Progress in Mathematics
Volume 196

$\hat{A}^{\times}_{\mathfrak{q}_i}/(\hat{A}^{\times}_{\mathfrak{q}_i})^{\ell} \approx \mathbb{Z}/\ell$. In this way, it follows not only that the natural map

$$A^{\times}/(A^{\times})^{\ell} \to \prod_{i=1}^{n} \hat{A}^{\times}_{\mathfrak{q}_i}/(\hat{A}^{\times}_{\mathfrak{q}_i})^{\ell}$$

is injective, but also that it is a linear map between two vector spaces over $\mathbb{F}_\ell$, given by an $r \times n$ matrix, where $r = r_2 + 1$ is given by the Dirichlet Unit Theorem. By finding an $r \times r$ invertible minor of this matrix we can choose in fact a finite set of prime ideals $\mathfrak{p}_1, \ldots, \mathfrak{p}_r$ among $\mathfrak{q}_1, \ldots, \mathfrak{q}_n$, such that the corestriction of the above map is an isomorphism as stated in our lemma.

□

3. The Etale Approximation to $BGL_n(A)$

The goal of this section is to describe the etale approximation $X_n(A)$ to $BGL_n(A)$ at the prime ℓ which comes together with a natural map

$$\chi_n : BGL_n(A) \to X_n(A)$$

After recalling the concepts of "etale topological type" and "etale approximation" we show that $X_n(A)$ can be expressed up to homotopy in terms of $X_n(\mathbb{F}_p)$'s for various residual fields $\mathbb{F}_p$'s of the ring A.

A simplicial object in a category $\mathcal{C}$ is a contravariant functor from the category Δ of standard simplices $\overline{n} = \{0, 1, 2, \ldots, n\}$, $n \geq 0$, to $\mathcal{C}$. With the natural transformations of functors as morphisms, the simplicial objects of $\mathcal{C}$ form a new category *simplicial-$\mathcal{C}$*. A pro-object in a category $\mathcal{C}$ is a functor from a small left filtering category to $\mathcal{C}$ ([3], III, 8.1). Similarly, the pro-objects of $\mathcal{C}$ form a new category *pro-$\mathcal{C}$*.

The *etale topological type* as defined in Friedlander [7], p. 36, is a covariant functor $(\ )_{et}$ from the category of *locally noetherian simplicial schemes* to the category of *pro-simplicial sets*. Recall that the category of schemes (as defined in [8], p. 74) contains the opposite category of commutative rings with identity. In particular, if a noetherian ring R is regarded as an affine scheme $X = Spec(R)$ and a scheme X as a simplicial scheme represented by the constant functor $\overline{n} \mapsto X$, we can apply $(\ )_{et}$ to the ring R. Also, the category of pro-simplicial sets can be regarded as a generalization of the category of CW-complexes [2]. In this context, we have the following basic examples ([7], 4.5, [2], p. 115, 125, [6], 3.2).

Example 3.1. *$pt \to (\mathbb{C})_{et}$ is a homotopy equivalence, where pt is a point.*

Example 3.2. *$(\mathbb{F}_q)_{et} \to (R)_{et}$ is a homotopy equivalence induced by the canonical map from a complete discrete valuation ring R to its residual finite field $\mathbb{F}_q$.*

Example 3.3. *$S^1 \to (\mathbb{F}_q)_{et}$ induces an isomorphism on mod ℓ cohomology if the finite field $\mathbb{F}_q$ has the characteristic $\neq \ell$ and the given map sends the generator of $\pi_1(S^1)$ to the Frobenius element of $\pi_1((\mathbb{F}_q)_{et})$. Here, S^1 is the one dimensional sphere.*

where ([11], p. 540)

$$E^i_\beta \approx \begin{cases} \mathbb{F}_l & \text{if } i=0 \\ \hat{\mathcal{O}}^\times_\beta/(\hat{\mathcal{O}}^\times_\beta)^\ell & \text{if } i=1 \\ 0 & \text{otherwise} \end{cases}$$

and ([11], p. 539)

$$H^i_{et}(Spec(\mathcal{O}),\mathbb{F}_\ell) \approx \begin{cases} \mathbb{F}_l & \text{if } i=0,3 \\ \mathcal{O}^\times/(\mathcal{O}^\times)^\ell & \text{if } i=2 \\ 0 & \text{otherwise} \end{cases}$$

in agreement with the fact that $Pic(\mathcal{O})$ has no ℓ-torsion and $\mathcal{O}$ contains $\sqrt[\ell]{1}$. Moreover, the map $E^0_\beta \to H^3_{et}(Spec(\mathcal{O}),\mathbb{F}_\ell)$ which appears in (6) is surjective (compare with [11], p. 241) and hence it immediately follows that $H^i_{et}(Spec(A),\mathbb{F}_\ell) = 0$ for $i > 2$. Finally, $\mathcal{O}^\times/(\mathcal{O}^\times)^\ell \approx {\mathbb{F}_\ell}^{r_2}$ and $\hat{\mathcal{O}}^\times_\beta/(\hat{\mathcal{O}}^\times_\beta)^\ell \approx {\mathbb{F}_\ell}^{2r_2+1}$ by the Dirichlet Unit Theorem and [13], p. 146. Hence, (6) and (5) give the following exact sequence

$$0 \to {\mathbb{F}_\ell}^{r_2+1} \to {\mathbb{F}_\ell}^{2r_2+1} \to {\mathbb{F}_\ell}^{r_2} \to H^2_{et}(Spec(A),\mathbb{F}_\ell) \to \mathbb{F}_\ell \to \mathbb{F}_\ell \to 0$$

from which it follows that $H^2_{et}(Spec(A),\mathbb{F}_\ell) = 0$ as it was stated. □

The second step is to apply the etale approximation functor X_n to all of the commutative diagrams (3) and to study the induced map

$$X_n(A) \to X_n(\hat{A}_{\mathfrak{p}_1},\dots,\hat{A}_{\mathfrak{p}_r};\mathbb{C}) \tag{7}$$

from $X_n(A)$ to the homotopy fibre product of all of the maps $X_n(\hat{A}_{\mathfrak{p}_i}) \to X_n(\mathbb{C})$, $i = 1,\dots,r$, denoted by $X_n(\hat{A}_{\mathfrak{p}_1},\dots,\hat{A}_{\mathfrak{p}_r};\mathbb{C})$. In order to do this, we can use the following known results:

Proposition 3.7 ([4] 2.7). *Let R be a notherian $\mathbb{Z}[\frac{1}{\ell}]$-algebra containing $\sqrt[\ell]{1}$ and let $\mathcal{U} \to (R)_{et}$ be a map of pro-simplicial sets which induces an isomorphism on mod ℓ cohomology. Then the induced etale approximation map*

$$X_n(R) \to X_n(\mathcal{U})$$

is a homotopy equivalence.

Proposition 3.8 ([1] 2.3). *There are natural homotopy equivalences of $\mathbb{F}_\ell$- complete simplicial sets*

$$(\mathbb{F}_\ell)_\infty BGL_n(\mathbb{F}_p) \to X_n(\hat{A}_\mathfrak{p})$$

$$(\mathbb{F}_\ell)_\infty BGL_n(\mathbb{C}_{top}) \to X_n(\mathbb{C})$$

where $\mathfrak{p}$ is any prime ideal of A and $\mathbb{F}_p$ is the residual field of A at $\mathfrak{p}$.

Here, $(\mathbb{F}_\ell)_\infty X$ means the $\mathbb{F}_\ell$- completion of the simplicial set X in the sense of Bousfield and Kan [3]. It follows that

Proposition 3.9. *The above map (7) is a homotopy equivalence.*

Proof. The idea of the proof is to factorize the map (7) as follows

$$X_n(A) \to X_n(\bigvee_{i=1}^{r}(\hat{A}_{\mathfrak{p}_i})_{et}) \to X_n(\hat{A}_{\mathfrak{p}_1}, ..., \hat{A}_{\mathfrak{p}_r}; \mathbb{C})$$

where the first map is the etale approximation map induced by (4), and according to 3.6 and 3.7 it is a homotopy equivalence, while the second map is given by the universal property of the homotopy fibre product.

All we have to prove is that the second map is also a homotopy equivalence. This fact can be immediately deduced by induction on r, the case $r = 2$ being shown in [1] (2.5). □

Corollary 3.10. *If χ_n induces an isomorphism on the mod ℓ cohomology then it induces a homotopy equivalence*

$$(\mathbb{F}_\ell)_\infty BGL_n(A) \to X_n(A)$$

Proof. According to 3.9 and [3], I, §5, it is enough to show that

$$X_n(\hat{A}_{\mathfrak{p}_1}, ..., \hat{A}_{\mathfrak{p}_r}; \mathbb{C})$$

is an $\mathbb{F}_\ell$-complete simplicial set. This fact follows at once by induction on r, the case $r = 2$ being shown in the proof of [1] 2.3. □

4. Proof of Theorem 1.1

Suppose that χ_n^* is an isomorphism for $n = n_0$. According to 3.10, it follows that χ_{n_0} induces a homotopy equivalence

$$[B\mu_m, (\mathbb{F}_\ell)_\infty BGL_{n_0}(A)] \approx [B\mu_m, X_{n_0}(A)] \tag{8}$$

where [,] stands for the homotopy classes of unpointed maps, m is any natural number, and μ_m is the cyclic group of order ℓ^m.

Next, observe that by 3.9 we have

$$[B\mu_m, X_{n_0}(A)] \approx [B\mu_m, X_{n_0}(\hat{A}_{\mathfrak{p}_1}, ..., \hat{A}_{\mathfrak{p}_r}; \mathbb{C})] \tag{9}$$

Moreover, we can think of the space of unpointed maps

$$Map(B\mu_m, X_{n_0}(\hat{A}_{\mathfrak{p}_1}, ..., \hat{A}_{\mathfrak{p}_r}; \mathbb{C}))$$

as the homotopy fibre product of all of the maps

$$Map(B\mu_m, X_{n_0}(\hat{A}_{\mathfrak{p}_i})) \to Map(B\mu_m, X_{n_0}(\mathbb{C}))$$

induced by (2).

Lemma 4.1. *The induced map*

$$[B\mu_m, X_{n_0}(\hat{A}_{\mathfrak{p}_1}, ..., \hat{A}_{\mathfrak{p}_r}; \mathbb{C})] \to \prod_{i=1}^{r}[B\mu_m, X_{n_0}(\hat{A}_{\mathfrak{p}_i})]$$

is injective.

Proof. By using the homotopy exact sequence associated to a homotopy fibre product, it is enough to check that each component of $Map(B\mu_m, X_{n_0}(\mathbb{C}))$ is simply connected and this was shown in [4], 3.6. □

Finally, by 3.8, we have

$$[B\mu_m, X_{n_0}(\hat{A}_{\mathfrak{p}_i})] \approx [B\mu_m, (\mathbb{F}_l)_\infty BGL_{n_0}(\mathbb{F}_{p_i})] \tag{10}$$

and also, by [14], p. 124, all of the groups $GL_{n_0}(A)$ and $GL_{n_0}(\mathbb{F}_{p_i})$, $i = 1, 2, ..., r$, satisfy the finiteness condition of the following

Theorem 4.2 (Lee [10]). *Let Φ be a group of virtually finite cohomological dimension, and let μ be a finite ℓ-group. Then the natural map*

$$Rep(\mu, \Phi) \to [B\mu, (\mathbb{F}_\ell)_\infty B\Phi]$$

is a bijection.

Here $Rep(\mu, \Phi)$ stands for the set of conjugacy classes of group homomorphisms $\mu \to \Phi$.

From (8), (9), 4.1, (10) and 4.2 we get a commutative square

$$\begin{array}{ccc} Rep(\mu_m, GL_{n_0}(A)) & \longrightarrow & \prod_{i=1}^r [B\mu_m, (\mathbb{F}_\ell)_\infty BGL_{n_0}(\mathbb{F}_{p_i})] \\ \simeq\downarrow & & \downarrow\simeq \\ [B\mu_m, (\mathbb{F}_\ell)_\infty BGL_{n_0}(A)] & \longrightarrow & \prod_{i=1}^r [B\mu_m, X_{n_0}(\hat{A}_{p_i})] \end{array}$$

in which the bottom horizontal arrow is injective. Hence, it follows that the map

$$\tau_{m,n_0}: \; Rep(\mu_m, GL_{n_0}(A)) \to \prod_{i=1}^r Rep(\mu_m, GL_{n_0}(\mathbb{F}_{p_i}))$$

induced by the canonical maps $A \to A/\mathfrak{p}_i \approx \mathbb{F}_{p_i}$ is injective for any $m \geq 0$. At this point, we conclude the proof of theorem 1.1 by referring to the following

Lemma 4.3. *There exists n_1 such that for every $n_0 \geq n_1$ we can find at least one m_0 for which the above map τ_{m,n_0} is not injective for $m = m_0$.*

Proof. We distinguish between two cases: Case 1. A is a principal ideal domain; and Case 2. A is not such a ring.

Case 1. Suppose by contradiction that for each n_1 there exists $n_0 \geq n_1$ such that τ_{m,n_0} is injective for all $m \geq 0$. Let $m \geq a$ be fixed for a moment, where $a \geq 1$ is the maximum exponent having the property that A contains the ℓ^a-th root of unity. Denote $A[\sqrt[\ell^m]{1}]$ by A_m and let $\mathcal{P}$ be any prime ideal of A_m not dividing $p_1 \cdot ... \cdot p_r$. Choose an embedding of μ_m in $A_m^\times$ and use it to define a module structure of $\mathcal{P}$, A_m over the groupal ring $A[\mu_m]$ by letting μ_m act on $\mathcal{P}$, A_m by multiplication. Let n_1 be the degree of the minimal polynomial of $\sqrt[\ell^m]{1}$ over A and $n_0 = n_1 + n'$, where $n' \geq 0$ is chosen such that τ_{m,n_0} is injective. Let $A^{n'}$ be the trivial $A[\mu_m]$-module of rank n' as a free module over A. Because A is a principal ideal domain, both $\mathcal{P}$ and A_m are free modules over A of the same rank n_1. Therefore, the $A[\mu_m]$-modules $M = \mathcal{P} \oplus A^{n'}$ and $N = A_m \oplus A^{n'}$ can be regarded as elements of $Rep(\mu_m, GL_{n_0}(A))$.

Also, for each i between 1 and r, $\mathcal{P}$ has an index in A_m relative prime to p_i, by the way $\mathcal{P}$ was chosen. Therefore, $\mathbb{F}_{p_i} \otimes_A \mathcal{P}$ is isomorphic to $\mathbb{F}_{p_i} \otimes_A A_m$ as $\mathbb{F}_{p_i}[\mu_m]$-modules, and hence $\tau_{m,n_0}(M) = \tau_{m,n_0}(N)$. Meanwhile, it is known that ℓ is divisible by $(1 - \sqrt[\ell^m]{1})$ in A_m and ℓ is a unit in A. Therefore, by choosing a generator σ of μ_m we have $(1-\sigma)M = \mathcal{P}$ and $(1-\sigma)N = A_m$. Because τ_{m,n_0} is injective, it follows that $\mathcal{P}$ is isomorphic to A_m as $A[\mu_m]$-modules. In other words, $\mathcal{P}$ is a principal ideal in A_m. By the Tchebotarev Theorem, we conclude that A_m must be a principal ideal domain. But this is not true for m sufficiently large, according to [9], p. 298, and [15], p. 44.

Case 2. Let $\mathcal{I}$ be a nonprincipal prime ideal of A not dividing $p_1 \cdot \ldots \cdot p_r$. The existence of such an ideal $\mathcal{I}$ follows from the Tchebotarev Theorem. Suppose that $\mathcal{I}^h$ is principal for some exponent $h > 1$. Consider $M := \mathcal{I}^{h-1} \oplus \mathcal{I}$ and $N := A \oplus A$ as $A[\mu_1]$-modules by letting μ_1 act by multiplication on the second and trivially on the first summands. Observe that M is a free module of rank 2 over A because A is a Dedekind ring, and in this way we can regard M and N as elements in $Rep(\mu_1, GL_2(A))$. Because $\mathcal{I}$ has index prime to p_i, we conclude that if $\tau_{1,2}$ is injective, then $\mathcal{I}$ must be a principal ideal, contradiction. □

References

[1] Anton, M.F.: *On a conjecture of Quillen at the prime 3*, to appear in J. Pure Appl. Algebra.

[2] Artin, M.; Mazur, B.: *Etale Homotopy*, Lecture Notes in Math. 100, Springer, Berlin, **1969**.

[3] Bousfield, A.K.; Kan, D.M.: *Homotopy Limits, Completions and Localizations*, Lecture Notes in Math. 304, Springer, Berlin, **1972**.

[4] Dwyer, W.G.: *Exotic cohomology for* $\mathbf{GL_n}(\mathbb{Z}[\frac{1}{2}])$, Proc. Amer. Math. Soc. **126** (1998), 2159–2167.

[5] Dwyer, W.G.; Friedlander, E.M.: *Algebraic and etale K-theory*, Trans. Amer. Math. Soc. **272** (1985), 247–280.

[6] Dwyer, W.G.; Friedlander, E.M.: *Topological models for arithmetic*, Topology **33** (1994), 1–24.

[7] Friedlander, E.M.: *Etale Homotopy of Simplicial Schemes*, Princeton Univ. Press, Princeton, **1982**.

[8] Hartshorne, R.: *Algebraic Geometry*, Springer, New York, **1993**.

[9] Lang, S.: *Cyclotomic Fields I and II*, Springer, New York, **1990**.

[10] Lee, C-N.: *On the unstable homotopy type of the classifying space of a virtually torsion-free group*, Math. Z. **224** (1997), 253–253.

[11] Mazur, B.: *Notes on etale cohomology of number fields*, Ann. Scient. Éc. Norm. Sup., 4e série, **6** (1973), 521–552.

[12] Milne, J.S.: *Etale Cohomology*, Princeton Univ. Press, Princeton, **1980**.

[13] Neukirch, J.: *Algebraische Zahlentheorie*, Springer, Berlin, **1992**.

[14] Serre, J-P.: *Cohomologie des groups discrets*, Princeton Univ. Press, Princeton, **1971**, pp.77–169.

[15] Washington, L.: *Introduction to Cyclotomic Fields*, Springer, Berlin, **1982**.

Department of Mathematics,
University of Notre Dame,
Notre Dame, IN 46556, U.S.A.
and Institute of Mathematics of the Romanian Academy,
Bucharest, Romania
E-mail address: `anton.4@nd.edu`

Progress in Mathematics, Vol. 196

On the Hurewicz map and Postnikov invariants of $K\mathbb{Z}$

Christian Ausoni

Abstract. The purpose of this note is to present a calculation of the Hurewicz homomorphism $h\colon K_*\mathbb{Z} \longrightarrow H_*(GL(\mathbb{Z});\mathbb{Z})$ on the elements of $K_*\mathbb{Z}$ known to generate direct summands. These results are then used to produce lower bounds for the Postnikov invariants of the space $K\mathbb{Z}$. Under extra hypotheses (compatible with the Quillen–Lichtenbaum conjecture for $\mathbb{Z}$), we give the exact p-primary part of the order of the latter invariants.

1. Introduction

D. Quillen defined, for any integer $n \geq 1$, the higher algebraic K-theory group K_nR of a ring R as the homotopy group $K_nR = \pi_n(BGL(R)^+)$. In this paper, we will calculate the Hurewicz homomorphism

$$h\colon K_*\mathbb{Z} \longrightarrow H_*(BGL(\mathbb{Z})^+;\mathbb{Z}) \cong H_*(GL(\mathbb{Z});\mathbb{Z})$$

on elements of $K_*\mathbb{Z}$ that are known to generate direct summands. One motivation for such a calculation is to obtain information on the homotopy type of the space $BGL(\mathbb{Z})^+$, which we will denote in the sequel by $K\mathbb{Z}$. Its weak homotopy type is uniquely determined by its homotopy groups $K_*\mathbb{Z}$ and by its Postnikov invariants, which are related to the Hurewicz homomorphism.

The Hurewicz homomorphism $h\colon K_*\mathbb{Z} \longrightarrow H_*(GL(\mathbb{Z});\mathbb{Z})$ has first been used by Borel [7] to calculate the rank of the finitely generated abelian group $K_m\mathbb{Z}$ for all $m \geq 1$: by the Milnor–Moore Theorem, the Hurewicz homomorphism induces an isomorphism from $K_*\mathbb{Z}\otimes\mathbb{Q}$ onto the primitives of $H_*(GL(\mathbb{Z});\mathbb{Q}) \cong \bigwedge_{\mathbb{Q}}(u_3, u_5, \dots)$, where $|u_i| = 2i-1$. Hence, if n is an odd integer ≥ 3, the group $K_{2n-1}\mathbb{Z}$ contains an infinite cyclic direct summand which injects in $H_{2n-1}(GL(\mathbb{Z});\mathbb{Z})$. How? In Theorem 3.2, we show that this injection is far from being split: it is multiplication by $(n-1)!$ (up to primes that do not satisfy Vandiver's Conjecture from number theory). Theorem 3.2 also gives the Hurewicz homomorphism on all 2-torsion classes of $K_*\mathbb{Z}$, and on the odd torsion classes of $K_*\mathbb{Z}$ corresponding to $\operatorname{Im} J$. We then apply these results to estimate the order of the Postnikov invariants of the space $K\mathbb{Z}$ (Theorem 4.1).

These calculations are made by comparing the p-adic completion $K\mathbb{Z}_p^\wedge$ of the space $K\mathbb{Z}$ to one of its topological models, called $JK\mathbb{Z}_p^\wedge$ and first defined by M. Bökstedt in [5]. We begin by reviewing some links between these spaces.

2. The model $JK\mathbb{Z}_p^\wedge$ for $K\mathbb{Z}_p^\wedge$

Let ℓ be an odd prime, and define $JK\mathbb{Z}(\ell)$ as the homotopy fibre of the composite map

$$BO \xrightarrow{\Psi_{\mathbb{R}}^{\ell}-1} BSpin \xrightarrow{c} BSU \tag{2.1}$$

where $\Psi_{\mathbb{R}}^{\ell}$ is the Adams operation ([1]), and where c is induced by the complexification of vector bundles. The homotopy groups of the space $JK\mathbb{Z}(\ell)$ are given by

$$\pi_n\big(JK\mathbb{Z}(\ell)\big) = \begin{cases} \mathbb{Z}/2 & \text{if } n=1 \text{ or if } n\equiv 2 \bmod (8),\\ \mathbb{Z}\oplus\mathbb{Z}/2 & \text{if } n\geq 9 \text{ and if } n\equiv 1 \bmod (8),\\ \mathbb{Z}/2(\ell^{\frac{n+1}{2}}-1) & \text{if } n\equiv 3 \bmod (8),\\ \mathbb{Z} & \text{if } n\equiv 5 \bmod (8),\\ \mathbb{Z}/(\ell^{\frac{n+1}{2}}-1) & \text{if } n\equiv 7 \bmod (8),\\ 0 & \text{otherwise.}\end{cases} \tag{2.2}$$

Let p be a prime number, and choose $\ell=3$ if $p=2$, ℓ a generator of the group of units of $\mathbb{Z}/p^2$ if p is odd. Following Bökstedt [5], let us then call $JK\mathbb{Z}_p^\wedge$ the space $JK\mathbb{Z}(\ell)_p^\wedge$. Here, $X_p^\wedge$ means the p-adic completion of a suitable space or group X. The homotopy group $\pi_n(JK\mathbb{Z}_p^\wedge)$ is isomorphic to $\pi_n\big(JK\mathbb{Z}(\ell)\big)\otimes\mathbb{Z}_p^\wedge$ and can be explicitly computed using (2.2) and the following formulas: if $n\equiv 3,7 \bmod (8)$ and if ℓ is chosen as above with respect to p, then

$$v_p(\ell^{\frac{n+1}{2}}-1) = \begin{cases} v_p(n+1)+1 & \begin{cases}\text{if } p\neq 2 \text{ and } 2(p-1)|n+1,\\ \text{or if } p=2 \text{ and } n\equiv 7 \bmod (8),\end{cases}\\ 3 & \text{if } p=2 \text{ and } n\equiv 3 \bmod (8),\\ 0 & \text{otherwise.}\end{cases} \tag{2.3}$$

Here v_p denotes the p-adic valuation.

Bökstedt showed that there is a map $\phi\colon K\mathbb{Z}_2^\wedge \longrightarrow JK\mathbb{Z}_2^\wedge$ which, after looping once, is a homotopy retraction (Theorem 2 of [5]). The recent calculation (in [18] and [15]) of the 2-primary part of $K_*\mathbb{Z}$ implies that the map

$$\phi\colon K\mathbb{Z}_2^\wedge \xrightarrow{\simeq} JK\mathbb{Z}_2^\wedge \tag{2.4}$$

is actually a homotopy equivalence.

When p is odd, the homotopy groups of $JK\mathbb{Z}_p^\wedge$ are isomorphic to direct summands of $(K_*\mathbb{Z})_p^\wedge$ (see [7] and [14]). If p is a regular prime, the Quillen–Lichtenbaum conjecture asserts that $K\mathbb{Z}_p^\wedge$ and $JK\mathbb{Z}_p^\wedge$ have same homotopy groups (see [10],

Corollary 2.3), while if p is irregular, there are p-torsion classes in $(K_*\mathbb{Z})^\wedge_p$ which do not appear in the homotopy groups of $JK\mathbb{Z}^\wedge_p$ (see [16]). It is not known in whole generality whether the group-level splitting

$$(K_*\mathbb{Z})^\wedge_p \;\cong\; \pi_*(JK\mathbb{Z}^\wedge_p) \;\oplus\; \dots$$

can be induced by a space level retraction $K\mathbb{Z}^\wedge_p \longrightarrow JK\mathbb{Z}^\wedge_p$ or not. However, it follows from the work of Quillen and Dwyer–Mitchell that this is the case when p is a Vandiver prime (Proposition 2.5), that is when p is an odd prime that does not divide the class number $h^+(\mathbb{Q}(\zeta_p))$ of the maximal real subfield of the cyclotomic field $\mathbb{Q}(\zeta_p)$. It is a conjecture by Vandiver that all primes verify this condition, and it is known to be true for $p < 4'000'000$ (see [17], page 158).

Proposition 2.5. *If p is a Vandiver prime, then $JK\mathbb{Z}^\wedge_p$ is a retract of $K\mathbb{Z}^\wedge_p$.*

Proof. If p is an odd prime, the space $BSU^\wedge_p$ splits as a product $BSU^\wedge_p \simeq BO^\wedge_p \times B(SU/SO)^\wedge_p$, thus induces a splitting $JK\mathbb{Z}^\wedge_p \simeq (\mathrm{F}\Psi^\ell_\mathbb{C})^\wedge_p \times (SU/SO)^\wedge_p$, where $(\mathrm{F}\Psi^\ell_\mathbb{C})^\wedge_p$ is the p-adic completion of the homotopy fibre $\mathrm{F}\Psi^\ell_\mathbb{C}$ of $\Psi^\ell_\mathbb{C} - 1\colon BU \longrightarrow BU$, or equivalently the homotopy fibre of $\Psi^\ell_\mathbb{R} - 1\colon BO^\wedge_p \longrightarrow BO^\wedge_p$ (because of the above choice of ℓ). However, the space $\mathrm{F}\Psi^\ell_\mathbb{C}$ is homotopy equivalent to $K\mathbb{F}_\ell$, and the reduction map $K\mathbb{Z}^\wedge_p \longrightarrow (K\mathbb{F}_\ell)^\wedge_p$ is a retraction according to [14].

On the other hand, W. Dwyer and S. Mitchell proved in [11], Theorem 9.3 and Example 12.2, that if p is a Vandiver prime, then $(U/O)^\wedge_p$ is a retract of $K\mathbb{Z}[\frac{1}{p}]^\wedge_p$. The space $(SU/SO)^\wedge_p$ is the universal cover of $(U/O)^\wedge_p$ and, by the localization exact sequence, $K\mathbb{Z}^\wedge_p$ is the universal cover of $K\mathbb{Z}[\frac{1}{p}]^\wedge_p$. This implies that $(SU/SO)^\wedge_p$ is a retract of $K\mathbb{Z}^\wedge_p$. The product of the above retractions

$$K\mathbb{Z}^\wedge_p \longrightarrow (\mathrm{F}\Psi^\ell_\mathbb{C})^\wedge_p \times (SU/SO)^\wedge_p \simeq JK\mathbb{Z}^\wedge_p$$

is then itself a retraction. □

3. The Hurewicz homomorphism for $K\mathbb{Z}$

Let us choose for all odd $n \geq 3$ a representative $b_n \in K_{2n-1}\mathbb{Z}$ of a generator of $K_{2n-1}\mathbb{Z}/(\mathrm{Torsion}) \cong \mathbb{Z}$, thus obtaining a decomposition $K_{2n-1}\mathbb{Z} \cong \langle b_n \rangle \oplus T_{2n-1}$, where T_{2n-1} is the (finite) torsion subgroup of $K_{2n-1}\mathbb{Z}$. Since the homomorphism $h\colon K_{2n-1}\mathbb{Z} \longrightarrow H_{2n-1}(GL(\mathbb{Z});\mathbb{Z})$ is injective after rationalization, there exists a generator v_n of an infinite cyclic summand of $H_{2n-1}(GL(\mathbb{Z});\mathbb{Z})$ and an integer $\mu_n > 0$ such that $h(b_n) \equiv \mu_n v_n$ modulo torsion elements. Equivalently, we may define μ_n as the order of the torsion subgroup of the cokernel of the homomorphism $h\colon K_{2n-1}\mathbb{Z} \longrightarrow H_{2n-1}(GL(\mathbb{Z});\mathbb{Z})/\{\mathrm{Torsion}\}$. On the other hand, if $n \geq 1$, it is known that $K_n\mathbb{Z}$ contains the following finite cyclic groups as direct summands:

$$\begin{cases} \mathbb{Z}/2 & \text{if } n \equiv 1,2 \bmod (8), \\ \mathbb{Z}/16 & \text{if } n \equiv 3 \bmod (8), \\ \mathbb{Z}/2^{v_2(n+1)+1} & \text{if } n \equiv 7 \bmod (8), \\ \mathbb{Z}/p^{v_p(n+1)+1} & \text{if } p \text{ is an odd prime and if } 2(p-1)|n+1. \end{cases} \tag{3.1}$$

We know, because of the equivalence $\phi\colon K\mathbb{Z}_2^\wedge \xrightarrow{\simeq} JK\mathbb{Z}_2^\wedge$ and of (2.2), that this is all the 2-torsion there is in $K_*\mathbb{Z}$. The odd torsion direct factors in (3.1) are given by [14] (see proof of Proposition 2.5). Let us choose a generator $\omega_{2,n}$ of the 2-torsion subgroup of $K_n\mathbb{Z}$ whenever $n \equiv 1,2,3,7 \bmod (8)$, and a generator $\omega_{p,n}$ of the p-torsion subgroup of $K_n\mathbb{Z}$ given by (3.1) whenever p is an odd prime with $2(p-1)|n+1$.

Theorem 3.2. *The Hurewicz homomorphism $h\colon K_*\mathbb{Z} \longrightarrow H_*(GL(\mathbb{Z});\mathbb{Z})$ has the following properties:*

(a) *If $p = 2$ or if p is a Vandiver prime, and if $n \geq 3$ is odd, then*

$$v_p(\mu_n) = v_p\big((n-1)!\big).$$

(b) *If p is an odd prime and if $(p,n) \neq (p, 2p-3), (3,11)$, then $\omega_{p,n}$ belongs to the kernel of h. The image $h(\omega_{p,2p-3})$ generates a direct summand of order p of $H_{2p-3}(GL(\mathbb{Z});\mathbb{Z})$, and $h(\omega_{3,11})$ is of order 3 in a direct summand of order 9 of $H_{11}(GL(\mathbb{Z});\mathbb{Z})$.*

(c) *If $n \neq 1,2,3,7,15$, then $\omega_{2,n}$ belongs to the kernel of h. If $n = 1$ or 2, then $K_n\mathbb{Z} \cong \mathbb{Z}/2$ and $h\colon K_n\mathbb{Z} \longrightarrow H_n(GL(\mathbb{Z});\mathbb{Z})$ is an isomorphism. The image $h(\omega_{2,3})$ generates the 2-torsion subgroup of $H_3(GL(\mathbb{Z});\mathbb{Z})$, which is of order 8. The image $h(\omega_{2,7})$ is of order 8 in a cyclic direct summand of order 16 of $H_7(GL(\mathbb{Z});\mathbb{Z})$, and $h(\omega_{2,15})$ is of order 2 in a cyclic direct summand of order 32 of $H_{15}(GL(\mathbb{Z});\mathbb{Z})$.*

To prove Theorem 3.2 we will need the equivalence (2.4), Proposition 2.5, as well as a computation of the Hurewicz homomorphism h' for $JK\mathbb{Z}(\ell)$. The next Lemma is the main ingredient of this computation.

For any $n \geq 2$, let us choose a generator ε_n of $\pi_{2n-1}(SU) \cong \mathbb{Z}$. Recall that there is an isomorphism of algebras $H_*(SU;\mathbb{Z}) \cong \bigwedge_{\mathbb{Z}}(x_2, x_3, \dots)$. Here x_i is a primitive class of degree $2i-1$, defined as the dual of the class $e_i = \sigma(c_i) \in H^{2i-1}(SU;\mathbb{Z})$, where σ is the cohomology suspension and $c_i \in H^{2i}(BSU;\mathbb{Z})$ is the ith Chern class. The Hurewicz homomorphism for SU was calculated by Douady ([9], Théorème 6), and is given by the rule

$$\varepsilon_n \longmapsto \pm(n-1)!\, x_n. \tag{3.3}$$

By looping the fibration

$$JK\mathbb{Z}(\ell) \xrightarrow{f} BO \xrightarrow{g} BSU, \tag{3.4}$$

where g is the composition (2.1), we get a map $\partial\colon SU \longrightarrow JK\mathbb{Z}(\ell)$ having the following properties.

Lemma 3.5. *Let ℓ be an odd prime and n an integer ≥ 2. The image of the element $x_n \in H_{2n-1}(SU;\mathbb{Z})$ under the homomorphism*

$$\partial_*\colon H_{2n-1}(SU;\mathbb{Z}) \longrightarrow H_{2n-1}(JK\mathbb{Z}(\ell);\mathbb{Z})$$

generates a direct summand in $H_{2n-1}(JK\mathbb{Z}(\ell);\mathbb{Z})$. This summand is of infinite order if n is odd, and of order $(\ell^n - 1)$ if n is even.

Proof. Suppose first n is odd. The integral cohomology algebra of SU is given by an isomorphism $H^*(SU;\mathbb{Z}) \cong \bigwedge_{\mathbb{Z}}(e_2, e_3, \dots)$, where e_i is the dual class of x_i. We must show that the class e_n is in the image of the homomorphism

$$\partial^*\colon H^*(JK\mathbb{Z}(\ell);\mathbb{Z}) \longrightarrow H^*(SU;\mathbb{Z}).$$

Consider the homotopy commutative diagram

$$\begin{array}{ccccccc} (1) & & SU & \xrightarrow{\partial} & JK\mathbb{Z}(\ell) & \xrightarrow{f} & BO \\ \varphi\downarrow & & \downarrow = & & \downarrow & & \downarrow g=(\Psi^\ell_{\mathbb{C}}-1)c \\ (2) & & SU & \longrightarrow & PBSU & \longrightarrow & BSU \end{array} \tag{3.6}$$

whose rows (1) and (2) are homotopy fibrations. Here $PBSU \longrightarrow BSU$ is the path fibration. For $i = 1$ or 2, let us call $(E_*^{*,*}(i), d_*^i)$ the Serre spectral sequence for $H^*(-;\mathbb{Z})$ associated to the fibration (i). The fibration morphism φ induces a morphism between these spectral sequences, which we will denote by $\varphi_*^{*,*}$. It suffices to verify that the element e_n in $E_2^{0,2n-1}(1) \cong H^{2n-1}(SU;\mathbb{Z})$ is a permanent cycle. Since the cohomology suspension $\sigma\colon H^{2n}(BSU;\mathbb{Z}) \longrightarrow H^{2n-1}(SU;\mathbb{Z})$ maps the n-th Chern class c_n to e_n, e_n is transgressive in $(E_*^{*,*}(2), d_*^2)$ and by naturality belongs to $E_{2n}^{0,2n-1}(1)$. Now $E_{2n}^{2n,0}(1)$ is a quotient of $H^{2n}(BO;\mathbb{Z})$, which contains only elements of order 2 since n is odd (see [6], Theorem 24.7 page 86). On the other hand, one can show by induction on n that $d_{2n}^1(e_n)$ is equal to $\varphi_{2n}^{2n,0}(c_n) = (\ell^n - 1)c^*(c_n)$ in $E_{2n}^{2n,0}(1)$, so is divisible by 2. Hence e_n is a permanent cycle in $E_*^{*,*}(1)$.

If n is even, we work with the Serre spectral sequences $(E^*_{*,*}(i), d_i^*)$ for $H_*(-;\mathbb{Z})$ of the fibrations $(i = 1, 2)$ of diagram (3.6). Using the homology suspension, one verifies that there is a primitive generator $p_n \in H_{2n}(BSU;\mathbb{Z}) = E^2_{2n,0}(2)$ that transgresses to $x_n \in H_{2n-1}(SU) = E^2_{0,2n-1}(2)$ at the $2n$-th stage. Since n is even, there is in $H_{2n}(BO;\mathbb{Z})$ an element $\bar{p}_n$ (the n-th Pontryagin class) that verifies $c_*(\bar{p}_n) = p_n$ (see [8], equation 61, page 19). Now $(\Psi^\ell_{\mathbb{C}} - 1)_*(p_n) = (\ell^n - 1)p_n$, so by naturality, the class $\bar{p}_n \in H_{2n}(BO;\mathbb{Z}) = E^2_{2n,0}(1)$ is transgressive in the spectral sequence $(E^*_{*,*}(1), d_1^*)$ and transgresses to $(\ell^n - 1)x_n$. It follows that $\partial_*(x_n)$ is of order $\ell^n - 1$ in $H_{2n-1}(JK\mathbb{Z}(\ell);\mathbb{Z})$. To verify that $\partial_*(x_n)$ indeed generates a direct summand, it is enough to check that the dual class of x_n in $H^{2n-1}(SU;\mathbb{Z}/(\ell^n - 1))$ is in the image of $\partial^*\colon H^{2n-1}(JK\mathbb{Z}(\ell);\mathbb{Z}/(\ell^n - 1)) \longrightarrow H^{2n-1}(SU;\mathbb{Z}/(\ell^n - 1))$. This can be proven using again a Serre spectral sequence argument of the same flavour as above. □

Remarks 3.7.

(a) Lemma 3.5, together with (3.3), allows one to compute the Hurewicz homomorphism h' of $JK\mathbb{Z}(\ell)$ on all elements of $\pi_*\big(JK\mathbb{Z}(\ell)\big)$ that are in the image of $\partial_*\colon \pi_*(SU) \longrightarrow \pi_*\big(JK\mathbb{Z}(\ell)\big)$. The only elements of $\pi_*\big(JK\mathbb{Z}(\ell)\big)$ that are not in this image are the 2-torsion elements in dimensions n with $n \equiv 1, 2 \bmod (8)$.

(b) A very similar argument to the one of the proof of Lemma 3.5 implies that, for any prime ℓ, the connecting map $\partial\colon SU \longrightarrow \mathrm{F}\Psi^{\ell}_{\mathbb{C}}$ has the following property: for any integer $n \geq 2$, the image of the element $x_n \in H_{2n-1}(SU;\mathbb{Z})$ under the homomorphism $\partial_*\colon H_{2n-1}(SU;\mathbb{Z}) \longrightarrow H_{2n-1}(\mathrm{F}\Psi^{\ell}_{\mathbb{C}};\mathbb{Z})$ generates a direct summand in $H_{2n-1}(\mathrm{F}\Psi^{\ell}_{\mathbb{C}};\mathbb{Z})$ of order $(\ell^n - 1)$.

(c) Notice that if $n \equiv 2 \bmod (4)$, the element $\partial_*(\varepsilon_n)$ is of order $2(\ell^n - 1)$ in $\pi_{2n-1}\big(JK\mathbb{Z}(\ell)\big)$, while $\partial_*(x_n)$ is of order $(\ell^n - 1)$ in $H_{2n-1}(JK\mathbb{Z}(\ell);\mathbb{Z})$.

Proof of Theorem 3.2.

(a) Let $p = 2$ and $\ell = 3$, or let p be a Vandiver prime and ℓ an odd prime that generates the units of $\mathbb{Z}/p^2$. We compare the Hurewicz homomorphisms of $K\mathbb{Z}$ and $JK\mathbb{Z}(\ell)$ by means of the following commutative diagram. Let us call $\psi\colon JK\mathbb{Z}^{\wedge}_p \longrightarrow K\mathbb{Z}^{\wedge}_p$ the inclusion as a summand given by Proposition 2.5, and choose an integer $k > \max\big\{v_p(\mu_n), v_p\big((n-1)!\big)\big\} + v_p(T)$, where T is the largest order of any p-torsion element in $K_{2n-1}\mathbb{Z}$ or $H_{2n-1}(K\mathbb{Z};\mathbb{Z})$.

$$\begin{array}{ccccccc}
\pi_*\big(JK\mathbb{Z}(\ell)\big) & \xrightarrow{\alpha_{JK\mathbb{Z}(\ell)}} & \pi_*(JK\mathbb{Z}^{\wedge}_p;\mathbb{Z}/p^k) & \xrightarrow{\psi_*} & \pi_*(K\mathbb{Z}^{\wedge}_p;\mathbb{Z}/p^k) & \xleftarrow{\alpha_{K\mathbb{Z}}} & \pi_*(K\mathbb{Z}) \\
{\scriptstyle h'}\downarrow & & {\scriptstyle \bar{h}'}\downarrow & & {\scriptstyle \bar{h}}\downarrow & & {\scriptstyle h}\downarrow \\
H_*(JK\mathbb{Z}(\ell);\mathbb{Z}) & \xrightarrow{\gamma_{JK\mathbb{Z}(\ell)}} & H_*(JK\mathbb{Z}^{\wedge}_p;\mathbb{Z}/p^k) & \xrightarrow{\psi_*} & H_*(K\mathbb{Z}^{\wedge}_p;\mathbb{Z}/p^k) & \xleftarrow{\gamma_{K\mathbb{Z}}} & H_*(K\mathbb{Z};\mathbb{Z}).
\end{array}$$

Here $\pi_*(-,\mathbb{Z}/p^k)$ and $\bar{h}$, $\bar{h}'$ are the mod p^k homotopy groups and Hurewicz maps (see Chapter 3 of [13]). The map α_X given in the diagram is the composite

$$\pi_*(X) \longrightarrow \pi_*(X) \otimes \mathbb{Z}/p^k \hookrightarrow \pi_*(X;\mathbb{Z}/p^k) \cong \pi_*(X^{\wedge}_p;\mathbb{Z}/p^k)$$

and the map γ_X is defined in a similar way. The assertion is proven by inspection of this diagram, using our knowledge of $h'\colon \pi_{2n-1}\big(JK\mathbb{Z}(\ell)\big) \longrightarrow H_{2n-1}(JK\mathbb{Z}(\ell);\mathbb{Z})$ (see Remark 3.7.a).

(b) If p is any odd prime and ℓ an odd prime that generates the units of $\mathbb{Z}/p^2$, the space $\mathrm{F}\Psi^{\ell}_{\mathbb{C}}$ splits off $K\mathbb{Z}$ after being localized at p. The element $\omega_{p,n}$ generates the factor $\pi_n\big((\mathrm{F}\Psi^{\ell}_{\mathbb{C}})_{(p)}\big)$ of $K_n\mathbb{Z}$, and is in the image of the homomorphism $\partial_*\colon \pi_n(SU) \longrightarrow \pi_n\big((\mathrm{F}\Psi^{\ell}_{\mathbb{C}})_{(p)}\big)$. It then follows from the rule (3.3) and the Remark 3.7.b that $h(\omega_{p,n}) = \big(\frac{n-1}{2}\big)!z_n$, where z_n is the generator of a direct summand of $H_n(K\mathbb{Z};\mathbb{Z})$ of order the p-primary part of $\ell^{\frac{n+1}{2}} - 1$ (see (2.3) for a description of it). The assertions then follow from the arithmetic behavior of $\big(\frac{n-1}{2}\big)!$ in $\mathbb{Z}/(\ell^{\frac{n+1}{2}} - 1)$ at p.

(c) If X is a simple space of finite type, p a prime and $\eta\colon X \longrightarrow X_p^\wedge$ the p-adic completion of X, the homomorphism $\eta_*\colon H_*(X;\mathbb{Z}) \longrightarrow H_*(X_p^\wedge;\mathbb{Z})$ restricts to an isomorphism from the p-torsion subgroup of $H_*(X;\mathbb{Z})$ onto the p-torsion subgroup of $H_*(X_p^\wedge;\mathbb{Z})$. The same is also true for homotopy groups. For us, this means that using the equivalence $\phi\colon K\mathbb{Z}_2^\wedge \longrightarrow JK\mathbb{Z}_2^\wedge$, we can just read off the Hurewicz map of $K\mathbb{Z}$ on 2-torsion elements from the Hurewicz map h' of $JK\mathbb{Z}(3)$.

The Eilenberg–Mac Lane space $K(\mathbb{Z}/2,1)$ splits off $JK\mathbb{Z}(3)$, so by the Hurewicz Theorem, h' must be an isomorphism in dimensions 1 and 2, and surjective in dimension 3.

The classes $\omega_{2,n}$ with $n \equiv 3 \bmod (4)$ correspond to classes of $\pi_n\big(JK\mathbb{Z}(3)\big)$ coming from $\pi_n(SU)$. Their image under the Hurewicz homomorphism can therefore be calculated as for the odd-p-torsion classes $\omega_{p,n}$ in part (b) of this proof.

Choose $n \geq 9$ satisfying $n \equiv 1,2 \bmod (8)$, and let us show that the class $\omega'_{2,n} \in \pi_n\big(JK\mathbb{Z}(3)\big)$ corresponding to $\omega_{2,n}$ is in the kernel of h'. Consider the following diagram

$$\begin{array}{ccc}
\pi_{n+1}(JK\mathbb{Z}(3);\mathbb{Z}/2) & \xrightarrow{d_*} & \pi_n\big(JK\mathbb{Z}(3)\big) \\
\bar{h}' \downarrow & & h' \downarrow \\
H_{n+1}(JK\mathbb{Z}(3);\mathbb{Z}/2) & \xrightarrow{d_*} & H_n(JK\mathbb{Z}(3);\mathbb{Z})
\end{array}$$

where d_* denotes the connecting homomorphism associated to the coefficient exact sequence $0 \to \mathbb{Z} \to \mathbb{Z} \to \mathbb{Z}/2 \to 0$. It is commutative (see [13], Lemma 3.2). The class $\omega'_{2,n} \in \pi_n\big(JK\mathbb{Z}(3)\big)$ is of order 2 and is in the image of d_*, so it suffices to show that the mod 2 Hurewicz homomorphism $\bar{h}'$ is trivial in dimension $n+1$.

Consider the mod 2 Moore space $P^{n+1}(2) = S^n/2$. By definition, an element α in $\pi_{n+1}(JK\mathbb{Z}(3);\mathbb{Z}/2)$ is the homotopy class of a map $\alpha\colon P^{n+1}(2) \longrightarrow JK\mathbb{Z}(3)$, and $\bar{h}(\alpha)$ is defined as $\alpha_*(e)$, where α_* is the homomorphism induced by α in mod 2 homology, and where e is the generator of $H_{n+1}(P^{n+1}(2);\mathbb{Z}/2) \cong \mathbb{Z}/2$. We claim that any such induced homomorphism α_* is zero. By duality, it is equivalent to prove the corresponding statement in mod 2 cohomology. There exists an isomorphism of Hopf algebras and of modules over the Steenrod algebra

$$H^*(JK\mathbb{Z}(3);\mathbb{Z}/2) \cong H^*(BO;\mathbb{Z}/2) \otimes H^*(SU;\mathbb{Z}/2)$$

(see [12], Remark 4.5). Recall the isomorphisms $H^*(BO;\mathbb{Z}/2) \cong \mathbb{Z}/2[w_1, w_2, \dots]$ and $H^*(SU;\mathbb{Z}/2) \cong \bigwedge_{\mathbb{Z}/2}(e_2, e_3, \dots)$, where w_i is the Stiefel–Whitney class of degree i, and e_i is primitive of degree $2i-1$. The action of the Steenrod algebra on these cohomology classes is well known. For instance, $Sq^1(w_i) = w_{i+1} + w_1 w_i$ and $Sq^2(w_i) = w_{i+2} + w_2 w_i$ if i is even, and $Sq^{2k} e_i = \binom{i-1}{k} e_{i+k}$. These relations, as well as the fact that $H^*(P^{n+1}(2);\mathbb{Z}/2)$ is concentrated in dimensions 0, n and $n+1$, force any induced homomorphism $H^{n+1}(JK\mathbb{Z}(3);\mathbb{Z}/2) \longrightarrow H^{n+1}(P^{n+1}(2);\mathbb{Z}/2)$ to be zero for the above choices of n. □

4. The order of the Postnikov invariants of $K\mathbb{Z}$

Let X be a connected simple space, for instance a connected H-space. For any integer $n \geq 1$, let us denote by $X \longrightarrow X[n]$ the nth-Postnikov section of X, and by k_X^{n+1} the $(n+1)$th-Postnikov invariant of X. Recall that k_X^{n+1} is an element of the cohomology group $H^{n+1}(X[n-1];\pi_n(X))$, which can be chosen canonically as the image of the fundamental class $u_X^{n+1} \in H^{n+1}(X[n-1],X;\pi_n(X))$ under the homomorphism induced by the inclusion of pairs $(X[n-1],\emptyset) \hookrightarrow (X[n-1],X)$. The Postnikov invariant k_X^{n+1} corresponds to a map $X[n-1] \longrightarrow K(\pi_n X, n+1)$ whose homotopy fiber is the nth-Postnikov section $X[n]$ of X.

If X is an H-space of finite type, all its Postnikov invariants are cohomology classes of finite order: this is the Arkowitz–Curjel Theorem ([2]). In particular, the Postnikov invariant of $K\mathbb{Z}$ are of finite order. The orders ρ_n of the Postnikov invariants $k_{K\mathbb{Z}}^{n+1}$ of $K\mathbb{Z}$ have previously been studied by Arlettaz and Banaszak in [3]. See especially their Proposition 5, which states that if $n \geq 5$ is an integer with $n \equiv 1 \bmod (4)$ and if $K_n\mathbb{Z}$ has no p-torsion, where p is an odd prime, then $v_p(\rho_n) \leq v_p\left(\frac{n-1}{2}!\right)$.

Theorem 4.1. *For any integer $n \geq 2$, the order ρ_n of the Postnikov invariant $k_{K\mathbb{Z}}^{n+1}$ of $K\mathbb{Z}$ verifies:*

(a)

$$v_2(\rho_n) = \begin{cases} 1 & \textit{if } n = 3, 7, \textit{ or if } n \geq 10 \textit{ and } n \equiv 2 \bmod (8), \\ v_2(\frac{n-1}{2}!) & \textit{if } n \equiv 1 \bmod (4), \\ 4 & \textit{if } n \geq 11 \textit{ and } n \equiv 3 \bmod (8), \textit{ or if } n = 15, \\ v_2(n+1)+1 & \textit{if } n \geq 23 \textit{ and } n \equiv 7 \bmod (8), \\ 0 & \textit{otherwise.} \end{cases}$$

(b) *If p is a Vandiver prime and $n \geq 5$ is an integer with $n \equiv 1 \bmod (4)$, then*

$$v_p(\rho_n) \geq v_p\left(\tfrac{n-1}{2}!\right),$$

and equality holds if the order e of the torsion subgroup of $K_n\mathbb{Z}$ verifies $v_p(e) \leq v_p\left(\frac{n-1}{2}!\right)$.

(c) *Let p be an odd prime. If $2(p-1)$ is a proper divisor of $n+1$, then*

$$v_p(\rho_n) \geq v_p(n+1)+1$$

(except if $p=3$ and $n=11$, where $v_3(\rho_{11}) \geq 1$ holds).

Proof. This is a consequence of Theorem 3.2, using the following general argument. If X is a connected simple space, n an integer ≥ 2, and ρ an integer ≥ 1, then the following statements are equivalent:

(1) The Postnikov invariant k_X^{n+1} verifies $\rho k_X^{n+1} = 0$ in $H^{n+1}\big(X[n-1], \pi_n(X)\big)$.

(2) There exists a homomorphism $g\colon H_n(X;\mathbb{Z}) \longrightarrow \pi_n(X)$ such that the composition $gh\colon \pi_n(X) \longrightarrow \pi_n(X)$ is multiplication by ρ, where $h\colon \pi_n(X) \longrightarrow H_n(X;\mathbb{Z})$ is the Hurewicz homomorphism. □

Remark 4.2. If p is a regular prime and if the p-adic Quillen–Lichtenbaum Conjecture for $\mathbb{Z}$ holds, then equality for $v_p(\rho_n)$ holds in the inequalities (b) and (c) of Theorem 4.1, and for other values of n, $v_p(\rho_n) = 0$.

Acknowledgement. The considerations in this note are part of my Ph.D. thesis [4]. It is a pleasure to thank my adviser Professor D. Arlettaz for his friendly help and support.

References

[1] J. F. Adams, *Vector fields on spheres,* Ann. of Math. **75** (1962), 603–632.

[2] M. Arkowitz and C. R. Curjel, *The Hurewicz homomorphism and finite homotopy invariants,* Trans. Amer. Math. Soc. **110** (1964), 538–551.

[3] D. Arlettaz and G. Banaszak, *On the non-torsion elements in the algebraic K-theory of rings of integers,* J. Reine Angew. Math. **461** (1995), 63–79.

[4] C. Ausoni, *Propriétés homotopiques de la K-théorie algébrique des entiers,* Thèse de Doctorat, Université de Lausanne, (1998).

[5] M. Bökstedt, *The rational homotopy type of* $\Omega\mathrm{Wh}^{\mathrm{Diff}}(*)$, Algebraic Topology, Aarhus 1982, Lectures Notes in Math. **1051**, Springer (1984), 25–37.

[6] A. Borel, *Topics in the homology theory of fibre bundles,* Lectures Notes in Math. **36**, Springer (1967).

[7] A. Borel, *Stable real cohomology of arithmetic groups,* Ann. sci. École Norm. Sup. 4ème série **7** (1974), 235–272.

[8] H. Cartan, *Démonstration homologique des théorèmes de périodicité de Bott, 2. Homologie et cohomologie des groupes classiques et de leurs espaces homogènes,* Sém. H. Cartan 1959/60, exposé 17, École Norm. Sup.

[9] A. Douady, *Périodicité du groupe unitaire,* Sém. H. Cartan 1959/60, exposé 11, École Norm. Sup.

[10] W. G. Dwyer and E. M. Friedlander, *Topological models for arithmetic,* Topology **33** (1994), 1–24.

[11] W. G. Dwyer and S. A. Mitchell, *On the K-theory spectrum of a ring of algebraic integers,* K-Theory **14** (1998), 201–263.

[12] S. A. Mitchell, *On the plus construction for* $BGL\mathbb{Z}[\frac{1}{2}]$ *at the prime 2,* Math. Z. **209** (1992), 205–222.

[13] J. Neisendorfer, *Primary homotopy theory,* Mem. Amer. Math. Soc. **232** (1980).

[14] D. Quillen, *Letter from Quillen to Milnor on* $\mathrm{Im}(\pi_i O \to \pi_i^s \to K_i\mathbb{Z})$, Algebraic K-theory, Evanston 1976, Lecture Notes in Math. **551**, Springer (1976), 182–188.

[15] J. Rognes and C. Weibel, *Two-primary algebraic K-theory of rings of integers in number fields,* to appear in J. Amer. Math. Soc.

[16] C. Soulé, *K-théorie des anneaux d'entiers de corps de nombres et cohomologie étale,* Inventiones Math. **55** (1979), 251–295.

[17] L. C. Washington, *Introduction to cyclotomic fields,* Graduate Texts in Math. **83**, Springer (1997).

[18] C. Weibel, *The 2-torsion in the K-theory of the integers,* C. R. Acad. Sci., Paris, Sér I **324, 6** (1997), 615–620.

D-Math HG G 27.1
ETH-Zentrum
CH-8092 Zurich, Switzerland
E-mail address: ausoni@math.ethz.ch

Progress in Mathematics, Vol. 196

Recognition principle for generalized Eilenberg–Mac Lane spaces

Bernard Badzioch

Abstract. We give a homotopy theoretical characterization of generalized Eilenberg–Mac Lane spaces which resembles the Γ-space structure used by Segal to describe infinite loop spaces.

1. Introduction

A generalized Eilenberg–Mac Lane space (GEM) is a space weakly equivalent to a product of Eilenberg–Mac Lane spaces $\prod_i K(\pi_i, i)$ with π_i abelian. The goal of this note is to prove the following characterization of GEMs:

Theorem 1.1. *Let R be a commutative ring with a unit and let* $\mathbf{FMod}_R$ *be the category of finitely generated free R-modules. If $H\colon \mathbf{FMod}_R \to \mathbf{Spaces}$ is a functor such that*

- *$H(0)$ is a contractible space,*
- *$H(R)$ is connected and for every n the projections $\mathrm{pr}_k\colon R^n \to R$ induce a weak equivalence $H(R^n) \xrightarrow{\simeq} H(R)^n$,*

then the space $H(R)$ is weakly equivalent to a product $\prod_{i=1}^{\infty} K(M_i, i)$ where M_i is an R-module.

Notation 1.2. By **Spaces** above and in the rest of this paper we denote the category of simplicial sets. Consequently, by 'space' we always mean an object of this category.

The above description of GEMs is modeled after the Γ-space structure introduced by Segal in [Se] to characterize infinite loop spaces. Just as for infinite loop spaces one gets the following corollary which is implicitly present in the work of Bousfield [Bo] and Dror [D1] who apply it to localization functors.

Corollary 1.3. *If F:* **Spaces** $\to$ **Spaces** *is a functor preserving weak equivalences and preserving products up to weak equivalence then F preserves GEMs.*

The rest of the paper is organized as follows. In Section 2 the Grothendieck construction on a diagram of small categories is recalled and some of its properties are stated. In Section 3 we give a description suitable for our purposes of the infinite symmetric product $\mathrm{Sp}^\infty X$ on a space X. In Section 4 the proof of Theorem 1.1 is presented. It essentially amounts to showing that for a functor $H\colon \mathbf{FMod}_R \to \mathbf{Spaces}$ satisfying the assumptions of the theorem the space $H(R)$ is a homotopy retract of $\mathrm{Sp}^\infty H(R)$. Since $\mathrm{Sp}^\infty H(R)$ is a GEM and the class of GEMs is closed under homotopy retractions the claim of the theorem will follow.

I am indebted to W. G. Dwyer for suggesting the topic of this work to me and for many discussions. The referee's comments were very helpful in improving the presentation of the paper. I also want to thank Sylwia Zabłocka for her support and friendship.

2. The Grothendieck construction

Definition 2.1. Let **Cat** denote the category of small categories and let **N** be the telescope category:

$$\mathbf{N} = (0 \xrightarrow{i_0} 1 \xrightarrow{i_1} \cdots \xrightarrow{i_{n-1}} n \xrightarrow{i_n} \cdots).$$

For a functor $P\colon \mathbf{N} \to \mathbf{Cat}$ the Grothendieck construction on P [Th] is the category $Gr(P)$ whose objects are pairs (n, c) where $n \in \mathbf{N}$ and $c \in P(n)$. A morphism $(n, c) \to (n', c')$ is a pair (i, φ) where i is the unique morphism $n \to n'$ in $\mathbf{N}$ and $\varphi \in \mathrm{Mor}_{P(n')}(P(i)c, c')$. The composition of $(i, \varphi)\colon (n, c) \to (n', c')$ and $(i', \varphi')\colon (n', c') \to (n'', c'')$ is defined to be the pair

$$(i' \circ i, \varphi' \circ P(i')\varphi)\colon (n, c) \to (n'', c'').$$

For every $n \in \mathbf{N}$ there is a functor

$$I_n\colon P(n) \to Gr(P), \quad I_n(c) = (n, c),$$

which lets us identify $P(n)$ with a subcategory of $Gr(P)$. It follows that any functor $F\colon Gr(P) \to \mathbf{C}$ defines a sequence of functors $F_n\colon P(n) \to \mathbf{C}$, $n = 0, 1, \ldots$. Moreover, for $n \in \mathbf{N}$ and $c \in P(n)$ let $\beta_{n,c}$ be the image under F of the morphism $(i_n, \mathrm{id}_{P(i_n)(c)}) \in \mathrm{Mor}_{Gr(P)}((n, c), (n+1, P(i_n)(c)))$. It is easy to check that the morphisms $\{\beta_{n,c}\}_{c\in P(n)}$ define a natural transformation of functors

$$\beta_n\colon F_n \to F_{n+1} \circ P(i_n).$$

The converse is also true [ChS, A.9]: any sequence of functors $\{F_n\colon P(n) \to \mathbf{C}\}_{n\geq 0}$ and natural transformations $\{\beta_n\colon F_n \to F_{n+1} \circ P(i_n)\}_{n\geq 0}$ can be used to define a functor $F\colon Gr(P) \to \mathbf{C}$ such that $F\,|_{P(n)} = F_n$.

Proposition 2.2. *For any functor $F\colon Gr(P) \to \mathbf{C}$ the natural morphism*

$$\mathrm{colim}_{\mathbf{N}}\, \mathrm{colim}_{P(n)} F_n \to \mathrm{colim}_{Gr(P)} F$$

is an isomorphism. Moreover, if $\mathbf{C} = \mathbf{Spaces}$ then the natural map

$$\mathrm{hocolim}_{\mathbf{N}}\, \mathrm{hocolim}_{P(n)} F_n \to \mathrm{hocolim}_{Gr(P)} F$$

is a weak equivalence.

Proof. The first statement follows directly from the definition of $Gr(P)$. The proof of the second can be found in [Sl, Prop. 0.2] or [ChS, Cor. 24.6]. □

3. Infinite symmetric products

Let Σ_n be the permutation group of the set $\{1, \ldots, n\}$. We will denote by $\mathbf{O}_{\Sigma_n}$ the orbit category of Σ_n whose objects are sets Σ_n/G for $G \subseteq \Sigma_n$ and whose morphisms are Σ_n-equivariant maps $\Sigma_n/G \to \Sigma_n/H$. Let $\mathbf{O}^{\mathrm{op}}_{\Sigma_n}$ be the opposite category. We can identify Σ_n with the subgroup of all elements of Σ_{n+1} which leave the element $n+1$ fixed. The inclusion $\Sigma_n \subset \Sigma_{n+1}$ induces a functor

$$J_n : \mathbf{O}^{\mathrm{op}}_{\Sigma_n} \to \mathbf{O}^{\mathrm{op}}_{\Sigma_{n+1}}, \quad J_n(\Sigma_n/G) = \Sigma_{n+1}/G.$$

This data in turn can be used to define a functor $O\colon \mathbf{N} \to \mathbf{Cat}$,

$$O(n) = \mathbf{O}^{\mathrm{op}}_{\Sigma_n}, \quad O(i_n) = J_n.$$

Let $\mathbf{Spaces}_*$ denote the category of pointed spaces and let $X \in \mathbf{Spaces}_*$. The group Σ_n acts on X^n by permuting the coordinates. As usual we have the fixed point functor

$$F_n X\colon \mathbf{O}^{\mathrm{op}}_{\Sigma_n} \to \mathbf{Spaces}_*$$

defined by $F_n X(\Sigma_n/G) = (X^n)^G$ —the fixed point set of the action of G on X^n.

Remark 3.1. For $G \subseteq \Sigma_n$ let $|\mathrm{orb}\, G|$ denote the number of orbits of the action of G on $\{1, \ldots, n\}$. Then there is a natural isomorphism $(X^n)^G \cong X^{|\mathrm{orb}\, G|}$.

Using the embedding $\Sigma_n \subset \Sigma_{n+1}$ one can think of $G \subseteq \Sigma_n$ as a subgroup of Σ_{n+1}. There is an obvious isomorphism

$$(X^{n+1})^G \cong (X^n)^G \times X$$

and since X is a pointed space, we have a map

$$\beta_{n,G}\colon (X^n)^G \to (X^{n+1})^G$$

$$(x_1, \ldots, x_n) \mapsto (x_1, \ldots, x_n, *).$$

One can check that the morphisms $\{\beta_{n,G}\}_{G \subseteq \Sigma_n}$ give a natural transformation

$$\beta_n\colon F_n X \to F_{n+1} X \circ J_n$$

and from the remarks made in Section 2 it follows that the functors $\{F_n\}_{n \geq 0}$ and natural transformations $\{\beta_n\}_{n \geq 0}$ can be assembled to define a functor

$$FX\colon Gr(O) \to \mathbf{Spaces}_*.$$

Lemma 3.2. $\mathrm{hocolim}_{Gr(O)} FX \simeq \mathrm{Sp}^\infty X$.

Proof. By Proposition 2.2 we have a weak equivalence

$$\mathrm{hocolim}_{Gr(O)} FX \simeq \mathrm{hocolim}_{\mathbf{N}}\, \mathrm{hocolim}_{\mathbf{O}^{\mathrm{op}}_{\Sigma_n}} F_n X.$$

Moreover, by [D1, Ch. 4, Lemma A.3],

$$\mathrm{hocolim}_{\mathbf{O}^{\mathrm{op}}_{\Sigma_n}} F_n X \simeq \mathrm{colim}_{\mathbf{O}^{\mathrm{op}}_{\Sigma_n}} F_n X$$

(this follows from the fact that $F_n X$ is a free diagram of spaces [D2, 2.7], and that for free diagrams homotopy colimits coincide with ordinary colimits). But $\mathrm{colim}_{\mathbf{O}^{\mathrm{op}}_{\Sigma_n}} F_n X \cong \mathrm{Sp}^n X$ and so we have

$$\mathrm{hocolim}_{\mathbf{N}}\, \mathrm{hocolim}_{\mathbf{O}^{\mathrm{op}}_{\Sigma_n}} F_n X \simeq \mathrm{hocolim}_{\mathbf{N}}\, \mathrm{Sp}^n X \simeq \mathrm{colim}_{\mathbf{N}} \mathrm{Sp}^n X \cong \mathrm{Sp}^\infty X,$$

where the second equivalence is a consequence of [BK, Ch. XII, 3.5]. □

4. Proof of Theorem 1.1

Let $H\colon \mathbf{FMod}_R \to \mathbf{Spaces}$ be a functor as in the theorem. One can assume that H takes its values in the category $\mathbf{Spaces}_*$ of pointed spaces (if not, replace H with $\widetilde{H}$, where $\widetilde{H}(M) = \mathrm{cofib}(H(0) \to H(M))$ for $M \in \mathbf{FMod}_R$). Moreover, once we know that the theorem holds for $R = \mathbb{Z}$, the ring of integers, the embedding $\mathbb{Z} \hookrightarrow R$ will induce a functor $\mathbf{FMod}_\mathbb{Z} \to \mathbf{FMod}_R$, and so the space $H(R)$ will have the structure of a GEM. Furthermore the action of the ring R on its free module $R \in \mathbf{FMod}_R$ via multiplications will induce an action of R on $H(R)$ and so the homotopy groups $\pi_i(H(R))$ will be R-modules as claimed. Therefore for the rest of this paper we will assume that $R = \mathbb{Z}$ and that $H\colon \mathbf{FMod}_\mathbb{Z} \to \mathbf{Spaces}_*$.

For a free abelian group on n generators $\mathbb{Z}^n \in \mathbf{FMod}_\mathbb{Z}$ the group Σ_n acts on $\mathbb{Z}^n$ by permuting the set of generators. For $G \subseteq \Sigma_n$ let $(\mathbb{Z}^n)^G$ be the subgroup of all elements of $\mathbb{Z}^n$ which are fixed by the action of G.

Remark 4.1. It is not difficult to check that, using the notation of 3.1, there is a natural isomorphism of groups $(\mathbb{Z}^n)^G \cong \mathbb{Z}^{|\mathrm{orb}\, G|}$.

For any $n \in \mathbf{N}$ we have a functor

$$F_n\mathbb{Z}\colon \mathbf{O}^{\mathrm{op}}_{\Sigma_n} \to \mathbf{FMod}_\mathbb{Z}, \quad F_n\mathbb{Z}(\Sigma_n/G) = (\mathbb{Z}^n)^G.$$

Arguments similar to those in Section 3 show that one can define a functor

$$F\mathbb{Z}\colon Gr(O) \to \mathbf{FMod}_\mathbb{Z}$$

such that $F\mathbb{Z}\,|_{\mathbf{O}^{\mathrm{op}}_{\Sigma_n}} = F_n\mathbb{Z}$. Observe that Σ_1 is a trivial group and so $F\mathbb{Z}\,|_{\mathbf{O}^{\mathrm{op}}_{\Sigma_1}} = F_1\mathbb{Z}$ is the constant functor with value $\mathbb{Z}$.

Lemma 4.2. *Let $\bar{Z}\colon Gr(O) \to \mathbf{FMod}_\mathbb{Z}$ be the constant functor with value $\mathbb{Z}$. There exists a natural transformation*

$$\theta\colon F\mathbb{Z} \to \bar{Z}$$

such that $\theta\,|_{\mathbf{O}^{\mathrm{op}}_{\Sigma_1}}$ is an isomorphism.

Proof. For $\Sigma_n/G \in \mathbf{O}^{\mathrm{op}}_{\Sigma_n} \subset Gr(O)$ define

$$\theta_{\Sigma_n/G}\colon (\mathbb{Z}^n)^G \to \mathbb{Z}$$
$$(k_1, k_2, \ldots, k_n) \mapsto \sum_i k_i.$$

It is easy to check that these maps give the required transformation of functors. □

Let $H\colon \mathbf{FMod}_{\mathbb{Z}} \to \mathbf{Spaces}_*$ be a functor satisfying the conditions of Theorem 1.1; that is, the projections $\mathbb{Z}^n \to \mathbb{Z}$ induce weak equivalences $H(\mathbb{Z}^n) \to H(\mathbb{Z})^n$.

Lemma 4.3. $\mathrm{hocolim}_{Gr(O)} H \circ F\mathbb{Z} \simeq \mathrm{Sp}^\infty H(\mathbb{Z})$.

Proof. It follows from 3.1 and 4.1 that for $G \subseteq \Sigma_n$ we have isomorphisms

$$H((\mathbb{Z}^n)^G) \cong H(\mathbb{Z}^{|\mathrm{orb}\, G|})$$

and

$$(H(\mathbb{Z})^n)^G \cong H(\mathbb{Z})^{|\mathrm{orb}\, G|}.$$

Their composition with the map $H(\mathbb{Z}^{|\mathrm{orb}\, G|}) \to H(\mathbb{Z})^{|\mathrm{orb}\, G|}$ induced by projections $\mathbb{Z}^{|\mathrm{orb}\, G|} \to \mathbb{Z}$ gives a map $\varphi_{n,G}\colon H((\mathbb{Z}^n)^G) \to (H(\mathbb{Z})^n)^G$ which, in view of the properties of H, is a weak equivalence. Moreover, the maps $\{\varphi_{n,G}\}_{n\geq 0,\, G\subseteq\Sigma_n}$ define a natural transformation of functors

$$\varphi\colon H \circ F\mathbb{Z} \to FH(\mathbb{Z}).$$

Therefore we have a weak equivalence

$$\mathrm{hocolim}_{Gr(O)} H \circ F\mathbb{Z} \xrightarrow{\simeq} \mathrm{hocolim}_{Gr(O)} FH(\mathbb{Z}).$$

But by Lemma 3.2 $\mathrm{hocolim}_{Gr(O)} FH(\mathbb{Z}) \simeq \mathrm{Sp}^\infty H(\mathbb{Z})$. □

To conclude the proof of the theorem observe that the natural transformation θ from Lemma 4.2 gives a transformation

$$H(\theta)\colon H \circ F\mathbb{Z} \to H \circ \bar{Z}$$

and so induces a map

$$\mathrm{hocolim}_{Gr(O)} H \circ F\mathbb{Z} \to \mathrm{hocolim}_{Gr(O)} H \circ \bar{Z} \to \mathrm{colim}_{Gr(O)} H \circ \bar{Z} \cong H(\mathbb{Z}).$$

On the other hand, the inclusion $\mathbf{O}^{\mathrm{op}}_{\Sigma_1} \subseteq Gr(O)$ gives a map

$$H(\mathbb{Z}) \simeq \mathrm{hocolim}_{\mathbf{O}^{\mathrm{op}}_{\Sigma_1}} H \circ F_1\mathbb{Z} \to \mathrm{hocolim}_{Gr(O)} H \circ F\mathbb{Z}$$

and since $\theta\,|_{\mathbf{O}^{\mathrm{op}}_{\Sigma_1}}$ is an isomorphism it is easy to see that the composition

$$H(\mathbb{Z}) \to \mathrm{hocolim}_{Gr(O)} H \circ F\mathbb{Z} \to H(\mathbb{Z})$$

has to be a weak equivalence. But, by 4.3, $\mathrm{hocolim}_{Gr(O)} H \circ F\mathbb{Z} \simeq \mathrm{Sp}^\infty H(\mathbb{Z})$, and so $H(\mathbb{Z})$ must be a GEM as a homotopy retract of a GEM (see [D1, Ch. 4, Thm. B.2]).

Remark 4.4. The above proof remains valid if we replace $\mathbf{FMod}_{\mathbb{Z}}$ by the category of free, finitely generated abelian monoids.

References

[Bo] A. K. Bousfield, *Unstable localization and periodicity*, Proceedings of the Barcelona Conference in Algebraic Topology 1994, Progress in Mathematics, vol. 136, Birkhäuser, 1996, pp. 33–50.

[BK] A. K. Bousfield, D. M. Kan, *Homotopy limits, completions and localizations*, Lecture Notes in Mathematics, vol. 304, Springer, 1972.

[ChS] W. Chachólski, J. Scherer, *Homotopy meaningful constructions: homotopy colimits*, Preprint #379, CRM Barcelona, 1998.

[D1] E. Dror Farjoun, *Cellular spaces, null spaces and homotopy localization*, Lecture Notes in Mathematics, vol. 1622, Springer, 1996.

[D2] E. Dror Farjoun, *Homotopy and homology of diagrams of spaces*, Algebraic Topology (Seattle, 1985), Lecture Notes in Mathematics, vol. 1286, Springer, 1987, pp. 93–134.

[Se] G. Segal, *Categories and cohomology theories*, Topology 13 (1974), pp. 293–312.

[Sl] J. Słomińska, *Homotopy colimits on E-I-categories*, Algebraic Topology Poznań 1989, Lecture Notes in Mathematics, vol. 1474, Springer, 1991, pp. 273–294.

[Th] R. W. Thomason, *Homotopy colimits in the category of small categories*, Math. Proc. Cambridge Philos. Soc. 85 (1979), pp. 91–109.

Department of Mathematics,
University of Notre Dame,
Notre Dame, IN 46556, USA
E-mail address: `badzioch.1@nd.edu`

Progress in Mathematics, Vol. 196

Groups with infinite homology

A. J. Berrick and P. H. Kropholler

1. Introduction

We consider the reduced homology of a group G with coefficients in the trivial module:

$$\widetilde{H}(G) = \bigoplus_{n=1}^{\infty} H_n(G;\mathbb{Z}).$$

A group is said to be acyclic if its reduced homology vanishes. Many interesting classes of groups have been discovered having this property ([2] is a useful survey). This is an indication that the reduced homology carries limited information.

Here we obtain information about $\widetilde{H}(G)$ for a class of groups G that includes all locally finite groups and all soluble groups of finite rank. We show that non-locally-finite groups in the class cannot be acyclic, and that in fact their reduced homology is infinite.

Our main result is as follows.

Theorem 1.1. *Let G be a group having a series of finite length whose factors are either infinite cyclic or locally finite. Then the reduced homology $\widetilde{H}(G)$ is infinite or zero. Indeed, if $\widetilde{H}(G)$ is a torsion group, then G is locally finite, and either:*

(i) *for some prime p occurring as the order of an element of G, and for infinitely many n, $H_n(G;\mathbb{Z})$ contains elements of order p; or*

(ii) *G is acyclic.*

The proof brings together some techniques from homotopy theory and some Euler characteristic arguments. There is an interesting dichotomy between the cases of torsion-free and non-torsion-free groups. This shows up in the choice of prime field that results in infinite homology. When G is torsion-free it has finite cohomological dimension, and in this case one must prove that at least one of its rational homology groups is nonzero. At the other extreme, when G is locally finite it may be acyclic —the McLain group $M(\mathbb{Q},\mathbb{F}_p)$ is an example. (The McLain group $M(\mathbb{Q},\mathbb{F}_p)$ may be thought of as the group of upper unitriangular matrices with entries in $\mathbb{F}_p$, but with rows and columns indexed by the rational numbers rather than the natural numbers; see [1].) Otherwise, we show that for some prime p that is the order of an element of G the integral homology contains elements of order p in arbitrarily high dimensions; this has long been known to be the case for finite G. The phrase "for some prime" is best possible, in that the example of the direct product of a locally finite acyclic group and a group of prime order shows that there may be a unique prime that is detected by integral homology.

Note that for groups in general it is possible to have the second sentence of the theorem hold but neither (i) nor (ii). For instance, by [3] there is a perfect, torsion-generated group with nth integral homology group of order n whenever n is prime, and zero otherwise.

2. Two ingredients from homotopy theory

Our arguments rest on two results, of independent interest, which have a homotopy theoretic pedigree. They sharpen results of [3] which consider only integral coefficients. Let k be a commutative ring. We write $\mathrm{thd}_k(G)$ for the trivial homological dimension of G over k; this is the largest integer m for which $H_m(G;k)$ is nonzero, or infinity in case there is no such integer. We also use similar notation with G replaced by a topological space. First, here is a lemma that is a useful ancillary result for applications of Miller's Theorem [5]. It is used implicitly in [3].

Lemma 2.1. *If X has the homotopy type of a CW-complex and $\mathrm{thd}_{\mathbb{Z}}(X)$ is finite, then the suspension ΣX has the homotopy type of a finite CW-complex.*

Proof. Write $k = \mathrm{thd}_{\mathbb{Z}}(X)$. When $k = 0$, X is acyclic, making ΣX contractible; so we may suppose that $k \geq 1$. Assuming also that X is actually a CW-complex, let $C(j)$ be the mapping cone of the inclusion $j\colon X^{(k)} \hookrightarrow X$ of its k-skeleton. Then the homology exact sequence of the pair $(X, X^{(k)})$ shows that $H_{k+1}(C(j))$ embeds in $H_k(X^{(k)})$, which is in turn a subgroup of the free abelian group of k-chains on $X^{(k)}$. Thus $C(j)$ has the homology of a wedge of $(k+1)$-spheres. Since $k \geq 1$, it follows from van Kampen's theorem that $C(j)$ is homotopy equivalent to $\bigvee S^{k+1}$. Therefore ΣX has the homotopy type of the mapping cone of some map $\bigvee S^{k+1} \to \Sigma X^{(k)}$, a CW-complex of dimension at most $k+2$. □

Theorem 2.2. *Let p be a fixed prime and let G be a group with $\mathrm{thd}_{\mathbb{F}_p}(G)$ finite. Then any homomorphism $G \to \mathrm{GL}(\mathbb{C})$ is trivial on all elements of p-power order.*

Proof. Since the argument involves only a tweaking of that given in [3] Theorem 3, we indicate it briefly. As in [3], for any cyclic subgroup C of p-power order in G, by change of basis the homomorphism can be made to restrict to a unitary representation on C, giving rise to the commuting diagram

$$\begin{array}{ccccc} BC & \longrightarrow & BU & \xrightarrow{\simeq} & \Omega^2 BU \\ \downarrow & & \downarrow\simeq & & \\ BG & \longrightarrow & B\,\mathrm{GL}(\mathbb{C}). & & \end{array}$$

We focus on the adjunction $\Sigma^2 BC \to \Sigma^2 BG \to BU$. Since $\Sigma^2 BG$ is simply-connected with $\mathrm{thd}_{\mathbb{F}_p}(\Sigma^2 BG)$ finite, by Theorems C, D of [5] $\mathrm{map}_*(BC, \Sigma^2 BG)$ is weakly contractible. This makes $[\Sigma^2 BC, \Sigma^2 BG] = 0$. Then the result follows as in [3] by appeal to Atiyah's embedding of the complex representation ring of C in $\mathbb{Z}\times[BC, BU]$. □

Theorem 2.3. *Let p be a fixed prime and let G be a locally finite group with $k = \mathrm{thd}_{\mathbb{F}_p}(G)$ finite. Then G is $\mathbb{F}_p$-acyclic, that is, $\mathrm{thd}_{\mathbb{F}_p}(G) = 0$.*

Proof. As ΣBG is simply-connected, we pass to its p-localization $(\Sigma BG)_{(p)}$, which is also simply-connected. By definition, it is $\mathbb{F}_q$-acyclic whenever $q \neq p$, while its homology with other prime field coefficients is that of ΣBG. Hence $\mathrm{thd}_{\mathbb{F}_p}((\Sigma BG)_{(p)})$ is finite. On the other hand, since G is locally finite it is the direct limit of its finite subgroups. So its homology is the direct limit of the homology of those finite subgroups; thus $\widetilde{H}((\Sigma BG)_{(p)}; \mathbb{Q}) = 0$. Therefore, by universal coefficients, $\mathrm{thd}_{\mathbb{Z}}((\Sigma BG)_{(p)})$ is finite, and the lemma thus implies that $(\Sigma BG)_{(p)}$ has the homotopy type of a finite-dimensional complex. It follows from Miller's Theorem that $[\Sigma BG, (\Sigma BG)_{(p)}] = 0$, whence by its universal property $(\Sigma BG)_{(p)}$ is contractible. This makes zero the reduced homology of ΣBG, hence of G, with coefficients in the local ring $\mathbb{Z}_{(p)}$. Since $\mathbb{F}_p$ is a $\mathbb{Z}_{(p)}$-module, it follows by universal coefficients that $\widetilde{H}(G; \mathbb{F}_p) = 0$ too. □

3. Relevant groups, and notation

For any group G, let $\tau(G)$ denote the unique largest normal locally finite subgroup (for example, see [6] p. 418). We consider the class of groups having a series of finite length whose factors are either infinite cyclic or locally finite. Then the number of infinite cyclic factors in such a series is an invariant of the group, known as the torsion-free rank or Hirsch length $h(G)$ (cf. [6] p. 407). Recall also that the Fitting subgroup, or nilpotent radical, of a group is defined to be the product of all its nilpotent normal subgroups. We shall use the following folklore characterization of this class.

Proposition 3.1. *Suppose that G admits a finite series*

$$1 = G_0 \triangleleft G_1 \triangleleft \cdots \triangleleft G_n = G$$

in which the factors G_i/G_{i-1} are either infinite cyclic or locally finite.

Let H denote the subgroup containing $\tau(G)$ such that $H/\tau(G)$ is the Fitting subgroup of $G/\tau(G)$. Let K denote the subgroup containing H such that $K/H = \tau(G/H)$. Then

1. *there is a finite dimensional $\mathbb{C}$-linear representation of G with kernel equal to $\tau(G)$,*
2. *$H/\tau(G)$ is torsion-free nilpotent and of finite Hirsch length,*
3. *K/H is finite, and*
4. *G/K is a Euclidean crystallographic group.*

Thus, schematically, G has the following decomposition.

$$\begin{array}{ll}
G & \\
\mid & \text{maximal abelian, by finite} \\
K & \\
\mid & \text{finite nilpotent} \\
H & \\
\mid & \text{torsion-free nilpotent} \\
\tau(G) & \\
\mid & \text{locally finite} \\
1 &
\end{array}
\quad \left.\begin{array}{l} \\ \\ \\ \\ \\ \end{array}\right\} \text{linear}$$

Proof. (Outline) Professor Wehrfritz has kindly indicated some steps in the argument here. Such a group is locally finite, by soluble with a finite series with abelian torsion-free factors of finite rank, by finite. The proof is by induction on the Hirsch length, noting that such G with no normal torsion are linear over the rationals [11], and such a G lying in $GL_n(\mathbb{Q})$ must be soluble by finite [9]. Now by a theorem of Mal'cev [6] p. 436, soluble groups linear over $\mathbb{Q}$ are torsion-free nilpotent by abelian by finite, while by Gruenberg [10] p. 102 their Fitting subgroups are nilpotent. Finally, the deduction that G/K is actually maximal abelian by finite, in other words crystallographic, may be seen from an argument of Zassenhaus [6] p. 435. □

Observe that this particular class of groups is closed under passage to subgroups, quotients and extensions.

For G in this class, an easy spectral sequence argument shows that the rational homology groups $H_n(G;\mathbb{Q})$ are all finite-dimensional over $\mathbb{Q}$, and that $\mathrm{thd}_{\mathbb{Q}}(G) \leq h(G) < \infty$ (cf. [8]). We shall use the notation $\bar{\chi}(G)$ for the naive Euler characteristic,

$$\bar{\chi}(G) = \sum_{n=0}^{\infty} (-1)^n \dim_{\mathbb{Q}} H_n(G;\mathbb{Q}).$$

This definition suits our purpose although it is not always easily or closely connected with the classical Euler characteristic. For example, the infinite dihedral group D_∞ has Euler characteristic zero whereas $\bar{\chi}(D_\infty) = 1$. We refer the reader to Brown's book [4] for a detailed account of Euler characteristics and cohomological methods.

4. Proof of the main theorem

Our strategy is to use the naive Euler characteristic to set up contradictions that reduce to consideration of locally finite groups. Throughout we assume that G satisfies the hypotheses of Proposition 3.1.

Lemma 4.1. *If G is nontrivial and torsion-free, then $\bar{\chi}(G) = 0$.*

Proof. Let H and K be the subgroups defined using Proposition 3.1. Since G is nontrivial and torsion-free, we know that $\tau(G)$ is trivial and H is infinite. Let L be any subgroup containing H as a normal subgroup of finite index. Now L is torsion-free and nilpotent-by-finite, of finite Hirsch length equal to the Hirsch length $h(H)$ of H. Let L_0 be a finitely generated subgroup of L that contains a transversal to H in L and which has full rank, that is $h(L_0) = h(H)$. Let H_0 denote $H \cap L_0$. Then the inclusion of H_0 into H induces isomorphisms in rational homology. Moreover the short exact sequence

$$H_0 \rightarrowtail L_0 \twoheadrightarrow H/L$$

embeds naturally into

$$H \rightarrowtail L \twoheadrightarrow H/L$$

and comparison of the associated Lyndon–Hochschild–Serre spectral sequences shows that the inclusion of L_0 into L induces isomorphisms in rational homology. Hence $\bar{\chi}(L_0) = \bar{\chi}(L)$. Since L_0 is torsion-free, virtually nilpotent and finitely generated, it is a Poincaré duality group of type FP and our naive Euler characteristic coincides with the genuine classical Euler characteristic which is zero in this case (see [4] pp. 201, 213, 224, [7]). Thus $\bar{\chi}(L)$ is zero.

Since G/K is a crystallographic group it admits a proper cocompact action on a Euclidean space X which we may suppose to be endowed with a CW-structure. In this way we have a contractible cocompact G-CW-complex X in which each stabilizer is a finite extension of K (see [12] (3.1.2), (3.1.3)). Consider the Leray spectral sequence ([4] VII.7.10)

$$E^1_{p,q} = \bigoplus_{\dim \sigma = p} H_q(G_\sigma; \mathbb{Q}) \qquad \Longrightarrow \qquad H_{p+q}(G; \mathbb{Q})$$

where σ runs through a set of orbit representatives of cells in X. Reading Euler characteristics, we have

$$\begin{aligned} \bar{\chi}(G) &= \sum_{p,q} (-1)^{p+q} \dim E^1_{p,q} \\ &= \sum_{\sigma} (-1)^{\dim \sigma} \left(\sum_q (-1)^q \dim H_q(G_\sigma; \mathbb{Q}) \right) \\ &= \sum_{\sigma} (-1)^{\dim \sigma} \, \bar{\chi}(G_\sigma). \end{aligned}$$

The stabilizers G_σ are all groups of the form L considered above and so their naive Euler characteristics are always zero. Hence $\bar{\chi}(G)$ is zero too. □

Lemma 4.2. *If $\widetilde{H}(G)$ is torsion then $\bar{\chi}(G/\tau(G)) = 1$.*

Proof. The universal coefficient theorem shows that

$$H_n(G;\mathbb{Q}) = H_n(G;\mathbb{Z}) \otimes \mathbb{Q} = 0$$

for all $n \geq 1$, and hence $\bar{\chi}(G) = 1$. The subgroup $\tau(G)$ makes no impact on the rational homology with trivial coefficients and therefore $\bar{\chi}(G/\tau(G)) = 1$ also. □

Lemma 4.3. *If* $\mathrm{thd}_{\mathbb{F}_p}(G) < \infty$ *for all primes p occurring as orders of elements of G, then $G/\tau(G)$ is torsion-free.*

Proof. Using Proposition 3.1 (1), there is a finite dimensional $\mathbb{C}$-linear representation ρ of G having kernel $\tau(G)$. Theorem 2.2 shows that every element of finite order in G belongs to the kernel of ρ, and hence $G/\tau(G)$ is torsion-free. □

To prove the main theorem, suppose now that $\widetilde{H}(G)$ is torsion, yet for all primes p occurring as orders of elements of G only finitely many $H_n(G;\mathbb{Z})$ contain p-torsion. Then by universal coefficients, for all such p, $\mathrm{thd}_{\mathbb{F}_p}(G) < \infty$. Bringing Lemmas 4.2 and 4.3 into play, we see that $G/\tau(G)$ is torsion-free and has naive Euler characteristic equal to 1. Applying Lemma 4.1 to this quotient shows that it is trivial, and so $G = \tau(G)$ is locally finite. Therefore, as in the proof of Theorem 2.3, its reduced rational homology is zero. By Theorem 2.3 and the universal coefficient theorem, we conclude that G is acyclic.

References

[1] A. J. Berrick, Two functors from abelian groups to perfect groups, *J. Pure Appl. Algebra* **44** (1987), 35–43.

[2] A. J. Berrick, A topologists's view of perfect and acyclic groups, *Topics in Geometry and Topology*, eds. M. Bridson & S. Salamon, Oxford Univ. Press (Oxford, 2001), to appear.

[3] A. J. Berrick and C. F. Miller III, Strongly torsion generated groups, *Math. Proc. Cambridge Philos. Soc.* **111** (1992), 219–229.

[4] K. S. Brown, *Cohomology of Groups*, Graduate Texts in Math. **87**, Springer (Berlin, 1982).

[5] H. Miller, The Sullivan conjecture on maps from classifying spaces, *Ann. of Math.* (2) **120** (1984), 39–87.

[6] D. J. S. Robinson, *An Introduction to the Theory of Groups*, Graduate Texts in Math. **80**, Springer (Berlin, 1982).

[7] S. Rosset, A vanishing theorem for Euler characteristics, *Math. Z.* **185** (1984), 211–215.

[8] U. Stammbach, On the weak (homological) dimension of the group algebra of solvable groups, *J. London Math. Soc. (2)* **2** (1970), 567–570.

[9] J. Tits, Free subgroups of linear groups, *J. Algebra* **20** (1972), 250–270.

[10] B. A. F. Wehrfritz, *Infinite Linear Groups,* Ergebnisse der Math. **76**, Springer (Berlin, 1973).

[11] B. A. F. Wehrfritz, On the holomorphs of soluble groups of finite rank, *J. Pure Appl. Algebra* **4** (1974), 55–69.

[12] J. A. Wolf, *Spaces of Constant Curvature*, Publish or Perish (Berkeley, CA, 1977), 4th edn.

Department of Mathematics,
National University of Singapore,
Kent Ridge 117543, Singapore
E-mail address: `berrick@math.nus.edu.sg`

School of Mathematical Sciences,
Queen Mary, University of London,
Mile End Road, London E1 4NS, UK
E-mail address: `P.H.Kropholler@qmw.ac.uk`

Progress in Mathematics, Vol. 196

Unstable splittings related to Brown-Peterson cohomology

J. Michael Boardman and W. Stephen Wilson

Abstract. We give a new and relatively easy proof of the splitting theorem of the second author for the spaces in the Omega spectrum for BP. We then give the first published proofs of our similar theorems for the spectra $P(n)$.

1. Introduction

In [Wil75], unstable splittings were constructed for the spaces in the Omega spectrum for Brown-Peterson cohomology, a cohomology theory with coefficient ring $BP^* \simeq \mathbf{Z}_{(p)}[v_1, v_2, \dots, v_n, \dots]$. This was done using the Postnikov decomposition and a multiple induction. The proof, from [Wil73] (again using Postnikov systems), that these spaces had no torsion was essential in this proof. In [RW77], the calculation of the homology of those spaces for BP was done as a Hopf ring, a great improvement, and this was used in [BJW95] to construct an unstable idempotent to get the splittings. Some of the splittings are not as H-spaces and so the full power of non-additive unstable operations was required. The splittings were generalized in [BW] to the spaces in the Omega spectrum for $P(n)$, $n > 0$, a theory with coefficient ring $P(n)^* \simeq BP^*/I_n$ where $I_n = (p, v_1, v_2, \dots, v_{n-1})$, after the calculation of the Hopf ring for these spaces in [RW96]. This calculation was done with the intent of getting an analogous splitting. The technique is again to construct an unstable idempotent to get the splittings.

Although the technique of constructing unstable idempotents is clearly the proper way to prove these results, it requires an immense amount of technical machinery which cannot be accused of being easily accessible. In fact, this difficulty has led to the present paper being published well before the paper with the first proof in it.

We wish to present a much more direct proof of these splittings which requires none of the machinery of unstable operations. In fact, the proofs could be done quite easily if one could just insert a few short paragraphs into the papers [RW77] and [RW96]. Unfortunately that option is not open to us. If we essentially reproduce those papers to insert what little extra is needed, then the proofs cease to be "easy." On the other hand, if we just create those paragraphs to be inserted,

1991 *Mathematics Subject Classification.* Primary 55N22; Secondary 55N20.

then the result remains obscure. We will try to walk a fine line between these two approaches. Our goal will be to write the necessary insertions in such a way that a rigorous proof has been accomplished when combined with the previous two papers but at the same time discuss the results in enough depth so the reader should be convinced of the result without having to consult the other papers.

First we need to establish some notation and state our results. We let $P(0)$ represent BP so BP is not an exceptional case. In fact it is both easier and quite different from the $n > 0$ case. There are also theories $BP\langle m\rangle$ with homotopy $\mathbf{Z}_{(p)}[v_1, \dots, v_m]$ and $P(n,m)$ with homotopy $BP\langle m\rangle_*/I_n$ ($I_0 = (0)$, so $BP\langle m\rangle = P(0,m)$) which are associative ring spectra by [SY76]. If E is a spectrum we denote the spaces in the Omega spectrum by $\{\underline{E}_k\}$. Define $g(n,m)$, for $n \leq m$ to be $2(p^n + p^{n+1} + \dots + p^m)$. We will show that for $k \leq g(n,m)$, $\underline{P(n,m)}_k$ splits off of $\underline{P(n)}_k$. Precisely:

Theorem 1.1 ([Wil75], and later [BJW95], for n=0 and also [BW] for n>0). *For $k \leq g(n,m)$ there is an unstable splitting*

$$\underline{P(n)}_k \simeq \underline{P(n,m)}_k \prod_{j>m} \underline{P(n,j)}_{k+2(p^j-1)}.$$

Remark 1.2. Once the bottom piece has been split off the rest of the splitting follows easily. If $k < g(n,m)$ then it follows that this is a splitting of H-spaces.

Remark 1.3. We do not recover the result of [BW] for $n > 0$ and $k = g(n,m)$ that this is still a splitting as H-spaces (only for odd primes). The fact that the homology splits off as Hopf algebras tells us nothing. If $k > g(n,m)$ then it is easy to see from our approach that the homology of the smaller piece no longer splits off and so there is no such homotopy splitting as well. One of the major attractions about our approach is that we don't have to worry in any way about the additivity of our splittings. For the $n = 0$, $k = g(0,m)$ case where the splittings are not additive, this is a major complication in the proof using unstable idempotents. Our approach is indifferent to such matters although it does show the non-additivity of this splitting because we see that the homology splitting cannot be as Hopf algebras.

Remark 1.4. Note that when $n = m$ the small bottom piece of the splitting is just a space in the spectrum for connective Morava K-theory. Furthermore, when $k < 2p^n - 2$, this is a space in the spectrum for periodic Morava K-theory.

Remark 1.5. Another major complication in [BW] for the $n > 0$ case is the prime 2. The theories involved are not homotopy commutative ring spectra and so the machinery for unstable idempotents must be extended and contorted to deal with this special case. These complications have led to significant delays in the publication of this work. One of the benefits of our approach is once again that we do not need to worry about such things. The $p = 2$ proof is identical to the odd prime proof and we need not be concerned whether any of the spectra are commutative ring spectra or not. The $p = 2$ version of the theorem is very important.

A major motivation for the theorem, and even for a second or third proof, is its important applications. First, it is easy to prove a generalization of Quillen's theorem from the splitting.

Theorem 1.6 (For $n=0$, [Qui71], and, for $n>0$, also [BW].). *For X a finite complex, $P(n)^*(X)$ is generated by non-negative degree elements.*

Proofs for the $n=0$ case using the splitting appear in [Wil75] and [BJW95]. In this last case a more general, purely algebraic, version about unstable modules is proven. Strickland has shown that Quillen's proof cannot be generalized to the $n>0$ case.

Second, all of the results of [RWY98] are unstable, and the only unstable input is this generalized Quillen theorem. Everything depends on it. Thus, we feel it is important to have a relatively simple and accessible proof, especially one with no complications associated with the prime two or the non-additive splittings.

The authors are particularly indebted to the organizers of the 1998 Barcelona Conference on Algebraic Topology and the Centre de Recerca Matemàtica for the conference and terrific work environment during the extended stay of the second author. Thanks are also due to our previous coauthors, David Johnson and Douglas Ravenel, for the work with them which led us to this paper.

2. Approach and simple parts of the proofs

In [RW77] and [RW96] the homology of the spaces in the Omega spectra were calculated by induction on degree using the bar spectral sequence

$$\mathrm{Tor}^{H_*(\underline{P(n)}_*)}(\mathbf{Z}/(p),\mathbf{Z}/(p)) \Rightarrow H_{*+1}(\underline{P(n)}_{*+1}).$$

The H_{n+1} is there to indicate that we gain a degree in this inductive calculation.

What was not noticed in those two papers was that we could easily have simultaneously calculated the homology of the bottom piece which splits off. We would just use the spectral sequence

$$\mathrm{Tor}^{H_*(\underline{P(n,m)}_*)}(\mathbf{Z}/(p),\mathbf{Z}/(p)) \Rightarrow H_{*+1}(\underline{P(n,m)}_{*+1}).$$

There is a map of spectra $P(n) \to P(n,m)$ which induces a map on the spectral sequences. The proof of the calculation of this spectral sequence is identical up to $k=g(n,m)$ with the exception of one slightly modified definition. It is easy to see that the calculation cannot go one step higher.

Once the homology is calculated it is easy to see that the Atiyah-Hirzebruch spectral sequence for the $P(n)$ and $P(n,m)$ homology of these spaces collapses and so they are free over the coefficient rings.

Corollary 2.1. *For $k \le g(n,m)$ the Atiyah-Hirzebruch spectral sequences*

$$H_*(\underline{P(n,m)}_k; P(n)_*) \text{ and } H_*(\underline{P(n,m)}_k; P(n,m)_*)$$

collapse and $P(n)_*(\underline{P(n,m)}_k)$ is $P(n)_*$ free and $P(n,m)_*(\underline{P(n,m)}_k)$ is $P(n,m)_*$ free.

This in turn allows us to calculate the cohomology theories as the dual and get:

Corollary 2.2. *For $k \leq g(n,m)$, the map*

$$P(n)^*(\underline{P(n,m)}_k) \to P(n,m)^*(\underline{P(n,m)}_k)$$

is surjective.

Proof of Theorem 1.1. To get our splitting we just note that the identity map of $\underline{P(n,m)}_k$, an element of $P(n,m)^k(\underline{P(n,m)}_k)$, has a lift to a map of $\underline{P(n,m)}_k$ to $\underline{P(n)}_k$, an element of $P(n)^k(\underline{P(n,m)}_k)$. That splits off the bottom piece. The other pieces are handled by taking the maps

$$\underline{P(n,j)}_{k+2(p^j-1)} \to \underline{P(n)}_{k+2(p^j-1)} \to \underline{P(n)}_k$$

where the first map is just the bottom piece splitting and the second map comes from the stable map of v_j. Using the H-space structure to add all of these maps up we see we have a homotopy equivalence due to the obvious isomorphism on homotopy groups.

This concludes the proof of the splitting from the calculation of the homology and the collapsing of the Atiyah-Hirzebruch spectral sequence. □

The summary given above constitutes a "proof" that if the splittings exist, then there must be an "easy" proof along the lines just outlined.

We now want to give a proof of the Quillen theorem.

Proof of Theorem 1.6. This proof is exactly the same as the proof of the original Quillen theorem given in [Wil75], but because it is of such major importance, and it is short, we reproduce it here. There are stable cofibrations

$$\Sigma^{2(p^m-1)}P(n,m) \xrightarrow{v_m} P(n,m) \longrightarrow P(n,m-1) \tag{2.1}$$

where $P(n,n-1)$ is the mod p Eilenberg–Mac Lane spectrum. These cofibrations give rise to long exact sequences in cohomology theories. Given a negative degree element $x \in P(n)^*(X)$ where X is a finite complex, we see that there is some $q \geq n$ for which x maps to zero in $P(n,q-1)^*(X)$ because mod p cohomology is zero in negative degrees. Using the above exact sequence there is an element $y_q \in P(n,q)^*(X)$ such that $v_q y_q$ is the image of x in $P(n,q)^*(X)$. Because x has negative degree, the element y_q, which is a map of X into the space $\underline{P(n,q)}_{|x|+2(p^q-1)}$, is in the range of the splitting Theorem 1.1 and so the element y_q can be lifted to $P(n)^*(X)$. We now look at the element $x - v_q y_q$. It will go to zero in $P(n,q)^*(X)$ and we can iterate this process. It is a finite process since the splitting Theorem 1.1, combined with the finiteness of X, tells us that for some large m, $P(n)^k(X)$ is the same as $P(n,m)^k(X)$ for a fixed $k = |x|$. At that point we have $x = \sum v_i y_i$, a finite sum. Thus any negative degree element is decomposable in this way and we have proven the generalized Quillen theorem. (For the $n = 0$ case we let $P(0,0)$ be the $\mathbf{Z}_{(p)}$ cohomology and we do not need the mod p cohomology.) □

3. The Homology

Let $H_*(-)$ be the standard mod p homology where p is the prime associated with the spectrum $P(n,m)$. Because $P(n,m)$ is a ring spectrum there are maps

$$\underline{P(n,m)}_i \times \underline{P(n,m)}_j \longrightarrow \underline{P(n,m)}_{i+j}$$

corresponding to cup product, in addition to the loop space product

$$\underline{P(n,m)}_i \times \underline{P(n,m)}_i \longrightarrow \underline{P(n,m)}_i.$$

These induce pairings

$$\circ : H_*(\underline{P(n,m)}_i) \otimes H_*(\underline{P(n,m)}_j) \to H_*(\underline{P(n,m)}_{i+j})$$

and

$$* : H_*(\underline{P(n,m)}_i) \otimes H_*(\underline{P(n,m)}_i) \to H_*(\underline{P(n,m)}_i).$$

Since $H_*(-)$ has a Kunneth isomorphism these pairings satisfy certain identities making $H_*(\underline{P(n,m)}_*)$ into a *Hopf ring*, [RW77], i.e., a ring object in the category of coalgebras.

There are special elements

$$\begin{aligned} e &\in P(n)_1(\underline{P(n,m)}_1), \\ a_{(i)} &\in P(n)_{2p^i}(\underline{P(n,m)}_1) && \text{for } 0 \le i < n, \\ [v_i] &\in P(n)_0(\underline{P(n,m)}_{-2(p^i-1)}) && \text{for } m \ge i \ge n, \quad i > 0, \quad \text{and} \\ b_{(i)} &\in P(n)_{2p^i}(\underline{P(n,m)}_2) && \text{for } i \ge 0, \end{aligned}$$

which have already been defined in [RW96] in $P(n)_*(\underline{P(n)}_*)$ and we get these by just pushing them down using the map from $P(n)$ to $P(n,m)$ to get them first in $P(n)_*(\underline{P(n,m)}_*)$ and then into $P(n,m)_*(\underline{P(n,m)}_*)$. They then push down non-trivially to $H_*(\underline{P(n,m)}_*)$.

A basic property which we need and which comes out of the construction of these elements, (this goes clear back to [Wil84]), is:

Proposition 3.1. *The elements e, $a_{(i)}$, $[v_i]$, and $b_{(i)}$ are permanent cycles in the Atiyah-Hirzebruch spectral sequence for $P(n)_*(-)$ and $P(n,m)_*(-)$.*

Other facts proven about these elements are in [Wil84, Proposition 1.1] and were repeated again in [RW96, Proposition 2.1, p. 1048] and are not repeated again here.

Let

$$e^\varepsilon a^I [v^K] b^J = e^\varepsilon \circ a_{(0)}^{\circ i_0} \circ \cdots \circ a_{(n-1)}^{\circ i_{n-1}} \circ [v_n^{k_n} v_{n+1}^{k_{n+1}} \cdots] \circ b_{(0)}^{\circ j_0} \circ b_{(1)}^{\circ j_1} \cdots$$

where $\varepsilon = 0$ or 1, $i_q = 0$ or 1, $k_q \ge 0$, and $j_q \ge 0$, (K and J finite), and if $n = 0$, $k_0 = 0$.

Definition 3.2. For $n > 0$ we say $e^\varepsilon a^I[v^K]b^J$ is *nm*-**allowable** if

$$J = p^n\Delta_{d_n} + p^{n+1}\Delta_{d_{n+1}} + \cdots + p^q\Delta_{d_q} + J'$$

where Δ_d has a 1 in the dth place and zeros elsewhere, $d_n \le d_{n+1} \le \cdots \le d_q$ and J' is non-negative implies $k_q = 0$. In other words,

$$[v_q] \circ b^{p^n\Delta_{d_n}+p^{n+1}\Delta_{d_{n+1}}+\cdots+p^q\Delta_{d_q}}$$

does not divide $e^\varepsilon a^I[v^K]b^J$ when $d_n \le d_{n+1} \le \cdots \le d_q$, $q < m$. We will denote the set of such (K, J) by $\mathcal{A}_{nm}$. If we eliminate the reference to m then we have the n-allowable of [RW96]. Because we do not want to use $[v_0]$ in the $n = 0$ case we set $\mathcal{A}_{0m} = \mathcal{A}_{1m}$. There is still a difference between the allowable elements because I is empty for $n = 0$ but not for $n = 1$.

We say $e^\varepsilon a^I[v^K]b^J$ is *nm*-**plus allowable** if $e^\varepsilon a^I[v^K]b^{J+\Delta_0}$ is *nm*-allowable. We will denote the set of such (K, J) by $\mathcal{A}^+_{nm}$. Note that $\mathcal{A}^+_{nm} \subset \mathcal{A}_{nm}$.

Define the shift operator s on J by

$$b^{s(J)} = b^{\circ j_0}_{(1)} \circ b^{\circ j_1}_{(2)} \circ \cdots . \tag{3.2}$$

Theorem 3.3. *Let $H_*(-)$ be the standard mod p homology with p the prime associated with $P(n, m)$.*

For $n > 0$ and $* \le g(n, m)$. $H_*(\underline{P(n,m)}_*)$ is the same as $H_*(\underline{P(n)}_*)$ stated in [RW96, Theorem 1.3, p. 1045] *except we replace the n and n-plus allowable with nm and nm-plus allowable. For $p = 2$ there is one more minor modification described in the appendix to this paper.*

For $n = 0$ and $* < g(0, m)$,

$$H_*(\underline{P(0,m)}_*) \simeq \bigotimes_{(K,J)\in\mathcal{A}_{0m}} E(e[v^K]b^J) \bigotimes_{(K,J)\in\mathcal{A}_{0m}} P([v^K]b^J).$$

For $n = 0$ and $k = g(0, m)$, as a coalgebra, $H_*(\underline{P(0,m)}_k)$ is the divided power coalgebra

$$\bigotimes_{(K,J)\in\mathcal{A}_{0m}} \Gamma([v^K]b^{J+\Delta_0}).$$

The elements $\gamma_{p^i}([v^K]b^{J+\Delta_0})$ represent $[v^K]b^{s^i(J+\Delta_0)}$.

Remark 3.4. Of course we insist that one only uses the elements which actually lie in the appropriate spaces. The element $e^\varepsilon a^I[v^K]b^J$ is in $H_s(\underline{P(n,m)}_k)$ where $s = \varepsilon + \sum 2p^{i_q} + \sum 2p^{j_q}$ and $k = \varepsilon + \sum i_q + 2\sum j_q - \sum 2(p^q - 1)k_q$.

Remark 3.5. For $n = 0$ and $* < g(0, m)$, this is the same as in [RW77], replacing allowable with $0m$-allowable.

Proof of Corollary 2.1 . The Atiyah-Hirzebruch spectral sequence respects the two products, $\circ$ and $*$, and all elements in the $P(n)_*(\underline{P(n,m)}_*)$ we are considering are constructed using these two products from the basic elements e, $a_{(i)}$, $[v_i]$, and $b_{(i)}$ which are all permanent cycles by Proposition 3.1. Thus the spectral sequence collapses. □

Primitives are calculated simultaneously as in [RW96, Theorem 1.4, p. 1046] and [RW77].

The $p = 2$ case deserves some discussion. The spectra $P(n)$ and $P(n, m)$ are not commutative ring spectra. However, the standard homology is still a commutative Hopf ring, see the explanation in [Wil84, pages 1030-31]. There are no concerns raised by this lack of commutativity in the collapse of the Atiyah-Hirzebruch spectral sequence, the use of duality to compute the cohomologies or their application to get the splittings. The lack of commutativity could make things very bad for some applications but it doesn't interfere in the slightest with what we are doing. It does make the proof of the splitting in [BW] much harder.

The proof of our theorem relies on being able to identify elements in the bar spectral sequence, compute differentials and solve multiplicative extension problems, all using Hopf ring techniques. The $n = 0$ case has no differentials but does have extension problems.

Let Q stand for the indecomposables.

Theorem 3.6. *In $QH_*(\underline{P(n,m)}_k)$, $k \leq g(n,m)$, any $e^\varepsilon a^I[v^K]b^J$ can be written in terms of nm-allowable elements.*

Proof. The construction and proof of an algorithm for the reduction of non-allowable elements is done on pages 273–275 of [RW77]. The proof applies with only notational modification to the case of nm-allowable when $I = 0$. We can then circle multiply by a^I to get our result. □

The homology and primitives are calculated simultaneously by induction on degree in the bar spectral sequence. Recall that for a loop space X with classifying space BX the bar spectral sequence converges to $H_*(BX)$, and its E^2-term is

$$\mathrm{Tor}_{*,*}^{H_*(X)}(\mathbf{Z}/(p), \mathbf{Z}/(p)).$$

When BX is also a loop space, we have a spectral sequence of Hopf algebras.

The E^2 term of the bar spectral sequence for $n > 0$ is calculated inductively from Theorem 3.3 just as in [RW96, Lemma 3.6, p. 1056]. For $n = 0$ the calculation is trivial as in [RW77]. Both cases must use the modified definition of allowable.

The complete behavior of the bar spectral sequence, using the modified definition of allowable, is given in [RW96, Theorem 3.7, p. 1057] for $n > 0$ and in [RW77] for $n = 0$. For $n > 0$ it is a gruesome description of all differentials and identification of elements in terms of the Hopf ring. There are no differentials for $n = 0$ and the identifications are much easier as well.

This does not complete the calculation of the homology but only the E^∞ term of the spectral sequence. Extension problems must be solved. However, first, since we claim that the proofs of all parts of the calculation are exactly the same as for the spaces in the Omega spectra for $P(n)$ and BP as in [RW96] and [RW77], some explanation is clearly needed in order to explain the differences between the two cases and to see why we cannot proceed up the Omega spectrum with $P(n, m)$. That difference comes about in the calculation of the differentials. For $n = 0$ there are no differentials so we do not see any difference at this step. For

$n > 0$ the differentials are not really calculated but inferred. Certain elements are shown to disappear and it is proven that they must be targets of differentials. They are counted and shown to be in one-to-one correspondence with possible sources of differentials. Thus, all possible sources must kill all necessary targets. The difference comes in the counting process. We use the following lemma and the difference is found in the proof, which we include.

Lemma 3.7. *Let $n > 0$. In $H_*\underline{P(n,m)}_{*<g(n,m)}$, there is a one-to-one correspondence between the set*

$$\{[v^{K+\Delta_n}]b^{J+(p^n-1)\Delta_0} : (K,J) \in \mathcal{A}_{nm}, j_0 = 0\}$$

and the set

$$\{[v^{K'}]b^{J'} : (K',J') \in \mathcal{A}_{nm} - \mathcal{A}^+_{nm}\}.$$

Remark 3.8. Recall that for $n = 0$ we have $0m$ and $1m$ give the same thing.

Proof. To see this, write

$$J = p^n\Delta_{d_n} + p^{n+1}\Delta_{d_{n+1}} + \cdots + p^q\Delta_{d_q} + J''$$

where q is maximal (this can be vacuous, i.e. $J = J''$, in which case we set $q = n-1$) and $d_n \leq d_{n+1} \leq \cdots \leq d_q$ and J'' is non-negative. Now let

$$J' = J'' + (p^n - 1)\Delta_0 + p^{n+1}\Delta_{d_n-1} + p^{n+2}\Delta_{d_{n+1}-1} + \cdots + p^{q+1}\Delta_{d_q-1}$$

and $K' = K + \Delta_{q+1}$. □

If we are not observant, we can find ourselves letting q get too large in this proof and creating a $[v_{m+1}]$, which does not exist. Looking closely, we find that the smallest space this could happen in is $k = g(n,m)$. However, the counting for the $n > 0$ case that matters here, is done for differentials and the targets all have an e with them which throws this problem up to the $k = g(n,m)+1$ space. We are not working in this range so this doesn't affect us. It does tell us that the splitting cannot be delooped though since it does say we cannot have as many differentials on our smaller space as we would need to get the size of the homology down to where it splits off. That is, all of our necessary targets are not there in the next space so some possible sources will survive. They do not survive in the space for $P(n)$ though, so the homology cannot split off.

The final problem which arises is the solving of the extension problems to give us the proper homology. Again, the proofs are the same for the $P(n,m)$ and $P(n)$ cases. The counting argument of the previous proof is used again in this proof. Here we need to solve various extension problems. First, we show that certain elements cannot be generators and then we show that the only thing that can prevent them from being generators is if they are p-th powers. We then show that they are in one-to-one correspondence with the only elements which could possibly have non-trivial p-th powers. We use the same counting Lemma as in the previous proof. This time, in the $n > 0$ case, we must always have an $a_{(i)}$ involved in the p-th powers which again throws it up to the $g(n,m) + 1$ space before we see the creation of an unwanted $[v_{m+1}]$. However, in the $n = 0$ case, we see our p-th power extensions can, and do, actually occur in the $g(0, m)$ space. Our result, in this case, is only correct as coalgebras and in the small space the homology is not a polynomial algebra and so cannot split off of the homology of the larger space as Hopf algebras, thus preventing the splitting from being as H-spaces. This completes the proof.

4. Appendix: $p = 2$

This paper requires the results of [RW96]. However, in that paper the results were not stated explicitly for $p = 2$. Because $p = 2$ is such an important part of the contribution of this paper, we must rectify that enough to do the $p = 2$ case here. The lack of precision with $p = 2$ in [RW96] originates in a similar vagueness in [Wil84]. First, we will straighten out the situation in [Wil84] and then we will do the same for [RW96].

The key to the solution is mentioned in the $p = 2$ comments in [Wil84, page 1030], namely, the element e must be included in the coproduct of the elements a_i. In particular, the Verschiebung is evaluated as $V(a_{(0)}) = e$. Using the mod 2 homology formula, $a_{(n-1)}^{*2} = a_{(0)} \circ [v_n] \circ b_{(0)}^{2^n-1}$, we can compute

$$(e \circ a_{(n-1)})^{*2} = a_{(0)} \circ (a_{(n-1)})^{*2} = a_{(0)} \circ a_{(0)} \circ [v_n] \circ b_{(0)}^{2^n-1}.$$

We note that

$$V(a_{(0)} \circ a_{(0)}) = V(a_{(0)}) \circ V(a_{(0)}) = e \circ e = b_1 = b_{(0)}.$$

Since $V(b_{(1)}) = b_{(0)}$ also and $b_{(1)}$ is the only element in degree 4 of this space, we must have $a_{(0)} \circ a_{(0)} = b_{(1)}$. Thus, we have $(e \circ a_{(n-1)})^{*2} = [v_n] \circ b_{(0)}^{2^n-1} \circ b_{(1)}$. For p odd, all elements with e in them were exterior. However, for $p = 2$, we need to look at all elements containing $e \circ a_{(n-1)}$ in both the cases we are considering.

For odd primes, we recall the result of [Wil84, Theorem 1] as

$$H_*\underline{K(n)}_* \simeq \bigotimes_{j_0 < p^n - 1} E(a^I b^J \circ e_1) \bigotimes_{\substack{I \neq I(1) \\ \text{if } i_0 = 1, \\ \text{then } j_0 < p^n - 1}} TP_{\rho(I)}(a^I b^J) \bigotimes_{\substack{I = I(1) \\ j_0 < p^n - 1}} P(a^I b^J)$$

where H is the mod p standard homology, $I(1)$ denotes the sequence of all ones and all $j_k < p^n$. The number $\rho(I) > 0$ is the smallest k with $i_{n-k} = 0$.

For $p = 2$ we must modify this to take into account the additional nontrivial squares which we have already identified. In this case our description is precise.

Theorem 4.1. *For $p = 2$, with the above constraints, $H_*\underline{K(n)}_* \simeq$*

$$\bigotimes_{\substack{j_0 < 2^n - 1 \\ i_{n-1} = 0}} E(a^I b^J \circ e_1) \quad \bigotimes_{\substack{j_0 < 2^n - 1 \\ i_{n-1} = 1}} TP_{\rho(s(I-\Delta_{n-1}))+1}(a^I b^J e_1)$$

$$\bigotimes_{\substack{I \neq I(1) \\ \text{if } i_0 = 1, \\ \text{then } j_0 < 2^n - 1 \\ \text{if } i_0 = 0 \text{ and } j_0 = 2^n - 1 \\ \text{then } j_1 = 0}} TP_{\rho(I)}(a^I b^J) \quad \bigotimes_{\substack{I = I(1) \\ j_0 < 2^n - 1}} P(a^I b^J)\,.$$

The essentials of the proof remain unchanged. We have taken the exterior generators which should be truncated polynomial generators and we have taken away the generators which are their squares. The size remains the same in either description. This is significantly easier than our next case because here we can solve our extension problems precisely.

It is tedious to reproduce the results of [RW96] here and then modify them slightly. We will keep the flavor of the rest of the paper and only produce the modifications.

Theorem 4.2. *For $p = 2$, the standard mod 2 homology, $H_*(\underline{P(n)}_*)$, fits in a short exact sequence of Hopf algebras with the associated graded algebra being given by* [RW96, Theorem 1.3] *(the odd prime answer). The quotient Hopf algebra is just the exterior algebra on generators $ea^I[v^K]b^J$ as in* [RW96, Theorem 1.3] *with $i_{n-1} = 1$. These elements, in the actual algebra, all have nontrivial squares which are contained in the set of generators of the subalgebra given by $a^I[v^K]b^J$ with $i_0 = 0$ and $(K, J) \in \mathcal{A}_n - \mathcal{A}_n^+$.*

Proof. In the Morava K-theory case we could evaluate the necessary squares directly. Here we cannot. It is essential that we know the squares are all nontrivial and linearly independent but it is not obvious how to do that directly. However, the proof in [RW96] need only be modified slightly. In particular, we can still work in the same bar spectral sequence with the same elements. We must consider the elements $a^I[v^K]b^J$ with $i_0 = 0$ and $(K, J) \in \mathcal{A}_n - \mathcal{A}_n^+$. After double suspension to $a^I[v^K]b^{J+\Delta_0}$ we know this is zero mod $*$ since $(K, J) \notin \mathcal{A}_n^+$. However, it cannot be a square because it has no $a_{(0)}$ in it and its degree is a multiple of 4. (If it was the square of an elements with e in it, thus making the $a_{(0)}$ unnecessary, then it would have to have degree 2 mod 4). All differential targets must be odd degree so our double suspended element must be zero which implies something happened to it in the previous spectral sequence as $ea^I[v^K]b^J$. Since it is odd degree it cannot be a square so it either is the target of a differential or it was already zero. Both happen. The counting argument of [RW96, page 1061-2] pairs these up with the potential source of differentials. For $p = 2$, the count, as given, uses $\gamma_2(\sigma ea^I[v^K]b^J)$ when $i_{n-1} = 1$ and $(K, J) \in \mathcal{A}_n^+$. However, at $p = 2$, this would have to be a d_1 differential which does not exist. We know that the elements which should be targets must be zero so the ones associated with these γ_2 must have already been zero. These are in 1-1 correspondence with our γ_2, or our $ea^I[v^K]b^J$ when $i_{n-1} = 1$ and $(K, J) \in \mathcal{A}_n^+$. These are precisely the elements we wish to have non-trivial squares! Thus, the only solution to our problem is that they have non-trivial squares, linearly independent, among the $a^I[v^K]b^J$ with $i_0 = 0$ and $(K, J) \in \mathcal{A}_n - \mathcal{A}_n^+$. Because they are squares, they never show up as $\sigma a^I[v^K]b^J$ in the next spectral sequence so they do not have to be killed there. Likewise, because $ea^I[v^K]b^J$ is not exterior, the γ_2 we worried about in the spectral sequence does not exist so those unwanted elements go away as well. This concludes the discussion of the differences for $p = 2$. □

References

[BJW95] J. M. Boardman, D. C. Johnson, and W. S. Wilson. Unstable operations in generalized cohomology. In I. M. James, editor, *The Handbook of Algebraic Topology*, chapter 15, pages 687–828. Elsevier, 1995.

[BW] J. M. Boardman and W. S. Wilson. $k(n)$-torsion-free H-spaces and $P(n)$-cohomology. In preparation.

[Qui71] D. G. Quillen. Elementary proofs of some results of cobordism theory using Steenrod operations. *Adv. in Math.*, 7:29–56, 1971.

[RW77] D. C. Ravenel and W. S. Wilson. The Hopf ring for complex cobordism. *Journal of Pure and Applied Algebra*, 9:241–280, 1977.

[RW96] D. C. Ravenel and W. S. Wilson. The Hopf ring for $P(n)$. *Canadian Journal of Mathematics*, 48(5):1044–1063, 1996.

[RWY98] D. C. Ravenel, W. S. Wilson, and N. Yagita. Brown-Peterson cohomology from Morava K-theory. *K-Theory*, 15(2):149–199, 1998.

[SY76] N. Shimada and N. Yagita. Multiplications in complex bordism theory with singularities. *Publications of Research Institute of Mathematical Sciences, Kyoto University*, 12:259–293, 1976.

[Wil73] W. S. Wilson. The Ω-spectrum for Brown-Peterson cohomology, Part I. *Commentarii Mathematici Helvetici*, 48:45–55, 1973.

[Wil75] W. S. Wilson. The Ω-spectrum for Brown-Peterson cohomology, Part II. *American Journal of Mathematics*, 97:101–123, 1975.

[Wil84] W. S. Wilson. The Hopf ring for Morava K-theory. *Publications of Research Institute of Mathematical Sciences, Kyoto University*, 20:1025–1036, 1984.

Johns Hopkins University, Baltimore, Maryland 21218, USA
E-mail address: `jmb@math.jhu.edu, wsw@math.jhu.edu`

Progress in Mathematics, Vol. 196

Stripping and splitting decorated mapping class groups

Carl-Friedrich Bödigheimer and Ulrike Tillmann*

Abstract. We study decorated mapping class groups, i.e., mapping class groups of surfaces with marked points and boundary components, and their behaviour under stabilization maps with respect to the genus, the number of punctures and boundary components. Decorated mapping class groups are non-trivial extensions of the undecorated mapping class group, and the first result states that the extension is homologically trivial when one stabilizes with respect to the genus. The second result implies that one also gets splittings of homology groups when stabilizing with respect to the number of punctures and boundary components.

1. Introduction and statement of results

Let $F^k_{g,n+1}$ be an oriented smooth surface of genus g with k marked points and $n+1$ boundary components. The mapping class group $\Gamma^k_{g,n+1}$ is the group of components of the space of diffeomorphisms of $F^k_{g,n+1}$ which fix the marked points and the boundary pointwise. If k and n are zero, they will be dropped from the notation. Gluing a torus with two boundary components to one of the boundary components of $F^k_{g,n+1}$, one gets a surface $F^k_{g+1,n+1}$. Extending diffeomorphisms by the identity induces a map of mapping class groups

$$\sigma\colon \Gamma^k_{g,n+1} \longrightarrow \Gamma^k_{g+1,n+1},$$

and we may define the associated stable mapping class group

$$\Gamma^k_{\infty,n} = \lim_{g\to\infty} \Gamma^k_{g,n+1}.$$

The n boundary components not used in this stabilization process will be called *free*. Consider now diffeomorphisms that may permute the punctures and free boundary components. To be precise, the boundary components should be thought of as having a parametrization and diffeomorphisms have to be compatible with these. The associated mapping class groups will be denoted by $\Gamma^{(k)}_{g,(n),1}$ and $\Gamma^{(k)}_{\infty,(n)}$. They are normal extensions

$$\Gamma^k_{g,n+1} \hookrightarrow \Gamma^{(k)}_{g,(n),1} \twoheadrightarrow \Sigma_k \times \Sigma_n, \qquad \Gamma^k_{\infty,n+1} \hookrightarrow \Gamma^{(k)}_{\infty,(n),1} \twoheadrightarrow \Sigma_k \times \Sigma_n$$

---

*Supported by an Advanced Fellowship of the EPSRC.

with quotients the product of the symmetric groups on k and n letters. Similarly, let $\Gamma^{(k)}_{g,n+1}$ and $\Gamma^{k}_{g,(n),1}$ denote the mapping class groups associated to the diffeomorphism groups that may permute the punctures respectively the free boundary components only. These extensions are important for the understanding of surface operads (see for example [G] and [GK]).

Theorem 1.1.
(1) $B(\Gamma^k_{\infty,n})^+ \simeq B\Gamma^+_\infty \times (\mathbb{C}P^\infty)^k$;
(2) $B(\Gamma^{(k)}_{\infty,(n)})^+ \simeq B\Gamma^+_\infty \times B\Sigma^+_n \times B(\Sigma_k \wr S^1)^+$.

These are two special cases of Theorem 3.1. Here $\mathbb{C}P^\infty$ denotes infinite dimensional complex projective space, Σ_k is the symmetric group on k letters, $\Sigma_k \wr S^1$ is the wreath product of the symmetric group Σ_k with the circle group, and X^+ denotes the Quillen plus construction of the space X with respect to the maximal perfect subgroup of its fundamental group. The most important ingredient for the proof of Theorem 1.1 is Harer's homology stability theorem for mapping class groups [H]. It allows one also to deduce results for the mapping class groups of closed surfaces and finite genus g. Theorem 1.1 has the following reinterpretation in homology, the first part of which has also been proved by Looijenga [L] and Morita [M].

Corollary 1.2.
(1) $H_*(\Gamma^k_{g,m}) = H_*(\Gamma_{g,1}) \otimes \mathbb{Z}[x_1, \ldots, x_k]$ *in dimensions* $* \leq g/2$ *when* $k+m \geq 1$; *here each* x_i *has degree* 2;
(2) $H_*(\Gamma^{(k)}_{g,(n),m}; \mathbb{F}) = H_*(\Gamma_{g,1}; \mathbb{F}) \otimes H_*(\Sigma_n; \mathbb{F}) \otimes H_*(\Sigma_k; \mathbb{F}[x_1, \ldots, x_k])$ *in dimensions* $* \leq g/2$ *when* $k+n+m \geq 1$; $\mathbb{F}$ *is any field.*

Remark The difficult question of determining the homotopy type of $B\Gamma^+_\infty$ is not addressed in this paper. Some progress has recently been made. In particular, $B\Gamma^+_\infty$ has been shown to be an infinite loop space [T]. Hence, by the above splitting, also $B(\Gamma^k_\infty)^+$ has an infinite loop space structure. It is curious to note that the methods of [T] however do not generalize to the case $k > 0$.

Instead of stabilizing with respect to the genus g, one can also consider stabilizations with respect to the number of boundary components or punctures:

$$\alpha\colon \Gamma^k_{g,n+1} \longrightarrow \Gamma^k_{g,n+2}, \qquad \beta\colon \Gamma^k_{g,n+1} \longrightarrow \Gamma^{k+1}_{g,n+1},$$

$$\tilde{\alpha}\colon \Gamma^{(k)}_{g,n+1} \longrightarrow \Gamma^{(k)}_{g,n+2}, \qquad \tilde{\beta}\colon \Gamma^{(k)}_{g,n+1} \longrightarrow \Gamma^{(k+1)}_{g,n+1},$$

$$\tilde{\tilde{\alpha}}\colon \Gamma^k_{g,(n),1} \longrightarrow \Gamma^k_{g,(n+1),1}, \qquad \tilde{\tilde{\beta}}\colon \Gamma^k_{g,(n),1} \longrightarrow \Gamma^{k+1}_{g,(n),1}.$$

For α, $\tilde{\alpha}$ and $\tilde{\tilde{\alpha}}$ glue a pairs of pants surface to the non-free boundary component of $F^k_{g,n+1}$. Similarly, for β, $\tilde{\beta}$ and $\tilde{\tilde{\beta}}$ glue a cylinder with one puncture to $F^k_{g,n+1}$. α and $\tilde{\alpha}$ have retractions on the group level induced by gluing a disk to one of the boundary components of the pairs of pants surface. β and $\tilde{\tilde{\beta}}$ have retractions on the group level induced by forgetting the $(k+1)$-st puncture. Such splittings on the level of groups cannot be found for $\tilde{\tilde{\alpha}}$ and $\tilde{\beta}$. However, using techniques from configuration spaces we show

Theorem 1.3. *The maps* $\tilde{\tilde{\alpha}}\colon B\Gamma^k_{g,(n),1} \to B\Gamma^k_{g,(n+1),1}$ *and* $\tilde{\beta}\colon B\Gamma^{(k)}_{g,n+1} \to B\Gamma^{(k+1)}_{g,n+1}$ *admit stable retractions.*

In particular, $\tilde{\tilde{\alpha}}$ and $\tilde{\beta}$ have splittings in homology. We note here that for infinite genus Theorem 1.3 follows from Theorem 1.1 and well-known stable splitting of the inclusion of Borel constructions

$$E\Sigma_n \times_{\Sigma_n} X^n \longrightarrow E\Sigma_{n+1} \times_{\Sigma_{n+1}} X^{n+1}$$

for connected X and any n (compare [BöM]).

2. Preliminaries on perfect subgroups and the plus construction

A group P is called *perfect* if every element can be written as a commutator, that is $P = [P, P]$. Any group G has a unique *maximal perfect* subgroup which we will denote by $P(G)$. As the homomorphic image of a perfect group is again perfect, $P(G)$ is a characteristic subgroup of G.

Let X be a connected space and let P be a perfect normal subgroup of its fundamental group $\pi_1(X)$. By attaching 2-cells and 3-cells, one can form a space X^+_P with the properties that the natural inclusion $q_P\colon X \to X^+_P$ induces (i) an epimorphism $\pi_1(X) \to \pi_1(X^+_P)$ with kernel P and (ii) an isomorphism $H_*(X;A) \to H_*(X^+_P;A)$ for any abelian group A with a $\pi_1(X^+_P)$-action. This construction is due to Quillen. We will always consider the plus-construction X^+ with respect to the maximal perfect subgroup of $\pi_1(X)$ and drop the subscript P from the notation. We recall a few functorial properties of the plus construction which will be needed later on [B; Sections 5 and 6].

(2.1) Given a map $f\colon X \to Y$ between two spaces there is a unique homotopy class of maps $f^+\colon X^+ \to Y^+$ making the following diagram commute

$$\begin{array}{ccc} X & \xrightarrow{f} & Y \\ {\scriptstyle q}\downarrow & & \downarrow{\scriptstyle q} \\ X^+ & \xrightarrow{f^+} & Y^+. \end{array}$$

In particular, X^+ is well-defined up to homotopy relative to X.

(2.2) As $P\pi_1(X \times Y) = P\pi_1(X) \times P\pi_1(Y)$, we have

$$(X \times Y)^+ = X^+ \times Y^+.$$

(2.3) If $F \to E \xrightarrow{p} B$ is a fibre sequence of connected spaces, then so is also $F^+ \to E^+ \to B^+$ provided either (i) $P\pi_1(B) = 1$ or (ii) p is quasi-nilpotent and F^+ is nilpotent.

A space X is *nilpotent* if its fundamental group acts nilpotently on its homotopy groups. A fibration p is *quasi-nilpotent* if the fundamental group $\pi_1(B)$ of the base acts nilpotently on the homology H_*F of the fiber.

3. Stripping decorations: The proof of Theorem 1.1

Let H_k and H_n be any subgroups of the symmetric groups Σ_k and Σ_n on k and n letters. Let $\Gamma^{H_k}_{g,H_n,1}$ be the subgroup of $\Gamma^{(k)}_{g,(n),1}$ that fits into the exact sequence

$$\Gamma^k_{g,n+1} \longrightarrow \Gamma^{H_k}_{g,H_n,1} \xrightarrow{\rho} H_k \times H_n.$$

Extending diffeomorphisms by the identity on an attached disk with an appropriate number of punctures and disks removed, or an attached torus with one boundary component defines the inclusion map

$$\text{incl}\colon \Gamma_{g,1} \longrightarrow \Gamma^{H_k}_{g,H_n,1}$$

and the stabilization map

$$\sigma\colon \Gamma^{H_k}_{g,H_n,1} \longrightarrow \Gamma^{H_k}_{g+1,H_n,1}.$$

These maps are illustrated in Figure 1 where the non-free boundary component of the surfaces is drawn as a rectangle, and gluing is defined by connected sum on this boundary component. (Note that this gluing is homotopy equivalent to the pairs of pants construction but strictly associative.) We may now also fill in the k punctures and n disks, and once again extend diffeomorphisms by the identity. This defines the forgetful map

$$\phi\colon \Gamma^{H_k}_{g,H_n,1} \longrightarrow \Gamma_{g,1}.$$

Clearly $\phi \circ \text{incl}$ is the identity homomorphism. As stabilization commutes with both incl and ϕ, these homomorphisms extend to the stable mapping class groups.

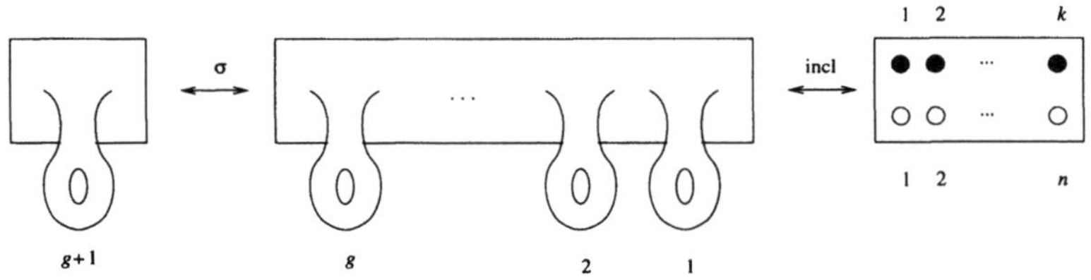

Figure 1: Stabilization and inclusion maps.

Theorem 3.1. $B(\Gamma^{H_k}_{\infty,H_n})^+ \simeq B\Gamma^+_\infty \times BH^+_n \times B(H_k \wr S^1)^+$.

Proof. Gluing n punctured disks onto the n free boundary components of $F_{g,n+1}$ commutes with stabilization, and hence defines a central extension

$$\mathbb{Z}^n \longrightarrow \Gamma_{\infty,n} \longrightarrow \Gamma^n_\infty.$$

This central extension gives rise to a map $\theta\colon \Gamma^n_\infty \to B\mathbb{Z}^n = (S^1)^n$. Recall that the connected components of the group of orientation preserving diffeomorphisms of a surface $F_{g,1}$ are contractible when $g \geq 2$ [ES]. Hence, identify $\Gamma^n_{g,1}$ up to homotopy with the group of diffeomorphisms of $F_{g,1}$ which fix the boundary as well as n marked points. Then the components θ_i of θ assign to a diffeomorphism ϕ the angle between $t_{\phi(i)}$ and $D\phi(t_i)$ where the t_i's are fixed tangent vectors at the marked points and $D\phi$ is the derivative of ϕ. θ is a group homomorphism and can be extended in an obvious way to a homomorphism $\rho \wr \theta\colon \Gamma^{(n)}_\infty \to \Sigma_n \wr S^1$. We have the following (up to homotopy) commutative diagram of fibrations.

$$\begin{array}{ccccc}
B\Gamma_{\infty,n} & \longrightarrow & B\Gamma^n_\infty & \xrightarrow{\theta} & (BS^1)^n \\
\| & & \downarrow & & \downarrow \\
B\Gamma_{\infty,n} & \longrightarrow & B\Gamma^{(n)}_\infty & \xrightarrow{\rho\wr\theta} & B(\Sigma_n \wr S^1) \\
\downarrow & & \downarrow & & \downarrow \\
* & \longrightarrow & B\Sigma_n & = & B\Sigma_n.
\end{array}$$

More generally, there is a fibration

$$B\Gamma_{\infty,k+n} \longrightarrow B\Gamma^{H_k}_{\infty,H_n} \xrightarrow{\rho\wr\theta} B(H_k \wr S^1) \times BH_n. \tag{3.2}$$

$B\Gamma_{\infty,n}$ has perfect fundamental group for any n, and therefore its plus construction is nilpotent. Furthermore, the fundamental group of the base space acts trivially on the homology of the fiber by Lemma 3.3 below. Hence, (3.2) is a quasi-nilpotent fibration. By (2.3) we conclude that therefore also its plus construction is a homotopy fibration:

$$B\Gamma^+_{\infty,k+n} \longrightarrow (B\Gamma^{H_k}_{\infty,H_n})^+ \xrightarrow{\rho\wr\theta} (B(H_k \wr S^1) \times BH_n)^+.$$

The forgetful map ϕ induces a map of fibrations from this into the trivial fibration

$$B\Gamma^+_\infty \longrightarrow B\Gamma^+_\infty \times (B(H_k \wr S^1) \times BH_n)^+ \longrightarrow (B(H_k \wr S^1) \times BH_n)^+.$$

The map of base spaces is the identity. By Harer's stability theorem [H], the homology of the stable mapping class group is independent of the number of free boundary components. Therefore the map of fibers is a homotopy equivalence, and hence so is the map of total spaces. An application of (2.2) finishes the proof. □

Lemma 3.3. *Σ_n acts trivially on $H_*B\Gamma_{g,n}$ for $* \le g/2$.*

Proof. This is trivially true for $n = 0$ and we may assume $n \ge 1$. By the stability theorem [H], [I], the Σ_n-equivariant map $H_*\Gamma_{g,n+1} \to H_*\Gamma_{g,n}$ induced by gluing a disk to the non-free boundary component induces an isomorphism in degrees $* \le g/2$. Hence it is enough to show that Σ_n acts trivially on $H_*\Gamma_{g,n+1}$ for $* \le g/2$.

Let x be an element in the k-th homology of $\Gamma_{g,n+1}$ with $k \le g/2$. As the inclusion incl: $\Gamma_{g,1} \to \Gamma_{g,n+1}$ induces an isomorphism on the k-th homology [H], [I], x can be represented by a closed chain z in the bar complex of $\Gamma_{g,1}$.

The action of Σ_n on the homology of $\Gamma_{g,n+1}$ is induced by conjugation. Let μ be an element of Σ_n. Then μ has a lift $\tilde{\mu}$ to $\Gamma_{g,(n),1}$ such that it may be represented by a diffeomorphism which is entirely supported on the surface $F_{0,n+1}$ (with $k = 0$) of Figure 1. The mapping classes appearing in the closed chain z, on the other hand, are represented by diffeomorphisms of the surface $F_{g,1}$. Diffeomorphisms with disjoint support commute, and hence $\tilde{\mu}^{-1} z \tilde{\mu} = z$. It follows that μ acts trivially on x. □

The lemma implies the following generalization of the stability theorem. Corollary 1.2 can be deduced immediately from this and the fact that by a theorem of Nakaoka [N], the Leray–Serre spectral sequence for the wreath product collapses at the E^2-term.

Proposition 3.4. $H_*\Gamma^{H_k}_{g,H_n,m} \simeq H_*\Gamma^{H_k}_{g+1,H_n,m}$ *for $* \le g/2$ and $k + n + m \ge 1$.*

Proof. The two groups are related by group homomorphisms

$$\Gamma^{H_k}_{g,H_n,m} \longleftarrow \Gamma^{H_k}_{g,H_n,m+1} \longrightarrow \Gamma^{H_k}_{g+1,H_n,m+1} \longrightarrow \Gamma^{H_k}_{g+1,H_n,m}.$$

Consider the induced maps of Serre spectral sequences for the fibration of type (3.2) associated to each of these groups. By the above lemma and the stability theorem [H], [I], the induced maps on the second term of the spectral sequences $E^2_{*,q}$ are isomorphisms for $q \le g/2$. The result thus follows by the Zeeman comparison theorem. □

We finish this section by identifying the maximal perfect subgroup of the decorated mapping class groups.

Proposition 3.5. $P(\Gamma^{H_k}_{g,H_n,1}) = \rho^{-1}(P(H_k \times H_n))$ *for $g > 2$, and in particular contains $\Gamma_{g,1}$.*

Proof. Let $g > 2$. Recall that for all n, $\Gamma_{g,n}$ is perfect [P]. There is a natural surjection $\Gamma_{g,n+k} \to \Gamma^k_{g,n}$. Hence, as the image of a perfect group is perfect, $\Gamma^k_{g,n}$ is perfect for all k and n. Now consider the general case. The kernel of ρ is $\Gamma^k_{g,n}$. As an extension of perfect groups is again perfect, we see $\rho^{-1}(P(H_k \times H_n))$ is perfect. It must be maximal as otherwise its image under ρ would be larger then $P(H_k \times H_n)$ and not perfect. But that is impossible. □

4. Stable retractions: The proof of Theorem 1.3

The purpose of this section is to prove Theorem 1.3. We will give details in the case of $\tilde{\beta}$. We will use a configuration space model for $B\Gamma^{(k)}_{g,n+1}$ and construct a map

$$R\colon\ \Omega^\infty\Sigma^\infty B\Gamma^{(k+1)}_{g,n+1} \longrightarrow \Omega^\infty\Sigma^\infty B\Gamma^{(k)}_{g,n+1} \tag{4.1}$$

such that $R\circ\Omega^\infty\Sigma^\infty\widetilde{\beta} \simeq \mathrm{Id}$. At the end of the proof, we will indicate the necessary changes to be made in the case of $\tilde{\tilde{\alpha}}$.

Fix a surface $F = F_{g,n+1}$ with a distinguished, non-free boundary curve, and let $C^k(F)$ denote the space of configurations $\mathcal{X} = \{x_1, \ldots, x_k\}$ of k distinct, unordered points x_i on the (interior of the) surface F. The group $D = \mathrm{Diff}^+(F;\partial)$ of orientation preserving diffeomorphisms of F which fix the boundary pointwise acts on $C^k(F)$.

Lemma 4.2. $E^k_{g,n+1} = ED \times_D C^k(F)$ *is a classifying space for the group* $\Gamma^{(k)}_{g,n+1}$.

Proof. This follows from the fibration $C^k(F) \to E^k_{g,n+1} \to BD \simeq B\Gamma_{g,n+1}$ and from the fact that $C^k(F)$ is a classifying space for the group $\pi_1 C^k(F)$, the group of braids with k strands in $F\times[0,1]$. The last assertion is proved using the covering space $\widetilde{C}^k(F)$ of $C^k(F)$, i.e., the space of ordered configurations of k points on F, which admits an inductive sequence of fibrations

$$\begin{array}{ccccccc}
\widetilde{C}^k(F_{g,n+1}) & \longleftarrow & \cdots & \longleftarrow & \widetilde{C}^2(F^{k-2}_{g,n+1}) & \longleftarrow & \widetilde{C}^1(F^{k-1}_{g,n+1}) = F^{k-1}_{g,n+1}. \\
{\scriptstyle P_0}\big\downarrow & & & & {\scriptstyle P_{k-2}}\big\downarrow & & \\
F_{g,n+1} & & & & F^{k-2}_{g,n+1} & &
\end{array}$$

Here P_i forgets the last point of a configuration; the total space of P_i is the fibre of P_{i-1}. Since all the base spaces are surfaces with at least one boundary curve, an inductive argument shows that $\widetilde{C}^k(F_{g,n+1})$ and thus $C^k(F_{g,n+1})$ has no higher homotopy. $\square$

The fibration $C^k(F) \to E^k_{g,n+1} \to B\Gamma_{g,n+1}$ now gives the group extension $\pi_1 C^k(F) \to \Gamma^{(k)}_{g,n+1} \to \Gamma_{g,n+1}$. The inclusion $\widetilde{\beta}$ is given by the inclusion of fibres

$$\begin{array}{ccc}
C^k(F) & \xrightarrow{\ b\ } & C^{k+1}(F) \\
\big\downarrow & & \big\downarrow \\
E^k_{g,n+1} & \xrightarrow{\ \widetilde{\beta}\ } & E^{k+1}_{g,n+1} \\
\big\downarrow & & \big\downarrow \\
B\Gamma^0_{g,n+1} & = & B\Gamma^0_{g,n+1},
\end{array}$$

which adds to a configuration $\mathcal{X} \in C^k(F)$ a new point in the following way. Let $S^1\times[-1,1]$ be a collar along the distinguished boundary curve. Define $b'\colon F \longrightarrow F$

as $b'(y,t) = \left(y, \frac{1}{2}t^2\right)$ for $y \in S^1, 0 \leqslant t \leqslant 1$, and extend b' smoothly to $S^1 \times [-1,0]$ and identically outside the collar. Set $\xi = \left(y_0, \frac{3}{4}\right)$ for some fixed $y_0 \in S^1$, and then define $b(\mathcal{X}) = \{b'(x_1), \ldots, b'(x_k), \xi\}$ for $\mathcal{X} = \{x_1, \ldots, x_k,\} \in C^k(F)$.

$C^k(F)$ is via b a closed subspace of $C^{k+1}(F)$; thus the same is true for $E^k_{g,n+1}$ in $E^{k+1}_{g,n+1}$; we denote the quotient by $\widetilde{E}^{k+1}_{g,n+1} = E^{k+1}_{g,n+1}/E^k_{g,n+1}$; and by $q_k\colon E^{k+1}_{g,n+1} \to \widetilde{E}^{k+1}_{g,n+1}$ the natural map. By V_k we denote the bouquet of these filtration quotients:

$$V_k = \bigvee_{i=1}^{k} \widetilde{E}^i_{g,n+1}.$$

Theorem 4.3. *There is a homotopy equivalence* $\Theta_k\colon \Omega^\infty\Sigma^\infty E^k_{g,n+1} \to \Omega^\infty\Sigma^\infty V_k$.

Proof. We replace (à la Kahn–Priddy, Quillen) $\Omega^\infty\Sigma^\infty Y$ for a connected space Y with base point $*$ by the labelled configuration space $C(\mathbb{R}^\infty; Y)$ of unordered configurations $z_1, \ldots, z_k$ of distinct points $z_i \in \mathbb{R}^\infty$ with (not necessarily distinct) labels $y_1, \ldots, y_k \in Y$; a point in $C(\mathbb{R}^\infty; Y)$ is thus a set $\{(z_1, y_1), \ldots, (z_k, y_k)\}$, where a pair (z_i, y_i) is deleted if $y_i = *$. A point in $E^k_{g,n+1}$ is given as a D-orbit of $\widetilde{e} = (e, \mathcal{X})$, where $e \in ED$ and $\mathcal{X} = \{x_1, \ldots, x_k\}$ a configuration on F. Set $\mathcal{X}_\alpha = \{x_i \in \mathcal{X} \mid i \in \alpha\}$ for any (non-empty) subset $\alpha \subset \{1, \ldots, k\}$. Regard $\mathcal{X}_\alpha$ as a point in $\mathbb{R}^\infty$ by choosing an embedding $\coprod_{l\geq 0} C^l(F) \hookrightarrow \mathbb{R}^\infty$. Then define

$$\Theta'_k\colon E^k_{g,n+1} \longrightarrow C(\mathbb{R}^\infty; V_k)$$

by $\Theta'_k(e, \mathcal{X}) = \{(\mathcal{X}_\alpha, q_l(e, \mathcal{X}_\alpha)) \mid \emptyset \neq \alpha \subset \{1, \ldots, k\},\ l = \#\alpha\}$. This Θ'_k is a continuous map which extends to a map Θ_k on $C\left(\mathbb{R}^\infty; E^k_{g,n+1}\right)$ since the latter is a free object in the category of infinite loop spaces generated by $E^k_{g,n+1}$. To prove that Θ_k is a homotopy equivalence, note first that Θ_1 is homotopic to the identity; then consider the diagram

$$\begin{array}{ccccc}
C\left(\mathbb{R}^\infty;\ E^k_{g,n+1}\right) & \xrightarrow{\ \widetilde{\beta}\ } & C\left(\mathbb{R}^\infty;\ E^{k+1}_{g,n+1}\right) & \xrightarrow{\ q_{k+1}\ } & C\left(\mathbb{R}^\infty;\ \widetilde{E}^{k+1}_{g,n+1}\right) \\
{\scriptstyle\Theta_k}\downarrow & & {\scriptstyle\Theta_{k+1}}\downarrow & & \| \\
C\left(\mathbb{R}^\infty;\ V_k\right) & \xrightarrow{\ \iota\ } & C\left(\mathbb{R}^\infty;\ V_{k+1}\right) & \xrightarrow{\ p_{k+1}\ } & C\left(\mathbb{R}^\infty;\ \widetilde{E}^{k+1}_{g,n+1}\right).
\end{array}$$

Here ι is induced by the inclusion $V_k \hookrightarrow V_{k+1}$, and p_{k+1} is induced by the projection onto the last leaf. It is commutative (up to homotopy): for the left-hand square, all $\left(\mathcal{X}_\alpha, q_l(e, \mathcal{X}_\alpha)\right)$ are deleted: if $k+1 \notin \alpha$ and $l = \#\alpha$, then $q_l(e, \mathcal{X}_\alpha)$ is contained in V_k; if $k+1 \in \alpha$ and $\mathcal{X} = b(\mathcal{X}')$, then $q_l(e, \mathcal{X}_\alpha)$ is the basepoint in V_{k+1} for all $l = 1, \ldots, k+1$. For the right-hand square only $\left(\mathcal{X}_\alpha, q_{k+1}(e, \mathcal{X}_\alpha)\right)$ for $\alpha = \{1, \ldots, k+1\}$ survives the projection q_{k+1}. Because the horizontal sequences are quasi-fibrations and all spaces are of the homotopy type of a CW-complex, it follows by induction that Θ_k is a homotopy equivalence. □

Under these homotopy equivalences Θ_k's, the retraction R of (4.1) corresponds to the projection onto the first $k-1$ leaves of the bouquet V_k. This proves Theorem 1.3 for the case $\tilde{\beta}$.

In the case of $\tilde{\tilde{\alpha}}$, we first note that on the level of mapping class groups, keeping a boundary component fixed is the same as fixing a point and the direction of a tangent vector at that point. Since we always have at least one boundary component the tangent bundle of F is trivial. Thus a tangent direction is nothing else but a label in S^1, and we need to consider the configuration space $C^n(F; S^1)$ of configurations of n points with labels in S^1 of the k-punctured surface $F = F^k_{g,1}$. The group D is now the group of orientation preserving diffeomorphisms $\mathrm{Diff}^+(F; \partial)$ which fix the (non-free) boundary component and the k punctures. The analogous statements of Lemma 4.2 and Theorem 4.3 can now be proved in a similar way.

We remark that the proof relies on the same idea used to stably split loop spaces of suspensions. For details see [Bö] or [CMT]. To reduce the essence of the proof even more: the main idea is that the classifying space of the k-th braid group of a surface (with labels) is a stable retract of the $(k+1)$-st braid group of the surface.

5. Appendix: Group pairs, H-action, and a general splitting result

The splitting of Theorem 3.1 can also be seen in a somewhat different context. The idea is that the retraction induced by the group homomorphisms incl and ϕ on the plus construction of the classifying spaces is *multiplicative*, and hence gives rise to a splitting of spaces. This uses the H-space structure of $B\Gamma^+_\infty$ which will be made explicit below.

Adopting arguments of [W] we consider the general situation: Two groups G and H form a *direct sum pair* if H is a subgroup of G and there is a group homomorphism $\oplus\colon H \times G \to G$. Furthermore, we will assume that for any $g_1, \dots, g_s \in G$ and $h_1, \dots, h_s \in H$ there exist elements $c \in P(G)$ and $d \in P(H)$ such that for all $i = 1, \dots, s$

$$1 \oplus g_i = c \cdot g_i \cdot c^{-1} \quad \text{and} \quad h_i \oplus 1 = d \cdot h_i \cdot d^{-1}. \tag{5.1}$$

Proposition 5.2. *BG^+ admits a left* H*-action by BH^+.*

This means there is a map $\mu\colon BH^+ \times BG^+ \to BG^+$ such that the restriction to the left factor is homotopic to the map induced by the inclusion incl: $H \to G$ and the restriction to the right factor is homotopic to the identity. When H is equal to G, then BG^+ is an H-space.

Proof. Note that, by (2.2), $(BH \times BG)^+ = BH^+ \times BG^+$. Thus the direct sum homomorphism $\oplus$ induces a map

$$m\colon BH^+ \times BG^+ \longrightarrow BG^+.$$

Let $*$ denote the basepoint of BG^+ and BH^+. The map $m(\_, *)\colon BH^+ \to BG^+$ is induced by the group homomorphism $\_ \oplus 1$. By property (5.1), $\_ \oplus 1$ factors through H. We want to prove that the induced map $f\colon BH^+ \to BH^+$ is a homotopy equivalence. $BP(H)$ is a regular cover of BH, and hence $BP(H)^+$ is the universal cover of BH^+. By property (5.1), as $d \in P(H)$, the map $BP(H)^+ \to BP(H)^+$ induced by f is the identity on homology (compare Lemma 1.3 [W]). Hence, by the Whitehead theorem it is a homotopy equivalence. But then the map $f\colon BH^+ \to BH^+$ is a homotopy equivalence as well. Similarly, $m(*, \_)$ is a homotopy equivalence of BG^+. Now choose homotopy inverses r and l for these two maps. Then $\mu = m \circ (r \times l)\colon BH^+ \times BG^+ \to BG^+$ defines an H-action. □

Corollary 5.3. *Assume G and H are as above, and that there is a splitting homomorphism $\phi\colon G \to H$. Then there is a splitting of spaces $BG^+ \simeq BH^+ \times F$.*

Proof. Let F be the homotopy fiber of the map $\phi\colon BG^+ \to BH^+$, and let $s\colon F \to BG^+$ denote the inclusion of the fiber. Define $BH^+ \times F \to BG^+$ by mapping (x, y) to $\mu(x, s(y))$. Because μ defines an H-action, this induces an isomorphism on homotopy groups and hence is a homotopy equivalence. □

$\Gamma^{H_k}_{\infty,H_n}$ and Γ_∞ form a direct sum pair with $\oplus\colon \Gamma_\infty \times \Gamma^{H_k}_{\infty,H_n} \to \Gamma^{H_k}_{\infty,H_n}$ defined by letting Γ_∞ act on the odd handles and $\Gamma^{H_k}_{\infty,H_n}$ act on the even handles. The map of the underlying surfaces is illustrated in Figure 2. To check property (5.1), let $h_1, \ldots, h_s \in \Gamma_\infty$. Then they are in the image of $\Gamma_{g,1}$ for some large g. Now choose an appropriate diffeomorphism of the surface $F_{2g,1} \subset F^k_{2g,n+1}$ which moves the even handles over the odd handles to the last g handles. Let $d \in \Gamma_\infty$ be its homotopy class. Similarly, define an element $c \in \Gamma_\infty$. By Proposition 3.5, c and d are in the maximal perfect subgroups of $\Gamma^{H_k}_{\infty,H_n}$ and Γ_∞ respectively. Hence, by the corollary, there exists a space F such that

$$(B\Gamma^{H_k}_{\infty,H_n})^+ \simeq B\Gamma^+_\infty \times F.$$

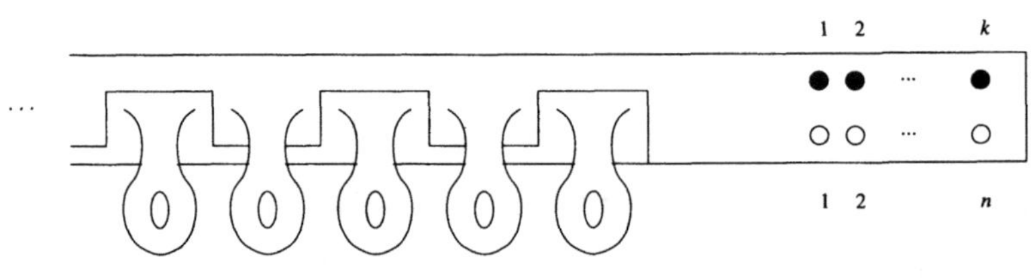

Figure 2: The direct sum homomorphism.

References

[B] A. J. Berrick, *An approach to algebraic K-theory*, Research Notes in Mathematics **56**, Pitman Publishing, 1982.

[Bö] C.-F. Bödigheimer, *Stable splittings of mapping spaces*, in Algebraic Topology, Lecture Notes in Math. **1286**, Springer-Verlag (1987), 174–188.

[BöM] C.-F. Bödigheimer, I. Madsen, *Homotopy quotients of mapping spaces and their stable splittings*, Quart. J. Math. Oxford **39** (1988), 401–409.

[CMT] F. R. Cohen, J. P. May, L. R. Taylor, *Splitting of certain spaces CX*, Math. Proc. Cambridge Philos. Soc. **84** (1978), 465–496.

[ES] C. J. Earle, A. Schatz, *Teichmüller theory for surfaces with boundary*, J. Diff. Geom. **4** (1970), 169–185.

[G] E. Getzler, *Batalin–Vilkovisky algebras and two-dimensional topological field theories*, Comm. Math. Phys. **159** (1994), 265–285.

[GK] E. Getzler, M. M. Kapranov, *Modular Operads*, Compositio Math. **110** (1998), 65–126.

[H] J. L. Harer, *Stability of the homology of the mapping class groups of orientable surfaces*, Ann. of Math. **121** (1985), 215–249.

[I] N. V. Ivanov, *Complexes of curves and the Teichmüller modular group*, Uspekhi Mat. Nauk **42** (1987), 55–107.

[L] E. Looijenga, *Stable cohomology of the mapping class group with symplectic coefficients and the universal Abel–Jacobi map*, J. Alg. Geom. **5** (1996), 135–150.

[M] S. Morita, *Characteristic classes of surface bundles*, Invent. Math. **90** (1987), 551–578.

[N] M. Nakaoka, *The infinite symmetric group*, Ann. of Math. **73** (1961), 229–257.

[P] J. Powell, *Two theorems on the mapping class group of a surface*, Proc. Amer. Math. Soc. **68** (3) (1978), 347–350.

[T] U. Tillmann, *On the homotopy of the stable mapping class group*, Invent. Math. **130** (1997), 257–275.

[W] J. B. Wagoner, *Delooping classifying spaces in algebraic K-theory*, Topology **11** (1972), 349–370.

Mathematisches Institut,
Beringstraße 1,
53155 Bonn, Germany
E-mail address: `boedigheimer@math.uni-bonn.de`

Mathematical Institute,
24–29 St. Giles Street,
Oxford, OX1 3LB, UK
E-mail address: `tillmann@maths.ox.ac.uk`

Progress in Mathematics, Vol. 196

Loop spaces of configuration spaces, braid-like groups, and knots

F. R. Cohen* and S. Gitler*

Abstract. The purpose of this note is to describe some relationships between the following topics: (1) higher dimensional variations of braids, (2) loop space homology, (3) Hopf algebras given by loop space homology, (4) natural groups attached to connected Hopf algebras, (5) analogues of Artin's (pure) braid group, (6) Alexander's construction of knots arising from loop spaces, and (7) Vassiliev's invariants of braids.

1. Introduction

The purpose of this article is to give higher dimensional analogues of braids as well as an analogue of "braiding" certain "pieces" of a manifold such as hyperplanes in Euclidean space or projective spaces in certain Lie groups. The main direction here is that one obtains a Hopf algebra via the homology of the loop space of a configuration space. These Hopf algebras then give groups which have properties that are analogous to Artin's (pure) braid group, and arise from classical algebraic topology. This article is a survey of how some of these results fit [7, 9, 11, 12, 13, 24]. A smattering of new results are included.

These Hopf algebras are special cases of a version of "braiding" which occur in a broader context. In addition to the "infinitesimal braid relations" to be defined below, there are additional relations depending on the underlying geometry of the manifold. The Lie algebras obtained in this way depend on certain naive features of the underlying manifold. Some examples are given below. One natural family of manifolds arises from the Lie groups $SU(n)$. The structures of the Hopf algebras, Lie algebras, and groups encountered here differ drastically in case of $SU(3)$ than those for $SU(n)$ with n not 3.

One of the Hopf algebras that occurs here is the universal enveloping algebra of the "universal Yang–Baxter Lie algebra" which satisfies relations that are sometimes called the "infinitesimal braid relations", or the "horizontal 4T relation" and "framing independence" to knot theorists. This Hopf algebra is the universal enveloping algebra of a graded Lie algebra which also occurs in the study of the

---

1991 *Mathematics Subject Classification.* Primary: 55P35, Secondary: 14D99.

*Partially supported by the NSF.

Vassiliev invariants of braids [3, 18, 19, 20, 26]. In particular, these algebras "account" for all of the Vassiliev invariants of braids as described below in Sections 4, 5 and 9.

The Hopf algebras encountered here give rise to natural groups. Namely, attached to any Hopf algebra with conjugation (or antipode) is the group of coalgebra maps from a fixed coalgebra to the Hopf algebra. With a natural fixed choice of coalgebra given below, this group is filtered, and the underlying set of the associated graded is the product of the underlying set of primitive elements in the Hopf algebra. In the special case for which the Hopf algebra is the homology of the loop space of the configuration space for $\mathbb{R}^{2n}$, the associated graded is a Lie algebra that is isomorphic to the Lie algebra associated to the descending central series for the pure braid group (where these Lie algebras are tensored with the rational numbers). There are further Lie algebras which take into account the underlying geometry of the manifold.

With a particular choice of coalgebra, the groups alluded to above are obtained by assembling the images of the Hurewicz homomorphism into a natural group which, as a set, is the product of the primitive elements in a Hopf algebra. One example of these groups of coalgebra maps is given by the Mal'cev completion of a free group [24, 16]. These completions assemble themselves in various "twisted" ways in the case where the Hopf algebra is the homology of the loop space of the configuration space for $\mathbb{R}^n$.

Namely, the associated Lie algebras are isomorphic (in characteristic zero) to the "universal Yang–Baxter Lie algebra", but the groups themselves have different structures. The point of this is that the Hopf algebras attached to these constructions "see" the same quadratic commutator relations as do the pure braid groups. The precise relations in the group of coalgebra maps also agrees with those of the pure braid group through "quadratic terms". However, the groups of coalgebra maps are filtered and are isomorphic on the level of associated graded modules, but there are terms of higher filtration which appear in the precise relations for groups of coalgebra maps, and the groups themselves are not actually isomorphic to the braid groups [24].

Regarding braids as equivalence classes of motions of distinct particles in the plane through time, there is an extension given by replacing points by other "pieces" of a manifold. There are natural "braid-like" groups which are defined for any manifold which reflect these motions, and arise as a target of the classical Hurewicz homomorphism. These "braid-like" groups satisfy relations given in the braid group up to "quadratic terms" as well as "extended Yang–Baxter relations" as described in Section 3 provided that the underlying manifold enjoys additional naive geometric properties. These structures are developed in Sections 5 and 6.

Constructions involving loop spaces of configuration spaces in turn inform on spaces of embeddings. For example, a classical result of J. W. Alexander gives that every "knot type" is given by "closing-up" a choice of braid [1]. Alexander's construction arises as a map from the loop space of a configuration space to the space of embeddings of S^1 in $\mathbb{R}^3$, and fits in a wider context. That is, there is a map out of the loop space of a configuration space to a space of embeddings of a circle in a manifold M which when $M = \mathbb{R}^3$ specializes to Alexander's map after applying π_0. One immediate observation in this context is that any component of the double loop space of smooth embeddings of S^1 in $\mathbb{R}^3$ splits as a product of the double loop space of the 3-sphere and some other (possibly contractible) space. (This result gives a possibly bizarre connection between the braid groups and knots as $\Omega^2 S^3$ is homotopy equivalent to the Quillen plus construction applied to the classifying space of the stabilized braid group.) A second related observation is a splitting for loop spaces of spaces of embeddings of circles in the product of a manifold $M \times \mathbb{R}^3$ as given in Section 7 here.

The calculations here fit naturally into a context of "quasi-embedding spaces" which are defined formally below. These "quasi-embedding spaces" are given by a subspace of the continuous functions from a circle to a manifold M which satisfy the property that the functions restricted to certain subspaces of the circle are embeddings of the subspaces, but not necessarily embeddings. The subspaces on which the functions are required to be embeddings are given by all cosets of the circle determined by the subgroup of all 2^r-th roots of unity for all $r > 0$. This is in contrast to the Vassiliev conditions for smooth maps which require embeddings except possibly at a finite number of double points.

In the case that the underlying manifold is $\mathbb{R}^n$, these "quasi-embedding spaces" are the inverse limit of spaces whose homology is given in terms of Hochschild homology of the universal enveloping algebras for the "universal Yang–Baxter Lie algebras". These spaces are the subject of Section 8.

Based on work of T. Kohno, there is a comparison of the Hopf algebras encountered here to those that give the Vassiliev invariants. This is carried out in Section 9.

The authors of this article would like to thank the referee of this article for pointing out other recent, and interesting work on pure braids, and their generalizations. Among these are work of Manin–Schechtman where the authors consider higher dimensional braids related to the KZ-equations by braiding 2-planes in 4-space. In addition, there is interesting work of Stefan Papadima, Simon Willerton, and Jacob Mostovoy that connect the Lie algebra relations that occur in the Lie algebra attached to the descending central series for the pure braid group, to the Vassiliev invariants as well as work of Kohno. Time constraints do not permit the opportunity to compare these interesting constructions with those considered here.

The first author would like to thank the CRM for providing a stimulating mathematical environment, and for their generous support. Part of the work here was done at the CRM.

Contents

1. Introduction 59
2. Braids and their generalizations to higher dimensions 62
3. "Universal extended Yang–Baxter Lie algebras" 63
4. Loop space homology of configuration spaces 64
5. Natural groups attached to connected Hopf algebras 66
6. Braid-like groups: analogues of Artin's (pure) braid group 69
7. On Alexander's construction of knots arising from loop spaces 71
8. Crude approximations to embedding spaces 73
9. Vassiliev invariants and the loop space homology of configuration spaces 76
References 77

2. Braids and their generalizations to higher dimensions

Recall Artin's braid group [2, 4, 22]. The so-called "pure braids" are those braids which leave the end-points unpermuted. A braid can be thought of as a path in the configuration space of k points in the plane $\mathbb{R}^2$. A loop in this configuration space specifies a pure braid. There is a naive construction of such a path given by a smooth function from a circle to the configuration space which represents these braids and which is described explicitly below. Higher dimensional analogues are given by "thickening" each strand, while explicit formulae are also given below.

Regard these thickenings as higher dimensional braids. These thickenings are partitioned into homotopy classes below, and naturally give rise to a group which is very much like Artin's braid group, but is not actually isomorphic to it. More precisely, this group is filtered with the property that the associated graded is a Lie algebra. Furthermore, this Lie algebra in characteristic zero, and apart from a formal shift in grading, is isomorphic to the "universal Yang–Baxter Lie algebra" to be defined below. By results in [14, 18], that Lie algebra is precisely the Lie algebra attached to the descending central series of the pure braid group.

In addition, the idea of higher dimensional braiding extends to braidings of other geometric objects in a manifold. Namely, a braid may be thought of as the locus obtained by k distinct points in the plane that move through time. One might replace points by other subspaces of a manifold, and then consider the object obtained by motions of distinct subspaces through time. Specific examples of these subspaces are given below by choices of hyperplanes in Euclidean space or projective spaces in Lie groups.

There is a resulting Hopf algebra, Lie algebra, and group. Furthermore, these constructions are in a naive way extensions of classical constructions with additional universal relations. These constructions are addressed below and give rise to the notion of the "extended Yang–Baxter Lie algebra relations", or "extended infinitesimal braid relations" [7].

Recall that the classical configuration space of ordered k-tuples of distinct points in a manifold M is given by

$$F(M,k) = \{(m_1, m_2, \ldots, m_k) \mid m_i \neq m_j \text{for } i \neq j\};$$

see [12]. The higher dimensional analogues of representations for generators of the pure braid groups are induced by maps given in [6, 7, 10, 11, 12] with $k \geq i > j \geq 1$:

$$A_{i,j} \colon S^{n-1} \to F(\mathbb{R}^n, k).$$

In the case of $n = 2$, these elements represent generators of $\pi_1(F(\mathbb{R}^2, k))$, the pure k-stranded braid group. Adjoints of these maps are

$$B_{i,j} \colon S^{n-2} \to \Omega F(\mathbb{R}^n, k).$$

A graded Lie algebra arises from these maps via the Samelson product in homotopy, the so-called homotopy Lie algebra which is discussed below. One feature of these graded Lie algebras is that, apart from gradings, they are isomorphic when n is even. The case when n is odd provides related, but slightly different stuctures. These Lie algebras will also be elucidated below.

The maps $A_{i,j}$ arise roughly by linking the diagonal in the i and j coordinates. Here fix points q_i in $\mathbb{R}^n$ given by $q_i = 4i(v)$ where v is the canonical unit vector $(1, 0, \ldots, 0)$. Regard z as a point of unit norm in $\mathbb{R}^n$, and define

$$A_{i,j}(z) = (x_1, x_2, \ldots, x_n)$$

where $x_t = q_t$ if t is not i, and $x_i = q_j + v$.

In addition, J. W. Alexander gave a well-known construction of knots and links from braids [1, 4]. Let $\mathrm{Emb}(S^1, M)$ denote the space of continuous embeddings of a circle in a fixed manifold M. In Section 7 of this note, there is a map

$$\Theta \colon \Omega F(M \times [0,1]^2, k) \to \mathrm{Emb}(S^1, M \times \mathbb{R}^3)$$

which specializes to Alexander's map on the level of π_0 when $M =$ point. This map then provides information about the space of embeddings $\mathrm{Emb}(S^1, M \times \mathbb{R}^3)$. There is an analogous version where one restricts to smooth embeddings and smooth loops. Some crude information concerning these spaces is given in Sections 6 and 7.

3. "Universal extended Yang–Baxter Lie algebras"

The "universal Yang–Baxter Lie algebras" are defined next [7, 11]. Fix a graded abelian group $V_k(n)$ with basis given by elements of degree n labelled by $B_{i,j}$ for $k \geq i > j \geq 1$. Next, consider the free graded Lie algebra $L[V_k(n)]$ generated by $V_k(n)$. Let $\mathcal{L}_k(n)$ denote the quotient Lie algebra obtained from the "graded infinitesimal braid relations":

1. $[B_{i,j}, B_{s,t}] = 0$ if $\{i,j\} \cap \{s,t\} = \phi$,
2. $[B_{i,j}, B_{i,t} + (-1)^n B_{t,j}] = 0$ for $1 \leq j < t < i \leq k$, and
3. $[B_{t,j}, B_{i,j} + B_{i,t}] = 0$ for $1 \leq j < t < i \leq k$.

In the case that the $B_{i,j}$ are of even degree, the relations above are both redundant, and equivalent to the relations given by Kohno in [18, 19]. There are further related relations in Lie algebras which are extensions of the $L[V_k(n)]$ where x_i for $1 \leq i \leq k$ are elements in a fixed choice of extension of the Lie algebra $L[V_k(n)]$.

The "extended infinitesimal braid relations" are defined as follows:

1. $[B_{i,j}, x_s] = 0$ if $\{i,j\} \cap \{s\} = \phi$,
2. $[B_{i,j}, x_i + x_j] = 0$.

A specific example is given next of a Lie algebra for which the "extended infinitesimal braid relations" are satisfied. Fix a graded Lie algebra $\mathcal{G}$ which is a free module over a commutative ring R together with the k-fold direct sum of $\mathcal{G}$ given by $\mathcal{G}^{\oplus k}$. Assume that (1) $\mathcal{L}_k(n)$ is an R-module (replacing $\mathcal{L}_k(n)$ by $\mathcal{L}_k(n) \otimes_{\mathbb{Z}} R$), and (2) $\mathcal{G}$ is R-free. Consider the coproduct in the category of graded Lie algebras (over R) $\mathcal{L}_k(n) \coprod \mathcal{G}^{\oplus k}$. Let

$$\mathcal{L}_k(n) \wr \mathcal{G}$$

denote the quotient of $\mathcal{L}_k(n) \coprod \mathcal{G}^{\oplus k}$ modulo the "extended infinitesimal braid relations" in which the elements x_i denote the element $a \oplus x \oplus b$ with x in $\mathcal{G}$, $a = 0^{\oplus i-1}$, and $b = 0^{\oplus n-i}$. This construction is reminiscent of the classical wreath product construction on the level of groups.

It was pointed out in [7] that $\mathcal{L}_k(n)$ is a torsion free Lie algebra which is finitely generated in each degree as a module over the integers. Thus by the Poincaré–Birkhoff–Witt theorem, $\mathcal{L}_k(n)$ embeds in its universal enveloping algebra.

4. Loop space homology of configuration spaces

In this section, consider the homology of the loop space of the configuration space $\Omega F(M,k)$. It is sometimes the case that these homology groups are torsion free. If not, restrict to field coefficients, $\mathbb{F}$. In addition, let $\operatorname{Prim} H_*(\Omega M; \mathbb{F})$ denote the Lie algebra of primitive elements in the Hopf algebra $H_*(\Omega M; \mathbb{F})$. The following theorem was proven in [11] and [7].

Theorem 4.1. [11, 7] *If $M = \mathbb{R}^n$ for $n > 2$, then the homology of the loop space $\Omega F(M,k)$ is isomorphic to the universal enveloping algebra of $\mathcal{L}_k(n-2)$ as a Hopf algebra. Furthermore, the elements $B_{i,j}$ are given by the Hurewicz image of the fundamental cycle for a sphere via the maps $B_{i,j}$ defined in Section* 1.

With some additional hypotheses on the underlying manifold M, there is an analogous theorem with field coefficients $\mathbb{F}$. A further naive idea of "braiding of a subspace A" of M is required. Namely, a manifold M is said to be A-dominated provided the following hold:

1. There exist embeddings $e_i \colon A \to M$, $i = 1, 2$, with disjoint images.
2. The map e_1 is isotopic to e_2.

3. The induced maps $\Omega(e_i)\colon \Omega(A) \to \Omega(M)$ (with respect to different base-points) induce surjections in homology (with $\mathbb{F}$-coefficients).

Notice that the subspace A plays a role analogous to points in $\mathbb{R}^2$ as follows. Braids correspond to motions of points in the plane through time. Analogously, one might consider subspaces A of a manifold and motions of disjoint copies of A through time. The homological property that the induced map $\Omega(A) \to \Omega(M)$ gives an epimorphism in homology provides a way to measure motions of disjoint copies of A through time. Furthermore, there is a "braid-like" group attached to these homological measures; these are given in Section 5.

Proposition 4.2. *Let $n > 3$ and $A = \Sigma\mathbb{C}P^{n-1}$. The manifolds* $\mathrm{SU}(n)$ *are A-dominated. Furthermore, the manifolds* $\mathrm{SU}(3)$ *and* $\mathrm{Sp}(2)$ *are not A-dominated for any A.*

Proof. Notice that $\Sigma\mathbb{C}P^{n-1}$ embeds in $\mathrm{SU}(n)$ [17]. Since $n > 3$, there are disjoint and isotopic copies of $\Sigma\mathbb{C}P^{n-1}$ embedded in $\mathrm{SU}(n)$ by general position [17]. Furthermore, the homology of the loop space of $\Sigma\mathbb{C}P^{n-1}$ surjects to that of the loop space of $\mathrm{SU}(n)$.

The analogous result fails for $\mathrm{SU}(3)$ and $\mathrm{Sp}(2)$. Here notice that if there were two disjoint embeddings of $\Sigma\mathbb{C}P^2$ in $\mathrm{SU}(3)$, then the cup product of the three class with the five class is zero in the cohomology of $\mathrm{SU}(3)$ by duality. This contradicts that the product of the three dimensional generator with the five dimensional generator gives the top class. A similar argument applies to $\mathrm{Sp}(2)$. □

The next theorem illustrates homological consequences of "braiding" of spheres and subspaces of certain choices of manifolds.

Theorem 4.3. [7] *Let M be a simply-connected m-dimensional manifold, $m > 2$, which satisfies*

1. *$M = N \setminus \{\text{point}\}$ for a manifold N,*
2. *$w_{m-1}(\tau(M)) = 0$ where $\tau(M)$ is the tangent bundle of M if $F = \mathbb{Z}/2\mathbb{Z}$,*
3. *the Euler class of $\tau(M)$ is zero if $\mathrm{char}(\mathbb{F}) \neq 2$,*
4. *the homology Hopf algebra $H_*(\Omega M;\mathbb{F})$ is a primitively generated Hopf algebra which is isomorphic to the universal enveloping algebra of the Lie algebra of primitive elements $\mathrm{Prim}\, H_*(\Omega M;\mathbb{F})$, and*
5. *M is A-dominated for some A.*

Then $H_*(\Omega F(M,k);\mathbb{F})$ is isomorphic to the universal enveloping algebra of $\mathcal{L}_k(n) \wr \mathcal{G}$, where $\mathcal{G}$ is the Lie algebra of primitive elements, $\mathrm{Prim}\, H_*(\Omega M;\mathbb{F})$.

The proof of the conclusion in the theorem above uses the hypothesis that M is A-dominated for some A. Furthermore, this conclusion is independent of the choice of A. In case the underlying manifold has a Euclidean factor, a similar conclusion follows next.

Theorem 4.4. [7] *Let M be a simply-connected m-dimensional manifold, where $m > 1$, and assume that the Hopf algebra $H_*(\Omega M;\mathbb{F})$ is primitively generated. Then $H_*(\Omega F(\mathbb{R} \times M,k);\mathbb{F})$ is isomorphic to the universal enveloping algebra of $\mathcal{L}_k(m-1) \wr \mathrm{Prim}\, H_*(\Omega M;\mathbb{F})$.*

In case $M = \mathrm{SU}(n+k)/\mathrm{SU}(n)$ and $\mathbb{F}$ is given by the rational numbers, then the homology algebra $H_*(\Omega F(M,k);\mathbb{F})$ is also described in [3]. The extensions with $M = \mathrm{SU}(3)$ are the most interesting.

There are analogous features of the above constructions given in work of Xicoténcatl [27] who considers configurations of orbits in a manifold with a free action of a group. Specific calculations apply to complex n-space minus the origin. He then obtains Lie algebras that support actions of cyclic groups, and which are analogous to the "universal Yang–Baxter Lie algebras" with additional symmetries.

5. Natural groups attached to connected Hopf algebras

Let $\mathrm{Hom}^{\mathrm{coalg}}(T[v], H)$ denote the set of coalgebra morphisms with source given by the tensor algebra over the integers with a single primitive algebra generator v in degree 1. Furthermore, the target H is a Hopf algebra with conjugation (antipode). Recall that this set is naturally a group with multiplication induced by the coproduct for the source and product for the target with inverses induced by the conjugation in H [23].

Throughout this section, assume that X is simply-connected. If $H_*(\Omega X)$ is torsion free, then it is naturally a coalgebra by the Künneth theorem. If homology is taken with field coefficients $\mathbb{F}$, then $H_*(\Omega X;\mathbb{F})$ is a Hopf algebra with conjugation. Thus if $\mathbb{F}$ is a field or if $H_*(\Omega X;\mathbb{Z})$ is torsion free, then for $R = \mathbb{Z}$ or $R = \mathbb{F}$,

$$\mathrm{Hom}^{\mathrm{coalg}}(T[v], H_*(\Omega X; R))$$

is always a group. Of course, this is a special case of the hom-sets in a category where the target is a group object in the category. It is unclear whether it is informative to look at this construction within the context here. Thus this note will be limited to the remarks below.

Next specialize to the group of pointed homotopy classes of maps $[\Omega S^2, \Omega X]$. As a set, this group is isomorphic to the direct product of all of the homotopy groups of ΩX in case X is simply-connected. This last assertion follows at once from the well-known fact that the single suspension of ΩS^2 is homotopy equivalent to the bouquet $\vee_{n\geq 2} S^n$.

Thus the group $[\Omega S^2, \Omega X]$ may be regarded as reassembling all of the homotopy groups of ΩX into a single group. It will be seen below that the group $[\Omega S^2, \Omega X]$ has additional structure that is not "seen" by the additive structure of all the homotopy groups of X. Indeed one of the groups that appears in this context is very close to Artin's braid group as will be seen in Section 6. In addition, the fact that the group $[\Omega S^2, \Omega X]$ gives the product of all of the homotopy groups for ΩX suggests that the choice of source has special features.

The groups $[\Omega S^2, \Omega X]$, and $\mathrm{Hom}^{\mathrm{coalg}}(H_*(\Omega S^2), H_*(\Omega X))$ are filtered groups via a decreasing filtration:

1. filtration j of $[\Omega S^2, \Omega X]$ is given by those maps which are null-homotopic when restricted to the $(j-1)$-skeleton;
2. filtration j of $\mathrm{Hom}^{\mathrm{coalg}}(H_*(\Omega S^2), H_*(\Omega X))$ is given by those maps which are trivial in homology in dimensions less than j.

Denote the associated graded groups by

$$E_0^j = F^j/F^{j+1}.$$

Let G denote the group $[\Omega S^2, \Omega X]$ or $\mathrm{Hom}^{\mathrm{coalg}}(H_*(\Omega S^2), H_*(\Omega X))$. Consider the commutator map $[-,-]\colon G\times G \to G$. Further, let ϕ denote the morphism which sends a map to the induced map on homology:

$$\phi\colon [\Omega S^2, \Omega X] \to \mathrm{Hom}^{\mathrm{coalg}}(H_*(\Omega S^2), H_*(\Omega X)).$$

Theorem 5.1. [9] *Assume that $H_*(\Omega X)$ is torsion free.*

1. *The function*

$$\phi\colon [\Omega S^2, \Omega X] \to \mathrm{Hom}^{\mathrm{coalg}}(H_*(\Omega S^2), H_*(\Omega X))$$

is a morphism of filtered groups.

2. *In case $G = [\Omega S^2, \Omega X]$, the associated graded E_0^j is isomorphic to $\pi_j(\Omega X)$.*
3. *In case $G = \mathrm{Hom}^{\mathrm{coalg}}(H_*(\Omega S^2), H_*(\Omega X))$, the associated graded E_0^j is isomorphic to the module of primitives concentrated in degree j,*

$$\mathrm{Prim}\, H_j(\Omega X).$$

4. *On the level of associated graded groups, the map*

$$E_0^j(\phi)\colon \pi_j(\Omega X) \to \mathrm{Prim}\, H_j(\Omega X)$$

is the Hurewicz homomorphism.

5. *The Hurewicz homomorphism $\pi_*(\Omega X) \to \mathrm{Prim}\, H_*(\Omega X)$ surjects to the module of primitives in $H_*(\Omega X)$ if and only if ϕ is a surjection.*
6. *The commutator map $[-,-]\colon G \times G \to G$ induces a homomorphism on the level of associated groups*

$$E_0^j \otimes E_0^k \to E_0^{j+k}$$

which endows the associated graded groups $\oplus_{j\geq 1} E_0^j = F^j/F^{j+1}$ with the structure of graded Lie algebra (with the usual exceptions for the primes 2 and 3 where it will be assumed for convenience that the primes 2 and 3 are units); this structure on the source is induced by the Samelson product $\langle -,-\rangle$. Namely the Lie bracket $[\alpha,\beta]$ is given by $\binom{j+k}{k}\langle\alpha,\beta\rangle$.

7. *The morphism $E_0^j(\phi)$ is a morphism of graded Lie algebras.*

8. *The associated graded module for the group* $\mathrm{Hom}^{\mathrm{coalg}}(H_*(\Omega S^2), H_*(\Omega X))$ *is given by*
$$\oplus_{j\geq 1} E_0^j = \oplus_{j\geq 1} \mathrm{Prim}\, H_j(\Omega X).$$
As a Lie algebra the Lie bracket $[a,b]$ *is given by* $\binom{j+k}{k}(a\otimes b-(-1)^{jk}b\otimes a)$ *for primitive elements* a *of degree* j*, and* b *of degree* k*.*

Remarks

1. The above theorem reflects features of Samelson products in homotopy theory as well as those of the Hurewicz map with the additional point that the classical Hurewicz map is the associated graded of a natural map. The result is a small modification of earlier classical results. The motivation for including these results is that they "explain" how the Lie algebras arising in the loop space homology of configuration spaces are associated gradeds for natural groups.
2. The graded module $\oplus_{j\geq 1} E_0^j = F^j/F^{j+1}$ is not, without additional assumptions, a graded Lie algebra, as $[x,x]$ and $[[x,x]x]$ may be non-zero. In case x is of even degree, and $[x,x]$ is non-zero, it is of order 2. In case x is of odd degree, and $[[x,x],x]$ is non-zero, it is of order 3. Furthermore, the binomial coefficients in part (8) above arise from the structure of the coproduct for $T[v]$.

In addition, a theorem of [23] concerning the structure of rational homotopy groups of ΩX gives information at once about the group $[\Omega S^2, \Omega X]$. The theorem in [23] which states that the rational homology of a loop space of a simply-connected space X is isomorphic to the universal enveloping algebra of the rational homotopy Lie algebra of ΩX is the associated graded version of the next theorem concerning filtered groups.

Theorem 5.2. [9] *Let* X *denote the rationalization of a simply-connected space of finite type. Then*
$$\phi\colon [\Omega S^2, \Omega X] \to \mathrm{Hom}^{\mathrm{coalg}}(H_*(\Omega S^2), H_*(\Omega X))$$
is an isomorphism of filtered groups.

The structure of the group $\mathrm{Hom}^{\mathrm{coalg}}(H_*(\Omega S^2), H_*(\Omega X))$ is frequently non-trivial as all of the homotopy groups assembles into a single group with non-trivial extensions. It is not yet clear what precisely determines these extensions as well as what the characteristic classes of the group extensions actually are. The following example gives one case where they are non-trivial and where the constructions in the above paragraph are useful.

Theorem 5.3. [23] *If* X *is a bouquet of* q *copies of the* 3*-sphere,* S *is a set of cardinality* q*, and* $F[S]_M$ *is the Mal'cev completion of* $F[S]$*, then there is an isomorphism of groups*
$$F[S]_M \to \mathrm{Hom}^{\mathrm{coalg}}(H_*(\Omega S^2), H_*(\Omega X; \mathbb{Q})).$$

It is the purpose of the next section to describe some explicit examples which use the above theorem as a further connection between the braid groups and the "Yang–Baxter Lie algebras". Namely, the subject of the next section uses the previous theorem as well as the next theorem where the ring R is assumed to be a PID.

Theorem 5.4. [9] *Let*

$$(\dagger): 1 \to A \to B \to C \to 1$$

be a short exact sequence of Hopf algebras with conjugation (antipode) that are free over R*, and which is split as coalgebras. Then for any fixed choice of coalgebra, there is a short exact sequence of groups*

$$(\ddagger): 1 \longrightarrow \mathrm{Hom}^{\mathrm{coalg}}(K,A) \longrightarrow \mathrm{Hom}^{\mathrm{coalg}}(K,B) \longrightarrow \mathrm{Hom}^{\mathrm{coalg}}(K,C) \longrightarrow 1.$$

Furthermore, if the short exact sequence $(\dagger)$ *is also multiplicatively split, then the extension* $(\ddagger)$ *is multiplicatively split and the group* $\mathrm{Hom}^{\mathrm{coalg}}(K,B)$ *is a semi-direct product of the groups* $\mathrm{Hom}^{\mathrm{coalg}}(K,A)$*, and* $\mathrm{Hom}^{\mathrm{coalg}}(K,C)$*.*

The next proposition is a remark which follows from the work in [8, 23].

Proposition 5.5. *Let* X *denote the localization at* p *of the* $(2n+1)$*-connected cover of the* $(2n+1)$*-sphere. Then the group* $[\Omega S^2, \Omega X]$ *has exponent*

1. p^n *if* p *is odd, and*
2. *bounded above by* $2^{(3n/2)+\epsilon}$ *for* $\epsilon = 0$ *or* 1*.*

Proof. The odd primary case follows at once from

1. the fact that S^{2n+1} localized at an odd prime is an H-space. This gives that $[\Omega S^2, \Omega S^{2n+1}]$ splits as a group $[\vee_{t\geq 2} S^t, \Omega X]$, and
2. the exponents given in [8].

The 2^q-th power map on ΩS^{2n+1} is homotopic to the looping of the degree 2^q map if $q > 1$ after localization at 2. Thus if $q > 1$, the effect of the 2^q-th power map on $[\Omega S^2, \Omega S^{2n+1}]$ is induced by the self-map of $[\Sigma\Omega S^2, S^{2n+1}]$ given by the degree 2^q map on the target which is then given by the induced map

$$[\vee_{t\geq 2} S^t, \Omega X] \to [\vee_{t\geq 2} S^t, \Omega X]$$

given by multiplication by degree 2^q on each homotopy group.

These bounds on exponents of homotopy groups are given by results in [25]. The proposition follows. □

6. Braid-like groups: analogues of Artin's (pure) braid group

The main point of this section is that the group of coalgebra maps in the previous section provide groups which have features that are analogous to Artin's (pure) braid group. Some of these "braid-like" properties are described below. The main theme is that these groups of coalgebra maps are filtered as above. The Lie algebras attached to these filtrations via Theorems 5.1 and 5.4 are isomorphic to the Lie algebra obtained from the descending central series for the pure braid group in the cases where the underlying manifolds are given by $\mathbb{R}^{2n}$ for $n > 1$ (and thus give the Vassiliev invariants of braids by work of Kohno [18, 19] as is pointed out in Section 8). There are generalizations of these "braid-like" groups for other choices of manifolds.

Consider the groups and maps defined in the last section where X is the configuration space of ordered k-tuples of distinct points in a manifold. The first result in this direction is to consider the group $\mathrm{Hom}^{\mathrm{coalg}}(H_*(\Omega S^2), H_*(\Omega X))$ as well as the filtration described above.

Theorem 6.1. *If $X = F(\mathbb{R}^n, k)$ for $n > 2$, and homology is taken with coefficients in the rational numbers, then the associated graded Lie algebra for the group of coalgebra morphisms $\mathrm{Hom}^{\mathrm{coalg}}(H_*(\Omega S^2), H_*(\Omega X))$ is isomorphic to the "universal Yang–Baxter Lie algebra" $\mathcal{L}_k(n-2) \otimes_{\mathbb{Z}} \mathbb{Q}$.*

The proof of this theorem follows at once from Theorems 5.3, and 5.4 above. Namely, the fibrations in [13] imply that the homology of $\Omega F(\mathbb{R}^n, k)$ is given by an iterated split extension as in Theorem 5.3 where the Hopf algebra kernels at each stage are given by primitively generated tensor algebras [7, 11]. The group extensions follow by direct calculation.

Notice that apart from a grading, the Lie algebras that occur in Theorem 6.1 when n is even are isomorphic to the Lie algebra obtained from the descending central series for Artin's braid group on k strands [14, 18]. This raises the question of whether the groups themselves are isomorphic at least after forming the nilpotent completion.

This turns out to not hold. These groups of coalgebra morphisms do not have the same relations. Namely, the relations in the pure braid group can be given in terms of commutators of weight 2 and weight 3. The relations in the group of coalgebra maps satisfy these through quadratic terms, and the groups on the level of associated graded modules agree. The actual relations in the group of coalgebra maps appear to be given by an infinite product of commutators of arbitrarily large length and is work in progress [24].

Since the relations for the pure braid group are not usually listed in terms of commutators, this presentation is recorded next. This formulation follows at once from the presentation listed in [22, p. 174], where the element $[g, h]$ in a group G is given by the commutator $ghg^{-1}h^{-1}$:

I. Generators for the k-stranded pure braid group PBr_k are given by $B_{j,i}$ for $1 \leq j < i \leq k$, and

II. A complete set of relations is given by

1. $[B_{r,s}, B_{i,k}] = 1$ for $s < i$ or $k < r$,
2. $[B_{k,s}, B_{i,k}] = [B_{i,s}{}^{-1}, B_{i,k}]$ for $i < k < s$, and
3. $[B_{r,s}, B_{i,k}] = [[B_{i,s}{}^{-1}, B_{i,r}{}^{-1}], B_{i,k}]$ for $1 < r < k < s$.

By construction, the map

$$\phi\colon [\Omega S^2, \Omega F(\mathbb{R}^n, k)] \to \mathrm{Hom}^{\mathrm{coalg}}(H_*(\Omega S^2), H_*(\Omega F(\mathbb{R}^n, k)))$$

is a surjection. The reason for this statement is that the Hurewicz homomorphism surjects to the module of primitives in $H_*(\Omega F(\mathbb{R}^n, k))$.

Theorem 6.2. *The group* $[\Omega S^2, \Omega F(\mathbb{R}^n, k)]$ *is the semi-direct product of*

$$\mathrm{Hom}^{\mathrm{coalg}}(H_*(\Omega S^2), H_*(\Omega F(\mathbb{R}^n, k)))$$

and the kernel of ϕ.

That the infinitesimal braid relations are satisfied in the associated graded for the group of coalgebra maps $\mathrm{Hom}^{\mathrm{coalg}}(H_*(\Omega S^2), H_*(\Omega F(\mathbb{R}^n, k)))$ suggests calling these groups "braid-like groups". If M is a manifold, there are similar "braid-like groups" given by

$$\mathrm{Hom}^{\mathrm{coalg}}(H_*(\Omega S^2), H_*(\Omega F(M, k))).$$

Theorem 6.3. *If M is a simply-connected m-dimensional manifold for $m > 2$, then the Lie algebra given by the associated graded for the group*

$$\mathrm{Hom}^{\mathrm{coalg}}(H_*(\Omega S^2), H_*(\Omega F(M \times \mathbb{R}, k); \mathbb{Q}))$$

is isomorphic to $\mathcal{L}_k(m-1) \wr \mathrm{Prim}\, H_*(\Omega M; \mathbb{Q})$.

In case M is $\mathrm{SU}(n)$, $\mathrm{SO}(n)$ or $\mathrm{Sp}(n)$, or certain associated homogeneous spaces, then similar results apply. The rank two cases of $\mathrm{SU}(3)$ and $\mathrm{Sp}(2)$ are more complicated than the cases of larger rank. Specific answers are given in [7].

7. On Alexander's construction of knots arising from loop spaces

By "closing up" pure braids, Alexander gave a procedure for constructing isotopy classes of smooth knots and links [1, 4]. This construction "fits" with the natural maps considered here, and there are similar constructions which apply to links by using paths. There is an explicit map given below, which for M = point corresponds to "cyclic closure" of pure braids and is induced by applying π_0 to a geometric map between function spaces given as

$$\Lambda_k\colon \Omega F(M \times [0,1]^2, k) \to \mathrm{Emb}(S^1, M \times \mathbb{R}^3).$$

Namely, an element in the loop space of the configuration space is a k-tuple of functions that do not simultaneously coincide. By (i) dividing the interval into $2k$ subintervals, (ii) "stringing out" these k functions over k alternating intervals via a time parameter, and (iii) "connecting" these strings in a compatible way on the complementary k alternating intervals, one gets an embedding of a circle. A formal description is as follows. Consider an element in the space $\Omega F(M \times [0,1]^2, k)$. That is a map from the circle to $F(M \times [0,1]^2, k)$ which consists of

1. a k-tuple of continuous functions $(f_1, f_2, \ldots, f_k)$,
2. $f_i\colon [0,1] \to M \times [0,1]^2$,

3. $(f_i)(0) = (f_i)(1) = x_i$ for fixed and distinct x_i, and
4. $(f_i)(t) \neq (f_j)(t)$ when $i \neq j$.

Next fix paths $p_i\colon [0,1] \to M \times \mathbb{R}^2 \times \mathbb{R}$ such that

1. $(p_i)(0) = (x_i, 1)$,
2. $(p_i)(1) = (x_{i+1}, 0)$ for $i < k$,
3. $(p_k)(1) = (x_1, 0)$,
4. the image of $p_i((0,1))$ is contained in the complement of $M \times [0,1]^2$ in $M \times \mathbb{R}^3$, and
5. the images $p_i([0,1])$ satisfy $p_i([0,1]) \cap p_j([0,1]) = \emptyset$ for $i \neq j$.

The construction of the map Λ_k follows directly: Consider the intervals

$$[i/2k, (i+1)/2k],$$

and define Λ_k in a piecewise manner as follows (which admits a "smooth analogue"):

1. $\Lambda_k(t) = (f_i(2k[t - 2i/2k]), 2k[t - 2i/2k])$ if $2i/2k \leq t \leq (2i+1)/2k$.
2. $\Lambda_k(t) = (p_i(2k[t - (2i+2)/2k])$ if $(2i+1)/2k \leq t \leq (2i+2)/2k$.

A result of Alexander [1] then admits a function space analogue which follows from his results together with the definition of the maps Λ_k.

Proposition 7.1. *In case $M =$ point, the smooth analogue*

$$\amalg_{k\geq 2}\{\pi_0\Lambda_k\}\colon\ \amalg_{k\geq 2}\{\pi_0\,\Omega^{\text{smooth}}F(\mathbb{R}^2,k)\} \to \pi_0\,\text{Emb}^{\text{smooth}}(S^1,\mathbb{R}^3)$$

induces a surjection of sets.

Proof. Notice that the definition of the maps Λ_k induce the map on the level of pure braids that is given by cyclic closure of pure braids. By Alexander's results [1], cyclic closure of pure braids gives all isotopy classes of knots for some choice of pure braid. The result follows. □

Consider the natural inclusion of $\text{Emb}(S^1, M)$ in the free loop space ΛM.

Proposition 7.2. *If M is simply-connected, the composite*

$$\Omega(M \times [0,1]^2) \to \text{Emb}(S^1, M \times \mathbb{R}^3) \to \Lambda(M \times \mathbb{R}^3)$$

is homotopic to the natural inclusion of $\Omega(M)$ in $\Lambda(M)$. Thus the loop space of $\text{Emb}(S^1, M \times \mathbb{R}^3)$ is homotopy equivalent to $\Omega^2 M \times X_M$ for some choice of a space X_M, and $\pi_*(\text{Emb}(S^1, M \times \mathbb{R}^3))$ contains $\pi_*(\Omega(M))$ as a direct summand.

Proof. The composite $\Omega(M \times [0,1]^2) \to \text{Emb}(S^1, M \times \mathbb{R}^3) \to \Lambda(M \times \mathbb{R}^3)$ is homotopic to the map which sends a loop to that loop plus the loop sum with a fixed loop. Since M is simply-connected, this map is homotopic to the natural inclusion of $\Omega(M)$ in $\Lambda(M)$. The result follows from the fact that the inclusion $\Omega(M) \to \Lambda(M)$ is split after looping once. □

Let $\mathrm{Emb}_\beta(S^1,\mathbb{R}^3)$ denote the path-component of an embedding β. An analogous standard observation is that the group SO(3) acts on the space of embeddings $\mathrm{Emb}(S^1,\mathbb{R}^3)$. Thus if β is a fixed embedding, there are maps

$$\{\beta\} \times \mathrm{SO}(3) \to \mathrm{Emb}_\beta(S^1,\mathbb{R}^3) \to S^2.$$

Proposition 7.3. *There is a homotopy equivalence*

$$\Omega^2\mathrm{Emb}_\beta(S^1,\mathbb{R}^3) \to \Omega^2 S^3 \times X_\beta$$

for some choice of a (possibly contractible) space X_β. Thus, the homotopy groups of the component of β, $\mathrm{Emb}_\beta(S^1,\mathbb{R}^3)$, are non-zero in arbitrarily large degrees. An analogous result holds in the case of smooth embeddings $\mathrm{Emb}_\beta^{\mathrm{smooth}}(S^1,\mathbb{R}^3)$.

Proof. Consider the composite of the maps

$$\{\beta\} \times \mathrm{SO}(3) \to \mathrm{Emb}_\beta(S^1,\mathbb{R}^3) \to S^2.$$

If α is an element of SO(3), this composite sends α to $\alpha(q)/||\alpha(q)||$ where $q = \beta(1) - \beta(-1)$. A choice of rotation sending q to $(1,0,0)$ in $\mathbb{R}^3$ gives that the above composite is homotopic to the standard map of SO(3) to S^2. Consider double looping the map $\mathrm{SO}(3) \to S^2$. Each connected component is homotopy equivalent to $\Omega^2 S^3$, and the induced map is an equivalence on the component of the base-point. Notice that if β is smooth, then the map g takes values in $\mathrm{Emb}_\beta^{\mathrm{smooth}}(S^1,\mathbb{R}^3)$. The proposition follows. □

One is led to wonder whether the spaces X_β have homotopy types that are independent of β. One consequence of a preprint due to A. Hatcher gives that the spaces X_β are contractible. A second question is whether there are function space interpretations of Markov moves for the maps given in Proposition 7.1. Proposition 7.3 gives yet another (weird?) connection between knots and the braid groups as the Quillen plus construction for the classifying space of the stable braid group gives $\Omega^2 S^3$.

8. Crude approximations to embedding spaces

The space of continuous embeddings of a circle in a manifold admits a natural embedding as a closed subspace of the space ΛM of all (free) continuous maps of S^1 to M. There is a natural tower of spaces that factors this natural inclusion and is described in more detail below.

Consider the subspace of continuous maps of a circle to the manifold which restrict to simultaneous embeddings on all cosets of the group G generated by all 2^k-th roots of unity in a circle for $k > 0$. Call this subspace $\mathrm{Emb}_G(S^1, M)$.

This subspace is the main object of study in this section. In particular, it is pointed out below that $\mathrm{Emb}_G(S^1, M)$ is the inverse limit of identifiable spaces which arose in the previous sections. One consequence is that when the manifold M is Euclidean n-space for $n > 2$, the rational homology of each space in this inverse limit is isomorphic to the Hochschild homology of the universal enveloping algebra for the "universal Yang–Baxter Lie algebra" as defined above and in [7, 11].

Let Γ_k denote the subgroup of the circle generated by the k-th root of unity ρ_k given by $e^{2\pi i/k}$. The inclusion of Γ_{2^j} in $\Gamma_{2^{j+1}}$ which sends ρ_{2^j} to $(\rho_{2^{j+1}})^2$ induces a tower of spaces

$$\begin{array}{ccc} X_{j+1} & \longrightarrow & F(M, 2^{j+2}) \\ & & \big\downarrow \pi_{2^{j+1}} \\ X_j & \longrightarrow & F(M, 2^{j+1}) \\ & & \big\downarrow \pi_{2^j} \\ & & F(M, 2^j) \end{array}$$

where

1. $\pi_{2^j} : F(M, 2^{j+1}) \to F(M, 2^j)$ is the projection induced by the inclusion of Γ_{2^j} in $\Gamma_{2^{j+1}}$, and
2. the fibre of π_{2^j} is X_j, the space $F(M \setminus Q(2^j), 2^{j+1} - 2^j)$ where $Q(2^j)$ is a subset of M having cardinality 2^j.

Consider the inverse limit $\varprojlim F(M, 2^j)$ of the spaces $F(M, 2^j)$. Notice that there is a map Θ: $\mathrm{Emb}(S^1, M) \to \Lambda \varprojlim F(M, 2^j)$ which is defined by the projection to $\Lambda F(M, k)$ given by

$$\Theta(f)(z) = (f(z), f(\rho_k(z)), f({\rho_k}^2(z)), \ldots, f({\rho_k}^{k-1}(z))).$$

Furthermore, each map $F(M, 2^{j+1}) \to F(M, 2^j)$ is the projection map in a fibre bundle and, if $M = \mathbb{R}^n \times N$, then this fibration has a cross-section [13]. Thus in the cases for which $M = \mathbb{R}^n \times N$ the homotopy groups of $\varprojlim F(M, 2^j)$ are given by the inverse limit of the homotopy groups of $F(M, 2^j)$ [5]. In addition, these maps have more structure some of which is described next.

Consider the subspace of $\Lambda F(M, k)$ given by the $\mathbb{Z}/k\mathbb{Z}$-equivariant maps from S^1 to $F(M, k)$ where $\mathbb{Z}/k\mathbb{Z}$ acts on S^1 by multiplication by $\rho_k = e^{2\pi i/k}$. Write $\Lambda^{\mathbb{Z}/k\mathbb{Z}} F(M, k)$ for the space of continuous $\mathbb{Z}/k\mathbb{Z}$-equivariant maps from S^1 to $F(M, k)$. The next remark is stated as a proposition.

Proposition 8.1. *The map Θ takes values in the subspace given by the inverse limit of the $\Lambda^{\mathbb{Z}/k\mathbb{Z}} F(M, k)$ for $k = 2^j$, $\varprojlim \Lambda^{\mathbb{Z}/k\mathbb{Z}} F(M, k)$, a subspace of $\Lambda \varprojlim F(M, 2^j)$.*

The rest of this section consists of some information concerning the inverse limit $\varprojlim \Lambda^{\mathbb{Z}/2^j\mathbb{Z}} F(M, 2^j)$ of the $\Lambda^{\mathbb{Z}/2^j\mathbb{Z}} F(M, 2^j)$. This inverse limit will be written

$$\Lambda^{\mathbb{Z}/2^\infty} F(M, \mathbb{Z}/2^\infty).$$

Notice that restriction to the subspace of maps which are embeddings on cosets of the group G defined above, $\mathrm{Emb}_G(S^1, M)$, is far from the space of embeddings. Namely, there are maps which have multiple points at irrational points of the circle.

Proposition 8.2.

1. *There is a commutative diagram*

$$\begin{array}{ccc} \mathrm{Emb}(S^1, M) & \longrightarrow & \Lambda^{\mathbb{Z}/2^\infty} F(M, \mathbb{Z}/2^\infty) \\ \downarrow 1 & & \downarrow \\ \mathrm{Emb}(S^1, M) & \xrightarrow{\text{inclusion}} & \Lambda M. \end{array}$$

2. *The space* $\Lambda^{\mathbb{Z}/2^\infty} F(M, \mathbb{Z}/2^\infty)$ *is homeomorphic to the subspace of* ΛM *given by the continuous maps of* S^1 *to* M *which are embeddings when restricted to cosets of* G *in* S^1, $\mathrm{Emb}_G(S^1, M)$.
3. *If* M *is a simply-connected manifold of dimension at least* 3, *then the rational homology of* $\mathrm{Emb}_G(S^1, M)$ *is not of finite type.*

Proof. Statement 1 is clear. Consider the space $\mathrm{Emb}_G(S^1, M)$. Notice that the map Θ extends to give a map $\Theta'\colon \mathrm{Emb}_G(S^1, M) \to \varprojlim \Lambda^{\mathbb{Z}/2^j Z} F(M, 2^j)$. Furthermore, the projection map $\pi\colon \varprojlim \Lambda^{\mathbb{Z}/2^j\mathbb{Z}} F(M, 2^j) \to \Lambda M$ takes values in the space $\mathrm{Emb}_G(S^1, M)$. Both natural composites $\pi \circ \Theta'$ and $\Theta' \circ \pi$ give the identity. Thus statement 2 follows.

To prove statement 3, notice that if M is $\mathbb{R}^n$ for $n > 2$, then the first non-vanishing homology group of $\varprojlim \Lambda^{\mathbb{Z}/2^j\mathbb{Z}} F(M, 2^j)$ is torsion free and not finitely generated. In case M is simply-connected, the first dimension where the homotopy is non-trivial gives that the first non-vanishing homotopy group of $\mathrm{Emb}_G(S^1, M)$ is not finitely generated. □

Let X be a space with a free $\mathbb{Z}/k\mathbb{Z}$-action, and $\Lambda^{\mathbb{Z}/k\mathbb{Z}} X$ the space of continuous $\mathbb{Z}/k\mathbb{Z}$-equivariant maps. The following proposition is well known.

Proposition 8.3. *If* X *is simply-connected, the natural inclusion of* $\Lambda^{\mathbb{Z}/k\mathbb{Z}} X$ *in* ΛX *is a rational homotopy equivalence.*

Proposition 8.4. *The space* $\Lambda^{\mathbb{Z}/2^j\mathbb{Z}} F(M, 2^j)$ *is rationally homotopy equivalent to* $\Lambda F(M, 2^j)$ *if* M *is a simply-connected manifold of dimension at least* 3.

There is a further interpretation of the above remarks. Define the space of almost embedded curves to be the homotopy orbit space

$$E\mathrm{SO}(2) \times_{\mathrm{SO}(2)} \mathrm{Emb}_G(S^1, M).$$

Consider reduced simplicial sets X. Let ΩX denote the simplicial (Kan) loop group of X, and $k[\Omega X]$ the simplicial group algebra over a field k as in [15] or [21, Corollary 7.3.14], and recall the following results.

1. The Hochschild homology of $k[\Omega X]$ is isomorphic to $H_*(\Lambda|X|; k)$, and
2. the cyclic homology of $k[\Omega X]$ is isomorphic to $H_*(E\mathrm{SO}(2) \times_{\mathrm{SO}(2)} \Lambda|X|; k)$.

Corollary 8.5. 1. *The rational homology of* $\Lambda^{\mathbb{Z}/k\mathbb{Z}}F(M,2^j)$ *is isomorphic to the Hochschild homology of the singular chain complex for* $\Omega F(M,2^j)$. *If* $M=\mathbb{R}^n$ *for* $n\geq 3$, *the rational homology of* $\Lambda^{\mathbb{Z}/k\mathbb{Z}}F(M,k)$ *is isomorphic to the Hochschild homology of the universal enveloping algebra of the "universal Yang–Baxter Lie algebra"* $\mathcal{L}_k(n-2)$.

2. *The space* $E\mathrm{SO}(2)\times_{\mathrm{SO}(2)}\mathrm{Emb}_G(S^1,M)$ *is homeomorphic to*

$$\varprojlim E\mathrm{SO}(2)\times_{\mathrm{SO}(2)}\Lambda^{\mathbb{Z}/k\mathbb{Z}}F(M,k).$$

The rational homology of $E\mathrm{SO}(2)\times_{\mathrm{SO}(2)}\Lambda^{\mathbb{Z}/k\mathbb{Z}}F(\mathbb{R}^n,k)$ *is given in terms of the cyclic homology of the universal enveloping algebra for the "universal Yang–Baxter Lie algebra"* $\mathcal{L}_k(n-2)$.

Loop spaces of configuration spaces also fit in a different context one of which is discussed in the next section.

9. Vassiliev invariants and the loop space homology of configuration spaces

The purpose of this section is to describe how the homology of the loop space of a configuration space "accounts" for all of the Vassiliev invariants of pure braids. The main additional input is given in work of T. Kohno [18, 19]. The information here is a direct comparison of those results with the results described above.

Let V_k^n be the vector space of Vassiliev invariants of order k for pure braids with n strands (as in [19, p. 130]). Let A_k^n be the complex vector space spanned by horizontal chord diagrams with n vertical strands modulo the "horizontal 4T relation" and the "framing independence relation"; these last two relations are precisely the "infinitesimal braid relations" of Section 3, and given by Kohno [19, Proposition 4.3] who proves that there is an isomorphism of complex vector spaces

$$V_k^n/V_{k-1}^n \to \mathrm{Hom}_{\mathbb{C}}(A_k^n,\mathbb{C}).$$

Consider the graded algebra A_*^n which in degree k is given by the complex vector space A_k^n, and is $\mathbb{C}$ in degree zero. Thus the Euler–Poincaré series for A_*^n is given by $\Sigma_{k\geq 0}\dim_{\mathbb{C}}(A_k^n)\,t^k$ in [19, Proposition 1.3.11].

Proposition 9.1. *Assume that* $q\geq 1$.

1. *If* $q\geq 1$, *then* $H_*(\Omega F(\mathbb{R}^{q+2},n);\mathbb{Z})$ *is torsion free of finite type. Moreover,*

$$\chi H_*(\Omega F(\mathbb{R}^{q+2},n);\mathbb{Z}) = [(1-t^q)(1-2t^q)\cdots(1-(n-1)t^q)]^{-1}, \textit{and}$$

$$\chi H_*(F(\mathbb{R}^{q+1},n);\mathbb{Z}) = (1+t^q)(1+2t^q)\cdots(1+(n-1)t^q).$$

2. *There is an equality of Euler–Poincaré series*

$$\chi H_*(\Omega F(\mathbb{R}^3,n);\mathbb{C}) = \Sigma_{k\geq 0}\dim_{\mathbb{C}}(A_k^n)\,t^k.$$

Proof. If $q \geq 1$, the space $\Omega F(\mathbb{R}^{q+2}, n)$ is homotopy equivalent to a product $\Pi_{1\leq i\leq n-1}\ \Omega(\vee_i S^{q+1})$ [7]. The formula in part 1 follows from the Euler–Poincaré series for a tensor algebra and the calculation of $H_*(F(\mathbb{R}^{q+1}, n); \mathbb{Z})$ made in [6]:

$$\chi H_*(\Omega F(\mathbb{R}^{q+2}, n); \mathbb{Z}) = [(1-t^q)(1-2t^q)\cdots(1-(n-1)t^q)]^{-1}, \text{ and}$$

$$\chi H_*(F(\mathbb{R}^{q+1}, n); \mathbb{Z}) = (1+t^q)(1+2t^q)\cdots(1+(n-1)t^q).$$

Notice that the Euler–Poincaré series

$$\Sigma_{k\geq 0} \dim_{\mathbb{C}}(A_k^n)\, t^k$$

is given by $[(1-t)(1-2t)\cdots(1-(n-1)t)]^{-1}$ by [19]. This is precisely the Euler–Poincaré series for $H_*(\Omega F(\mathbb{R}^3, n); \mathbb{C})$, and the proposition follows. □

Remark A straightforward geometric interpretation of Proposition 9.1 is given by considering the ordered pairs of distinct pure braids α and β, and using the classical theory of Samelson products in $H_*(\Omega F(\mathbb{R}^3, n); \mathbb{Z})$ to "measure" $\alpha\beta^{-1}$.

References

[1] J. W. Alexander, A lemma on systems of knotted curves, Proc. Nat. Acad. Science USA **9** (1923), 93–95.

[2] E. Artin, The theory of braids, Ann. of Math. **48** (1947), 101–126.

[3] D. Bar Natan, On the Vassiliev invariants, Topology **34** (1995), 423–472.

[4] J. Birman, Braids, Links, and Mapping Class Groups, Ann. of Math. Studies 66 (1971).

[5] A. K. Bousfield, and D. M. Kan, Homotopy Limits, Completions and Localizations, Lecture Notes in Math. 304, Springer-Verlag (1972).

[6] F. R. Cohen, The homology of C_{n+1}-spaces, Lecture Notes in Math. 533, Springer-Verlag (1976), 207–351.

[7] F. R. Cohen, S. Gitler, On loop spaces of configuration spaces, preprint.

[8] F. R. Cohen, J. C. Moore, and J. A. Neisendorfer, The double suspension and exponents of the homotopy groups of spheres, Ann. of Math. **110** (1979), 549–565.

[9] F. R. Cohen, T. Sato, On the group of homotopy groups, student topology seminar, Univ. of Rochester, Spring 1998, preprint.

[10] F. R. Cohen, L. R. Taylor, On the representation theory associated to the cohomology of configuration spaces, Contemp. Math. 146 (1993), 91–109.

[11] E. Fadell and S. Husseini, Geometry and Topology of Configuration Spaces, in preparation.

[12] E. Fadell and S. Husseini, The space of loops on configuration spaces and the Majer–Terracini index, Topological Methods in Nonlinear Analysis, Journal of the Julius Schauder Center (1996).

[13] E. Fadell, and L. Neuwirth, Configuration Spaces, Math. Scand. **10** (1962), 111–118.

[14] M. Falk, and R. Randell, The lower central series of a fiber-type arrangement, Invent. Math. **82** (1985), 77–88.

[15] T. G. Goodwillie, Cyclic homology, derivations, and Hochschild homology, Topology **24** (1985), 187–215.

[16] R. Hain, Infinitesimal presentations of the Torelli group, J. Amer. Math. Soc. **10** (1997), 597–691.

[17] I. M. James, Stiefel Manifolds, London Mathematical Society Lecture Note Series 78 (1965).

[18] T. Kohno, Linear represenations of braid groups and classical Yang–Baxter equations, Contemp. Math. 78 (1988), 339–363.

[19] T. Kohno, Vassiliev invariants and de Rham complex on the space of knots, Contemp. Math. 179 (1994), 123–138.

[20] M. Kontsevich, Vassiliev's knot invariants, Advances in Soviet Mathematics **16** (1993), 137–150.

[21] J.-L. Loday, Cyclic Homology, Grundlehren der mathematischen Wissenschaften 301 (1992), Springer-Verlag, Berlin.

[22] W. Magnus, A. Karrass, D. Solitar, Combinatorial Group Theory (1976), Dover.

[23] J. W. Milnor, and J. C. Moore, On the structure of Hopf algebras, Ann. of Math. **81** (1965), 211–264.

[24] T. Sato, On free groups and morphisms of coalgebras, student topology seminar, Univ. of Rochester, Spring 1998, preprint.

[25] P. S. Selick, 2-primary exponents for the homotopy groups of spheres, Topology **23** (1984), 97–99.

[26] V. Vassiliev, Cohomology of knot spaces, Theory of singularities and its applications, Amer. Math. Soc. (1992).

[27] M. Xicoténcatl, Orbit configuration spaces, Thesis, Univ. of Rochester, Spring 1996.

Department of Mathematics,
University of Rochester,
Rochester, NY 14627, USA
E-mail address: `cohf@math.rochester.edu`, `gitler@math.rochester.edu`

Progress in Mathematics, Vol. 196

On the homotopy type of infinite stunted projective spaces

Frederick R. Cohen* and Ran Levi

1. Introduction

Consider the space $X_n = \mathbb{R}\mathbf{P}^\infty/\mathbb{R}\mathbf{P}^{n-1}$ together with the boundary map in the Barratt–Puppe sequence

$$X_n \longrightarrow \Sigma\mathbb{R}\mathbf{P}^{n-1}.$$

Francis Sergeraert and Vladimir Smirnov [6] have considered the low dimensional homotopy groups of X_n as well as their loop space homology. Their results are quite interesting, and their questions fit with several results that appeared previously in the literature [3, 8].

It is the purpose of this note to elaborate on a few of these remarks, as well as pointing out some natural associated questions and their connection to a recent result of Jie Wu [8], where he shows that the 2-torsion in the homotopy of the 3-sphere is a summand of the homotopy of $\Sigma\mathbb{R}\mathbf{P}^2$. That result is a consequence of the structure considered here. Moreover, it will be shown here that the homotopy type of the spaces X_n is closely related to that of certain finite complexes described below, where all spaces are tacitly assumed to be localized at the prime 2.

Here are some concrete results, the first of which is an observation that was also pointed out by Broto [2].

Theorem 1.1. *There is a fibration*

$$S^3 \longrightarrow X_2 \longrightarrow K(\mathbb{Z}, 2),$$

which is split after looping.

Notice that the calculation of loop space homology for X_2 follows trivially from this splitting result and the known loop space homology for S^3.

Let A_3 denote the 6-skeleton of the Lie group G_2. Notice that the homotopy type of A_3 is given by a 3-cell complex with cells in dimension 3, 5, and 6, which is determined by its cohomology. The identification of the relationship between A_3, and G_2 is not specifically used in the theorem below, but is pointed out as G_2 appears in several interesting contexts.

* Partially supported by the NSF.

Theorem 1.2. *There is a 2-local fibration*

$$A_3 \longrightarrow X_3 \longrightarrow BS^3,$$

where A_3 is the 6-skeleton of the Lie group G_2. Furthermore, there is a splitting

$$\Omega_0^4 X_3 \simeq \Omega_0^3 S^3 \times \Omega_0^4 A_3,$$

where Ω_0^n denotes the component of the constant map in an iterated loop space.

The calculation of loop space homology for X_3 is a bit more involved. A work of David Anick [1] gives the homology of ΩA_3. Let V denote the 5-skeleton of the Lie group G_2. Then A_3 is obtained from V by attaching a single 6-cell. It is easy to see that $H = H_*(\Omega V, \mathbb{F}_2)$ (in fact with any coefficients) is a tensor algebra on two generators a and b of dimensions 2 and 4 respectively. The obvious inclusion induces an H-module structure on $H_*(\Omega A_3, \mathbb{F}_2)$. Let $H\langle t\rangle$ denote the free associative H-algebra on one generator t in dimension 5. Thus $H\langle t\rangle$ is isomorphic to a tensor algebra on a, b and t. Define a differential d on $H\langle t\rangle$ by $d(t) = a^2$ and $d(a) = d(b) = 0$. This turns $H\langle t\rangle$ into a differential graded algebra and Anick's theorem now gives

Theorem 1.3. *The mod-2 loop space homology of A_3 is isomorphic as an H-module to the homology of the differential graded algebra $(H\langle t\rangle,\ d)$ defined above. Moreover the Poincaré series for $H_*(\Omega A_3, \mathbb{F}_2)$ is given by the formula*

$$P_{\Omega A_3}(t) = \frac{1 - t^4}{1 - t^2 - t^4 - t^7}.$$

The fibration

$$\Omega A_3 \longrightarrow \Omega X_3 \longrightarrow S^3$$

is not split because the connecting map $S^3 \longrightarrow A_3$ is essential. However, it is degree 2 on the bottom cell and hence trivial on mod-2 homology. Since the fibration is multiplicative, the mod-2 Serre spectral sequence collapses at the E^2-page and one obtains

Corollary 1.4. *There is an isomorphism of $H_*(\Omega A_3, \mathbb{F}_2)$-modules*

$$H_*(\Omega X_3, \mathbb{F}_2) \cong H_*(\Omega A_3, \mathbb{F}_2) \otimes E[x_3].$$

Remark 1.5. The low dimensional homotopy of A_3 is quite easy to compute, as the natural map $A_3 \longrightarrow K(Z, 3)$ induces a homology isomorphism through dimension 6, and the homotopy fibre is easy to handle through low dimensions, as demonstrated below.

Let $P^n(q)$ denote the n-dimensional mod-q Moore space, namely, the cofibre of the degree q map on the $(n-1)$-sphere. Recall the following theorem of J. Wu, where $P^n(k)$ denotes the cofibre of a degree k map on the $(n-1)$-sphere.

Theorem 1.6 (Wu). *Let $\gamma\colon P^3(2) \longrightarrow BSO(3)$ denote the map given by inclusion to the bottom skeleton. Let Y denote its homotopy fibre. Then*

1. *Y is homotopy equivalent to $\Sigma(\mathbb{R}\mathbf{P}^4/\mathbb{R}\mathbf{P}^1) \vee P^6(2)$ and*

2. *after looping 4 times and restricting to components of the constant map* $\Omega^4\gamma$ *has a right homotopy inverse.*

The boundary map in the Barratt–Puppe sequence

$$X_n \longrightarrow \Sigma\mathbb{R}\mathbf{P}^{n-1}$$

together with Theorem 1.2 and naturality gives a new proof of part 2 of Wu's theorem. This theorem gives a way of computing the homotopy of $P^3(2)$ through a considerable range, as it is given terms of the homotopy groups of the 3-sphere and Y.

Again, consider the map

$$X_n \longrightarrow \Sigma\mathbb{R}\mathbf{P}^{n-1}$$

together with the natural map

$$\Sigma\mathbb{R}\mathbf{P}^3 \longrightarrow BS^3$$

which induces an isomorphism on homology in dimension four. The composite of these two maps gives a map

$$X_4 \longrightarrow BS^3$$

with homotopy theoretic fibre A_4.

Theorem 1.7. *There is a fibration*

$$S^7 \vee P^6(2) \longrightarrow X_4 \longrightarrow BS^3,$$

and so A_4 *is homotopy equivalent to* $S^7 \vee P^6(2)$. *Furthermore, this fibration splits after looping.*

There are infinitely many elements of order 8 in the homotopy groups of $P^6(2)$, and thus in the homotopy groups of X_4 [3]. It seems reasonable to conjecture that there does not exist an element of order 16.

The maps described above fit in a more systematic context. Namely, consider the natural maps $g_n\colon \mathbb{R}\mathbf{P}^{n-1} \longrightarrow SO(n)$, which induce surjections in cohomology. Using these maps one gets the following

Theorem 1.8. *There are maps*

$$\alpha_n\colon X_n \longrightarrow BSpin(n),$$

such that the fibre F_n *of* α_n *is a finite complex.*

One might be tempted to conjecture that the spaces X_n have homotopy exponents. In addition, one might wonder how the splittings of the loop space of the suspension of X_n impinge on features of the degree 2 map on spheres.

Indeed, the maps

$$X_n \longrightarrow \Sigma\mathbb{R}\mathbf{P}^{n-1}$$

yield factorisations of several useful maps related to the degree 2 map. In particular when n is even, the degree 2 map on S^n factors through X_n and $\Sigma\mathbb{R}\mathbf{P}^{n-1}$, and when n is odd the map

$$P^{n+1}(2) \longrightarrow S^n,$$

given by collapsing onto the top cell followed by η factors through X_n and $\Sigma\mathbb{R}\mathbf{P}^{n-1}$. A discussion on how to obtain these factorisations and a speculation on their possible utility is in the last section of the paper.

The work described in this note started when the authors were both visiting the CRM during the emphasis semester in spring 1998. Both authors take the pleasure of expressing their thanks to the CRM for its kind and generous hospitality.

2. The spaces X_n

Consider the natural map from $\mathbb{R}\mathbf{P}^\infty$ to $K(\mathbb{Z},2)$, given by the first non-vanishing integral cohomology class. Since $K(\mathbb{Z},2)$ is simply-connected this map factors through X_2. Thus there is a fibration

$$F \longrightarrow X_2 \longrightarrow K(\mathbb{Z},2).$$

Pulling this fibration back once, one obtains a principal fibration with fibre S^1 and base space X_2. Inspection of the integral cohomology Serre spectral sequence for this fibration gives that the cohomology of the fibre is isomorphic to the cohomology of the 3-sphere. Since F is obviously simply-connected, the first statement of Theorem 1.1 follows.

To see that the fibration in the theorem splits after looping, observe that the connecting map

$$\Omega K(\mathbb{Z},2) = S^1 \longrightarrow F = S^3$$

is null-homotopic for the obvious reason. Thus the projection from ΩX_2 to S^1 has a section and the fibration splits.

Corollary 2.1. *The torsion in the homotopy of X_2 has an exponent at any prime p.*

Next analyse X_3 and X_4. Consider the inclusion i of $\mathbb{Z}/2\mathbb{Z}$ as the centre of the Lie group S^3. This homomorphism induces a map

$$Bi\colon \mathbb{R}\mathbf{P}^\infty = B\mathbb{Z}/2\mathbb{Z} \longrightarrow BS^3.$$

In cohomology this map takes the generator $u_4 \in H^*(BS^3,\mathbb{F}_2)$ to z^4, where $z \in H^*(B\mathbb{Z}/2\mathbb{Z},\mathbb{F}_2)$ is the generator. Since BS^3 is 3-connected, Bi factors through both X_3 and X_4. Thus there are fibrations for $i = 3,4$

$$A_i \longrightarrow X_i \longrightarrow BS^3. \tag{1}$$

The cohomology of A_i in both cases is easy to compute using the Serre spectral sequence for the pulled back fibrations

$$S^3 \longrightarrow A_i \longrightarrow X_i. \tag{2}$$

This can be done either with mod-2 coefficients or integrally. In either case the integral homology of the spaces A_i have only 2-torsion and their mod-2 cohomology is given by the proposition below.

Proposition 2.2. *The mod-2 cohomology of A_3 is generated additively by classes a_3, b_5 and a_3^2, with $Sq^2 a = b$ and $Sq^1 b = a^2$.*

The mod-2 cohomology of A_4 is generated additively by classes x_5, y_6 and z_7, with $Sq^1 x = y$.

The explicit calculation is straight-forward, and the non-trivial reduced integral homology groups are as follows:

$$H_3(A_3) = \mathbb{Z},\quad H_5(A_3) = \mathbb{Z}/2\mathbb{Z},\quad H_5(A_4) = \mathbb{Z}/2\mathbb{Z},\quad \text{and}\quad H_7(A_4) = \mathbb{Z}.$$

The analysis of A_4 follows directly: It is clear by the cohomological calculation that its 6-skeleton is given by the Moore space $P^6(2)$. The fibre of the obvious map $P^6(2) \longrightarrow K(\mathbb{Z}/2\mathbb{Z}, 5)$ gives the 5-connected cover of $P^6(2)$ and makes it visible that $\pi_6(P^6(2))$ is isomorphic to $\mathbb{Z}/2\mathbb{Z}$ and is generated by η on the bottom cell. Attaching a 7-cell to $P^6(2)$ by this unique non-trivial element would have resulted in a Sq^2 connecting the 5 and 7 cells in the mod-2 cohomology of cofibre. The cohomological structure of A_4 thus implies the structure claimed in Theorem 1.7. That the fibration in the theorem splits after looping follows by pulling it back once and observing the connecting map from S^3 to the fibre A_4 is null-homotopic for connectivity reasons.

Next analyse the homotopy type of A_3. The 5-skeleton of this space is seen to be given by $Y = S^3 \cup_\eta e^5$ by inspection of cohomology. The natural map $Y \longrightarrow K(\mathbb{Z}, 3)$ induces an isomorphism on mod-2 cohomology up to dimension 5 and its fibre is the 4-connected cover of Y. By inspection of the associated Serre spectral sequence, $\pi_5(Y) = \mathbb{Z}$ and the cohomological structure implies that the attaching map is given by a multiple of the generator by an integer divisible by 2 exactly once. One can in fact verify by doing an integral Serre sequence calculation that the attaching map is exactly twice the generator and so the structure of A_3 is thus determined. Notice that it has been shown here that the homotopy type of a simply-connected space with the mod-2 and integral cohomology of A_3 is determined uniquely. Thus it follows that A_3 is homotopy equivalent to the 6-skeleton of the Lie group G_2 as claimed.

An old theorem of J. Harper [4] shows that the 2-local homotopy type of the Lie group G_2 is determined by its mod-2 cohomology. The result here by contrast is obtained by very elementary methods and does not overlap with Harper's calculation, namely in his work the main issue is analysing the attaching map for the top cell (14-dimensional) in G_2.

The splitting result claimed in Theorem 1.7 follows immediately from the fact that the connecting map $S^3 \longrightarrow A_4 = P^6(2) \vee S^7$ is null-homotopic for connectivity reasons.

For the splitting result claimed in Theorem 1.2 one needs the following

Lemma 2.3. *Let Z be a 2-connected space and let $g\colon S^3 \longrightarrow Z$ be any map. If $\pi_4(g)$ is trivial then $\Omega^2(2g)$ is null-homotopic when restricted to $\Omega^2 S^3\langle 3\rangle$ at the prime* 2.

Proof. Let f denote $2g$. Thus f is given by the composite

$$S^3 \xrightarrow{2} S^3 \xrightarrow{g} Z.$$

Recall that the loops on the second Hilton–Hopf invariant $\Omega h_2\colon \Omega^2 S^3 \longrightarrow \Omega^2 S^5$ has order 2 in the group $[\Omega^2 S^3, \Omega^2 S^5]$. Thus the composition

$$\Omega^2 S^3 \xrightarrow{2} \Omega^2 S^3 \xrightarrow{\Omega h_2} \Omega^2 S^5$$

is null-homotopic.

This remains true if all spaces are replaced by their 1-connected covers and so the composition

$$\Omega^2(S^3\langle 3\rangle) \xrightarrow{2} \Omega^2(S^3\langle 3\rangle) \xrightarrow{\Omega h_2} \Omega^2 S^5$$

is null and 2 on $\Omega^2(S^3\langle 3\rangle)$ lifts to the fibre of Ωh_2, which is given by ΩS^3. Thus after passing to 1-connected covers $\Omega^2 f$ is homotopic to a composition

$$\Omega^2(S^3\langle 3\rangle) \xrightarrow{l} \Omega S^3 \xrightarrow{\Omega j} \Omega^2(S^3\langle 3\rangle) \xrightarrow{\Omega^2 g} \Omega^2(Z\langle 3\rangle).$$

But notice that j takes the fundamental class in $\pi_3(S^3)$ to $\eta \in \pi_3(\Omega(S^3\langle 3\rangle))$, which under Ωg is taken, by hypothesis, to 0. Hence $\Omega^2 g\Omega j$ is null-homotopic and so $\Omega^2 f$ is null-homotopic, restricted to 1-connected covers. □

Notice that in the fibration

$$S^3 \xrightarrow{\delta} A_3 \longrightarrow X_3$$

the fibre inclusion δ is degree 2 on the bottom cell. Also the map $A_3 \longrightarrow K(\mathbb{Z},3)$ corresponding to the bottom cohomology class induces an isomorphism on cohomology up to dimension 5. Hence $\pi_4(A_3) = 0$ and the conditions of Lemma 2.3 are satisfied. Thus looping twice and passing to universal covers, the map

$$\Omega^2\delta\colon \Omega^2 S^3 \longrightarrow \Omega^2 A_3$$

is null-homotopic, implying the splitting result of Theorem 1.2.

Finally analyse X_n for higher values of n. Let $g_n\colon \mathbb{RP}^{n-1} \longrightarrow SO(n)$ be a map which induces the inclusion of generators in homology. Then there is an induced diagram where the left column is a cofibration and the right column is a fibration,

$$\begin{array}{ccc} \mathbb{RP}^{n-1} & \longrightarrow & SO(n) \\ \downarrow & & \downarrow \\ \mathbb{RP}^{\infty} & \xrightarrow{=} & \mathbb{RP}^{\infty} \\ \downarrow & & \downarrow \\ X_n & \xrightarrow{\alpha_n} & BSpin(n)\,. \end{array}$$

Consider the fibration

$$\mathbb{RP}^{\infty} \longrightarrow BSpin(n) \longrightarrow BSO(n).$$

Using either the well known calculation of the cohomology of $BSpin(n)$ [5] or inspection of the Serre spectral sequence of this fibration, one observes that for some minimal positive integer r depending on n the class $z^{2^r} \in H^{2^r}(RP^{\infty}, \mathbb{F}_2)$ is an infinite cycle. Let $u_{2^r} \in H^{2^r}(BSpin(n), \mathbb{F}_2)$ be any class restricting to z^{2^r}.

Lemma 2.4. *Let F_n denote the fibre of α_n. Then there is an isomorphism*

$$H^*(F_n, \mathbb{F}_2) \cong H^*(X_n, \mathbb{F}_2)/(z^{2^r}) \otimes \mathrm{Tor}_{H^*(BSpin(n),\mathbb{F}_2)/(u_{2^r})}(\mathbb{F}_2, \mathbb{F}_2)$$

of $H^*(X_n, \mathbb{F}_2)$-modules. In particular F_n has the homotopy type of a finite complex.

Proof. Observe that α_n^* takes any choice of u_{2^r} to the cohomology class in $H^*(X_n, \mathbb{F}_2)$ which maps to z^{2^r} and that any class in $H^{2^r}(BSpin(n), \mathbb{F}_2)$ which is in the image of the map from $H^*(BSO(n), \mathbb{F}_2)$ is sent to 0 by α_n^*. The calculation now becomes direct by using the Eilenberg–Moore spectral sequence for the fibration

$$F_n \longrightarrow X_n \longrightarrow BSpin(n)$$

and Smith's "big collapse theorem" [7]. □

3. Some applications and sample calculations

Proposition 3.1. *For any positive integer n the degree 2 map on S^{2n} factors through X_{2n}. Also if n is odd then the map*

$$P^{2n}(2) \longrightarrow S^{2n-1},$$

given by collapsing onto the top cell followed by η, factors through X_{2n+1}.

Proof. Let $i\colon S^{2n} \longrightarrow X_{2n}$ denote the inclusion of the bottom cell and let $\delta\colon X_{2n} \longrightarrow \Sigma\mathbb{R}\mathbf{P}^{2n-1}$ denote the connecting map in the Barratt–Puppe sequence. Then by projecting to the top cell, one gets a map $q\colon X_{2n} \longrightarrow S^{2n}$. It is immediate that δ is of degree 2 in dimension $2n$. Thus the first part of the proposition follows.

For any n the $2n$-skeleton of X_{2n-1} is given by $P^{2n}(2)$. By composing with δ and collapsing to the top cell, one gets a map $P^{2n}(2) \longrightarrow S^{2n-1}$. Restricted to the bottom cell, this map is immediately seen to be of degree 0 and is hence null-homotopic. Thus it factors through the top cell via a map $\alpha\colon S^{2n} \longrightarrow S^{2n-1}$. The map α can be either η or the null map.

Consider the commutative diagram of cofibrations

$$\begin{array}{ccccc}
P^{2n}(2) & \longrightarrow & X_{2n-1} & \longrightarrow & X_{2n+1} \\
\Big\downarrow{=} & & \Big\downarrow & & \Big\downarrow{f} \\
P^{2n}(2) & \longrightarrow & \Sigma\mathbb{R}\mathbf{P}^{2n-2} & \longrightarrow & C \\
\Big\downarrow & & \Big\downarrow & & \Big\downarrow{h} \\
* & \longrightarrow & \Sigma\mathbb{R}\mathbf{P}^{\infty} & \longrightarrow & \Sigma\mathbb{R}\mathbf{P}^{\infty}.
\end{array}$$

Notice that C is $(2n+1)$-dimensional whereas X_{2n+1} is $2n$-connected. Since $\Sigma\mathbb{R}\mathbf{P}^{2n-2}$ is $(2n-1)$-dimensional, f induces the zero map on mod-2 cohomology. Thus h^* surjects in all dimensions. Notice that additively, $H^*(C, \mathbb{F}_2)$ is isomorphic to that of $H^*(\Sigma\mathbb{R}\mathbf{P}^{2n}, \mathbb{F}_2)$. Thus h^* is in fact an isomorphism through dimension $2n+1$. Now, let $\sigma x^{2n-2} \in H^{2n-1}(\Sigma\mathbb{R}\mathbf{P}^{\infty}, \mathbb{F}_2)$ denote the generator. Then

$$Sq^2(\sigma x^{2n-2}) = \sigma Sq^2(x^{2n-2}) = \sigma(Sq^1(x^{n-1})^2).$$

Thus $Sq^2(\sigma x^{2n-2}) = 0$ if n is even and is equal to σx^{2n} if n is odd. The claim now follows by commutativity of the diagram

$$\begin{array}{ccccc}
P^{2n}(2) & \longrightarrow & \Sigma\mathbb{R}\mathbf{P}^{2n-2} & \longrightarrow & C \\
\Big\downarrow & & \Big\downarrow & & \Big\downarrow \\
S^{2n} & \xrightarrow{\alpha} & S^{2n-1} & \longrightarrow & K
\end{array}$$

using the fact that η is detected by Sq^2. □

A remark related to the last proposition is the following. By suspending the factorisation of the degree 2 map on an even sphere one obtains a factorisation for the same map on an odd sphere,

$$S^{2n+1} \longrightarrow \Sigma X_{2n} \longrightarrow \Sigma^2 RP^{2n-1} \longrightarrow S^{2n+1}.$$

Barratt's distributivity formula for the degree 2 map on an odd sphere states that

$$\Omega[2] = 2 + (\Omega[\iota_{2n+1}, \iota_{2n+1}] \circ H_2),$$

where the second summand is the second Hilton–Hopf invariant composed with the loops on the Whitehead square

$$\Omega S^{2n+1} \xrightarrow{H_2} \Omega S^{4n+1} \xrightarrow{\Omega[\iota_{2n+1},\iota_{2n+1}]} \Omega S^{2n+1}.$$

The identity above holds in the non-abelian group $[\Omega S^{2n+1}, \Omega S^{2n+1}]$. One may wonder if this last map is null-homotopic after looping $2n$ times more.

The splittings here do not seem to inform on the problem, but factoring the degree 2 map through the double suspension of the projective space might be useful.

Next consider the space A_3 obtained as the fibre of the map $X_3 \longrightarrow BS^3$, constructed above. The calculation of loop space homology becomes easy using Anick's technique described in [1]. Specifically, A_3 is obtained from $V = S^3 \cup_\eta e^5$ by attaching a 6-cell. The cohomology of A_3 determines the attaching map $f\colon S^5 \longrightarrow V$. One observes easily that on mod-2 homology the adjoint $\mathrm{ad}(f)\colon S^4 \longrightarrow \Omega V$ takes the generator to the element $a^2 \in H_4(\Omega V, \mathbb{F}_2)$. In this situation Anick's theorem [1, 3.7] applies and the first claim of Theorem 1.3 follows. One also obtains the formula

$$\frac{1}{P_{\Omega A_3}(t)} = \frac{1+t}{P_N(t)} - \frac{t}{P_{\Omega V}(t)} - t^5,$$

where N is defined to be the quotient of the algebra $H_*(\Omega V, \mathbb{F}_2)$ by the two sided ideal generated by a^2. The calculation of the Poincaré series for ΩA_3 thus follows from knowing the series for the algebra N.

Notice that there is a short exact sequence of graded vector spaces

$$0 \longrightarrow \Sigma^4 T[a,b] \longrightarrow T[a,b] \longrightarrow N \longrightarrow 0.$$

From this exact sequence one obtains that

$$P_N(t) = P_{\Omega V}(t) - t^4 P_{\Omega V}(t) = \frac{1-t^4}{1-t^2-t^4}.$$

Plugging this into the formula above and simplifying, the proof of Theorem 1.3 is complete.

Wu's splitting theorem is also implied by the observations made here. We recall the relevant part of the theorem.

Theorem 3.2 (Wu). *Let $\gamma\colon P^3(2) \longrightarrow BSO(3)$ denote the map given by inclusion to the bottom skeleton. Let Y denote its homotopy fibre. Then after looping four times and restricting to components of the constant map $\Omega_0^4\gamma$ has a right homotopy inverse.*

Proof. Consider the following commutative diagram of fibrations:

$$\begin{array}{ccccc} A_3 & \longrightarrow & X_3 & \longrightarrow & BS^3 \\ \big\downarrow & & \big\downarrow & & \big\downarrow \\ Y & \longrightarrow & P^3(2) & \xrightarrow{\gamma} & BSO(3) \end{array}$$

where the vertical map in the centre is the connecting map in the Barratt–Puppe sequence in the cofibration defining X_3 and the right vertical map is the obvious one. Looping this diagram twice and taking connected components the right vertical map becomes an equivalence.

Thus looping four times and taking connected components there is a diagram

$$\begin{array}{ccccccc} \Omega_0^4 A_3 & \longrightarrow & \Omega_0^4 X_3 & \longrightarrow & \Omega_0^3 S^3 & \longrightarrow & \Omega_0^3 A_3 \\ \big\downarrow & & \big\downarrow & & \big\downarrow {=} & & \big\downarrow \\ \Omega_0^4 Y & \longrightarrow & \Omega_0^4 P^3(2) & \xrightarrow{\Omega_0^4\gamma} & \Omega_0^3 S^3 & \longrightarrow & \Omega_0^3 Y\,. \end{array}$$

By Theorem 1.2 the top right horizontal map is null-homotopic. Hence the same applies to the bottom right horizontal map. Thus the map $\Omega_0^4\gamma$ admits a cross-section up to homotopy. Since a multiplicative fibration with a cross-section is trivial, the result follows. □

Finally, the following low dimensional homotopy calculations are obvious from the splitting results presented here and well known facts on the homotopy groups of the spaces involved. These calculations are motivated by some computer oriented questions of Sergeraert. The program developed by Sergeraert and his coauthors however can certainly handle spaces much more general than the specific idiosyncratic cases here. The remarks here give a corroboration of those calculations.

Proposition 3.3. *The 2-primary low dimensional homotopy of X_3 is given as follows:*

$$\pi_i(X_3) = \begin{cases} 0 & i \le 2 \\ \mathbb{Z}/2\mathbb{Z} & i = 3 \\ 0 & i = 4 \\ \mathbb{Z}/2\mathbb{Z} & i = 5, 6 \\ \mathbb{Z}/4\mathbb{Z} \oplus \mathbb{Z}/2\mathbb{Z} & i = 7. \end{cases}$$

The 2-primary low dimensional homotopy of X_4 is given as follows:

$$\pi_i(X_4) = \begin{cases} 0 & i \leq 3 \\ \mathbb{Z} & i = 4 \\ \pi_{i-1}(S^3) \oplus \mathbb{Z}/2\mathbb{Z} & i = 5, 6 \\ \pi_6(S^3) \oplus \mathbb{Z} \oplus \mathbb{Z}/4\mathbb{Z} & i = 7 \\ \pi_{i-1}(S^3) \oplus (\mathbb{Z}/2\mathbb{Z})^3 & i = 8, 9 \\ \pi_9(S^3) \oplus (\mathbb{Z}/8\mathbb{Z})^2 & i = 10. \end{cases}$$

Proof. That the space X_3 is 2-connected and $\pi_3(X_3) \cong \mathbb{Z}/2\mathbb{Z}$ follows at once by looking at the cohomology of X_3. By Theorem 1.2, there is a splitting

$$\Omega_0^4 X_3 \simeq \Omega^3 S^3 \times \Omega_0^4 A_3,$$

where Ω_0^n denotes the zero component and A_3 is the 6-skeleton of the Lie group G_2. Thus it suffices to compute the homotopy of A_3 through the specified range. Notice that A_3 is also the 6-skeleton of $K(\mathbb{Z}, 3)$. Hence the inclusion induces an isomorphism on integral homology though dimension 5 and so $\pi_i(A_3) = 0$ for $i = 4, 5$. The fibre of this inclusion is the 3-connected cover of A_3, which is seen by inspection of the Serre spectral sequence to be 6-connected with π_7 given by $\mathbb{Z}/2\mathbb{Z}$. It follows that $\pi_4(X_3) = 0$ and $\pi_i(X_3)$ are given by $\pi_{i-1}(S^3)$ for $i = 5, 6$. Finally for $i = 7$ there is an exact sequence

$$\pi_7(A_3) \longrightarrow \pi_7(X_3) \longrightarrow \pi_6(S^3),$$

which is split because of Theorem 1.2. The result for X_3 follows.

The calculation for X_4 uses the splitting of Theorem 1.7

$$\Omega X_4 \simeq S^3 \times \Omega(S^7 \vee P^6(2)).$$

Thus

$$\pi_i(X_4) \cong \pi_{i-1}(S^3) \oplus \pi_i(S^7 \vee P^6(2)) \cong \pi_{i+1}(S^3) \oplus \pi_i(S^7) \oplus \pi_i(P^6(2))$$

through dimension 10. The right hand side equality follows from looking at the fibre of the inclusion $S^7 \vee P^6(2) \longrightarrow S^7 \times P^6(2)$, given by $\Sigma(\Omega S^7 \wedge \Omega P^6(2))$ which is 10-connected. The result for X_4 follows from the known homotopy of S^7 and $P^6(2)$ in this range. □

The homotopy of the 3-sphere is of course known through a larger range than is dealt with here. With some extra effort one can work out a few more homotopy groups for A_3 as well as the cross terms that occur in bigger dimensions arising from the homotopy groups of the wedge $S^7 \vee P^6(2)$.

Corollary 3.4. *There are infinitely many summands of $\mathbb{Z}/8\mathbb{Z}$ in the homotopy of X_4.*

This follows from a calculation by J. Wu and the first author [3] for the homotopy of mod-2 Moore spaces. Explicit dimensions of an infinite family of $\mathbb{Z}/8\mathbb{Z}$ summands are listed in this article.

Conjecture 3.5. *There do not exist any elements of order* 16 *in the homotopy of* X_4.

The conjecture fits with the Barratt conjecture for the Moore space. The 2-primary component of the homotopy of S^3 has exponent 4. The best known exponent for the 2-primary homotopy of S^7 is 32, but a conjecture of Barratt and Mahowald gives 8 as an upper bound.

References

[1] D. Anick, *A counterexample to a conjecture of Serre*, Ann. of Math. **115** (1982), 1–33.

[2] C. Broto, Private communication.

[3] F. R. Cohen and J. Wu, *A remark on the homotopy groups of* $\Sigma^n\mathbb{R}\mathbf{P}^2$, Contemp. Math. **181** (1995), 65–81.

[4] J. R. Harper, Unpublished.

[5] D. Quillen, *The mod* 2 *cohomology rings of extra-special* 2*-groups and the Spinor groups*, Math. Ann. **194**, Springer-Verlag (1971), 197–212.

[6] F. Sergeraert and V. Smirnov, To appear.

[7] L. Smith, *Homological algebra and the Eilenberg–Moore spectral sequence*, Trans. Amer. Math. Soc. **129** (1967), 58–93.

[8] J. Wu, *A product decomposition of* $\Omega_0^3\Sigma\mathbf{R}P^2$, Topology **37** (1998), No. 5, 1025–1032.

Department of Mathematics,
University of Rochester,
Rochester, New York, 14627, USA
E-mail address: `cohf@math.rochester.edu`

Department of Mathematical Sciences,
University of Aberdeen,
Aberdeen, AB24 3UE, United Kingdom
E-mail address: `ran@maths.abdn.ac.uk`

Progress in Mathematics, Vol. 196

Stable splittings of $\Omega SU(n)$

M.C. Crabb and J.R. Hubbuck

1. Introduction

In his Northwestern PhD thesis [5], M. Hopkins considered the stable decomposability of the space of based loops $\Omega SU(n)$ and proved a splitting theorem when a generator of π_2 is inverted, thereby providing evidence in support of a conjecture of M. Mahowald. Later W. Richter proved that $\Omega SU(n)$ splits stably as an infinite wedge of spaces which have the homotopy types of finite complexes, extending work of S. Mitchell (and confirming Mahowald's conjecture). We consider the stable indecomposability of some of these finite complexes.

Our model for ΩG is the space of based maps $\mathrm{Map}_*(S^1, G)$ with $S^1 \subset \mathbb{C}$ the unit circle. Let $L \in P(\mathbb{C}^n)$ be a line in $\mathbb{C}^n$, and $\pi_L : \mathbb{C}^n \to \mathbb{C}^n$ be orthogonal projection onto L. H. Hopf noticed that for each $\lambda \in S^1$

$$\rho_L(\lambda) = \lambda\pi_L + (1 - \pi_L)$$

lies in $U(n)$ and therefore $\rho_L \in \Omega U(n)$. We use these elements to define subspaces of $\Omega U(n)$ by setting

$$\Omega^{\{k\}}U(n) = \{f \in \Omega U(n) : f = \rho_{L_1} \cdot \rho_{L_2} \cdot \ldots \cdot \rho_{L_k} \text{ for some } L_i \in P(\mathbb{C}^n)\}$$

where the product is taken in $U(n)$. Taking the standard subgroup $U(1) \subset U(n)$ and projecting $\Omega^{\{k\}}U(n)$ from $U(n)$ to $SU(n) = U(n)/U(1)$, we obtain a "Mitchell filtration" of $\Omega SU(n)$,

$$P(\mathbb{C}^n) = R^1SU(n) \subset R^2SU(n) \subset \ldots \subset R^{k-1}SU(n) \subset R^kSU(n) \subset \ldots \subset \Omega SU(n).$$

More extensive accounts of filtrations on $U(n)$ and $\Omega SU(n)$ can be found in [2, 11].

Let $W_k(n) = R^kSU(n)/R^{k-1}SU(n)$. The stable splitting theorem of Richter referred to above [10, 3] establishes that there is a stable equivalence of spectra

$$\Omega SU(n) \simeq \bigvee_{k\geq 1} W_k(n).$$

In the notation of [11], $F_{n,k} = R^kSU(n)$. It is shown there in Theorem A that there is a homotopy commutative diagram

$$\begin{array}{ccc} R^kSU(n) & \longrightarrow & BU(k) \\ \downarrow & & \downarrow \\ \Omega SU(n) & \longrightarrow & \Omega SU \simeq BU. \end{array}$$

The vertical maps and the top map are inclusions.

Our homology theories will be unreduced; except where the notation denotes otherwise. Let b_i denote both the generator of $H_{2i}(P(\mathbb{C}^\infty), \mathbb{Z})$ dual to the i-th power of the Euler class of the Hopf line bundle, and its inclusion in $H_{2i}(BU, \mathbb{Z})$; we use the same notation for these standard homology generators whether the coefficients be $\mathbb{Z}$, $\mathbb{Z}/2^q\mathbb{Z}$ or $\mathbb{Z}_{(2)}$.

The homotopy commutative diagram induces a commutative diagram:-

$$\begin{array}{ccc} \mathbb{Z}\langle\{b_i, 1 \le i \le n-1, \le k\}\rangle & \rightarrow & \mathbb{Z}\langle\{b_i, i \ge 1, \le k\}\rangle \\ \downarrow & & \downarrow \\ \mathbb{Z}[b_i]_{1 \le i \le n-1} & \longrightarrow & \mathbb{Z}[b_i]_{i \ge 1} \end{array}$$

where $S\langle\{b_i,\ r,\ l\}\rangle$ denotes the free S-module with generators b_i for i in the range r, of length l.

The projection of the tensor to the symmetric power on generators b_0 $(= 1)$, $b_1, \dots, b_{n-1}$ induces a surjection

$$\bigotimes^k H_*(P(\mathbb{C}^n), \mathbb{Z}) \to H_*(R^kSU(n), \mathbb{Z}),$$

whereas the projection from the tensor power to the symmetric power on generators $b_1, \dots, b_{n-1}$ induces a surjection

$$\bigotimes^k \tilde{H}_*(P(\mathbb{C}^n), \mathbb{Z}) \to \tilde{H}_*(W_k(n), \mathbb{Z}).$$

So we identify $\tilde{H}_*(W_k(n), \mathbb{Z})$ with $\mathbb{Z}\langle\{b_i, 1 \le i \le n-1,\ k\}\rangle$.

When $n = 2$, the splitting echoes the classical splitting of ΩS^3 and $W_k(2) = S^{2k}$.

Proposition 1.1. *The space $W_k(3)$ is stably indecomposable for all $k \ge 1$ (at the prime 2).*

The proof is a routine argument using mod 2 homology, the relation $Sq_2(b_2) = b_1$ and the Cartan formula.

The first interesting case occurs when $n = 4$.

Theorem 1.2. *The space $W_k(4)$ is stably indecomposable for all $k \ge 1$ (at the prime 2).*

The theorem is established in Section 2.

We now fix k and seek results for all n. It is well known that $W_1(n) = P(\mathbb{C}^n)$ is stably indecomposable at the prime 2 for all n.

Theorem 1.3. *The space $W_2(n)$ is stably indecomposable for all finite n (at the prime 2).*

The strategy of proof is similar to that used in [4, 6, 7]. We replace the anti-symmetric algebras used in [4] with symmetric algebras. Also, as in Theorem 1.1 of [4], finite dimensionality is quite crucial for Theorem 1.3.

When n becomes infinite, $R^k SU(\infty) = BU(k)$ and the Richter splitting retrieves a well known theorem of V. Snaith [16], $BU(k) \simeq \bigvee_{1 \leq i \leq k} MU(i)$, where $MU(i) = W_i(\infty)$.

Theorem 1.4. *The spectrum $MU(2)$ splits stably as a wedge of two indecomposable spectra, provided one inverts the prime 3.*

This is established in Section 5 of the paper. A proof of the existence of a 2-complete splitting can be found in Theorem D of [12].

2. Notations and the proof of Theorem 1.2

Let $k_*(X)$ denote 2-local connective complex K-theory of the complex X with coefficients $\mathbb{Z}_{(2)}[v]$. So

$$k_*(\Omega SU(n)) = \mathbb{Z}_{(2)}[v][\beta_1, \beta_2, \dots, \beta_{n-1}]$$

where the $\beta_i \in k_{2i}(P(\mathbb{C}^\infty))$, $i \geq 0$, form a basis dual to the j-th powers of the Bott-Euler class of the canonical line bundle over $P(\mathbb{C}^\infty)$; these also lie in $k_{2i}(\Omega SU(n))$ under the inclusion. The descriptions of $k_*(R^k SU(n))$ and $\tilde{k}_*(W_k(n))$ are completely analogous with those in homology, replacing the ground ring $\mathbb{Z}$ by $\mathbb{Z}_{(2)}[v]$ and the generators b_i by β_i.

In the Pontrjagin ring $k_*(P(\mathbb{C}^\infty)) \otimes \mathbb{Q}$, let $\tilde{\beta}_i = \beta_1^i/i! = \beta_i + \ldots + v^{i-1}\beta_1/i!$, where

$$\beta_i = \beta_1(\beta_1 - v)\ldots(\beta_1 - (i-1)v)/i!.$$

We will soon need the explicit expressions $\tilde{\beta}_2 = \beta_2 + v\beta_1/2$ and $\tilde{\beta}_3 = \beta_3 + v\beta_2 + v^2\beta_1/6$. The Thom map

$$\mathrm{Th} : k_{2q}(\Omega SU(n)) \to H_{2q}(\Omega SU(n), \mathbb{Z}_{(2)})$$

satisfies $\mathrm{Th}(\beta_i) = b_i$ and $\mathrm{Th}(v) = 0$.

We will use also the integral Chern character of Adams

$$\chi : k_{2q}(X) \longrightarrow \bigoplus_{i \geq 0} H_{2(q-i)}(X, \mathbb{Z}_{(2)}),$$

extending the Thom map.

Proposition 2.1. *Let Y be a $(2t-1)$-connected pointed complex. Then*

$$\chi(\tilde{k}_{2q}(Y)) \supset 2^{q-t} \bigoplus_{i \geq 0} \tilde{H}_{2q-2i}(Y, \mathbb{Z}_{(2)}).$$

The proposition is known but does not appear to be recorded explicitly in the literature; it is closely connected with work of L. Smith in [15]. We outline the proof in Section 6.

When $X = P(\mathbb{C}^\infty), \chi(\beta_1) = b_1$. As $b_i = b_1^i/i!$ in $H_*(P(\mathbb{C}^\infty),\mathbb{Q})$, rationally $\chi(\tilde{\beta}_s) = b_s$.

We begin the proof of Theorem 1.2. The group $\tilde{H}_*(W_k(4),\mathbb{Z})$ is spanned by monomials of length k in b_1, b_2 and b_3. We prove Theorem 1.2 by assuming that $W_k(4)$ splits stably as a non-trivial wedge $X_1 \vee X_2$ where the lowest positive dimensional class in $\tilde{H}_*(X_1,\mathbb{Z})$ is b_1^k and so X_2 is $(2k+1)$-connected and obtain a contradiction.

Lemma 2.2. *A lowest positive dimensional non-zero class in $\tilde{H}_*(X_2,\mathbb{Z}/2\mathbb{Z})$ must be of the form $b_1^\sigma b_3^\tau$ where $\tau > 0$.*

This lemma is an elementary consequence of the relation $Sq_2(b_2) = b_1$ and the Cartan formula.

Proof of Theorem 1.2. We consider $b_2^\sigma b_3^\tau \in \tilde{H}_{4\sigma+6\tau}(W_k(4),\mathbb{Z}/2\mathbb{Z})$ as in Lemma 2.2. Then

$$Sq_{2^\sigma}(b_2^\sigma b_3^\tau) = b_1^\sigma b_3^\tau \in \tilde{H}_{2\sigma+6\tau}(X_2,\mathbb{Z}/2\mathbb{Z}).$$

Therefore there exists $\tilde{w} \in \tilde{H}_{4\sigma+6\tau}(X_2,\mathbb{Z}/2\mathbb{Z})$ with $Sq_{2^\sigma}(\tilde{w}) = b_1^\sigma b_3^\tau$. As $Sq_{2i}(b_3) = 0$ for all $i > 0$, we must have $\tilde{w} = b_2^\sigma b_3^\tau +$ distinct monomials. So there is a representative class $w \in \tilde{H}_{4\sigma+6\tau}(X_2,\mathbb{Z}_{(2)})$ with

$$w = b_2^\sigma b_3^\tau + a_1 b_1 b_2^{\sigma-2} b_3^{\tau+1} + \ldots + a_l b_1^l b_2^{\sigma-2l} b_3^{\tau+l}$$

where $a_i \in \mathbb{Z}_{(2)}$. As X_2 is $(2\sigma+6\tau-1)$-connected, we set $t = \sigma+3\tau$ in Proposition 2.1. So there exists $u \in \tilde{k}_{4\sigma+6\tau}(X_2)$ with $\chi(u) = 2^\sigma(0,\ldots,w)$. In $\tilde{k}_{4\sigma+6\tau}(X_2) \bigotimes \mathbb{Q}$,

$$u = 2^\sigma\{\tilde{\beta}_2^\sigma \tilde{\beta}_3^{tau} + a_1\tilde{\beta}_1\tilde{\beta}_2^{\sigma-2}\tilde{\beta}_3^{\tau+1} + \ldots + a_l\tilde{\beta}_1^l\tilde{\beta}_2^{\sigma-2l}\tilde{\beta}_3^{\tau+l}\}.$$

As $u \in \mathbb{Z}_{(2)}[v,\beta_1,\beta_2,\beta_3]$, the coefficient of $v^{\sigma+2\tau}\beta_1^{\sigma+\tau} \in \mathbb{Z}_{(2)}$. But this coefficient has the form

$$\begin{aligned} &2^\sigma\{2^{-\sigma}6^{-\tau} + a_1 2^{-\sigma+2}6^{-\tau-1} + \ldots + a_l 2^{-\sigma+2l}6^{-\tau-l}\} \\ &\quad = \{2^{-\tau}3^{-\tau} + a_1 2^{-\tau+1}3^{-\tau-1} + \ldots + a_l 2^{-\tau+l}3^{-\tau-l}\}. \end{aligned}$$

This is only possible if $\tau = 0$ and therefore $\sigma = k$. This contradicts the assumption that X_2 is $(2k+1)$-connected and completes the proof of Theorem 1.2.

3. Symmetrisation

A homogeneous polynomial of degree M in k-variables $p(\underline{\xi})$ is called a numerical M-form over $\mathbb{Z}_{(2)}$ if $p(\underline{\mu}) \in \mathbb{Z}_{(2)}$ whenever it is evaluated at $\underline{\mu} \in \mathbb{Z}_{(2)}^k$. Any such form can be written as

$$p(\underline{\xi}) = \sum a_{i_1,i_2,\dots,i_k} \xi_1^{i_1}.\xi_2^{i_2}\dots\xi_k^{i_k}/i_1!.i_2!\dots i_k!$$

where $\sum i_j = M, \underline{\xi} = (\xi_1, \xi_2, \dots, \xi_k)$ and the coefficients $a_{i_1,i_2,\dots i_k}$ lie in $\mathbb{Z}_{(2)}$. It is shown in [7] that there exists a smallest integer $K_2(k, 2^s)$ such that whenever $M > K_2(k, 2^s)$, all the coefficients of a numerical M-form are divisible by 2^s. The values of $K_2(k, 2^s)$ are known for $k = 2$, at least implicitly, by work of L. Schwartz [13, 1]. In particular $K_2(2, 2) = 4$ and we will show that $K_2(2, 4) = 10$. The upper bound for $K_2(k, 2^s)$ obtained in [7] is too large to be useful in computations.

In $k_*(P(\mathbb{C}^\infty)) \otimes \mathbb{Q} = \mathbb{Q}[v][\xi]$, let $\xi = \beta_1$ and so

$$\beta_i = \xi(\xi - v)\dots(\xi - v(i-1))/i!.$$

The fixed points of the Adams operator ψ_3 in $k_*(P(\mathbb{C}^\infty)) \otimes \mathbb{Q}$ are scalar multiples of ξ^i.

Let $BT^k = P(\mathbb{C}^\infty) \times P(\mathbb{C}^\infty) \times \dots \times P(\mathbb{C}^\infty)$, k-factors, and $k_*(BT^k) \otimes \mathbb{Q} = \mathbb{Q}[v][\xi_1, \xi_2, \dots, \xi_k]$, where $\xi_i = 1 \otimes \dots \otimes 1 \otimes \xi \otimes 1 \dots \otimes 1$ with ξ in the i-th position. Basic properties of the Adams operator imply that the fixed points of ψ_3 in $k_{2q}(BT^k)$ coincide with the numerical q-forms $p(\underline{\xi})$ which lie in $k_{2q}(BT^k)$, or more precisely lie in $k_{2q}(BT^k) \cap \mathbb{Q}[\xi_1, \xi_2, \dots \xi_k] \subset k_{2q}(BT^k) \otimes \mathbb{Q}$. In homology one has in a similar manner, $H_*(P(\mathbb{C}^\infty), \mathbb{Q}) = \mathbb{Q}[x]$ and we set $x = b_1$ and so $b_i = x^i/i!$. Then $H_*(BT^k, \mathbb{Q}) = \mathbb{Q}[x_1, x_2, \dots, x_k]$ and $\mathrm{Th}(p(\underline{\xi})) = p(\underline{x})$.

All homology groups in this note are free of integral torsion, but at times we need different coefficients. When coefficients change, any missing homomorphism will be the standard coefficient reduction from $\mathbb{Z}$ or $\mathbb{Z}_{(2)}$. We state part of the discussion above as a lemma.

Lemma 3.1. *The natural transformation*

$$\mathrm{Th} : k_{2s}(BT^k) \longrightarrow H_{2s}(BT^k, \mathbb{Z}/2^q\mathbb{Z})$$

maps fixed points of ψ_3 to zero, when $s > K_2(k, 2^q)$.

Let

$$t_H : \tilde{H}_*(W_k(n), \mathbb{Z}_{(2)}) \longrightarrow H_*(BT^k, \mathbb{Z}_{(2)})$$

and

$$t_k : \tilde{k}_*(W_k(n)) \longrightarrow k_*(BT^k)$$

be the symmetrisation maps determined by

$$t(c_{i_1}.c_{i_2}...c_{i_k}) = \sum_{\sigma \in S_k} c_{i_{\sigma(1)}} \otimes c_{i_{\sigma(2)}} \otimes \dots \otimes c_{i_{\sigma(k)}},$$

where $c_i = b_i$ or β_i.

Under the natural inclusions of modules,

$$H_*(W_k(n), \mathbb{Z}_{(2)}) \subset H_*(W_k(\infty), \mathbb{Z}_{(2)}) \subset H_*(BU(k), \mathbb{Z}_{(2)})$$

and $\tilde{k}_*(W_k(n)) \subset \tilde{k}_*(W_k(\infty),\mathbb{Z}) \subset k_*(BU(k))$, the above are restrictions of the symmetrisation maps in the commutative square.

$$\begin{array}{cccc} t_k: & k_*(BU(k)) & \longrightarrow & k_*(BT^k) \\ & \Big\downarrow \text{Th} & & \Big\downarrow \text{Th} \\ t_H: & H_*(BU(k),\mathbb{Z}) & \longrightarrow & H_*(BT^k,\mathbb{Z}). \end{array}$$

In Section 5 we will use the fact that t_H and t_k are induced by the transfer of the inclusion $BT^k \to BU(k)$.

Theorem 3.2. *The natural transformation*

$$\text{Th}: k_{2s}(BU(k)) \longrightarrow H_{2s}(BU(k),\mathbb{Z}/2\mathbb{Z})$$

maps the fixed points of ψ_3 *to zero if* $s > K_2(k,2^q)$, *where* $q = 1+\nu_2(k!)$.

This follows from Lemma 3.1 and the diagram immediately above with $\mathbb{Z}/2^q\mathbb{Z}$ coefficients in homology. One adds the facts that under the symmetrisation map, $t_H(b_{i_1}b_{i_2}\dots b_{i_r})$ is non zero in $H_*(BT^k,\mathbb{Z}/2^q\mathbb{Z})$ for $r \le k$ (where $\nu_2(k!)$ denotes the exponent of the highest power of 2 which divides $k!$) and that t_k commutes with ψ_3.

We remark that it is shown in [14] that the subspace of $k_{2s}(BU(k))$ fixed by ψ_3 corresponds to the numerical s-forms $p(\underline{\xi})$ which are invariant under S_k such that $p(\underline{\mu})$ is divisible by the order of the stabiliser of $\underline{\mu} \in \mathbb{Z}^k_{(2)}$.

Corollary 3.3. *Let* $W_k(n)$ *be stably homotopic to a non-trivial wedge* $X_1 \vee X_2$ *where* X_2 *is* $(2s-1)$*-connected,* $s > k$. *Then* $s \le K_2(k,2^q)$, *where* $q = 1+\nu_2(k!)$.

Proof. Let $S^{2s} \to X_2$ be the inclusion of a cell of smallest positive dimension which we compose with the inclusion into $W_k(n)$ Let $\tilde{k}_{2s}(S^{2s}) = \mathbb{Z}_{(2)}[v]\langle u\rangle$. The image of u in $H_{2s}(BU(k),\mathbb{Z}/2\mathbb{Z})$ under the homomorphism

$$\tilde{k}_{2s}(S^{2s}) \to \tilde{H}_{2s}(S^{2s},\mathbb{Z}/2\mathbb{Z}) \to H_{2s}(W_k(n),\mathbb{Z}/2\mathbb{Z}) \to H_{2s}(BU(k),\mathbb{Z}/2\mathbb{Z})$$

is non zero. This coincides with its image under the composition

$$\tilde{k}_{2s}(S^{2s}) \to \tilde{k}_{2s}(W_k(n)) \subset k_{2s}(BU(k)) \to H_{2s}(BU(k),\mathbb{Z}/2\mathbb{Z}).$$

As $\psi_3(u) = u$, by Theorem 3.2, $s \le K_2(k,2^q)$.

4. The proof of Theorem 1.3

In this section we consider only $k=2$ for $n \le \infty$. We determine the classes in $\tilde{H}_*(BU(2),\mathbb{Z}/2\mathbb{Z})$, and therefore $\tilde{H}_*(W_2(n),\mathbb{Z}/2\mathbb{Z})$, which can be lowest dimensional classes of a wedge summand; we call these "splitting classes".

We recall the work of Schwartz referred to earlier in [13] in more detail. The Pontrjagin ring $H_*(BT^2,\mathbb{Z}_{(2)})$ is a divided polynomial algebra on two 2-dimensional generators x and y say, where $t_H(b_ib_j) = (x^i/i!)(y^j/j!)+(x^j/j!)(y^i/i!)$. The subring of numerical forms is generated by x, y and classes q_i, $i \ge 1$, defined inductively by $q_1 = x.y^2/2! - x^2/2!.y$ and $q_k = (q_k^2 - h^{2^{k-1}}q_k)/2$, $k \ge 1$, where $h =$

$x^3 - x^2.y + y^3$, [13, 1]. We will write x_i, y_i for $x^i/i!$, $y^i/i!$, $i \geq 1$. Direct computation shows that the image of the space of numerical forms in $H_*(BT^2, \mathbb{Z}/4\mathbb{Z})$ is spanned by:-

Degree	Generators
0	1.
2	x_1, y_1.
4	$2x_2$, x_1y_1, $2y_2$.
6	$2x_3$, $q_1 = x_1.y_2 - x_2.y_1$, $2x_2.y_1$, $2y_3$.
8	$2xy_3$, $xq_1 = 2x_2y_2 + x_3y_1$, $yq_1 = 2x_2y_2 - x_1y_3$.
10	$x^2q_1 = 2x_3y_2$, $y^2q_1 = 2x_2y_3$.
12	$q_1^2 = 2x_3y_3$, $q_2 = 2\{x_1y_5 + x_2y_4 + x_4y_2 + x_5y_1\}$.
14	$xq_2 = 2\{x_3y_4 + x_5y_2\}$, $yq_2 = 2\{x_2y_5 + x_4y_3\}$.
16	$xyq_2 = 2\{x_3y_5 + x_5y_2\}$.
18	$q_1q_2 = 2\{x_3y_6 + x_6y_3\}$.
20	$xq_1q_2 = 2x_7y_3$, $yq_1q_2 = 2x_3y_7$.

We deduce that $K_2(2,4) = 10$.

Proposition 4.1. *The image of the fixed subspace of* ψ_3 *under the Thom map* Th : $k_*(BU(2)) \to H_*(BU(2), \mathbb{Z}/2\mathbb{Z})$ *has basis* $\{b_0^2, b_0b_1, b_1^2, b_1b_3, b_3^2\}$.

Proof. Theorem 3.2 and the computation of $K_2(2,4)$ above imply that the classes in the image lie in $H_{2s}(BU(2), \mathbb{Z}/2\mathbb{Z})$, $0 \leq s \leq 10$. The fixed points of ψ_3 in $k_*(BU(2))$ are mapped by t_k to fixed points of ψ_3 in $k_*(BT^2)$. These in turn are mapped by Th to numerical forms in $H_*(BT^2, \mathbb{Z}_{(2)})$. Those fixed points in $k_*(BU(2))$ whose images are non-zero in $H_*(BU(2), \mathbb{Z}/2\mathbb{Z})$ give rise to symmetric numerical forms with non-zero images in $H_*(BT^2, \mathbb{Z}/4\mathbb{Z})$. By inspection of the table above, the only possible classes in $H_*(BU(2), \mathbb{Z}/2\mathbb{Z})$ are b_0^2, corresponding to 1, b_0b_1, corresponding to $x_1 + y_1$, b_1^2, corresponding to $2x_1y_1$, b_1b_2, corresponding to $x_1y_2 + x_2y_1$, b_1b_3, corresponding to $x_1y_3 + x_3y_1$ and b_3^2, corresponding to $2x_3y_3$.

Of these, the class b_1b_2 is exceptional; $t_H(b_1b_2) \in H_6(BT^2, \mathbb{Z}/2\mathbb{Z})$ corresponds to a fixed point of ψ_3 in $k_6(BT^2)$, but not to a fixed point in $k_6(BU(2))$. Rationally the class in $k_6(BU(2)) \otimes \mathbb{Q}$ would have to be a $a\tilde{\beta}_1\tilde{\beta}_2 = a\beta_1(\beta_2 + v\beta_1/2)$, where a is a unit in $\mathbb{Z}_{(2)}$, This does not lie in $k_6(BU(2))$. One can check directly that b_3^2 does arise from a fixed point of ψ_3. This establishes the proposition.

Corollary 4.2. *Only* b_1^2 *and* b_1b_3 *can be splitting classes in* $\tilde{H}_*(W_2(n), \mathbb{Z}/2\mathbb{Z})$, *for* $n \leq \infty$.

Proof. We need to show that b_3^2 is not a splitting class.

If $n = 4$, the result follows from Theorem 1.2 and the approach in general is similar to the proof of that result. For $n > 4$, we use the relations

$$Sq_2(b_4) = b_3,\ Sq_4(b_4) = b_2,\ Sq_2(b_3) = 0,\ Sq_2(b_2) = b_1,$$

and assuming the generators exist,

$$Sq_4(b_5) = b_3,\ Sq_2(b_5) = 0,\ Sq_4(b_6) = 0 \quad \text{and} \quad Sq_2(b_6) = b_5.$$

So in $\tilde{H}_{14}(W_2(n), \mathbb{Z}/2\mathbb{Z})$, $Sq_2(b_4b_3) = b_3^2$. If Y is a stable summand of $W_2(n)$ with splitting class b_3^2, there exists a class

$$c = \alpha b_4 b_3 + \beta b_5 b_2 + \gamma b_6 b_1 \in \tilde{H}_{14}(Y, \mathbb{Z}/2\mathbb{Z})$$

with $Sq_2(c) = b_3^2$ and $Sq_4(c) = 0$. Thus $\alpha b_3^2 + \beta b_5 b_1 + \gamma b_5 b_1 = b_3^2$ and $\alpha b_2 b_3 + \beta b_3 b_2 = 0$. This implies that $\alpha = \beta = \gamma = 1$ and that $n \geq 6$. Therefore if b_3^2 is a splitting class, there exists an element

$$w = (lb_4 b_3 + mb_5 b_2 + nb_6 b_1) \in \tilde{H}_{14}(Y, \mathbb{Z}_{(2)})$$

where l, m and n are units in $\mathbb{Z}_{(2)}$. We apply Proposition 2.1 with $q = 7$ and $t = 6$ and deduce that $2w = \chi(u)$ for some $u \in \tilde{k}_{14}(Y)$. By a similar argument to that used in Theorem 1.2, in $\tilde{k}_{14}(Y) \otimes \mathbb{Q}$,

$$u = 2(l\tilde{\beta}_4\tilde{\beta}_3 + m\tilde{\beta}_5\tilde{\beta}_2 + n\tilde{\beta}_6\tilde{\beta}_1)$$

where as before $\tilde{\beta}_i = \beta_i + \ldots + v^{i-1}\beta_1/i!$. But as $u \in \tilde{k}_{14}(Y)$, the coefficient of $v^5\beta_1^2$ in $\mathbb{Q}[v, \beta_1, \ldots, \beta_{n-1}]$ must lie in $\mathbb{Z}_{(2)}$. As this coefficient is $2\{l/(4!)(3!) + m/(5!)(2!) + n/(6!)\}$, we have a contradiction. Thus b_3^2 is not a splitting class.

Proof of Theorem 1.3. When $n = 2$ or 3, the result is routine and the case $n = 4$ follows from Theorem 1.2. So we can assume that $n > 4$.

We have

$$\tilde{k}_4(W_2(n)) = \mathbb{Z}_{(2)}\langle\beta_1^2\rangle, \quad \tilde{k}_6(W_2(n)) = \mathbb{Z}_{(2)}\langle\beta_1\beta_2\rangle \oplus \mathbb{Z}(2)\langle v\beta_1^2\rangle,$$
$$\tilde{k}_8(W_2(n) = \mathbb{Z}_{(2)}\langle\beta_1\beta_3\rangle \oplus \mathbb{Z}_{(2)}\langle\beta_2^2\rangle \oplus \mathbb{Z}_{(2)}\langle v\beta_1\beta_2\rangle \oplus \mathbb{Z}_{(2)}\langle v^2\beta_1^2\rangle.$$

If $W_2(n)$ splits non-trivially as $X_1 \vee X_2$, we can assume by Corollary 4.2 that

$$H_4(X_1, \mathbb{Z}_{(2)}) = \mathbb{Z}_{(2)}\langle b_1^2\rangle, \quad H_8(X_1, \mathbb{Z}_{(2)}) = \mathbb{Z}_{(2)}\langle b_2^2 + lb_1b_3\rangle,$$
$$H_4(X_2, \mathbb{Z}_{(2)}) = 0, \quad H_6(X_2, \mathbb{Z}_{(2)}) = 0$$

and

$$H_8(X_2, \mathbb{Z}_{(2)}) = \mathbb{Z}_{(2)}\langle b_1b_3 + 2mb_2^2\rangle$$

where $l, m \in \mathbb{Z}_{(2)}$. So

$$\tilde{k}_8(X_1) = \mathbb{Z}_{(2)}\langle\beta_2^2 + l\beta_1\beta_3\rangle \oplus \mathbb{Z}_{(2)}\langle v\beta_1\beta_2\rangle \oplus \mathbb{Z}_{(2)}\langle v^2\beta_1^2\rangle$$

and

$$\tilde{k}_8(X_2) = \mathbb{Z}_{(2)}\langle\tilde{\beta}_1\tilde{\beta}_3 + 2m\tilde{\beta}_2^2\rangle,$$

where the rationally defined class

$$\tilde{\beta}_1\tilde{\beta}_3 + 2m\tilde{\beta}_2^2 = \beta_1\beta_3 + 2m\beta_2^2 + (1 + 2m)v\beta_1\beta_2 + 6^{-1}(1 + 3m)v^2\beta_1^2$$

(and so $m \in \mathbb{Z}_{(2)}$ is a unit).

Now $H_{4n-6}(W_2(n), \mathbb{Z}_{(2)}) = \mathbb{Z}_{(2)}\langle b_{n-2}b_{n-1}\rangle$. The subspace of $\tilde{k}_{4n-6}(W_2(n))$ fixed by ψ_3 is $\mathbb{Z}_{(2)}\langle\zeta\rangle$, where ζ is a rational multiple of $\tilde{\beta}_{n-2}\tilde{\beta}_{n-1}$. A routine computation verifies that

$$t!\tilde{\beta}_t = v^{t-1}\beta_1 + v^{t-2}2\beta_2 + v^{t-3}(1 - (-1)^t)\beta_3 \bmod 4$$

and so $\zeta = v^{2n-4}\beta_1^2 + 2v^{n-6}\beta_1\beta_3 \bmod 4$. Reducing mod 2, $\zeta = v^{2n-4}\beta_1^2$ and so $\zeta \in \tilde{k}_*(X_1)$. But this implies that $v^{n-6}\beta_1\beta_3 \bmod 2$ lies in $\tilde{k}_*(X_1)$, which is false.

We deduce that $W_2(n)$ is indecomposable at the prime 2.

5. Splitting $BU(2)$

Theorem 5.1. *There is an equivalence of spectra*

$$BU(2)_+ \simeq S^0 \vee MU(1) \vee X_1 \vee X_2,$$

if the prime 3 is inverted, where the summands on the right hand side are indecomposable and have splitting classes $b_0^2, b_0b_1, b_1^2, b_1b_3$ respectively.

Theorem 1.3 is an immediate corollary.

We construct idempotents using the inclusion $i : T^2 \to U(2)$. It induces both $i_* : \tilde{H}_*(BT_+^2, \mathbb{Z}) \to \tilde{H}_*(BU(2)_+, \mathbb{Z})$, corresponding to the projection from the tensor to the symmetric square on the generators $b_i, i \geq 0$, and also the (stable) transfer $i^! : BU(2)_+ \to BT_+^2$ inducing $i_*^! : \tilde{H}_*(BU(2)_+, \mathbb{Z}) \to \tilde{H}_*(BT_+^2, \mathbb{Z})$ (which also realises the symmetrisation maps t_H and t_k of Section 3). The composition $i_*i_*^!$ is multiplication by $\chi(U(2)/T^2) = 2$ in homology. The composition $i_*^!.i_*$ is $1 + \tau_*$, where τ is induced from the involution of the torus which interchanges the factors and so corresponds to the matrix $\left[\begin{smallmatrix} 0 & 1 \\ 1 & 0 \end{smallmatrix}\right]$.

Let $\sigma : BT^2 \to BT^2$ denote the map induced by $\left[\begin{smallmatrix} -1 & 1 \\ -1 & 0 \end{smallmatrix}\right]$. Then $\sigma^3 = 1$ and $\tau\sigma\tau = \sigma^{-1}$.

Lemma 5.2. *The stable map $e = 3^{-1}(1 + i\sigma i^!) : BU(2)_+ \to BU(2)_+$ acts as an idempotent on $\tilde{H}_*(BU(2)_+, \mathbb{Z}_{(1/3)})$.*

Proof. As $i_*^!i_* = 1 + \tau$, we have

$$(1 + i_*\sigma_*i_*^!)^2 = 1 + 2(i_*\sigma_*i_*^!) + i_*\sigma_*(1 + \tau_*)\sigma_*i_*^!.$$

But $i_*\tau_* = i_*$ and $\tau_*i_*^! = i_*^!$, as τ is given by an inner automorphism of $U(2)$ and so induces a map homotopic to the identity on $BU(2)$. So $i_*\sigma_*^2i_*^! = i_*\tau_*\sigma_*^2i_*^! = i_*\sigma_*\tau_*i_*^! = i_*\sigma_*i_*^!$ and $i_*\sigma_*\tau_*\sigma_*i_*^! = i_*\tau_*i_*^! = i_*i_*^! = 2$. Therefore $(1 + i_*\sigma_*i_*^!)^2 = 3(1 + i_*\sigma_*i_*^!)$ and $e_*^2 = e_*$.

Using a well known technique, written in detail in Section 4 of [4], e decomposes the spectrum $BU(2)_+$ into a wedge of two spectra.

To identify the image, Y say, of the idempotent we compute its Poincaré series. Let G be the subgroup of $GL(2, \mathbb{Z})$ generated by σ and τ, which has order 6. The rational homology of Y can be identified with the invariants $H_*(BT^2, \mathbb{Q})^G$. that is, with $\mathbb{Q}[x, y]^G$, in the notation of the section above. Using a superscript to denote homogeneous polynomials of a given degree, we give the Poincaré series in terms of an indeterminate q of dimension 2 as :-

$$\begin{aligned}&\frac{1}{\#G}\sum_{g\in G}\sum_{j\geq 0}\mathrm{tr}(g_*:(\mathbb{Q}[x,y])^{(j)}\to(\mathbb{Q}[x,y])^{(j)})q^j\\ =&\frac{1}{\#G}\sum_{g\in G}(\det(1-qg))^{-1}\quad\text{(Molien's theorem)}\\ =&\frac{1}{6}((1-q)^{-2}+2(1+q+q^2)^{-1}+3(1-q^2)^{-1})\\ =&(1-q^2)^{-1}(1-q^3)^{-1}.\end{aligned}$$

So we can split $Y=S^0\vee X_1$, where the splitting class of X_1 is b_1^2 in dimension 4. As the Poincare series of $BU(2)_+$ is $(1-q)^{-1}(1-q^2)^{-1}$, the Poincaré series of the summand complementary to Y is $q(1-q)^{-1}(1-q^3)^{-1}$. But $MU(1)$ with Poincaré series $q(1-q)^{-1}$ and splitting class b_0b_1 cannot lie in Y and so the complementary summand must split as $MU(1)\vee X_2$. This completes the proof of the theorem.

In the proof above we have quoted Snaith's result that $MU(1)$ is a stable summand of $BU(2)$. A strategy similar to that above can be used to establish this (without inverting 3). One must take care with base points. Let $P=\left[\begin{smallmatrix}1&0\\0&0\end{smallmatrix}\right]$ induce $\pi:BT^2\to BT^2$ and the zero matrix induce $\theta:BT^2\to BT^2$. Then $i_*(\pi_*-\theta_*)i_*^!$ is a suitable idempotent.

6. Appendix

We describe the proof of Proposition 2.1, which we restate for the case $t=0$ as

Proposition 6.1. *Suppose Y is a finite pointed complex. Then the cokernel of the 2-local Chern character in degree $2q$*

$$\chi:\tilde{k}_{2q}(Y)\to\bigoplus_{i\geq 0}\tilde{H}_{2(q-i)}(Y,\mathbb{Z}_{(2)})$$

is annihilated by 2^q.

The general case is established by replacing Y by the (stable) desuspension $\Sigma^{-2t}Y$.

We shall need a number of properties of the Chern character χ, for which the basic references are the papers of Mahowald and Milgram [9] and Smith [15].

First we give an easy proof under the assumption (satisfied in our application) that each Thom map $\tilde{k}_{2(q-i)}(Y)\to\tilde{H}_{2(q-i)}(Y,\mathbb{Z}_{(2)})$, $i\geq 0$, is surjective. Consider the long exact sequence

$$\cdots\to\tilde{k}_{2(q-1)}(Y)\xrightarrow{v\cdot}\tilde{k}_{2q}(Y)\xrightarrow{\mathrm{Th}}\tilde{H}_{2q}(Y,\mathbb{Z}_{(2)})\to\cdots$$

relating connective k-theory and homology. The proof follows by induction on q, using the facts that $\chi(v)=2$ and χ is multiplicative.

In general the proof runs as follows. We use the notation $P(\mathbb{R}^r)^H$ for the Thom space of the Hopf line bundle H over the real projective space $P(\mathbb{R}^r)$. The Chern character χ fits into a long exact sequence

$$\cdots \to \tilde{k}_{2q}(Y) \xrightarrow{\chi} \bigoplus_{i \geq 0} \tilde{H}_{2(q-i)}(Y) \to \tilde{k}_{2q-1}(Y \wedge P(\mathbb{R}^\infty)^H) \to \cdots$$

By connectivity, the inclusion map induces an isomorphism $\tilde{k}_{2q-1}(Y \wedge P(\mathbb{R}^{2q})^H) \to \tilde{k}_{2q-1}(Y \wedge P(\mathbb{R}^\infty)^H)$. In stable homotopy, the identity map on $P(\mathbb{R}^{2q})^H$ has order 2^q or 2^{q+1} (depending upon q mod 4); in k-theory its order is exactly 2^q. (To be precise, let us write $k^0\{X; Y\}$ for the k-theory maps between pointed complexes X and Y, that is, $[\mathbf{X}; \mathbf{Y} \wedge \mathbf{k}]$, where $\mathbf{X}$ and $\mathbf{Y}$ are the suspension spectra of X and Y and $\mathbf{k}$ is the k-theory spectrum. Then $1 \in k^0\{P(\mathbb{R}^{2q})^H; P(\mathbb{R}^{2q})^H\}$ has order 2^q. This can be seen by induction, using the cofibre sequence

$$P(\mathbb{R}^{2(q-1)})^H \to P(\mathbb{R}^{2q})^H \to P(\mathbb{R}^2)^{(2q-1)H} = \Sigma^{2(q-1)} P(\mathbb{R}^2)^H,$$

to reduce to the case of $P(\mathbb{R}^2)^H$.) The conclusion of Proposition 6.1 thus follows from the exact sequence.

Finally, we isolate the corollary which appeared as the essential step in the proofs of Theorem 1.2 and Corollary 4.2.

Corollary 6.2. *Suppose further that χ is injective on $\tilde{k}_{2q}(Y)$, and let y be an element of $\tilde{H}_{2q}(Y, \mathbb{Z}_{(2)})$. Then there exists a lift $x \in \tilde{k}_{2q}(Y)$ of $2^q y$ which is fixed by ψ_3.*

We lift $(0, \ldots, 0, 2^q y) \in \bigoplus_{i \geq 0} \tilde{H}_{2(q-i)}(Y, \mathbb{Z}_{(2)})$ to a class $x \in \tilde{k}_{2q}(Y)$. The Adams operation ψ_3 ($= \psi^{1/3}$) corresponds, under χ, to 1 on $\tilde{H}_{2q}(Y, \mathbb{Z}_{(2)})$ and 3^{-i} on $\tilde{H}_{2(q-i)}(Y, \mathbb{Z}_{(2)})$.

References

[1] A. Baker, F. Clarke, N. Ray and L. Schwartz, *On the Kummer congruence and the stable homotopy of BU,* Trans. Amer. Math. Soc., **316** (1989), 385–432.

[2] M.C. Crabb, *On the stable splitting of $U(n)$ and $\Omega U(n)$,* in Algebraic Topology, Barcelona 1986, Lecture Notes in Mathematics, **1298** (1987), 35–53.

[3] M.C. Crabb and S.A. Mitchell, *The loops on $U(n)/O(n)$ and $U(2n)/Sp(n)$,* Math. Proc. Camb. Phil. Soc., **104** (1988), 95–103.

[4] M.C. Crabb, J.R. Hubbuck and J.A.W. McCall, *Stable summands of $U(n)$,* Proc. Roy. Soc. Edin., **127A** (1997), 963–973.

[5] M.J. Hopkins, *Ph.D. thesis,* Northwestern University, Evanston, 1984.

[6] J.R. Hubbuck, *Some stably indecomposable loop spaces,* in Homotopy and related topics, Lecture Notes in Mathematics, **1418** (1989), 70–77.

[7] J.R. Hubbuck, *Numerical Forms,* J. Lond. Math. Soc., **55** (1997), 65–75.

[8] J.R. Hubbuck, *The Hurewicz homomorphism and numerical forms,* in Homotopy and Geometry, Banach Center Publications, Institute of Mathematics, Polish Academy of Sciences, Warszawa, **45** (1998), 235–240.

[9] M. Mahowald and R.J. Milgram, *Operations which detect Sq^4 in connective K-theory and their applications,* Quart. J. Math., **27** (1976), 415–432.

[10] M. Mahowald and W. Richter, *$\Omega SU(3)$ splits after four suspensions, but not after two,* Math. Z., **208** (1991), 597–615.

[11] S.A. Mitchell, *A filtration of the loops on $SU(n)$ by Schubert varieties,* Math. Z., **193** (1986), 347–362.

[12] S.A. Mitchell and S. Priddy, *Stable splittings derived from the Steinberg module,* Topology, **22** (1983), 285–298.

[13] L. Schwartz, *K-théorie et homotopie stable,* Thesis, Université de Paris-VII, 1983.

[14] L. Schwartz, *K-théorie des corps finis et homotopie stable du classifiant d'un groupe de Lie,* J. Pure Appl.Algebra, **34** (1984), 291–300.

[15] L. Smith, *The Todd character and the integrality theorem for the Chern character,* Ill. J. Math., **17** (1973), 301–310.

[16] V. Snaith, *Algebraic K-theory and localised stable homotopy theory,* Mem. Amer. Math. Soc. 43, Amer. Math. Soc., Providence R.I., 1983.

Department of Mathematical Sciences,
University of Aberdeen,
Aberdeen AB24 3UE, Scotland, UK.
E-mail address: `m.crabb@maths.abdn.ac.uk` `j.hubbuck@maths.abdn.ac.uk`

Progress in Mathematics, Vol. 196

Structure of mod p H-spaces with finiteness conditions

Juan A. Crespo

Abstract. The aim of this paper is to prove in full generality that a 1-connected mod p H-space with noetherian mod p cohomology is, up to p-completion, the total space of a principal fibration with base a mod p finite H-space and fibre a product of a finite number of copies of $\mathbb{C}P^\infty$. The case $p = 2$ was considered in [5].

1. Introduction

A finite H-space is an H-space whose underlying space is homotopy equivalent to a CW-complex with a finite number of cells. In the localized version, a mod p finite H-space stands for an H-space which is finite up to p-completion or equivalently for an H-space which mod p cohomology ring is finite dimensional, where by p-completion we understand Bousfield-Kan p-completion [4]. An H-space, being simple, is p-good in the sense of Bousfield-Kan, so we will assume without loss of generality that our mod p H-spaces are p-complete.

The three connected cover of a finite H-space is not homotopy equivalent to a finite CW-complex. In fact the mod p cohomology ring is no longer finite. For example the three connected cover of the three dimensional sphere S^3, obtained as the fibre of the fibration $S^3\langle 3\rangle \longrightarrow S^3 \longrightarrow K(\hat{\mathbb{Z}}_p, 3)$, has not a finite cohomology ring. It is however a noetherian ring, $H^*(S^3\langle 3\rangle; \mathbb{F}_p) \cong \mathbb{F}_p[x_{2p}] \otimes E[\beta x_{2p}]$.

In order to construct the 3-connected cover of a mod p finite H-space X, we firstly consider its universal cover, $\tilde{X}$, which is again mod p finite. Following [8] it is 2-connected and $\pi_3(\tilde{X})$ is torsion free. Choose a map $\tilde{X} \to K(\hat{\mathbb{Z}}_p{}^m, 3)$ that induces an isomorphism between the three dimensional homotopy groups. The 3-connected cover $X\langle 3\rangle$ is defined as the homotopy fibre of that map, thus it fits in a principal fibration

$$((\mathbb{C}P^\infty)^\wedge_p)^m \to X\langle 3\rangle \to \tilde{X} .$$

A spectral sequence argument shows now that the mod p cohomology ring of $X\langle 3\rangle$ is not finite but finitely generated; that is, noetherian. Other known examples of H-spaces with noetherian mod p cohomology ring are $(\mathbb{C}P^\infty)^\wedge_p$ and $B\mathbb{Z}/p^r$ for any positive integer r.

The author is partially supported by DGES grant PB97-0203.

Lin [14, Question 2.4] asks whether a simply connected H-space with finitely generated mod p cohomology algebra has the same mod p cohomology as a product of $K(\mathbb{Z},2)$'s with 3-connective covers of finite H-spaces and finite H-spaces.

In this paper we obtain

Theorem 1.1. *If X is a 1-connected mod p H-space with noetherian mod p cohomology algebra, then there exists a mod p finite H-space $F = F(X)$ and a principal H-fibration*

$$((\mathbb{C}P^\infty)^\wedge_p)^n \to X \to F(X) . \tag{1}$$

It is important to remark that the basis and the fiber in the fibration (1) can be obtained in a functorial way from X. In [9], Dror Farjoun gives a nice construction of localization functors with respect to maps. In particular the nullification functor for $B\mathbb{Z}/p$, $L_{B\mathbb{Z}/p}$ was investigated by Neisendorfer [16] who defined F as the composition of the functor $L_{B\mathbb{Z}/p}$ and the Bousfield-Kan p-completion and gave the relationship $F(X\langle n\rangle) \simeq X^\wedge_p$ for 1-connected finite complexes X with finite $\pi_2(X)$ and for any positive integer n. $F(X)$ in the principal fibration coincides with the functor introduced by Neisendorfer. On the other hand, the map $((\mathbb{C}P^\infty)^\wedge_p)^n \longrightarrow X$ coincides with $CW_{B\mathbb{Z}/p^\infty}(X) \longrightarrow X$ where CW_A is the colocalization functor introduced by Dror Farjoun in [9].

Corollary 1.2. *A mod p H-space is the three connected cover of a mod p finite H-space if and only if its mod p cohomology ring is three connected and noetherian.*

These results answer positively the question of Lin. In fact we show that not only such an H-space has the same cohomology but it fits in a sort of central extension of those. Namely, the H-fibration (1).

Here we present the odd prime version of the Theorem 1.1. In [5] was exposed the even prime version. From now p will thus always denote an odd prime.

In fact we will work under more general conditions that those mentionned in Theorem 1.1. We will consider H-spaces satisfying the finiteness conditions

- $(F1)$ $H^*(X;\mathbb{F}_p)$ is of finite type.
- $(F2)$ $H^*(X;\mathbb{F}_p)$ has a finite number of polynomial generators.
- $(F3)$ The module of the indecomposables $QH^*(X;\mathbb{F}_p)$ is locally finite as module over the Steenrod algebra.

Recall that a module over the Steenrod algebra is called locally finite provided any submodule generated by a single element is finite (cf. [17]).

It is clear that if an H-space has finitely generated cohomology then it satisfies conditions $(F1)$, $(F2)$ and $(F3)$. Under these less restrictive conditions one has a more general version of Theorem 1.1.

Theorem 1.3. *Let X be a 1-connected mod p H-space satisfying conditions $(F1)$, $(F2)$ and $(F3)$. Then there exists a principal H-fibration*

$$((\mathbb{C}P^\infty)^\wedge_p)^n \times B\mathbb{Z}/p^{k_1} \times \ldots \times B\mathbb{Z}/p^{k_s} \to X \to F(X) .$$

where $F(X)$ is an H-space such that $H^*(F(X);\mathbb{F}_p)$ is locally finite.

The typical examples of H-spaces with locally finite cohomology are provided by loop spaces of finite H-spaces. For example

Example 1.4. *Consider ΩS^3, the loop space of S^3. Its mod p cohomology is locally finite but not finite. In fact $H^*(\Omega S^3;\mathbb{F}_p) \cong \Gamma[\gamma_2]$, that is, it is an infinitely generated Hopf algebra. Classify the 2-dimensional class of $H^*(\Omega S^3;\mathbb{F}_p)$ by a map $\Omega S^3 \longrightarrow K(\mathbb{Z}/p,2)$. The homotopy fibre of this map, Y, is again an H-space that fits in a fibration*

$$B\mathbb{Z}/p \longrightarrow Y \longrightarrow \Omega S^3.$$

An easy computation with the SSS gives us that $H^*(Y;\mathbb{F}_p) \cong \mathbb{F}_p[x_2] \otimes \Gamma(y_{2p}) \otimes E[z_{2p-1}]$. The space Y satisfies conditions $(F1), (F2)$ and $(F3)$ but its cohomology is not noetherian.

We will denote by $\mathcal{A}$, the mod p Steenrod algebra. It is generated by the Steenrod powers P^i, $i \geq 0$ and the Bockstein operator β, subject to the Adem relations. The Bockstein β and the powers P^{p^n} form a system of algebra generators. We will use the notation

$$P^{\Delta_n} = P^{p^n} P^{p^{n-1}} \dots P^p P^1$$

for $n \geq 0$ and formally $P^{\Delta_{-1}} = P^0 = id$. With this notation we can express the mod p cohomology of $B^2\mathbb{Z}/p$ as the algebra

$$H^*(B^2\mathbb{Z}/p;\mathbb{F}_p) \cong \mathbb{F}_p[\iota, \beta P^1\beta\iota, \dots, \beta P^{\Delta_n}\beta\iota, \dots] \otimes E[\beta\iota, P^1\beta\iota, \dots, P^{\Delta_n}\beta\iota, \dots]$$

where $\iota \in H^2(B^2\mathbb{Z}/p;\mathbb{F}_p)$ is the fundamental class.

Theorem 1.5. *For any mod p H-space X satisfying the conditions* (F1), (F2) and (F3) *and any polynomial generator $x \in H^*(X;\mathbb{F}_p)$ of degree $\deg x > 1$, there exists a finite subquotient of $H^*(B^2\mathbb{Z}/p;\mathbb{F}_p)$ of the form, either*

$$M_n^{\mathrm{I}} = \left\langle (\beta P^1\beta\iota)^{p^{m_n}}, \dots, (\beta P^{\Delta_{n-1}}\beta\iota)^{p^{m_1}}, P^{\Delta_n}\beta\iota, \beta P^{\Delta_n}\beta\iota \right\rangle_{\mathbb{F}_p}$$

with $n \geq 0$, $m_0 = 0$ and $m_{k-1} \leq m_k \leq m_{k-1}+1$ or

$$M_n^{\mathrm{II}} = \left\langle \beta P^1\beta\iota)^{p^{m_n}}, \dots, (\beta P^{\Delta_{n-1}}\beta\iota)^{p^{m_1}}, P^{\Delta_n}\beta\iota, \beta P^{\Delta_n}\beta\iota, \iota^{p^m} \right\rangle_{\mathbb{F}_p}$$

with $n \geq 0$, $m_0 = 0$, $m_{k-1} \leq m_k \leq m_{k-1}+1$, $m \geq 0$, and an epimorphism of unstable $\mathcal{A}$-modules:

$$\tilde{\tau} : \Sigma QH^*X \longrightarrow\!\!\!\!\rightarrow M_n^\bullet$$

with $\tilde{\tau}(x) = P^{\Delta_n}\beta\iota$, where $M_n^\bullet$ denotes either M_n^{I} or M_n^{II}.

In case that $H^*(X;\mathbb{F}_p)$ is noetherian and 1-connected only M_n^{I} can occur.

This shows that a polynomial generator cannot live in a large dimension unless linked by Steenrod operations to other nilpotent generators in a way codified by the unstable $\mathcal{A}$-modules $M_n^\bullet$. In particular if the generator has dimension bigger than 2, then the module is $M_n^\bullet$ with $n \geq 0$ an hence this class will have non-trivial Bockstein. Hence polynomial rational classes can only happen in dimension 2.

We can give examples of H-spaces whose cohomology realize the information codified by M_n^{I}.

Example 1.6.

$n=0$: *$M_0^{\mathrm{I}}=\{P^1\beta\iota,\beta P^1\beta\iota\}$. This implies that the space realizing it, has to have a polynomial class in dimension $2p$ and an exterior class in dimension $2p+1$, related by a Bockstein. This is the cohomology of $S^3\langle 3\rangle$.*

$n=1$: *There are two different possibilities for the module M_1^{I}, these are given in Example (3.10). For the case $M_1^{\mathrm{I}}=\{P^{\Delta_1}\beta\iota,\beta P^{\Delta_1}\beta\iota,(\beta P^1\beta\iota)^p\}$ that we can represent as*

$$\circ \xrightarrow{\beta} \bullet \xrightarrow{P^1} \bullet$$

we have that an H-space candidate to realize the information codified by this M_1^{I} will have a polynomial class in dimension $2p^2$ linked by the above operations. This corresponds to the cohomology of the three connected cover of the Harper H-space, K_p:

$$H^*(K_p\langle 3\rangle;\mathbb{F}_p)\cong\mathbb{F}_p[x_{2p^2}]\otimes E[y_{2p^2+1},z_{2p^2+2p-1}]$$

with the relations $\beta x=y$ and $P^1y=z$.

The paper is organized as follows. In Section 2 we compute the Nil-localization of Hopf algebras. Using the coaugmentation of this functor we construct a sequence of H-fibrations $B\mathbb{Z}/p\longrightarrow X\longrightarrow E\longrightarrow B^2\mathbb{Z}/p$ for a given H-space X whose cohomology is noetherian. The map $B\mathbb{Z}/p\longrightarrow X$ detects a prescribed polynomial generator. Section 3 contains information about the structure of $H^*(B^2\mathbb{Z}/p;\mathbb{F}_p)$ and Section 4 about differential Hopf algebras. In Section 5 we compute the SSS for the H-fibrations constructed in Section 2, and in Section 6 the convergence of this SSS. Finally in Section 7 we show Theorems 1.1 and 1.3.

We will denote by $\mathcal{U}$ and $\mathcal{K}$ the categories of unstable modules and algebras over the Steenrod algebra respectively. For an element $x\in M$, M an object either of $\mathcal{U}$ or $\mathcal{K}$, we will use the notation

$$P_0x=\begin{cases}P^{\frac{|x|}{2}}=x^p \text{ if } |x|\equiv 0(2)\\ \beta P^{(|x|-1)/2}x \text{ if } |x|\equiv 1(2)\end{cases}$$

I would like to thank Carles Broto for his help and observations. I also would like to thank Jérôme Scherer the accurate reading of the original version and a lot of useful comments. Finally I would like to express my gratitude to the *Centre de Recerca Matemàtica* for its kind and generous hospitality during the Topology Semester in 1998.

2. Nil-localization of Hopf algebras

In this section we use techniques of Nil-localization [6, 17], to construct $\mathcal{A}$-maps between $H^*(X;\mathbb{F}_p)$ and the cohomology of elementary abelian p-groups V, compatible with the Hopf algebra structure of the spaces involved. Using Lannes' results [12] these maps can be realized by geometric maps, that is, maps $BV\longrightarrow X$.

An (F)-isomorphism in the category $\mathcal{U}$ is a morphism f such that its kernel and cokernel are nilpotent in the following sense: an unstable $\mathcal{A}$-module M is said nilpotent, in the category $\mathcal{U}$, if for any $x \in M$, ${P_0}^{n_x} x = 0$ for n_x big enough. That is, if $M \in \mathcal{K}$, it is said to be nilpotent if $x^{p^{s_x}} = 0$ for any element of M^{even}.

Nil-localization is a functor that associates to every unstable $\mathcal{A}$-module M a nil-closed $\mathcal{A}$-module $\mathcal{N}(M)$ together with an (F)-isomorphism $M \to \mathcal{N}(M)$. These properties characterize the Nil-localization of M up to isomorphism. It will not be necessary to remember here the definition of a nil-closed $\mathcal{A}$-module, but just recall that the mod p cohomology of elementary abelian p groups are the most important examples. In [6] it is shown that if M is an $\mathcal{A}$-algebra then $\mathcal{N}(M)$ is also an object of $\mathcal{K}$ and $M \longrightarrow \mathcal{N}(M)$ is a morphism in $\mathcal{K}$. Similar arguments show that if M is an $\mathcal{A}$-Hopf algebra then so is $\mathcal{N}(M)$ and $M \longrightarrow \mathcal{N}(M)$ is an $\mathcal{A}$-Hopf algebra map.

We will consider A, an $\mathcal{A}$-Hopf algebra satisfying conditions $(F1)$ and $(F2)$, that is A is of finite type and A has a finite number of polynomial generators. We can write $A \cong P \otimes N$ where P is a polynomial algebra concentrated in even degrees and N is nilpotent. Then, if we consider $\sqrt{0}$ the radical of A, $\dfrac{A}{\sqrt{0}}$ coincides with P. Obviously P satisfies the conditions of the Adams and Wilkerson theorem in [1], hence we have an inclusion of P in the cohomology of BT^n, the classifying space of an n-dimensional torus where n is the transcendence degree of A.

Proposition 2.1 ([2]). *Let A be a connected $\mathcal{A}$-Hopf algebra satisfying conditions (F1) and (F2). Then the inclusion*

$$j : \frac{A}{\sqrt{0}} \hookrightarrow H^*(BT^n; \mathbb{F}_p)$$

given by the Adams-Wilkerson embedding identifies $\dfrac{A}{\sqrt{0}}$ with a sub-$\mathcal{A}$-Hopf algebra of $H^*(BT^n; \mathbb{F}_p)$. In particular it is an (F)-isomorphism.

Proof. By the Borel decomposition theorem of Hopf algebras we know that $\dfrac{A}{\sqrt{0}}$ is a polynomial algebra on n generators. Remember that in [2] it is shown that the inclusion j is an $\mathcal{A}$-Hopf algebra map and hence $\operatorname{Im} j$ is a sub-$\mathcal{A}$-Hopf algebra of $H^*(BT^n; \mathbb{F}_p)$ of maximal degree. These are of the form $\mathbb{F}_p[u_1^{p^{r_1}}, \dots, u_n^{p^{r_n}}]$ where $r_i \geq 1$.

Now to prove the result we have to show that $\ker j$ and $\operatorname{coker} j$ are nilpotent. Obviously, $\ker j$ is trivial, and

$$\operatorname{coker} j \cong \frac{\mathbb{F}_p[u_1, \dots, u_n]}{\mathbb{F}_p[u_1^{p^{r_1}}, \dots, u_n^{p^{r_n}}]} .$$

Take a class $\overline{y} \in \operatorname{coker} j$, represented by $y \in \mathbb{F}_p[u_1, \dots, u_n]$. As $\operatorname{coker} j$ is concentrated in even degree, to check that this element is nilpotent, in the sense of the Steenrod algebra, we have to check that $\overline{y}^{p^s} = 0 \in \operatorname{coker} j$ for s big enough. But taking $s \geq \max\{r_1, \dots, r_n\}$ one has that $y^{p^s} \in \mathbb{F}_p[u_1^{p^{r_1}}, \dots, u_n^{p^{r_n}}]$ so that $\overline{y}^{p^s}$ is trivial in coker j as we wanted to prove. □

Now we are going to compute the Nil-localization [6] of $\dfrac{A}{\sqrt{0}}$ and $H^*(BT^n; \mathbb{F}_p)$.

Let A be an unstable algebra over $\mathcal{A}$. Consider the forgetful functor $O : \mathcal{K}' \longrightarrow \mathcal{K}$ where $\mathcal{K}'$ consists of the unstable $\mathcal{A}$-algebras concentrated in even degrees. Its left adjoint is described in [13] as

$$\tilde{O}A = \{x \in A^{even} \; ; \; \theta x = 0 \, \forall \, \theta \in \beta\mathcal{A}\}.$$

There is an obvious inclusion $O\tilde{O}A \overset{i}{\hookrightarrow} A$ which is an $\mathcal{A}$-algebra map.

Lemma 2.2. *The inclusion* $i : O\tilde{O}A \hookrightarrow A$ *is an* (F)*-isomorphism, that is*

$$\mathcal{N}(A) \cong \mathcal{N}(O\tilde{O}A).$$

Proof. One has just to check that $\dfrac{A}{O\tilde{O}A}$ is nilpotent in $\mathcal{U}$, that is, $P_0^{n_x} x = 0$ for n_x big enough and for any element x in coker i. Remember that for any $x \in M$, $M \in \mathcal{K}$, one has by instability that $\beta P_0 x = 0$. Thus, using Adem relations one has that $P_0 x \in O\tilde{O}A$ for any element $x \in A$. This shows that $\dfrac{A}{O\tilde{O}A}$ is nilpotent and hence the lemma follows. □

In particular if one has A a nil-closed algebra then $\mathcal{N}(A) \cong A$, and hence

$$\mathcal{N}(H^*(BT; \mathbb{F}_p)) \cong H^*(BV; \mathbb{F}_p),$$

where V is an elementary abelian group of rank n.

Now consider the map $A \overset{\pi}{\longrightarrow} \dfrac{A}{\sqrt{0}}$, this is not, in general, an $\mathcal{A}$-map because the action of the Bockstein is not internal in $\sqrt{0}$. However, the composite $O\tilde{O}A \overset{\pi \circ i}{\longrightarrow} \dfrac{A}{\sqrt{0}}$ is an $\mathcal{A}$-map.

Lemma 2.3. *The morphism of unstable* $\mathcal{A}$*-algebras* $f = \pi \circ i$ *is an* (F)*-isomorphism.*

Proof. On the one hand $\ker f = \{x \in O\tilde{O}A \; ; \; x^n = 0 \text{ for } n \text{ big enough}\}$. In particular we have $x^{p^s} = 0$ for some s. As $O\tilde{O}A$ is concentrated in even degrees this implies $P_0^s x = 0$ for all $x \in O\tilde{O}A$. That is, every element in $\ker f$ is nilpotent.

On the other hand, take an element $\overline{x} \in \operatorname{coker} f$ and x a representant of this class in A. We have to check that $P_0^n x = 0$ in coker f for n big enough, or equivalently $P_0^n x \in O\tilde{O}A$, that is $\theta P_0^n x = 0$ for all $\theta \in \beta A$. In fact, this is true for $n = 1$, because $\beta P_0 x = 0 \, \forall \, x$ and

$$P^i P_0 x = \begin{cases} P_0 P^{\frac{i}{p}} x & if \; i \equiv 0(p) \\ 0 & \text{otherwise.} \end{cases}$$ □

The two lemmas above prove,

Proposition 2.4. *Let A be a connected $\mathcal{A}$-algebra. Then we have a diagram*

$$\begin{array}{ccccc} A & \longleftarrow & O\tilde{O}A & \longrightarrow & \frac{A}{\sqrt{0}} \\ \downarrow & & \downarrow & & \downarrow \\ \mathcal{N}(A) & \xleftarrow{\cong} & \mathcal{N}(O\tilde{O}A) & \xrightarrow{\cong} & \mathcal{N}(\frac{A}{\sqrt{0}}) \end{array}$$

where the bottom horizontal arrows are isomorphisms.

Lemma 2.5. *The sub-Hopf algebras of maximal transcendence degree in*

$$H^*(BV;\mathbb{F}_p) \cong \mathbb{F}_p[v_1,\dots,v_n] \otimes E[u_1,\dots,u_n]$$

with $|u_i| = 1$ and $\beta u_i = v_i$, are of the form

$$\mathbb{F}_p[v_1,\dots,v_k,v_{k+1}^{p^{m_{k+1}}},\dots,v_n^{p^{m_n}}] \otimes E[u_1,\dots,u_j] \tag{2}$$

where $0 \le j \le k$.

Theorem 2.6. *Let A be a connected $\mathcal{A}$-Hopf algebra, satisfying conditions (F1) and (F2). Then the Nil-localization of A is an $\mathcal{A}$-Hopf algebra map*

$$l : A \longrightarrow H^*(BV;\mathbb{F}_p)$$

where V is an elementary abelian group of rank n, the transcendence degree of A.

Proof. Notice that Nil-localization applied to the map

$$j : \frac{A}{\sqrt{0}} \longrightarrow H^*(BT^n;\mathbb{F}_p)$$

of Proposition 2.1 provides the diagram

$$\begin{array}{ccc} \frac{A}{\sqrt{0}} & \xrightarrow{j} & H^*(BT^n;\mathbb{F}_p) \\ \downarrow & & \downarrow \\ \mathcal{N}(\frac{A}{\sqrt{0}}) & \xrightarrow{\cong} & H^*(BV;\mathbb{F}_p). \end{array}$$

This together with Proposition 2.4 provides the map l. Nil-localization preserves the structure of Hopf algebra, so $\mathcal{N}(A)$ is a Hopf algebra isomorphic as an $\mathcal{A}$- algebra to $H^*(BV;\mathbb{F}_p)$. The uniqueness of the $\mathcal{A}$-Hopf algebra structure of $H^*(BV;\mathbb{F}_p)$ gives us the fact that l is an $\mathcal{A}$-Hopf algebra map. □

Corollary 2.7. *Let A be a connected $\mathcal{A}$-Hopf algebra satisfying conditions (F1) and (F2). Then there is a system of generators for A as an algebra such that we can write*

$$A \cong E[y_1,\dots,y_s,\dots] \otimes \mathbb{F}_p[\beta y_1,\dots,\beta y_k,x_{k+1}\dots,x_n] \otimes \frac{\mathbb{F}_p[z_1,\dots,z_t,\dots]}{(z_1^{p^{m_1}},\dots,z_t^{p^{m_n}},\dots)}$$

and such that, up to ordering,

$$l(y_j) = \begin{cases} u_j & 1 \leq j \leq r \qquad r \leq k \\ 0 & i > k \end{cases}$$

$$l(x_i) = v_i^{p^{m_i}}$$

Proof. We have constructed the map l as the coaugmentation of the Nil-localization of A. Then $l(A)$ is a sub-Hopf algebra of $H^*(BV; \mathbb{F}_p)$ of maximal transcendence degree that is of the form (2). As $l(A)$ is of maximal transcendence degree in $H^*(BV; \mathbb{F}_p)$ the condition for the image of the polynomial generators is clear for a determined basis. Finally, the condition on the generators y_j, $\beta y_s = x_s$ if $l(y_s) = u_s$ is necessary because the map l is an $\mathcal{A}$-map. □

What makes this result interesting is the case when $A = H^*(X; \mathbb{F}_p)$ for X a connected H-space. Using the results of Lannes [12], we can realize the algebraic map l by a geometric map

$$f : BV \longrightarrow X$$

with $f^* = l$. Moreover, since l is a map of Hopf algebras, it commutes with the diagonal and geometrically this means that f is an H-map.

In particular this map determines the degrees where a polynomial generator could appear in the mod p cohomology of such an H-space.

Proposition 2.8. *Let X be a connected mod p H-space whose cohomology satisfies conditions $(F1)$ and $(F2)$. Then the polynomial generators in $H^*(X; \mathbb{F}_p)$ appear in degrees $2p^i$.*

Notice that we cannot find other restrictions about the value of i, because Aguadé and Smith gave in [3] examples of spaces realizing the cohomologies $E \otimes \mathbb{F}_p[x_{2p^i}]$, for all i, E being an exterior algebra. These are the cohomologies of $Sp(k)\langle 3\rangle$, the 3-dimensional cover of the Lie group $Sp(k)$ with $k = (p^{i-1} + 1)/2$.

Assume now that V' is an arbitrary elementary abelian p-group and $f' : BV' \longrightarrow X$ any map. By universality of the coaugmentation, l, of the Nil-localization, the induced map $f'^* : H^*(X; \mathbb{F}_p) \longrightarrow H^*(BV'; \mathbb{F}_p)$ factors as a composition

$$H^*(X; \mathbb{F}_p) \xrightarrow{l} H^*(BV; \mathbb{F}_p) \longrightarrow H^*(BV'; \mathbb{F}_p)$$

and then f' itself factors as $BV' \longrightarrow BV \xrightarrow{f} X$ for a certain homomorphism of groups $V' \longrightarrow V$. Finally f' is also an H-map.

Remark 2.9. *Notice that if f'^* maps trivially all the polynomial generators of $H^*(X; \mathbb{F}_p)$ then it is trivial, because f' factors through l.*

The same arguments as in [5, section 2] apply and we obtain

Theorem 2.10. *Let X be a connected mod p H-space satisfying conditions $(F1)$, $(F2)$, and $(F3)$, and let x be a polynomial generator of $H^*(X;\mathbb{F}_p)$. Then, there exists an H-map*

$$f : B\mathbb{Z}/p \longrightarrow X$$

with $f^*(x) = v^{p^n}$, where v is a 2-dimensional generator of $H^*(B\mathbb{Z}/p;\mathbb{F}_p)$, and a sequence of H-fibrations

$$B\mathbb{Z}/p \xrightarrow{f} X \xrightarrow{g} E \xrightarrow{h} B^2\mathbb{Z}/p.$$

3. PGA-ideals of $H^*(B^2\mathbb{Z}/p;\mathbb{F}_p)$.

The motivation for this section is the study of certain ideals in the mod p cohomology of $B^2\mathbb{Z}/p$. Among those we will find the possible kernels in mod p cohomology of the projection map p of an H-fibration $F \xrightarrow{j} E \xrightarrow{p} B^2\mathbb{Z}/p$.

Definition 3.1. *Let R be an $\mathcal{A}$-Hopf algebra. An ideal of R is called a Primitively Generated $\mathcal{A}$-ideal (PGA-ideal for short) if it is an $\mathcal{A}$-ideal generated by a sequence of primitive elements.*

Recall that

$$H^*(B^2\mathbb{Z}/p;\mathbb{F}_p) \cong \mathbb{F}_p[\iota, \beta P^1\beta\iota, \ldots, \beta P^{\Delta_n}\beta\iota, \ldots] \otimes E[\beta\iota, P^1\beta\iota, \ldots, P^{\Delta_n}\beta\iota, \ldots]$$

is a primitively generated Hopf algebra. Thus the primitives are the elements ι^{p^s} of degree $2p^s$, $P^{\Delta_n}\beta\iota$ of degree $2p^{n+1}+1$ and $(\beta P^{\Delta_n}\beta\iota)^{p^s}$ of degree $2p^s(p^n+1)$, for all $s \geq 0$ and $n \geq -1$. Observe that in each degree $P(H^*(B^2\mathbb{Z}/p;\mathbb{F}_p))$ is at most one dimensional.

Lemma 3.2. *The following equalities hold in $H^*(B^2\mathbb{Z}/p;\mathbb{F}_p)$*

$$P^t P^{\Delta_r}\beta\iota = 0 \; for \; 0 < t < p^{r+1} \tag{3}$$

$$P^{p^j}\beta P^{\Delta_{j-1}}\beta\iota = \beta P^{\Delta_j}\beta\iota\,. \tag{4}$$

Using this lemma we know how acts an element of type P^{p^n} over a polynomial generator $\beta P^{\Delta_j}\beta\iota$.

Lemma 3.3. *For $n \geq 0$ and $j \geq 0$*

$$P^{p^n}(\beta P^{\Delta_j}\beta\iota) = \begin{cases} (\beta P^{\Delta_{j-1}}\beta\iota)^p & n = 0 \\ 0 & 0 < n \leq j \\ \beta P^{\Delta_{j+1}}\beta\iota & n = j+1 \\ 0 & \textit{otherwise.} \end{cases}$$

Proof. We start with the case $n = 0$. Applying Adem relations

$$P^{p^0}(\beta P^{\Delta_j}\beta\iota) = P^1(\beta P^{\Delta_j}\beta\iota) = \lambda_0\beta P^{1+p^j}P^{\Delta_{j-1}}\beta\iota + \mu_0 P^{1+p^j}\beta P^{\Delta_{j-1}}\beta\iota.$$

Observe that the first term of this sum is zero because $P^{\Delta_{j-1}}\beta\iota$ has degree $2p^j+1 < 2(p^j+1)$. Now the second term, $\beta P^{\Delta_{j-1}}\beta\iota$, has degree $2(p^j+1)$ so if we apply the operation P^{1+p^j} we obtain $(\beta P^{\Delta_{j-1}}\beta\iota)^p$. Observe that

$$\mu_0 = \binom{\alpha}{0} = 1.$$

Let us see what happens if $0 < n < j+1$. Applying again Adem relations

$$P^{p^n}\beta P^{p^j} = \sum_{k=0}^{p^{n-1}} \lambda_k \beta P^{p^n+p^j-k}P^k + \sum_{k=0}^{\frac{p^{n-1}}{p}} \mu_k P^{p^n+p^j-k}\beta P^k$$

where λ_k , $\mu_k \in \mathbb{F}_p$. Then evaluating on $P^{\Delta_{j-1}}\beta\iota$ and by the previous lemma one obtains

$$\lambda_k\beta P^{p^n+p^j-k}(P^kP^{\Delta_{j-1}}\beta\iota) = 0$$

$$\mu_k P^{p^n+p^j-k}\beta(P^kP^{\Delta_{j-1}}\beta\iota) = 0$$

when $0 < k < p^{j+1}$. So we have only to look at the case $k = 0$:

$$\lambda_0\beta P^{p^n+p^j}P^0P^{\Delta_{j-1}}\beta\iota = \lambda_0\beta P^{p^n+p^j}P^{\Delta_{j-1}}\beta\iota = 0$$

because $|P^{\Delta_{j-1}}\beta\iota| = 2p^j+1 < 2(p^n+p^j)$ the degree of the operation that we apply. On the other hand, $\mu_0 P^{P^n+p^j}\beta P^{\Delta_{j-1}}\beta\iota = 0$ because $|\beta P^{\Delta_{j-1}}\beta\iota| = 2(p^j+1) < 2(p^j+p^n)$ in case that $n > 0$. Then we have shown the lemma for $0 \le n < j+1$. The case $n = j+1$ is the second result of the previous lemma. Finally, if $n > j+1$, the result is obvious by instability. □

It is also easy to check

Lemma 3.4. *Let x be any element of an unstable algebra over the Steenrod algebra. Then*

$$P^{p^s}x^p = (P^{p^{s-1}}x)^p.$$

By gluing these two lemmas together we have

Lemma 3.5.

$$P^{p^i}(\beta P^{\Delta_j}\beta\iota)^{p^s} = \begin{cases} (\beta P^{\Delta_{j-1}}\beta\iota)^{p^{s+1}} & i = s \\ (\beta P^{\Delta_{j+1}}\beta\iota)^{p^s} & i = s+j+1 \\ 0 & \textit{otherwise.} \end{cases}$$

With this information we can deduce what the PGA-ideals of $H^*(B^2\mathbb{Z}/p;\mathbb{F}_p)$ are. First we have,

Lemma 3.6. *The minimal PGA-ideal of $H^*(B^2\mathbb{Z}/p;\mathbb{F}_p)$ containing $P^{\Delta_n}\beta\iota$ is*

$$J_{minimal}(P^{\Delta_n}\beta\iota) = \Big((\beta P^1\beta\iota)^{p^n}, \dots, (\beta P^{\Delta_{n-1}}\beta\iota)^p, \\ P^{\Delta_n}\beta\iota, \beta P^{\Delta_n}\beta\iota, \dots, P^{\Delta_{n+k}}\beta\iota, \beta P^{\Delta_{n+k}}\beta\iota, \dots\Big).$$

In the same way,

Lemma 3.7. *The minimal PGA-ideal of $H^*(B^2\mathbb{Z}/p;\mathbb{F}_p)$ containing $(\beta P^{\Delta_n}\beta\iota)^{p^s}$, with $n \geq 0$, $s \geq 0$, is*

$$J_{minimal}((\beta P^{\Delta_n}\beta\iota)^{p^s}) = \Big((\beta P^{\Delta_1}\beta\iota)^{p^{s+n}}, \dots, (\beta P^{\Delta_{n-1}}\beta\iota)^{p^{s+1}}, \\ (\beta P^{\Delta_n}\beta\iota)^{p^s}, \dots, (\beta P^{\Delta_{n+r}}\beta\iota)^{p^s}, \dots\Big).$$

¿From these two descriptions its easy to obtain a description of all PGA-ideals of $H^*(B^2\mathbb{Z}/p;\mathbb{F}_p)$.

Lemma 3.8. *The PGA-ideals of $H^*(B^2\mathbb{Z}/p;\mathbb{F}_p)$ are either 0, (ι^{p^m}), $\tilde{H}^*(B^2\mathbb{Z}/p;\mathbb{F}_p)$ or one of the following types were $n \geq -1$ and $s \geq 0$,*

Type Is:

$$J_{n+1}(s) = \Big((\beta P^1\beta\iota)^{p^{m_n}}, \dots, (\beta P^{\Delta_{n-1}}\beta\iota)^{p^{m_1}}, \\ \beta P^{\Delta_n}\beta\iota, \dots, \beta P^{\Delta_{n+s}}\beta\iota, \dots, \beta P^{\Delta_{n+s+k}}\beta\iota, \dots \\ P^{\Delta_{n+s}}\beta\iota, \dots, P^{\Delta_{n+s+k}}\beta\iota, \dots\Big),$$

were $s \geq 0$, $m_1 = 1$ and $m_k = m_{k-1} + \epsilon$, $\epsilon = 0, 1$. Here s just determines the subindex $n+s$ where $P^{\Delta_{n+s}}\beta\iota$ and its Bockstein $\beta P^{\Delta_{n+s}}\beta\iota$ appear.

Type IIs:

$$J_{n+1}(s) = \Big(\iota^{p^m}, (\beta P^1\beta\iota)^{p^{m_n}}, \dots, (\beta P^{\Delta_{n-1}}\beta\iota)^{p^{m_1}}, \\ \beta P^{\Delta_n}\beta\iota, \dots, \beta P^{\Delta_{n+s}}\beta\iota, \dots, \beta P^{\Delta_{n+s+k}}\beta\iota, \dots \\ P^{\Delta_{n+s}}\beta\iota, \dots, P^{\Delta_{n+s+k}}\beta\iota, \dots\Big),$$

where $m > 1$, $s \geq 0$, $m_1 = 1$ and $m_k = m_{k-1} + \epsilon$, $\epsilon = 0, 1$.

Type IIIs:

$$J_{n+1}(s) = \Big((\beta P^{\Delta_1}\beta\iota)^{p^{m_n}}, \dots, (\beta P^{\Delta_{n-1}}\beta\iota)^{p^{m_1}}, \\ (\beta P^{\Delta_n}\beta\iota)^{p^s}, \dots, (\beta P^{\Delta_{n+r}}\beta\iota)^{p^s}, \dots\Big),$$

where $m_0 = s$, $m_1 = s+1$ and $m_k = m_{k-1} + \epsilon$, $\epsilon = 0, 1$ and $s \geq 0$.

Type IVs:

$$J_{n+1}(s) = \Big(\iota^{p^m}, (\beta P^{\Delta_1}\beta\iota)^{p^{m_n}}, \dots, (\beta P^{\Delta_{n-1}}\beta\iota)^{p^{m_1}}, \\ (\beta P^{\Delta_n}\beta\iota)^{p^s}, \dots, (\beta P^{\Delta_{n+r}}\beta\iota)^{p^s}, \dots\Big),$$

where $m_0 = s$,$m_1 = s+1$ and $m_k = m_{k-1} + \epsilon$, $\epsilon = 0, 1$, $m \geq 1$ and $s \geq 0$.

Proof. Suppose that the ideal J is non trivial, so it has to contain some primitive of $H^*(B^2\mathbb{Z}/p;\mathbb{F}_p)$, namely $P^{\Delta_n}\beta\iota$ or $(\beta P^{\Delta_n}\beta\iota)^{p^s}$. If it contains any of this elements it has to contain one of the ideals that we have described in 3.6 and 3.7. Obviously that ideals could be bigger, that is, it could contain other primitives such as, for example, roots of the elements there described. Suppose first that it contains one of the ideals $J_{minimal}(P^{\Delta_n}\beta\iota)$ described in 3.6, that is, it contains $P^{\Delta_n}\beta\iota$ but not $P^{\Delta_{n-1}}\beta\iota$. If one of the generators of the ideal is $(\beta P^{\Delta_{n-k}}\beta\iota)^{p^{m_k}}$ then, just applying the operation $P^{p^{m_k}}$ the element $(\beta P^{\Delta_{n-k-1}}\beta\iota)^{p^{m_k+1}}$ has also to be contained in the $\mathcal{A}$-ideal. Now, without contradiction, we can have the root of this element $(\beta P^{\Delta_{n-k-1}}\beta\iota)^{p^{m_k}}$. On the other hand, it is not possible that our ideal has $(\beta P^{\Delta_{n-k-1}}\beta\iota)^{p^{m_k-1}}$ as a generator because in this case we will not have $(\beta P^{\Delta_{n-k}}\beta\iota)^{p^{m_k}}$ as a generator but $(\beta P^{\Delta_{n-k}}\beta\iota)^{p^{m_k-1}}$, which could be obtained from $(\beta P^{\Delta_{n-k-1}}\beta\iota)^{p^{m_k-1}}$ applying $P^{p^{m_k-1}}$. This is the reason for the choice of the exponents m_i in the cases Is and IIs. Obviously our ideal can have ι^{p^m} as a generator for $m \geq 1$; in case $m = 0$ the ideal coincides with $\tilde{H}^*(B^2\mathbb{Z}/p;\mathbb{F}_p)$. In case that the ideal contains a minimal ideal of type $J_{minimal}((\beta P^{\Delta_n}\beta\iota)^{p^s})$, this being the biggest ideal included, that is, $(\beta P^{\Delta_n}\beta\iota)^{p^s}$ appears but not $P^{\Delta_n}\beta\iota$, then the argument is the same as the previous one. We have to impose $m_1 = p^{s+1}$ because if not $J_{minimal}((\beta P^{\Delta_{n-1}}\beta\iota)^{p^s})$ would be contained in our ideal, which is bigger than $J_{minimal}((\beta P^{\Delta_n}\beta\iota)^{p^s})$, contradicting the assumption made above. □

For an ideal J of type Is, IIs, III$_0$ or IV$_0$ as described in 3.8 we define the sub $\mathcal{A}$-module $\Sigma\hat{L}$ as
$\Sigma\hat{L} = \langle P^{\Delta_{n+s+1}}\beta\iota, \beta P^{\Delta_{n+s+1}}\beta\iota, \ldots, P^{\Delta_{n+s+k}}\beta\iota, \beta P^{\Delta_{n+s+k}}\beta\iota, \ldots\rangle_{\mathbb{F}_p}$ if J is either of type Is or IIs, and
$\Sigma\hat{L} = \langle \beta P^{\Delta_{n+1}}\beta\iota, \ldots \beta P^{\Delta_{n+k}}\beta\iota, \ldots\rangle_{\mathbb{F}_p}$ if the ideal J is either of type III$_0$ or IV$_0$.

In case that $J = \tilde{H}^*(B^2\mathbb{Z}/p;\mathbb{F}_p)$ we define

$$\Sigma\hat{L} = \langle P^1\beta\iota, \beta P^1\beta\iota, \ldots, P^{\Delta_n}\beta\iota, \beta P^{\Delta_n}\beta\iota, \ldots\rangle_{\mathbb{F}_p}.$$

For a better understanding of these ideals we recommend to look at the graphics given in Appendix 8. The corresponding quotients in $H^*(B^2\mathbb{Z}/p;\mathbb{F}_p)$ are of relevance in the argument of this work.

Definition 3.9. *We define the unstable $\mathcal{A}$-modules $M_n^\bullet = J/\Sigma\hat{L}$, where J is a primitively generated $\mathcal{A}$-ideal of $H^*(B^2\mathbb{Z}/p;\mathbb{F}_p)$ of one of the types* Is, IIs, III$_0$ *or* IV$_0$. *Alternatively we can describe the $\mathcal{A}$-modules $M_n^\bullet$ as the sub quotients of $H^*(B^2\mathbb{Z}/p;\mathbb{F}_p)$ described as vector spaces by*

$$\begin{aligned} M_{n+s}^{\mathrm{I}s} = \Big\langle & (\beta P^1\beta\iota)^{p^{m_n}}, \ldots, (\beta P^{\Delta_{n-1}}\beta\iota)^{p^{m_1}}, \\ & \beta P^{\Delta_n}\beta\iota, \ldots, P^{\Delta_{n+s}}\beta\iota, \beta P^{\Delta_{n+s}}\beta\iota\Big\rangle_{\mathbb{F}_p}, \end{aligned}$$

where $m_0 = 0,\ m_1 = 1,\ m_k = m_{k-1} + \epsilon,\ \epsilon = 0, 1,\ s \geq 0.$

$$M_{n+s}^{\mathrm{II}s} = \Big\langle (\beta P^1 \beta\iota)^{p^{m_n}}, \dots, (\beta P^{\Delta_{n-1}} \beta\iota)^{p^{m_1}}, \beta P^{\Delta_n} \beta\iota, \dots, P^{\Delta_{n+s}} \beta\iota, \beta P^{\Delta_{n+s}} \beta\iota, \iota^{p^m} \Big\rangle_{\mathbb{F}_p},$$

where $m > 0, m_0 = 0,\ m_1 = 1,\ m_k = m_{k-1} + \epsilon,\ \epsilon = 0, 1,\ s \geq 0.$

$$M_n^{\mathrm{III}} = \Big\langle (\beta P^1 \beta\iota)^{p^{m_n}}, \dots, (\beta P^{\Delta_{n-1}} \beta\iota)^{p^{m_1}}, \beta P^{\Delta_n} \beta\iota \Big\rangle_{\mathbb{F}_p},$$

being $m_0 = 0,\ m_1 = 1,\ m_k = m_{k-1} + \epsilon,\ \epsilon = 0, 1.$

$$M_n^{\mathrm{IV}} = \Big\langle (\beta P^1 \beta\iota)^{p^{m_n}}, \dots, (\beta P^{\Delta_{n-1}} \beta\iota)^{p^{m_1}}, \beta P^{\Delta_n} \beta\iota, \iota^{p^m} \Big\rangle_{\mathbb{F}_p},$$

with $m > 0,\ m_0 = 0,\ m_1 = 1,\ m_k = m_{k-1} + \epsilon,\ \epsilon = 0, 1.$

These ideals are better understood graphically. We are going to give some examples of them for the case M_n^{I} that will be the case of interest for us.

Example 3.10.

$n = -1$: $M_{-1}^{\mathrm{I}} = \{\beta\iota\}$

$n = 0$: $M_0^{\mathrm{I}} = \{P^1 \beta\iota, \beta P^1 \beta\iota\}.$

$$\circ \xrightarrow{\beta} \bullet$$

$n = 1$: *There are two possibilities*

$M_1^{\mathrm{I}} = \{P^{\Delta_1} \beta\iota, \beta P^{\Delta_1} \beta\iota, (\beta P^1 \beta\iota)^p\}$

$$\circ \xrightarrow{\beta} \bullet \xrightarrow{P^1} \bullet$$

$M_1^{\mathrm{I}} = \{P^{\Delta_1} \beta\iota, \beta P^{\Delta_1} \beta\iota, \beta P^1 \beta\iota\}.$

$$\circ \xrightarrow{\beta} \bullet \xleftarrow{P^p} \bullet$$

$n = 2$: *Now we have four possibilities*

$M_2^{\mathrm{I}} = \{P^{\Delta_2} \beta\iota, \beta P^{\Delta_2} \beta\iota, (\beta P^{\Delta_1} \beta\iota)^p, (\beta P^1 \beta\iota)^{p^2}\}.$

$$\circ \xrightarrow{\beta} \bullet \xrightarrow{P^1} \bullet \xrightarrow{P^p} \bullet$$

$M_2^{\mathrm{I}} = \{P^{\Delta_2} \beta\iota, \beta P^{\Delta_2} \beta\iota, (\beta P^{\Delta_1} \beta\iota)^p, (\beta P^1 \beta\iota)^p\}.$

$$\circ \xrightarrow{\beta} \bullet \xrightarrow{P^1} \bullet \xleftarrow{P^{p^2}} \bullet$$

$M_2^{\mathrm{I}} = \{P^{\Delta_2} \beta\iota, \beta P^{\Delta_2} \beta\iota, , \beta P^{\Delta_1} \beta\iota, (\beta P^1 \beta\iota)^p\}$

$$\circ \xrightarrow{\beta} \bullet \xleftarrow{P^{p^2}} \bullet \xrightarrow{P^1} \bullet$$

$$M_2^{\mathrm{I}} = \{P^{\Delta_2}\beta\iota, \beta P^{\Delta_2}\beta\iota, \beta P^{\Delta_1}\beta\iota, \beta P^1\beta\iota\}.$$

$$\circ \xrightarrow{\beta} \bullet \xleftarrow{P^{p^2}} \bullet \xleftarrow{P^p} \bullet$$

4. Differential Hopf algebras over $\mathbb{F}_p$

Here we study the structure of some differential Hopf algebras related to the Serre spectral sequence of an H-fibration $F \longrightarrow X \longrightarrow B^2\mathbb{Z}/p$. The general model for the SSS of an H-fibration is a bigraded differential Hopf algebra, (E, d), where the differential d is of bidegree $(n, 1-n)$.

Definition 4.1. *We will say that a bigraded differential Hopf algebra E satisfies the Kudo condition if for all $x \in E^{0,2n}$ with $d(x) = a \neq 0 \in E^{2n+1,0}$ then the element $a \otimes x^{p-1}$ is a non trivial class in $H(E)$.*

We will define the following differential Hopf algebras that there will be relevant in following sections.

Definition 4.2. *A bigraded differential Hopf algebra is of*

Type N: *if $E \cong A \otimes B$, with $E^{s,0} \cong A^s$ and $E^{0,t} \cong B^t$ and if it satisfies the Kudo condition, and of*

Type K: *if $E \cong A \otimes B \otimes E[c]$, with $E^{s,0} \cong A^s$, $E^{0,t} \cong B^t$, and $E[c]$, an exterior algebra primitively generated by a single generator in bidegree (s,t) with $s, t \neq 0$, and if it satisfies the Kudo condition.*

Definition 4.3. *The transgression for E is the restriction of the differential*

$$\tau_{|E^{0,n-1}} : E^{0,n-1} \longrightarrow E^{n,0},$$

that is, $\tau : B^{n-1} \longrightarrow A^n$.

These algebras are classified in the following types according to the possible transgression

Type N_0: E of type N and $d \equiv 0$

Type N_1: E of type N with transgression $\tau : B^{2m} \longrightarrow A^{2m+1}$ non-trivial.

Type N_2: E of type N and $d(x) = a$ with transgression $\tau : B^{2m-1} \longrightarrow A^{2m}$ non-trivial.

Type K_0: E of type K and $d \equiv 0$

Type K_1: E of type K and $d(c) = 0$ with non-trivial transgression.

Type K_2: E of type K and $d(c) \neq 0$.

We will assume that B is of finite type. According to the Borel's theorem we can write

$$B \cong \frac{\mathbb{F}_p[x_1, x_2, \dots, x_s, \dots]}{(x_1^{p^{\alpha_1}}, x_2^{p^{\alpha_2}}, \dots, x_r^{p^{\alpha_s}}, \dots)} \otimes E[y_1, \dots, y_r, \dots]$$

where α_i can be infinite.

We will assume in this section that A does not contain zero divisors in even degrees and also that $P^n(A)$ is at most 1-dimensional for every n.

Lemma 4.4. ([5, lemma 4.1])*Let $\{E, d\}$ be a differential Hopf algebra of type N. Then*

1. $d(B^m) \subset P^n(A) \otimes B^{m-n+1}$.
2. *If the transgression* $\tau : B^{n-1} \longrightarrow P^n(A)$ *is trivial then* $d \equiv 0$.

Lemma 4.5. *Let (E, d) be a bigraded differential Hopf algebra of type N_i, $i \neq 0$. That is $E \cong A \otimes B$, and*

$$B \cong \frac{\mathbb{F}_p[z_1, z_2, \ldots, z_s, \ldots]}{(z_1^{p^{\alpha_1}}, z_2^{p^{\alpha_2}}, \ldots, z_s^{p^{\alpha_s}}, \ldots)} \otimes E[y_1, \ldots, y_r, \ldots]\,.$$

Denote for simplicity by x_i the generators of B whether they are of the form z_i or y_i. Let $d(x_1) = a \in P^n(A)$, x_1 with minimal height in $d^{-1}(a)$. Then there is a system of generators $\{x_1, \tilde{x}_2, \ldots, \tilde{x}_r, \ldots\}$ of B as an algebra such that

1. *For $j > 1$, $\tilde{x}_j = x_j - bx_1$ and $d(\tilde{x}_j) = 0$ where b is a polynomial in the generators x_k in degrees less than $m = deg(x_j)$.*
2. $x_j^{p^{\alpha_j}} = 0 \Longrightarrow \tilde{x}_j^{p^{\alpha_j}} = 0$, $x_k^2 = 0 \Longrightarrow \tilde{x}_k^2 = 0$, *hence*
$$B \cong \frac{\mathbb{F}_p[z_1, \tilde{z}_2, \ldots, \tilde{z}_s]}{(z_1^{p^{\alpha_1}}, \tilde{z}_2^{p^{\alpha_2}}, \ldots, \tilde{z}_r^{p^{\alpha_s}}, \ldots)} \otimes E[\tilde{y}_1, \ldots, \tilde{y}_r, \ldots]$$
if $x_1 = z_1$ that is, is in even degree, or
$$B \cong \frac{\mathbb{F}_p[\tilde{z}_1, \tilde{z}_2, \ldots, \tilde{z}_s]}{(z_1^{p^{\alpha_1}}, \tilde{z}_2^{p^{\alpha_2}}, \ldots, \tilde{z}_r^{p^{\alpha_s}}, \ldots)} \otimes E[y_1, \ldots, \tilde{y}_r, \ldots]$$
in case that $x_1 = y_1$.

Proof. Let us consider x_j the first generator different from x_1 in minimal degree such that $d(x_j) \neq 0$. According to Lemma 4.4, and because $P^n(A)$ is 1-dimensional, we have that $d(x_j) = ba$ where b is a polynomial in the generators x_k in degrees less than the degree of x_j. We are going to distinguish the cases in which the dimension of the transgressive element $|x_1|$ is even or odd.

Suppose first that $|x_1|$ is odd, then $|a|$ is even and thus is not a zero divisor. We have

$$0 = d(ba) = d(a)b \pm ad(b) = ad(b) = 0$$

and then $d(b) = 0$.

If we consider $\tilde{x}_j = x_j - x_1 b$, applying the differential one obtains $d(\tilde{x}_j) = ab - (ab + x_1 d(b)) = ab - ab = 0$. Moreover, in this case to prove (2) one only has to observe that $x_1^{p^{\alpha_j}} = 0$ because x_1 is exterior and then the same holds for $(x_1 b)^{p^{\alpha_j}} = 0$.

Let us consider now the case $|x_1|$ even. Then $d(x_j) = ab$ where $b = \sum b_k x_1^k$ with the possibility $k = 0$ for some k and with b_k polynomials in generators in degree less than $|x_j|$. If we take $\tilde{x}_j = x_j - \sum 1/(k+1) b_k x_1^{k+1}$ we will have $d(\tilde{x}_j) = 0$ when $k + 1 \not\equiv 0(p)$, because $d(\sum 1/(k+1) b_k x_1^{k+1}) = a \sum b_k x_1^k$ if $k + 1 \not\equiv 0(p)$.

Then

$$d(\tilde{x}_j) = d(x_j - \sum 1/(k+1) b_k x_1^{k+1}) = a \sum b_k x_1^k - d(\sum 1/(k+1) b_k x_i^{k+1}) = 0\,.$$

We are going to see that the exponent k with $k+1 \equiv 0(p)$ is not possible. Let us consider first

$$\Delta(d(x_j)) = \Delta(a)\Delta(\sum b_k x_1^k)$$
$$= \sum \Delta(a)\Delta(b_k)\Delta(x_1^{kp-1}) \oplus \sum \Delta(a)\Delta(b_k)\Delta(x_1^k)\,.$$

Observe that in particular we have summands of type $ax_1^{kp-1} \otimes b_{pk-1} + b_{pk-1} \otimes ax_1^{kp-1}$. On the other hand we have

$$d(\Delta(x_j)) = d(x_j \otimes 1 + 1 \otimes x_j \pm \sum x'_j \otimes x''_j)\,.$$

Then applying the differential

$$d(\Delta(x_j)) = d(x_j) \otimes 1 + 1 \otimes d(x_j) + \sum (d(x'_j) \otimes x''_j \pm x'_j \otimes d(x''_j))\,,$$

that is, any term in this expression is of the form either $d(\alpha)\otimes b$ or $b\otimes d(\alpha)$ where α is an element of B. Observe, that any element of B can not have differential $a \otimes x^{rp-1}$ because these elements are product of the type $a \otimes x^{p-1} \otimes x^{tp}$, and by the Kudo condition these elements are non trivial classes in $H(E)$. So, any of these expressions cannot have these elements and then we can write $d(x_j) = a\sum b_k x_1^k$, where the sum is taken over $k \not\equiv p-1(p)$.

To prove (2) in case that dimension of x_1 is even consider

$$\Delta(x_j) = x_j \otimes 1 + 1 \otimes x_j + x_1 \otimes y + \sum z \otimes y' + \text{ terms in different degrees}$$

where z is a polynomial in the same degree as x_1 that does not contain x_1. Hence,

$$d(\Delta(x_j)) = ab \otimes 1 + 1 \otimes ab + a \otimes y + \text{ terms in different degrees}$$

On the other hand,

$$\Delta(d(x_j)) = \Delta(a)\Delta(b) = (a \otimes 1 + 1 \otimes a)(b \otimes 1 + 1 \otimes b + \sum b'_k \otimes b''_k) =$$
$$ab \otimes 1 + 1 \otimes ab + a \otimes b + \text{ terms in different degrees} \quad (5)$$

comparing both expressions we see that $b = y$. Then $\Delta(x_j) = x \otimes 1 + 1 \otimes x + x_1 \otimes b +$ terms in different degrees .

As $x_j^{p^{\alpha_j}} = 0$ also $\Delta(x_j^{p^{\alpha_j}}) = (\Delta(x_j))^{p^{\alpha_j}} = 0$. This equality implies that the term $(x_1 \otimes b)^{p^{\alpha_j}} = (x_1 \otimes \sum x_1^k b_k)^{p^{\alpha_j}} = 0$. As we have shown $k \not\equiv p-1(p)$ then this implies either $x_1^{p^{\alpha_j}} = 0$ or $(\sum x_1^k b_k)^{p^{\alpha_j}} = 0$. Observe that both possibilities give $(\tilde{x}_j)^{p^{\alpha_j}} = (x_j - \sum 1/(k+1) b_k x_1^{k+1})^{p^{\alpha_j}} = 0$ and the lemma follows. □

This result asserts that, under the above conditions if an algebra is of type N_i, $i = 1, 2$ then we can choose a system of algebra generators such that just one generator has non trivial transgression. From here it is easy to compute the homology of differential Hopf algebras of type N_1 or N_2.

Corollary 4.6. *Let (E, d) of type N, that is $E^{s,t} \cong A^s \otimes B^t$, and let B be of finite type. Suppose $d \not\equiv 0$. Then there exists a system of transgressive generators of B as an algebra such that,*

$$B \cong \frac{\mathbb{F}_p[x_1, \dots, x_r, \dots]}{(x_1^{p^{\alpha_1}}, \dots, x_r^{p^{\alpha_r}}, \dots)} \otimes E[y_1, \dots, y_r, \dots]$$

and then

- *If E is of type N_1, $d(x_1) = a$*
  - $H(E) = \dfrac{A}{(a)} \otimes \dfrac{\mathbb{F}_p[x_1^p, \dots, x_r, \dots]}{(x_1^{p^{\alpha_1}}, \dots, x_r^{p^{\alpha_r}} \dots)} \otimes E[y_1, \dots, y_r, \dots] \otimes E[a \otimes x^{p-1}]$.
- *If E is of type N_2, $d(y_1) = a$*
  - $H(E) = \dfrac{A}{(a)} \otimes \dfrac{B}{(y_1)}$.

We deal now with the case of Hopf algebras of type K.

Lemma 4.7. *Let (E, d) be a differential Hopf algebra of type K_1, that is, $E = A \otimes B \otimes E[c]$ where $E^{*,0} \cong A$, $E^{0,*} \cong B$, and $d(c) = 0$. Then d is internal on $A \otimes B$.*

Proof. Let x_1 be the generator in minimal degree of B with non trivial differential. We can assume that $x_1 \in B$ because d is a derivation and c has trivial differential. Applying the DHA lemma $d(x_1)$ has to be primitive in E. Remember that $P(E) = P(A) + P(B) + \mathbb{F}_p c$. As we are assuming the Kudo condition, $c \notin \operatorname{Im} d$. So that $d(x_1) \in P(A)$.

Let us consider now $b \in B$ the next generator in minimal degree such that $d(b) \neq 0$. Suppose that the differential of this element has any factor on c. We can write, in these conditions, $d(b) = \lambda c + \mu$, where λ and μ are polynomials in $A \otimes B$.

On the one hand one has $d(\Delta(b)) = d(b) \otimes 1 + 1 \otimes d(b) + \sum d(b'_k) \otimes b''_k + b'_k \otimes d(b''_k)$. As b has been chosen in minimal degree with $d(b) \neq 0$, then $d(b'_k)$ and $d(b''_k)$ are zero unless for the case in which b'_k or b''_k have any factor in x_1, that is, are of the form $x_1^{n_k} \tilde{b}_k$. Then $d(x_1^{n_k} \tilde{b}_k) = n_k x_1^{n_k - 1} a \tilde{b}_k$ that does not contain any factor on c. Then $d(\Delta(b))$ could contains factors on c just in $d(b) \otimes 1 + 1 \otimes d(b)$.

On the other hand,

$$\Delta(d(b)) = \Delta(\lambda c + \mu) = \Delta(\lambda)\Delta(c) + \Delta(\mu) =$$
$$(\lambda \otimes 1 + 1 \otimes \lambda + \sum \lambda'_k \otimes \lambda''_k)(c \otimes 1 + 1 \otimes c) + \Delta(\mu) . \quad (6)$$

Then this expression cannot contains factors in c except in $E^{*,0} \otimes E^{0,*}$ but in particular in the expression (6) we have $\lambda \otimes c$. Then $\lambda = 0$. Applying now this argument inductively over the degree of the following generators, one obtains that their differential cannot have any factor on c. $\square$

Corollary 4.8. *Let $(E \cong A \otimes B \otimes E[c], d)$ be a differential Hopf algebra of type K_1. Then there is a system $\{x_1, \dots, x_r, \dots\}$ of generators of B as an algebra, such that $d(x_1) = a \in P^n(A)$ and $d(x_r) = 0$, for $r > 1$.*

Proof. Lemma 4.7 asserts that d is internal on $A \otimes B$, thus the same argument used in 4.4 and 4.5 applies here. □

Now combining the above results

Corollary 4.9. *Let $(E \cong A \otimes B \otimes E[c], d_n)$ be a differential Hopf algebra of type K_i with $i = 0, 1$ and $d_n(c) = 0$. Assume that there are no primitive elements in $P^n(A)$, then $d \equiv 0$.*

The conditions of this lemma will hold for the differential algebras of type $(E \cong A \otimes B \otimes E[c], d_{2n+1})$ that will appear in the next sections. Thus the differential will always be trivial. Hence we are going to concentrate our calculations on differential algebras of type K where $d = d_{2n}$.

Corollary 4.10. *Let $(E \cong A \otimes B \otimes E[c], d_{2n})$ be a differential Hopf algebra of type K_1, with B of finite type and c a primitive element. Suppose that the differential is non trivial on this page and that $P^{2n}(A)$ is 1-dimensional and does not contain zero divisors. Then there is a system of generators of B as an algebra such that we can write*

$$B \cong \frac{\mathbb{F}_p[x_1, \dots, x_r, \dots]}{(x_1^{p^{\alpha_1}}, \dots, x_r^{p^{\alpha_r}}, \dots)} \otimes E[y_1, \dots, y_r, \dots]$$

and such that $d(y_1) = a$ and

$$H(E, d) \cong \frac{A}{(a)} \otimes \frac{\mathbb{F}_p[x_1, \dots, x_r, \dots]}{(x_1^{p^{\alpha_1}}, \dots, x_r^{p^{\alpha_r}}, \dots)} \otimes E[y_2, \dots, y_r, \dots] \otimes E[c] .$$

Finally we are going to compute the homology of differential Hopf algebras of type K_2. We are going to assume that in this case there are no transgressive elements. This will be so if we assume that there are no primitives in certain dimensions. The reason for this assumption is that in the cases that we will study in the next section there will be no primitive elements in the dimensions of those pages of the SSS in which we kill the exterior generator produced by the Kudo transgression theorem.

Proposition 4.11. *Let $(E \cong A \otimes B \otimes E[c], d)$ be a differential Hopf algebra of type K_2, with B of finite type and such that $d(c) = a \in P^{s+n}(A)$. Suppose that $P^{s+n}(A)$ is 1-dimensional and a is not a zero divisor. Then if $P^n(A) = 0$ we have*

1. $d(B) = 0$
2. $H(E) \cong \dfrac{A}{(a)} \otimes B$.

Proof. To prove (1) we consider an element of B with non trivial differential and show that its differential lies in $A \otimes B$. Let b be the generator in minimal degree such that $d(b) \neq 0$. We can write $d(b) = \mu + \lambda c$ were λ and μ belong to $A \otimes B$ and are polynomials in generators in degrees less than that of b. Then

$$0 = d^2(b) = d(\mu) + d(\lambda \otimes c) = \lambda \otimes d(c) = \lambda \otimes a$$

because the election of b as the generator of B in minimal degree with non trivial differential implies $d(\lambda) = d(\mu) = 0$. As a is not a zero divisor the expression $a \otimes \lambda = 0$ implies $\lambda = 0$ and then $d(b) \in A \otimes B$.

Now we will see that the differential of this element has to be a primitive element. The proof is the same as that of the DHA lemma in [10]. We write $d(b) = \sum a_k b_k$, then $\Delta(d(b)) = d(\Delta(b)) = d(b \otimes 1 + 1 \otimes b + \sum b'_k \otimes b''_k)$. As b is the generator in minimal degree of B with non trivial differential, $d(b'_k)$ and $d(b''_k)$ are zero for all k and then we see that $d(b)$ is primitive.

Therefore $d(b) \in P(A \otimes B) = P(A) + P(B)$, so $d(b) \in P^n(A) = 0$. That is $d(b) = 0$.

The second part of the result is trivial once one knows (1). □

5. Serre spectral sequence for H-fibrations over $B^2\mathbb{Z}/p$

We are interested now in the behaviour of the Serre spectral sequence of an H-fibration $F \longrightarrow X \longrightarrow B^2\mathbb{Z}/p$:

$$E_2^{*,*} \cong H^*(B^2\mathbb{Z}/p; \mathbb{F}_p) \otimes H^*(F; \mathbb{F}_p) \Longrightarrow H^*(X; \mathbb{F}_p) \tag{7}$$

We will use in this section the Kudo transgression theorem that will assert that Kudo condition in the previous section is satisfied by our fibrations. We recall here the theorem:

Theorem 5.1 ([11, 15]). *Let*

$$F \xrightarrow{i} E \xrightarrow{q} B$$

be a fibration with B a connected CW-complex. Let p be a prime. If p is odd, assume that B is simply connected. Consider the SSS

$$E_2^{n,t} = H^n(B; \mathbb{F}_p) \otimes H^t(F; \mathbb{F}_p) \Longrightarrow H^*(E; \mathbb{F}_p).$$

Suppose that $x \in H^t(F; \mathbb{F}_p)$ survives to $E_{t+1}^{0,t}$ and

$$d_{t+1}(x) = y \in E_{t+1}^{t+1,0} = H^{t+1}(B; \mathbb{F}_p).$$

(a)If $p = 2$ and $0 \le n \le t$ then $Sq^n(x) \in H^{t+n}(F; \mathbb{F}_2)$ survives to $E_{t+n+1}^{0,t+n}$ and

$$d_{t+n+1}(Sq^n(x)) = Sq^n(y) \in E_{t+n+1}^{t+n+1,0} = H^{t+n+1}(B; \mathbb{F}_2).$$

(b) If p is odd and $0 \le n \le t$ then $P^n(x) \in H^{t+2n(p-1)}(F; \mathbb{F}_p)$ survives to $E_{t+2n(p-1)+1}^{0,t+2n(p-1)}$ and

$$d_{t+2n(p-1)+1}(P^n(x)) = P^n(y) \in E_{t+2n(p-1)+1}^{t+2n(p-1)+1,0} = H^{t+2n(p-1)+1}(B; \mathbb{F}_p).$$

If $0 \le n \le t$ then $\beta P^n(x) \in H^{t+2n(p-1)+1}(B; \mathbb{F}_p)$ survives to $E_{t+2n(p-1)+2}^{0,t+2n(p-1)+1}$ and

$$d_{t+2n(p-1)+2}(\beta P^n(x)) = \beta P^n(y) \in E_{t+2n(p-1)+2}^{t+2n(p-1)+2,0} = H^{t+2n(p-1)+2}(B; \mathbb{F}_p).$$

If $\deg(x) = 2k$ *is even then the class* $y \otimes x^{p-1} \in E_2^{2k+1,2k(p-1)}$ *survives to* $E_{2k(p-1)+1}^{2k+1,2k(p-1)}$ *and*

$$d_{2k(p-1)+1}(y \otimes x^{p-1}) = -\beta P^k(y) \in E_{2k(p-1)+1}^{2kp+2,0} = H^{2kp+2}(B; \mathbb{F}_p).$$

We will call the last differential in the Theorem a **Kudo differential** or Kudo transgression.

Our first step will be determine which are the possible differentials for the generators of the finite part of the fibre. By the results of Section 4 we can assume all the generators are transgressive.

Lemma 5.2. *Let* $X \longrightarrow Y \longrightarrow B^2\mathbb{Z}/p$ *be an H-fibration with* $H^*(X;\mathbb{F}_p)$ *of finite type. Let* x *be a nilpotent generator of* $H^{even}(X;\mathbb{F}_p)$. *Then the transgression of* x *is trivial.*

Proof. We have a commutative diagram

$$\begin{array}{ccccc} X & \longrightarrow & Y & \longrightarrow & B^2\mathbb{Z}/p \\ {\scriptstyle f}\uparrow & & \uparrow & & \| \\ B\mathbb{Z}/p & \longrightarrow & * & \longrightarrow & B^2\mathbb{Z}/p \end{array}$$

By naturality of the SSS this diagram induces a map, denoted again by f, between the SSS of our fibration and that of the universal fibration, such that $\tau(x) = \tau(f(x))$ using τ for both transgressions. As f factors via the Nil-localization of $H^*(X;\mathbb{F}_p)$, we have that it is trivial over the finite part of $H^*(X;\mathbb{F}_p)$. Hence $f(x) = 0$ and then x has to transgress to an odd dimensional element representing the zero class in the corresponding page of the SSS of the universal fibration, that is, it is either a decomposable in $H^*(B^2\mathbb{Z}/p;\mathbb{F}_p)$ or just zero (recall that the image of a Kudo differential lies in even dimension). Notice that by DHA lemma, the transgression of any element in $H^*(X;\mathbb{F}_p)$ has to be a primitive element in the basis, that is in $H^*(B^2\mathbb{Z}/p;\mathbb{F}_p)$, but there are no decomposable primitives in $H^{odd}(B^2\mathbb{Z}/p;\mathbb{F}_p)$. Hence the transgression $\tau(x)$ of x has to be trivial. □

In the same way we obtain the transgressions of the generators of $H^*(X;\mathbb{F}_p)$ in odd degrees.

Lemma 5.3. *Let* x *be a generator of* $H^{odd}(X;\mathbb{F}_p)$ *in dimension bigger than 1 with non trivial transgression. Then its transgression* $\tau(x)$ *is either*

$$\begin{cases} (\beta P^{\Delta_n}\beta\iota)^{p^s} & \text{where } s \geq 0 \\ \text{or} & \\ (\iota)^{p^m} & \text{where } m > 0\,. \end{cases}$$

Proof. The same argument as in 5.2 shows that x restricts trivially to the cohomology of $B\mathbb{Z}/p$ and therefore the transgression of x has to be an element representing the zero class in the corresponding page of the SSS associated to the universal fibration. Since the transgression of any element in $H^*(X;\mathbb{F}_p)$ has to be a primitive element by DHA lemma $\tau(x)$ has to be either a decomposable primitive in even dimension or $\beta P^{\Delta_n}\beta\iota$ because these elements are images of previous Kudo differentials in the SSS for the universal fibration. Remember that the decomposable primitives in even degrees $H^*(B^2\mathbb{Z}/p;\mathbb{F}_p)$ are of one of the types described in the lemma. □

Theorem 5.4. *Let $X \longrightarrow Y \longrightarrow B^2\mathbb{Z}/p$ be an H-fibration with X satisfying conditions $(F1)$, $(F2)$ and $(F3)$. Then there is a system of transgressive generators for $H^*(X;\mathbb{F}_p)$*

$$H^*(X;\mathbb{F}_p) \cong \mathbb{F}_p[x_1,\dots,x_t] \otimes \frac{\mathbb{F}_p[z_1,\dots,z_r,\dots]}{(z_1^{p^{m_1}},\dots,z_r^{p^{m_r}}\dots)} \otimes E[y_1,\dots,y_m,\dots]$$

such that the pages of the SSS are of one of the types N_0, N_1, N_2, K_0, K_1 or K_2 of the previous section.

Proof. We will proceed by induction on s, the number of the page we are working. For $s=2$, it is clear that E_2 is of the type N_0 or N_2, the subindex being determined if d is trivial or not. Let E_s be the first page of the SSS that the differential is non trivial, that is $d_k = 0$ for $k < s$ and then E_k is of type N_0 for $k < s$. It could be E_2. Suppose first that s is even. That is, the page is of type N_2. Applying the results of 4.6, there is a system of generators for $H^*(X;\mathbb{F}_p)$ such that E_{s+1} is once again of type N.

Suppose now that there is a differential in an odd page such that there exists x with $d(x) \neq 0$. Using lemma 5.2, x, that lies in even degree in $H^*(X;\mathbb{F}_p)$, has to be a polynomial generator of $H^*(X;\mathbb{F}_p)$. Then following 2.8, x has degree $2p^{n+1}$ for some n and the page in which we are working is of type $E_{2p^{n+1}+1}$. The transgression of x has to be a primitive in this dimension, that is $d(x) = P^{\Delta_n}\beta\iota$.

The Kudo transgression theorem implies that Kudo's condition holds: this means that the page is of type N_1, and then we are under the hypothesis of 4.6. We can thus choose a system of transgressive generators for $E^{0,*}$ such that any generator different from x has trivial differential and then $H(E)$ is of type K with $c = P^{\Delta_n}\beta\iota \otimes x^{p-1}$, the generator of the exterior algebra in the description.

Claim 5.5. *Let $a = P^{\Delta_n}\beta\iota$. The class of $a \otimes x^{p-1}$ is primitive in the following pages of the SSS.*

Proof. Denote $c = a\otimes x^{p-1}$. $\Delta(c) = \Delta(x_1^{p-1})\Delta(a) = (x_1^{p-1}\otimes 1 + 1\otimes x_1^{p-1} + \sum x_i'\otimes x_i'')^{p-1}(a\otimes 1 + 1\otimes a)$. The element a is a trivial class in E_{s+1} and this page just contains factors in a if they are of the form $a\otimes x_1^{p-1}$. Then those terms in the expression of $\Delta(c)$ that do not contain $a\otimes x_1^{p-1}$ are zero in E_{s+1} so that

$$\Delta(c) = \Delta(a\otimes x_1^{p-1}) = (ax_1^{p-1}\otimes 1 + 1\otimes ax_1^{p-1})\,. \qquad \square$$

Suppose then that we have obtained a page of type K via an odd transgression $d_{2p^{n+1}+1}$. Kudo's theorem implies that the class $c = P^{\Delta_n}\beta\iota\otimes x^{p-1}$ survives until $E_{2(p^{n+2}-p^{n+1})+1}$ and in this page one has $d_{2(p^{n+2}-p^{n+1})+1}(c) = \beta P^{\Delta_{n+1}}\beta\iota$.

It is very important to remark the following facts:

1. $H^*(B^2\mathbb{Z}/p;\mathbb{F}_p)$ does no contain primitive elements in odd dimension in the degrees between $2p^{n+1}+2$ and $2(p^{n+2}-p^{n+1})$.
2. There are no non trivial primitive elements in degree $2(p^{n+2}-p^{n+1})+1$ in $H^*(B^2\mathbb{Z}/p;\mathbb{F}_p)$.

By the first observation we can apply the corollaries 4.9 and 4.10 to ensure that all the pages with $d(c)=0$ are again of type K, moreover it will be of type K_0 for odd pages by the first observation. Notice that in those pages it is not possible to obtain a new exterior generator with bidegree (i,j), $i,j\neq 0$, because they appear when a generator applies over a primitive generator in odd degree.

Finally the page $E_{2(p^{n+2}-p^{n+1})+1}$ is of type K_2 and the hypothesis of 4.11 hold, basically we need that there are no primitives in degree $2(p^{n+2}-p^{n+1})+1$. Then the following page, applying Lemma 4.11, is of type N, as it was at the beginning, and we start the process again. □

Remark 5.6. *This theorem asserts that the exterior generators of pages in the SSS that are not on the axis, are not permanent cycles. They always disappear in a posterior page of the SSS.*

Notice that the transgression is determined by the primitive elements of $H^*(B^2\mathbb{Z}/p;\mathbb{F}_p)$ and the whole SSS.

More precisely, we can give, at this moment, the analogous result to [5, 6.3].

Proposition 5.7. *For any mod p H-space X satisfying the conditions* F1, F2 and F3 *and any polynomial generator $x\in H^*(X;\mathbb{F}_p)$, there exists a finite subquotient of $H^*(B^2\mathbb{Z}/p;\mathbb{F}_p)$, $M_n^\bullet$, and an epimorphism of unstable $\mathcal{A}$-modules:*

$$\tilde{\tau}:\Sigma QH^*X \twoheadrightarrow M_n^\bullet$$

with $\tilde{\tau}(x)=P^{\Delta_n}\beta\iota$, where $M_n^\bullet$ denotes either M_n^{I} or M_n^{II}. Moreover, x could be completed to a system of generators where $\tilde{\tau}(y)=0$ for any polynomial generators $y\neq x$.

Proof. Given $x\in H^{even}(X;\mathbb{F}_p)$ a polynomial generator of $H^*(X;\mathbb{F}_p)$, we can construct, following Section 2, H-fibrations $B\mathbb{Z}/p\longrightarrow X\longrightarrow E\longrightarrow B^2\mathbb{Z}/p$ such that the map $B\mathbb{Z}/p\longrightarrow X$ detects x in cohomology. We are going to compute the SSS for the fibration $X\longrightarrow E\longrightarrow B^2\mathbb{Z}/p$.

By naturality of the SSS one has that x is transgressive with transgression $P^{\Delta_n}\beta\iota$ (look at the proof of 5.4). Then the p-powers of this element, x^{p^s}, have transgression $P^{\Delta_{n+s}}\beta\iota$, for all s, which are odd primitives in bigger dimensions. Up to a change of basis for $H^*(X;\mathbb{F}_p)$, those are the only elements in $H^{even}(X;\mathbb{F}_p)$ with non trivial transgression. On the other hand, Kudo's transgression theorem determines that the transgressions, (that are all non trivial!) of the elements of type $P^{\Delta_{n+s}}\beta\iota\otimes x^{p^s-1}$ are $\beta P^{\Delta_{n+s+1}}\beta\iota$. Moreover the action of the Steenrod algebra determines other possible different transgressions codified by one of the ideals $J_n(s)$ of the section 3.

Only elements in $H^{odd}(X;\mathbb{F}_p)$ could have non trivial transgression to a decomposable primitive of $H^*(B^2\mathbb{Z}/p;\mathbb{F}_p)$, as we said in Lemma 5.3, but always in a compatible way with the transgression of Kudo. That is, if an element y is transgressive, then for any Steenrod operation θ, θy transgresses to $\theta(\tau y)$. This determines an ideal of primitive elements of $H^*(B^2\mathbb{Z}/p;\mathbb{F}_p)$ that are transgressions of elements in $H^*(X;\mathbb{F}_p)$. This ideal has to be compatible with the fact that $P^{\Delta_{n+s}}\beta\iota$ and $\beta P^{\Delta_{n+s+1}}\beta\iota$, for $s \geq 0$ are transgressions determined by $\tau(x)$.

This compatibility was studied in section 3 and is codified by the modules $M_n^\bullet$, $\bullet = \text{I}, \text{II}$. The cases $\bullet = \text{III}, \text{IV}$ cannot occur since the primitive element $P^{\Delta_n}\beta\iota$ does not appear. □

Lemma 5.8. *Let $X \xrightarrow{j} Y \xrightarrow{p} B^2\mathbb{Z}/p$ be an H-fibration where X satisfies conditions $(F1), (F2)$ and $(F3)$. Then either the fibration is trivial or the PGA-ideal $I = \ker p^*$ is of type* Is, IIs *or $\tilde{H}^*(B^2\mathbb{Z}/p;\mathbb{F}_p)$.*

Proof. According to Proposition 3.8, the PGA-ideals of $H^*(B^2\mathbb{Z}/p;\mathbb{F}_p)$ are either 0, $(\iota)^{p^m}$, $\tilde{H}^*(B^2\mathbb{Z}/p;\mathbb{F}_p)$ or of one of the types I, II, III or IV. Assume that $I = \ker p^*$ is a PGA-ideal of type III or IV , $(\iota)^{p^m}$ or just zero. These ideals do not contain primitive generators in odd dimensions so polynomial generators have to transgress trivially. This implies that they map trivially along $B\mathbb{Z}/p \xrightarrow{f} F$. Then, by 2.9 this map is trivial in cohomology and therefore null-homotopic [12]. Now we can show applying Zabrodsky's lemma that $E \simeq F \times B\mathbb{Z}/p$ (see [5, lemma 6.1]). □

Once again the result obtained suggests the analogous definition to that introduced in [5, def 6.2]

Definition 5.9. *A non trivial H-fibration is said to be of type* I *(resp.* II*) if the transgression maps, in an epimorphic way, to a module of type* I *(resp.* II*).*

6. Convergence of the SSS

Now we study the convergence of SSS of fibrations either of type I or II. The results obtained are the p analogous to those obtained in [5]. Most of them were valid by just changing 2 by p in the proofs. Notice that when one has described the transgression map and the different pages of the SSS the convergence is determined by results independent on the prime in where we are working.

Consider a non trivial H-fibration of type I

$$X \longrightarrow Y \longrightarrow B^2\mathbb{Z}/p$$

where X is 1-connected and $H^*(X;\mathbb{F}_p)$ of finite type.

There is a system of transgressive generators such that

$$H^*(X;\mathbb{F}_p) \cong \frac{\mathbb{F}_p[x_1,\dots,x_r,x_{r+1},\dots,x_s,\dots]}{(x_{r+1}^{p^{m_1}},\dots,x_s^{p^{m_s}},\dots)} \otimes E[y_0,\dots,y_m,\dots]$$

and transgression

$$\tilde{\tau} : \Sigma QH^*(X;\mathbb{F}_p) \twoheadrightarrow M_n^{\text{I}}$$

where

$$M_n^{\mathrm{I}} \cong \langle P^{\Delta_n}\beta\iota, \beta P^{\Delta_n}\beta\iota, (\beta P^{\Delta_{n-1}}\beta\iota)^{p^{m_1}}, \dots, (\beta P^1\beta\iota)^{p^{m_n}}\rangle_{\mathbb{F}_p}.$$

We can suppose

$$\tilde{\tau}(x_1) = P^{\Delta_n}\beta\iota, \quad \tilde{\tau}(y_0) = \beta P^{\Delta_n}\beta\iota,$$
$$\tilde{\tau}(y_1) = (\beta P^{\Delta_{n-1}}\beta\iota)^{p^{m_1}}, \dots, \tilde{\tau}(y_n) = (\beta P^{\Delta_0}\beta\iota)^{p^{m_n}}$$

and $\tilde{\tau}$ trivial otherwise. These formulas determine the transgression and then the SSS. E_∞ in our SSS is $E_\infty \cong A_\infty \otimes B_\infty$ where

$$A_\infty \cong \mathbb{F}_p[\iota] \otimes N',$$
$$N' \cong \frac{\mathbb{F}_p[\beta P^{\Delta_0}\beta\iota, \dots, \beta P^{\Delta_{n-1}}\beta\iota]}{\left((\beta P^{\Delta_0}\beta\iota)^{p^{m_n}}, \dots, (\beta P^{\Delta_{n-1}}\beta\iota)^{p^{m_1}}\right)} \otimes \Lambda[P^{\Delta_{-1}}\beta\iota, \dots, P^{\Delta_{n-1}}\beta\iota]$$

and

$$B_\infty \cong \mathbb{F}_p[x_2, \dots, x_r] \otimes N'',$$
$$N'' \cong \frac{\mathbb{F}_p[x_{r+1}, \dots, x_s, \dots]}{(x_{r+1}^{p^{m_1}}, \dots, x_n^{p^{m_n}}, \dots)} \otimes \Lambda[y_{n+1}, \dots, y_m, \dots].$$

With this notation we have:

Proposition 6.1. *For a fibration of type* I *and 1-connected fiber*

1. $H^*(Y;\mathbb{F}_p) \cong \mathbb{F}_p[x_1', x_2', \dots, x_r'] \otimes N$ *where* $x_1' = h^*(\iota)$ *and* $g^*(x_i') = x_i$ *if* $i \geq 2$.
2. *N is a nilpotent Hopf algebra of finite type that fits in a exact sequence of Hopf algebras*

$$1 \longrightarrow N' \xrightarrow{h^*} N \xrightarrow{g^*} N'' \longrightarrow 1.$$

For type II fibrations one obtains the correspondent version. Consider namely an H-fibration of type II. We can write

$$H^*(X;\mathbb{F}_p) \cong \frac{\mathbb{F}_p[x_1, \dots, x_r, x_{r+1}, \dots, x_s, \dots]}{(x_{r+1}^{p^{m_1}}, \dots, x_s^{p^{m_s}}, \dots)} \otimes E[y_0, \dots, y_m, \dots],$$

where the generators form a good system of transgressive generators and the transgression is determined by

$$\tilde{\tau} : \Sigma QH^*(X;\mathbb{F}_p) \twoheadrightarrow M_n^{\mathrm{II}}$$

where

$$M_n^{\mathrm{II}} \cong \langle P^{\Delta_n}\beta\iota, \beta P^{\Delta_n}\beta\iota, (\beta P^{\Delta_{n-1}}\beta\iota)^{p^{m_1}}, \dots, (\beta P^1\beta\iota)^{p^{m_n}}, \iota^{p^m}\rangle_{\mathbb{F}_p}$$

in such a way that for this system of generators

$$\tilde{\tau}(y_0) = \beta P^{\Delta_n}\beta\iota, \quad \tilde{\tau}(y_1) = (\beta P^{\Delta_{n-1}}\beta\iota)^{p^{m_1}}, \dots,$$
$$\tilde{\tau}(y_n) = (\beta P^{\Delta_0}\beta\iota)^{p^{m_n}}, \quad \tilde{\tau}(y_{n+1}) = \iota^{p^m},$$

$\tilde{\tau}$ being trivial over all other generators.

Then $E_\infty \cong A_\infty \otimes B_\infty$ where

$$A_\infty \cong N' \cong \frac{\mathbb{F}_p[\iota, \beta P^{\Delta_0}\beta\iota, \dots, \beta P^{\Delta_{n-1}}\beta\iota]}{(\iota^{p^m}, (\beta P^{\Delta_0}\beta\iota)^{p^{m_n}}, \dots, (\beta P^{\Delta_{n-1}}\beta\iota)^{p^{m_1}})} \otimes E[\beta\iota, \dots, P^{\Delta_{n-1}}\beta\iota]$$

and

$$B_\infty \cong \mathbb{F}_p[x_2, \dots, x_r] \otimes N'' \quad N'' \cong \frac{\mathbb{F}_p[x_{r+1}, \dots, x_s, \dots]}{(x_{r+1}^{p^{m_1}}, \dots, x_n^{p^{m_n}}, \dots)} \otimes E[y_{n+1}, \dots, y_m, \dots]$$

We obtain now

Proposition 6.2. *For a fibration of type* II *with 1-connected fiber*

1. $H^*(Y;\mathbb{F}_p) \cong \mathbb{F}_p[x_2', \dots, x_r'] \otimes N$ *where* $g^*(x_i') = x_i$ *if* $i \geq 2$.
2. N *is a nilpotent Hopf algebra that fits in an exact sequence of Hopf algebras*

$$1 \longrightarrow N' \xrightarrow{h^*} N \xrightarrow{g^*} N'' \longrightarrow 1\,.$$

In this case N *has a new 2-dimensional nilpotent class.*

7. The results

Now we apply the iterative process of [5, section 8], obtaining the following result.

Theorem 7.1. *Let* X *be a 1-connected mod* p H*-space that satisfies conditions* $(F1)$, $(F2)$ *and* $(F3)$. *Then*

1. *There is a sequence of mod* p H*-spaces*
$$X = X_0 \longrightarrow X_1 \longrightarrow \dots \longrightarrow X_n = F(X)$$
where all X_i *satisfy also conditions* $(F1)$, $(F2)$ *and* $(F3)$, *the depth of* X_i *is the depth of* X_{i-1} *minus one and* $X_n = F(X)$ *is* $L_{B\mathbb{Z}/p}$*-local.*
2. *The maps*
$$X_i \longrightarrow X_{i+1}$$
are principal H*-fibrations with fibre either* $(\mathbb{C}P^\infty)^\wedge_p$ *or* $B\mathbb{Z}/p^k$ *for some* $k \geq 1$.
3. *The composition* $X \longrightarrow F(X)$ *is also a principal fibration with fibre the product of the fibres of the maps* $X_i \longrightarrow X_{i+1}$.

Assume furthermore that $H^*(X;\mathbb{F}_p)$ *is noetherian, then*

4. *As algebras,* $H^*(X;\mathbb{F}_p) \cong P \otimes N$, *where* P *is a polynomial algebra and* N *is a 2-connected finite Hopf algebra.*
5. *In the above sequence of mod* p H*-spaces*
$$X = X_0 \longrightarrow X_1 \longrightarrow \dots \longrightarrow X_n = F(X)$$
all X_i *have noetherian mod* p *cohomology.*
6. $X_n = F(X)$ *is a mod* p *finite* H*-space.*
7. *For all* $0 \leq i \leq n$ *the fibrations involved in the construction of*
$$X_i \longrightarrow X_{i+1}$$
are of type I *and* $X_i \longrightarrow X_{i+1}$ *are principal* H*-fibrations with fibre* $(\mathbb{C}P^\infty)^\wedge_p$.

8. *The composition $X \longrightarrow F(X)$ is also a principal fibration:*

$$((\mathbb{C}P^\infty)^\wedge_p)^n \longrightarrow X \longrightarrow F(X).$$

Proof. [Sketch proof] The idea of the proof given in [5] is the following. Given a 1-connected H-space satisfying the finiteness conditions one can detect one of the polynomial generators with a map $B\mathbb{Z}/p \longrightarrow X$. We construct the H-space E_1, Borel construction of this map, and using the results of the previous sections we compute its cohomology. It will depend just on the module $M_n^\bullet$ that classifies the fibration. If the module is of type M_n^{II} one obtains that E_1 is an H-space with one polynomial generator less than the original X but with a new nilpotent class in dimension 2. If the fibration is of type I, one obtains a new H-space with the same number of generators as X but with the original polynomial generator detected replaced by one in dimension 2. We iterate this process with the new H-space stopping if the fibration obtained is of type II in any step. If the process is either finite or infinite one obtains at the end, that the limit space, $\operatorname{hocolim} E_k$, is an H-space ([5, proposition 8.3]) satisfying again the finiteness conditions but with one polynomial generator less. We call this space X_1 and repeat this process now with X_1. The fibre of the map $X \longrightarrow X_1$ will be $(\mathbb{C}P^\infty)^\wedge_p$ if the process has been infinite and $B\mathbb{Z}/p^k$ if we were done in k steps. When we do that n-times, the number of polynomial generators of the original H-space X, we will have obtained an H-space X_n that satisfies again the finiteness conditions but without polynomial generators. That is X_n is a 1-connected mod p H-space with $H^*(X_n;\mathbb{F}_p)$ nilpotent and $QH^*(X_n;\mathbb{F}_p)$ locally finite as $\mathcal{A}$-module, hence $\operatorname{map}(B\mathbb{Z}/p, X_n) \simeq X_n$. In other words, X_n satisfies the Sullivan conjecture or equivalently X_n is $L_{B\mathbb{Z}/p}$-local, that is $F(X_n) \simeq X_n$ where F is the functor defined in the introduction.

In case one assumes $H^*(X;\mathbb{F}_p)$ noetherian, the final H-space $F(X)$ is a finite simply connected H-space. If any of the fibrations involved in the construction was of type II we have added a nilpotent class in dimension 2, which survives during all the process. So $F(X)$ is 1-connected but not 2-connected. This is impossible [8, Theorem 6.11]. Thus all fibrations have to be of type I and the the fibres of $X_i \longrightarrow X_{i+1}$ are always $(\mathbb{C}P^\infty)^\wedge_p$. □

Proof. [Proof of Theorem 1.1] It follows from Theorem 7.1(7) for the case in which $H^*(X;\mathbb{F}_p)$ is noetherian. □

Proof. [Proof of Theorem 1.3] In case that the H-space satisfies conditions $(F1)$, $(F2)$ and $(F3)$ but the cohomology is not necessary noetherian the fibrations of the construction of the spaces X_i could be of type II. This is possible because if the cohomology of our H-space X is not finitely generated the final H-space $F(X)$ in theorem 7.1 will be nilpotent but not finite and then $F(X)$ will be not finite. Then [8, Theorem 6.11] does not applies and we cannot conclude that the fibration is of type I (see [5]). If any of the fibrations involved in the construction of X_i are of type II, the process is finite and the fibre is not $(\mathbb{C}P^\infty)^\wedge_p$ but $B\mathbb{Z}/p^k$. The theorem follows applying 7.1(2) and (3). □

8. Appendix

We are going to give here a graphic with the relations between the primitive generators of $H^*(B^2\mathbb{Z}/p;\mathbb{F}_p)$ that is very useful to understand the ideals described in section 3.

Firstly for a primitive generator of type $P^{\Delta_n}\beta\iota$ one has

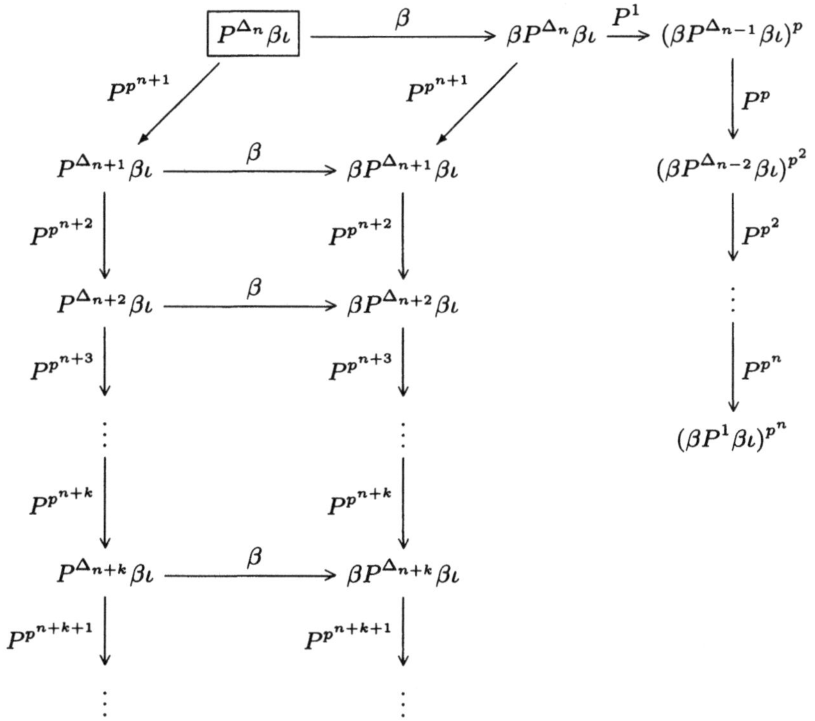

while for a primitive generator of type $(\beta P^{\Delta_n}\beta\iota)^{p^s}$ we will have

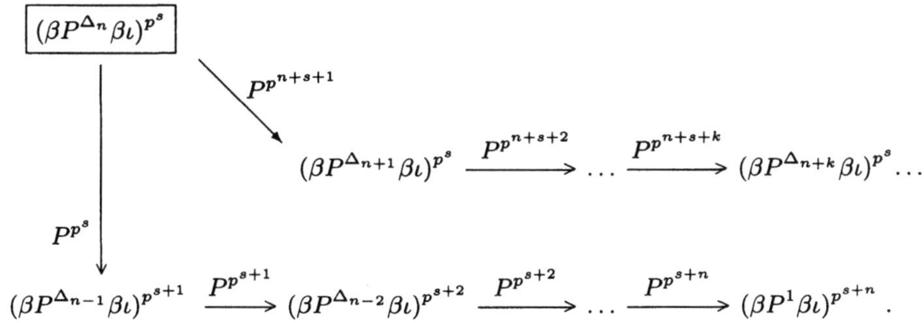

References

[1] J. F. Adams, C. W. Wilkerson, *Finite H-Spaces and algebras over the Steenrod algebra*, Ann. of Math. **111** (1980), 95–143.

[2] J. Aguadé, L. Smith, *On the mod p Torus Theorem of John Hubbuck*, Math. Z. **191** (1986), 325–326.

[3] J. Aguadé, L. Smith, *Modular cohomology algebras*, Amer. J. Math. **107** (1985), 507–530.

[4] A.K. Bousfield, D. Kan, *Homotopy Limits, Completions and Localizations*, Lect. Not. Math. **304**, Springer-Verlag 1972.

[5] C. Broto, J. A. Crespo, *H-spaces with noetherian mod two cohomology algebra*, Topology **38** **(2)**(1999), 353–386.

[6] C. Broto, S. Zarati, *Nil-localization of unstable algebras over the Steenrod algebra*, Math. Z. **199** (1988), 525–537.

[7] C. Broto, S. Zarati, *On sub-$\mathcal{A}_p^*$-algebras of H^*V*, in *Algebraic Topology, Homotopy and Group Cohomology, Proceedings, Barcelona 1990*, Lect. Not. Math. **1509**, Springer Verlag (1992), 35–49.

[8] W. Browder, *Torsion in H-spaces*, Ann. of Math. **74** (1961),24–51.

[9] E. Dror Farjoun, *Cellular Spaces, Null spaces and homotopy Localization*, Lect. Not. Math. **1622**, Springer Verlag (1996).

[10] R. M. Kane, *The Homology of Hopf Spaces*, North-Holland Mathematical Library. **40**, (1988).

[11] T. Kudo, *A transgression theorem* , Mem. Fac. Sci. Kyûsyû Univ. Ser. A **9** (1956), 79–81.

[12] J. Lannes *Sur les espaces functionneles dont la source est le classifiant d'un p-groupe abelien elementaire*, Publ. Math. IHES **75** (1992), 135–244.

[13] J. Lannes, S. Zarati, *Sur les $\mathcal{U}$-injectifs*, Ann. Scient. Ec. Norm. Sup. **19** (1986), 1–31.

[14] J. P. Lin, *H-spaces with Finiteness Conditions*, Chapter 22 in *Handbook of Algebraic Topology*, I. M. James (editor), Elsevier Science B.V. (1995).

[15] J. McCleary, *User's Guide to Spectral Sequences*, Math. Lecture Series **12**, Publish or Perish, Inc. (1985).

[16] J. Neisendorfer, *Localizations and connected covers of finite complexes*, Contemp. Math. **181** (1995).

[17] L. Schwartz, *Unstable Modules over the Steenrod Algebra and Sullivan's Fixed Point Set Conjecture*, Chicago Lectures in mathematics, The University of Chicago Press (1994).

Centre de Recerca Matemàtica
Institut d'Estudis Catalans
E-08193 Bellaterra, Spain
E-mail address: chiqui@mat.uab.es

Progress in Mathematics, Vol. 196

Composition methods in the homotopy groups of ring spectra

Brayton Gray

1.

Progress in calculating the homotopy groups of spheres has seen two major breakthroughs. The first was Toda's work, culminating in his book [11] in which the EHP sequences of James and Whitehead were used inductively; "composition methods" were used to construct elements and evaluate homomorphisms. The second was the Adams spectral sequence. Each method has advantages and disadvantages. Toda's method has the advantage that unstable homotopy groups are calculated along with stable groups. It has the disadvantage that it applies only to spheres and in particular, naturality under maps between spaces cannot be applied. The Adams method has the advantage that much of the bookkeeping work is accomplished in advance during the calculation of the Ext groups. The disadvantage that it does not calculate unstable groups has been eliminated, for certain nice spaces, by the work of [3]. This work implies that for certain spaces, the unstable homotopy is as accessible as the stable homotopy. It is the purpose of this work to examine how, in certain cases, the methods of Toda can be used for spaces other than spheres. We will begin by summarizing the methods used by Toda. We will discuss the possibility of using these methods for other spectra and work out the example of the Moore space spectrum $S^0 \cup_{p^r} e^1$ for $p > 3$.

2.

The main tool Toda used was the EHP sequence. Localized at 2, this is a long exact sequence for each $n \geq 1$

$$\cdots \longrightarrow \pi_{r+2}(S^{2n+1}) \xrightarrow{P}$$
$$\pi_r(S^n) \xrightarrow{E} \pi_{r+1}(S^{n+1}) \xrightarrow{H} \pi_{r+1}(S^{2n+1}) \xrightarrow{P} \pi_{r-1}(S^n) \longrightarrow \cdots$$

By induction first on the stem $\sigma = r - n$ and then on n, the determination of $\pi_{r+1}(S^{n+1})$ above is done when the other 4 groups are known. Compositions are used, and formulas for P and H on compositions allow one to calculate the groups involved. As an example, we cite the following formulas:

Proposition 2.1.

a) $H(\alpha \circ E\beta) = H(\alpha) \circ E\beta$.

b) $H(E\alpha \circ \beta) = E(\alpha \wedge \alpha) \circ H(\beta)$.

c) $P(\alpha \circ E^2\beta) = P(\alpha) \circ \beta$.

Proposition 2.2 (Barratt-Hilton formula). *If $\alpha \in \pi_r(S^n)$ and $\beta \in \pi_s(S^m)$ then $E^m\alpha \circ E^r\beta = (-1)^{(r-n)(s-m)} E^n\beta \circ E^s\alpha$. (In other words, on S^{n+m}, $\alpha\beta = (-1)^{(r-n)(s-m)}\beta\alpha$.)*

We will demonstrate how to calculate the 4 stem with these methods. We first need knowledge of previous stems.

Proposition 2.3.

	group	*generators*	*relations*
$\pi_{n+1}(S^n) =$	$\mathbf{Z},\ n=2$	η	
	$\mathbf{Z}/2,\ n>2$	η	2η
$\pi_{n+2}(S^n) =$	$\mathbf{Z}/2$	η^2	
$\pi_{n+3}(S^n) =$	$\mathbf{Z}/2,\ n=2$	η^3	
	$\mathbf{Z}/4,\ n=3$	ω	$2\omega = \eta^3$
	$\mathbf{Z}\oplus\mathbf{Z}/4,\ n=4$	ν, ω	
	$\mathbf{Z}/8,\ n\geq 5$	ν	$2\nu = \omega$.

Now the Hopf map $\eta\colon S^3 \longrightarrow S^2$ induces an isomorphism in homotopy:

$$\pi_r(S^3) \xrightarrow{\cong} \pi_r(S^2)$$

when $r > 2$, so $\pi_6(S^2) \simeq \mathbf{Z}/4$ generated by the composition $\eta\omega$. The map $\pi_5(S^5) \xrightarrow{P} \pi_3(S^2)$ generates the kernel of E so $P(\iota) = 2\eta$. We wish to evaluate

$$\pi_8(S^5) \xrightarrow{P} \pi_6(S^2)$$

$\pi_8(S^5)$ is generated by ν which is not a double suspension. However $2\nu = \omega$ is a double suspension so we apply 2.1 c)

$$P(2\nu) = P(\omega) = P(\iota) \circ \omega = 2\eta \circ \omega = \eta \circ 2\iota \circ \omega = \eta \circ \omega \circ 2\iota = 2(\eta\omega)$$

since $2\iota \circ \omega = \omega \circ 2\iota$ stably by 2.2 and hence on S^3 since $\pi_6(S^3) \longrightarrow \pi_{6+n}(S^{3+n})$ is a monomorphism. But $P(2\nu) = 2(\eta\omega)$ implies that $P(\nu) = \pm\eta\omega$ since P is a homomorphism. It follows that $E(\eta\omega) = 0$ and $0 \longrightarrow \pi_7(S^3) \longrightarrow \pi_7(S^5)$ is exact. We can define two compositions in $\pi_7(S^3)$: $\omega\eta$ and $\eta\nu$. We apply 2.1 (a) in the first case to get $H(\omega\eta) = H(\omega) \circ \eta = \eta^2$ and 2.1 (b) in the second case to get $H(\eta \circ \nu) = \eta^2 \circ H(\nu) = \eta^2$. Thus $\omega\eta = \eta\nu$ generates $\pi_7(S^3)$ and has order 2. Since ν has Hopf invariant 1, the EHP sequence splits when $n = 3$ and $\pi_{n+1}(S^4) \simeq \pi_n(S^3) \oplus \pi_{n+1}(S^7)$; in particular $\pi_8(S^4) \simeq \mathbf{Z}/2 \oplus \mathbf{Z}/2$ generated by $\omega\eta = \eta\nu$ and $\nu\eta$. By 2.2, on $S^5, \omega\eta = \eta\omega = 0$, so in the sequence

$$\pi_{10}(S^9) \xrightarrow{P} \pi_8(S^4) \xrightarrow{E} \pi_9(S^5) \xrightarrow{H} \pi_9(S^9)$$

$P(\eta) = \omega\eta$. Since $\pi_9(S^5)$ is finite, $H = 0$ and $\pi_9(S^5) = \mathbf{Z}/2$ generated by $\nu\eta$. By 2.2, on S^6 $\nu\eta = \eta\nu = \omega\eta = 0$, so $E(\pi_9(S^5)) = 0$. Furthermore, $\pi_{10}(S^{11}) = 0$ so $\pi_{10}(S^6) = 0$ and $\pi_{n+4}(S^n) = 0$ for $n \geq 6$. The interested reader will find that the calculation of the 5 stem is just as easy. There is one other useful result which we did not need here.

Proposition 2.4 (Barratt-Toda formula). *If $\alpha \in \pi_r(S^n)$ and $\beta \in \pi_s(S^m)$, the difference*

$$[\alpha, \beta] = E^{m-1}\alpha \circ E^{r-1}\beta - (-1)^{(r-n)(s-m)}E^{n-1}\beta \circ E^{s-1}\alpha\colon S^{r+s-1} \longrightarrow S^{m+n-1}$$

is equal to $P(EH(\alpha) \wedge H(\beta))$.

Using this one can see, for example, that

$$\nu\eta = \eta\nu + \nu\eta = P(\iota) \text{ on } S^5$$

$$\omega\eta = \omega\eta + \eta\omega = P(\eta) \text{ on } S^4 (\text{since } H(\omega) = \eta).$$

This method would continue very successfully if the homotopy groups of spheres were finitely generated. They clearly are not and one soon runs out of compositions. As a partial remedy to this Toda defined "secondary compositions", or what are commonly called Toda brackets and proved formulas similar to 2.1 for these operations.

When localized at $p > 2$, there are two types of EHP sequences. Define S_n by the formula

$$S_n = \begin{cases} S^{2k+1} & \text{if } n = 2k+1 \\ \widehat{S}^{2k} = J_{p-1}(S^{2k}) & \text{if } n = 2k \end{cases}$$

where J_m is the mth stage of the James construction.

Then the EHP sequences are defined by fibrations:

$$S_{2n} \xrightarrow{E} \Omega S_{2n+1} \xrightarrow{H_p} \Omega S_{2np+1}$$

$$S_{2n-1} \xrightarrow{E} \Omega S_{2n} \xrightarrow{H} \Omega S_{2np-1}$$

where H_p is the pth James Hopf invariant and H is the Toda-Hopf invariant ([7], [11]). Note that in case $p = 2$, these both degenerate into a single fibration

$$S_n \longrightarrow \Omega S_{n+1} \xrightarrow{H_2} \Omega S_{2n+1}$$

and $S_n = S^n$. Composition methods have been applied for $p > 2$ [6] with similar success. Proposition 2.1 generalizes directly while 2.2 requires some consideration. We define a modified suspension σ as follows: if $f\colon S^r \longrightarrow S_n$ let us write $\sigma(f)$ for the adjoint of $E \circ f\colon S^r \longrightarrow \Omega S_{n+1}$. We will write $SX = X \wedge S^1$ for the classical suspension of X and if $f\colon X \longrightarrow Y$, we write $Sf\colon SX \longrightarrow SY$ for the suspension of f.

Corresponding to Proposition 2.2 (at $p = 2$) we have

Theorem 2.5 (p local Barratt-Hilton Theorem). *Suppose that $\alpha \in \pi_r(S_n)$ and $\beta \in \pi_s(S_m)$. Then*

$$\sigma^m\alpha \circ \sigma^r\beta = (-1)^{(r-n)(s-m)}\sigma^n\beta \circ \sigma^s\alpha$$

in $\pi_*(S_{n+m})$.

Proof. We consider 3 cases:

a) if $n = 2k+1$ and $m = 2\ell+1$, the formula holds on $S^{2k+2\ell+2}$; since the suspension $E\colon \pi_{r+s}(S^{2k+2\ell+1}) \longrightarrow \pi_{r+s+1}(S^{2k+2\ell+2})$ is a monomorphism, this formula holds on $S^{2k+2\ell+1} = S_{m+n-1}$ and hence on S_{m+n}

b) if $n = 2k$ and $m = 2\ell + 1$, apply case a) to $\sigma\alpha$ and β to demonstrate the formula on $S^{2k+2\ell+1} = S_{m+n}$

c) if $n = 2k$ and $m = 2\ell$, new arguments are needed.

Let $\iota\colon S^{2n} \longrightarrow \widehat{S}^{2n}$ be the inclusion. To discuss case c) we rely on

Lemma 2.6. *Suppose $g\colon S^r \longrightarrow \widehat{S}^{2n}$. Then the diagram:*

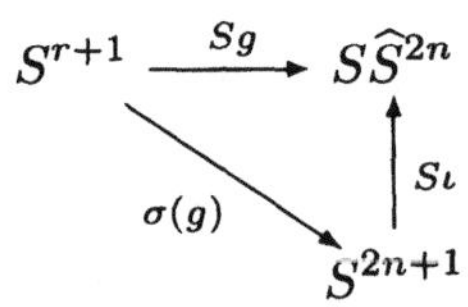

commutes up to homotopy.

Proof. Let $K_n = S^{2n-1} \cup_{w_n} e^{2np-2}$ be the $2np-2$ skeleton of $\Omega^2 S^{2n+1}$. According to [7], [10], there is a map $\gamma\colon SK_n \longrightarrow \widehat{S}^{2n}$ such that $\Omega\gamma$ has a right homotopy inverse. In particular, there is a lifting of g to a map $g'\colon S^r \longrightarrow SK_n$

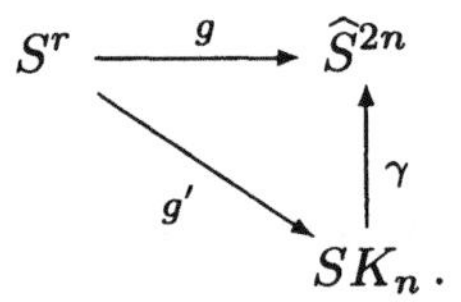

Write $\sigma(\gamma)\colon S^2K_n \longrightarrow S^{2n+1}$ for the adjoint of $E \circ \gamma\colon SK_n \longrightarrow \Omega S^{2n+1}$. We now show that

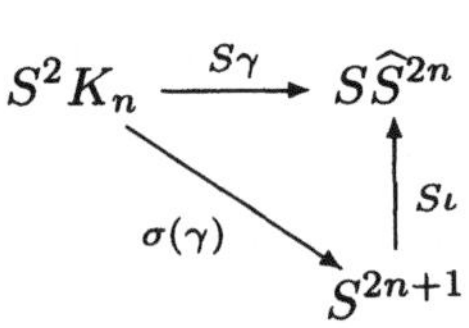

commutes up to homotopy. The Lemma then follows by combining these two diagrams. □

Now $S\widehat{S}^{2n} \simeq S^{2n+1} \vee S^{4n+1} \vee \cdots \vee S^{2(p-1)n+1}$; let $H_i\colon \widehat{S}^{2n} \longrightarrow S^{2n}_\infty \longrightarrow S^{2ni}_\infty$ be the restriction for the ith James-Hopf invariant for $1 \leq i \leq p-1$. Combining, we get a map:

$$\theta\colon \widehat{S}^{2n} \longrightarrow \prod_{i=1}^{p-1} S^{2ni}_\infty \simeq \Omega(S^{2n+1} \times \cdots \times S^{2n(p-1)+1}) \longrightarrow \Omega(S\widehat{S}^{2n}),$$

where the last map is the splitting map from the loops on a product to the loops on a wedge. Since ΩH_i has order prime to p for $1 < i \leq p-1$,

$$H_i \circ \gamma\colon SK_n \longrightarrow \widehat{S}^{2n} \longrightarrow S^{2ni}_\infty$$

is null homotopic if $1 < i \leq p-1$. Therefore $\theta \circ \gamma$ factors through ΩS^{2n+1}. Since the adjoint of θ is an equivalence, $S\gamma$ factors through S^{2n+1}. This factorization can be identified as $\sigma(\gamma)$ by applying the inclusion $\widehat{S}^{2n} \longrightarrow \Omega S^{2n+1}$ and taking adjoints.

To finish the proof of 2.5, choose a map $\mu\colon \widehat{S}^{2k} \wedge \widehat{S}^{2\ell} \longrightarrow \widehat{S}^{2k+2\ell}$ which is degree 1 in dimension $2k+2\ell$. Then consider the diagram:

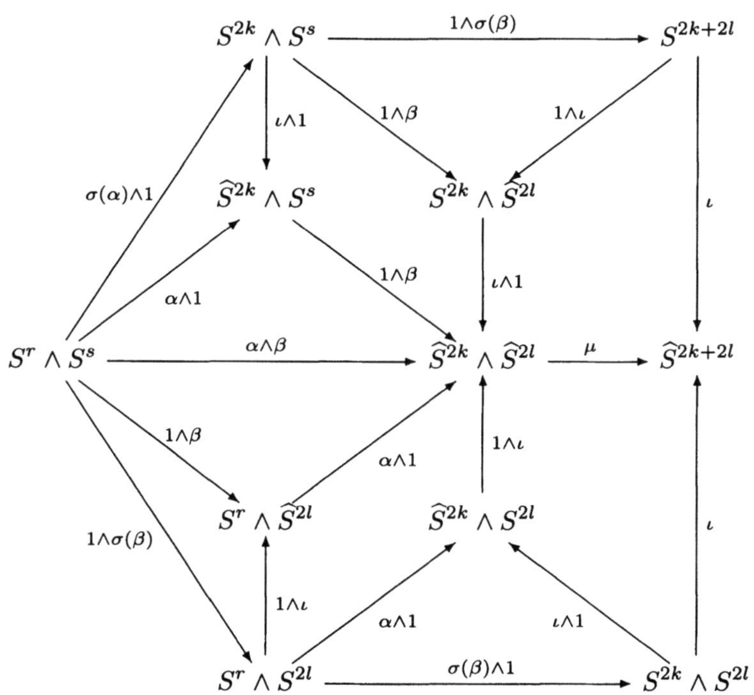

where the lemma is applied in 4 of the 6 triangular regions. If we identify $\sigma^t(f)$ with $1 \wedge f$, we introduce signs $(-1)^{(s-1)(r-2k)}$ and $(-1)^{(r-2k)(2l-1)}$ which combine to give $(-1)^{rs}$. □

3.

The utility of the EHP method applies when what you are studying is not a single space, but a sequence of spaces $\{X_n\}$ together with a suspension homomorphism $\pi_r(X_n) \xrightarrow{\sigma} \pi_{r+1}(X_{n+1})$. If σ is induced by a map $X_n \xrightarrow{e_n} \Omega X_{n+1}$, what we are in fact dealing with is a spectrum $X = \{X_n, e_n\}$. The utility of an EHP approach depends on the extent to which one has control over the fiber of e_n:

$$F_n \longrightarrow X_n \xrightarrow{e_n} \Omega X_{n+1}$$

From this point of view, it is clear that one can form two stably equivalent spectra $\{X_n, x_n\} \simeq \{Y_n, y_n\}$ with the spaces X_n and Y_n vastly different. We think of the sequence of spaces $\{X_n\}$ as an unstable development of the spectrum X. We seek a favorable unstable development of X. Ideally, one might seek an unstable development in which $F_n = \Omega^{k(n)} X_{\ell(n)}$ for some functions $k(n), \ell(n)$. This is explored in [9] and it appears formally that the spectra $V(m)$ of Smith and Toda are good candidates for this. In particular, $V(-1) = S^0$ and $V(0) = S^0 \cup_p e^1$ for $p > 2$. Considerable attention to this spectrum and the related spectra $S^0 \cup_{p^r} e^1$ will be deferred to section 4.

To apply "composition methods," unstable classes need to be composed. This suggests a ring structure in $\pi_*^S(X)$ which is associative (composition is associative) and commutative (if we wish to have a Barratt-Hilton formula). Thus one might begin by considering a homotopy associative homotopy commutative ring spectrum X. For Toda brackets we may need to consider higher homotopies of associativity and commutativity. However, in order to have a successful unstable composition theory it is first necessary to develop a stable composition theory. To do this we will find certain self maps of a ring spectrum X which correspond to elements in the stable homotopy of X in such a way that the product in $\pi_* X$ corresponds to composition.

Suppose now that X is a ring spectrum. Using the multiplication $X \wedge X \xrightarrow{\mu} X$ we can think of X as an "X-module spectrum".

Definition 3.1. *Call a map $\phi\colon S^r X \longrightarrow X$ right modular if there is a commutative diagram:*

$$\begin{array}{ccc} S^r X \wedge X & \xrightarrow{\phi \wedge 1} & X \wedge X \\ {\scriptstyle S^r\mu}\downarrow & & \downarrow{\scriptstyle \mu} \\ S^r X & \xrightarrow{\phi} & X . \end{array}$$

Note that the composition of modular maps is modular. Write $\mathrm{Mod}_r(X)$ for the group of homotopy classes of right modular maps $\rho\colon S^r X \longrightarrow X$. Then $\{\mathrm{Mod}_r(X)\}$ is a graded ring with unit under composition.

Theorem 3.2. *If X is homotopy associative, $\pi_*^S(X) \cong Mod_*(X)$ as graded rings.*

Proof. Define $F\colon \pi_r^S(X) \longrightarrow \mathrm{Mod}_r(X)$ by $F(\alpha) = \widehat{\alpha}$ where $\hat{\alpha}$ is the composition:

$$S^r X \xrightarrow{\alpha\wedge 1} X \wedge X \xrightarrow{\mu} X$$

to check modularity, consider the diagram:

$$\begin{array}{ccccc} S^r X \wedge X & \xrightarrow{\alpha\wedge 1\wedge 1} & X \wedge X \wedge X & \xrightarrow{\mu\wedge 1} & X \wedge X \\ \downarrow{\scriptstyle S^r\mu} & & \downarrow{\scriptstyle 1\wedge\mu} & & \downarrow{\scriptstyle \mu} \\ S^r X & \xrightarrow{\alpha\wedge 1} & X \wedge X & \xrightarrow{\mu} & X. \end{array}$$

Now define $G\colon\ \mathrm{Mod}_r(X) \longrightarrow \pi_r^S(X)$ by $G(\phi) = \phi \circ S^r\iota$

$$S^r = S^r \wedge S^0 \xrightarrow{S^r\iota} S^r \wedge X \xrightarrow{\phi} X.$$

Clearly $G \circ F = Id$. To see that $F \circ G = Id$, consider the diagram:

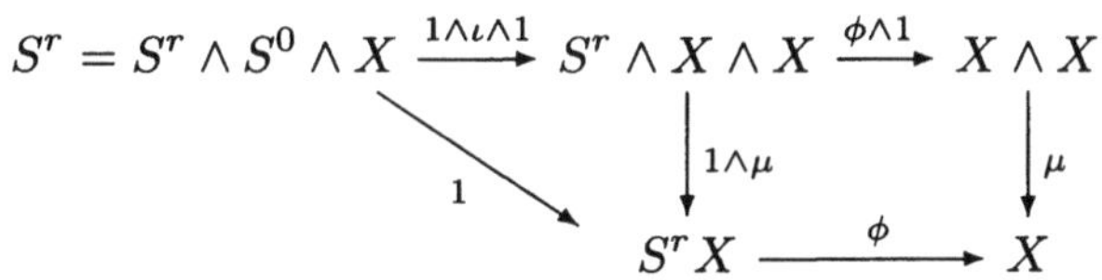

where $(F \circ G)(\phi)$ is the upper right hand composite. Furthermore, F is a ring homomorphism. To see this consider the following diagram where $\alpha \in \pi_r^S(X)$ and $\beta \in \pi_s^S(X)$. The top right hand composition represents $F(\alpha \circ \beta)$ while the lower left hand composition represents $F(\alpha) \circ F(\beta)$

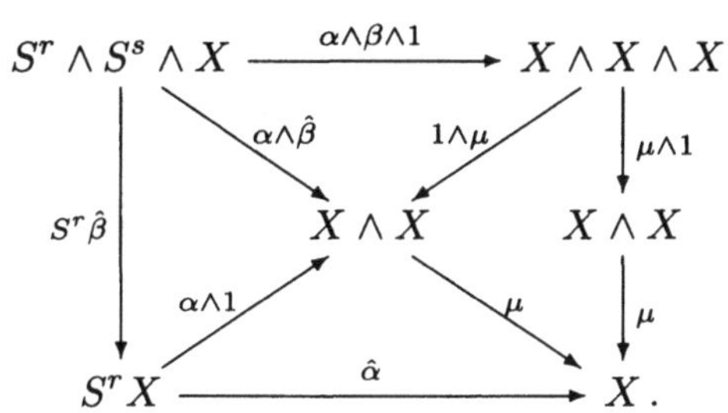

This completes the proof. □

Now suppose in addition that X is homotopy commutative. Then by 3.2, composition is graded commutative, i.e.;

$$\widehat{\alpha} \circ S^r\widehat{\beta} \sim (-1)^{rs}\widehat{\beta} \circ S^s\widehat{\alpha}.$$

4.

In this section we will develop the special case of the Moore space spectrum $S^0 \cup_{p^r} e^1$, $p > 2$. By the results of [9], this spectrum can be represented as $\{T_m, \sigma_n\}$, where $T_{2n} = S^{2n+1}\{p^r\}$, the fiber of the degree p^r map on S^{2n+1} and T_{2n-1} is a space constructed by Anick [1] and developed further in [2] and [12]. T_{2n-1} is the total space in a fibration

$$S^{2n-1} \longrightarrow T_{2n-1} \longrightarrow \Omega S^{2n+1}$$

where the connecting map $\Omega^2 S^{2n+1} \longrightarrow S^{2n-1}$ has degree p^r. We will write $T_n(p^r)$ when we wish to keep the exponent in mind. There are EHP fibrations:*

$$T_{2n-1} \xrightarrow{E} \Omega T_{2n} \xrightarrow{H} BW_n$$

$$T_{2n} \xrightarrow{E} \Omega T_{2n+1} \xrightarrow{H} BW_{n+1}\,.$$

The spaces T_m are a homotopically simple unstable representation of the Moore space spectrum and we seek an unstable version of modularity. To do this we need to find a functorial extension of a map $S^m \xrightarrow{\alpha} T_n$ to a "modular map" $T_m \xrightarrow{\widehat{\alpha}} T_n$.

This is accomplished in 2 steps:

a) extend α to a map $P_m = S^m \cup_{p^r} e^{m+1} \xrightarrow{\alpha'} T_n$ so that it is "modular" in the appropriate sense,
b) extend α' to $\widehat{\alpha}$ so that it is an H map.

The idea behind point a) is to use the fact that $p^r\pi_*(T_n) = 0$ to define an extension. This has some technical complexity. Part b) is easily obtained from

Theorem 4.1 ([9, 2, 12]). *a) T_n is a Homotopy commutative and homotopy associative H space and $p^r \circ \pi_*(T_n) = 0$.*
b) Let $\alpha'\colon P_m \longrightarrow X$ where X is a homotopy commutative and homotopy associative H space and $p^r\pi_r(X) = 0$. Then there is a unique H map $\widehat{\alpha}\colon T_m \longrightarrow X$ extending α'.

Proof. The case n even was worked out in [9] and the existence but not the uniqueness when n is odd appear in [2]. The uniqueness and H property in case n is odd appears in [12]. □

Proposition 4.2. *Given an H map $\phi\colon T_m \longrightarrow T_n$, there is a unique H map $\sigma\phi\colon T_{m+1} \longrightarrow T_{n+1}$ such that the diagram:*

$$\begin{array}{ccc} T_m & \xrightarrow{\phi} & T_n \\ {\scriptstyle E}\downarrow & & \downarrow{\scriptstyle E} \\ \Omega T_{m+1} & \xrightarrow{\Omega\sigma\phi} & \Omega T_{n+1} \end{array}$$

is homotopy commutative.

*Note that it is an open conjecture [8] that $BW_n = \Omega T_{2np-1}(p)$ making these sequences formally similar to those in section 2.

Proof. There is a unique H extension of the composite:

$$SP_m \longrightarrow ST_m \xrightarrow{\Sigma\phi} ST_n \longrightarrow T_{n+1}$$

to an H map $\sigma(\phi)\colon T_{m+1} \longrightarrow T_{n+1}$. It then follows that the diagram in question commutes when restricted to P_r. Since all maps are H maps, it commutes to homotopy by 4.1 b). □

Definition 4.3. *Suppose X, Y, Z are H spaces, and $f\colon X \wedge Y \longrightarrow Z$. We will call f an H map in two variables (or an H map for short) if the adjoints of f:*

$$X \longrightarrow Z^Y$$

$$Y \longrightarrow Z^X$$

are H maps with the induced H space structure on the function space.

Proposition 4.4. *There is a unique H map in two variables*

$$\mu\colon T_r \wedge T_s \longrightarrow T_{r+s}$$

extending the composition $P_r \wedge P_s \xrightarrow{\pi} P_{r+s} \xrightarrow{\iota} T_{r+s}$ where $\pi_*(1 \otimes 1) = 1$ in homology.

Proof. The adjoint map $P_r \longrightarrow T_{r+s}^{P_s}$ has a unique extension to an H map by 4.1. This gives a map $T_r \longrightarrow T_{r+s}^{P_s}$ and hence $T_r \wedge P_s \longrightarrow T_{r+s}$. The extension to a map $T_r \wedge T_s \longrightarrow T_{r+s}$ is similar.

Clearly we have commutative diagrams:

$$\begin{array}{ccc} T_r \wedge T_s & & \\ \Big\downarrow{\scriptstyle \tau} & \searrow^{\mu} & \\ & (-1)^{rs} & T_{r+s} \\ & \nearrow_{\mu} & \\ T_s \wedge T_r & & \end{array}$$

and a similar associativity diagram if $p > 3$ or $r > 1$. □

Given spaces A, B, we will often encounter the map $A \wedge \Omega B \xrightarrow{e} \Omega(A \wedge B)$ which is adjoint to the evaluation map. e is an H map in the second variable. In particular we can then extend

$$\begin{array}{ccccc} P_r \wedge \Omega T_s & \xrightarrow{e} & \Omega(P_r \wedge T_s) & \xrightarrow{\Omega\mu} & \Omega T_{r+s} \\ \Big\downarrow & & & \nearrow_{e'} & \\ T_r \wedge \Omega T_s & & & & \end{array}$$

so that e' is an H map in two variables, and

$$\begin{array}{ccc} T_r \wedge \Omega T_s & \xrightarrow{e'} & \Omega T_{r+s} \\ {\scriptstyle 1\wedge E}\uparrow & & \uparrow{\scriptstyle E} \\ T_r \wedge T_{s-1} & \xrightarrow{\mu} & T_{r+s-1} \end{array}$$

commutes up to homotopy, and similarly in the other variable.

In order to define modularity, we require the following.

Lemma 4.5. *There is a map* $\epsilon\colon P_1 \wedge \Omega T_n \longrightarrow T_n$ *so that the diagram:*

$$\begin{array}{ccc} P_1 \wedge \Omega T_n & \xleftarrow{1\wedge\sigma} & P_1 \wedge T_{n-1} \\ \uparrow & \searrow^{\epsilon} & \downarrow{\scriptstyle \mu} \\ S^1 \wedge \Omega T_n & \xrightarrow[ev]{} & T_n \end{array}$$

commutes up to homotopy.

Proof. Since the identity map of T_n has order p, there is a lifting $T_n \longrightarrow T_n\{p\}$ where $T_n\{p\}$ is the fiber of the pth power map:

$$\cdots \longrightarrow T_n\{p\} \longrightarrow T_n \xrightarrow{p} T_n\,.$$

However $\Omega T_n\{p\}$ is homotopy equivalent to the fiber of the map $\Omega T_n \longrightarrow \Omega T_n$ induced by the degree p map on S^1; i.e., $\Omega T_n\{p\} \simeq T_n^{P_1}$ as H spaces. It follows that there is an H map $\epsilon'\colon \Omega T_n \longrightarrow T_n^{P_1}$ with the composite

$$\Omega T_n \xrightarrow{\epsilon'} T_n^{P_1} \longrightarrow T_n^{S^1} \simeq \Omega T_n$$

homotopic to the identity. Furthermore, the restriction

$$T_{n-1} \xrightarrow{E} \Omega T_n \xrightarrow{\epsilon'} T_n^{P_1}$$

is an H map and hence is the adjoint to $\mu\colon P_1 \wedge T_{n-1} \longrightarrow T_n$. Taking adjoints yields the lemma. □

Our next step is to define, for each homotopy class $\alpha \in \pi_r(T_n)$ an H map $T_r \xrightarrow{\widehat{\alpha}} T_n$ extending α. It suffices to construct a map $P_r \longrightarrow T_n$ and this is accomplished as the composition:

$$P_r = P_1 \wedge S^{r-1} \xrightarrow{1\wedge d^*} P_1 \wedge \Omega T_n \xrightarrow{\epsilon} T_n$$

$\widehat{\alpha}$ is then the unique H extension. Consequently we have defined a homomorphism

$$F_\epsilon\colon \pi_r(T_n) \longrightarrow [T_r, T_n]_H$$

by $F_\epsilon(\alpha) = \widehat{\alpha}$.

Clearly we have a left inverse

$$G\colon [T_r, T_n]_H \longrightarrow \pi_r(T_n)$$

by restriction. We wish to determine the image of F_ϵ as the maps which are in some sense "modular." We will call a map ϕ k-modular if the diagram:

$$\begin{array}{ccc} T_k \wedge T_r & \xrightarrow{1\wedge\phi} & T_k \wedge T_n \\ {\scriptstyle\mu}\big\downarrow & & \big\downarrow{\scriptstyle\mu} \\ T_{k+r} & \xrightarrow{\sigma^k(\phi)} & T_{k+n} \end{array}$$

commutes. Clearly ϕ is k-modular iff the composition $d_k(\phi)$:

$$P_{k+r+1} \xrightarrow{\nabla} P_k \wedge P_r \longrightarrow P_k \wedge T_r \xrightarrow{1\wedge\phi} P_k \wedge T_n \xrightarrow{\mu} T_{k+n}$$

is null homotopic. Now $\sigma\big(d_k(\phi)\big) = d_{k+1}(\phi)$, so if ϕ is k-modular, it is $(k+1)$-modular.

Strong Conjecture 4.6. *There is a choice of ϵ so that the diagram:*

$$\begin{array}{ccccc} P_1 \wedge P_1 \wedge \Omega T_n & & \xrightarrow{\quad 1\wedge\epsilon \quad} & & P_1 \wedge T_n \\ {\scriptstyle\mu\wedge 1}\big\downarrow & & & & \big\downarrow{\scriptstyle\mu} \\ P_2 \wedge \Omega T_n & \xrightarrow{ev} & P_1 \wedge T_n & \xrightarrow{\mu} & T_{n+1} \end{array}$$

commutes up to homotopy.

Proposition 4.7. *The strong conjecture implies that $\widehat{\alpha}$ is k-modular for each $k \geq 1$.*

Proof. We prove that $\widehat{\alpha}$ is 1-modular by factoring $d_1(\widehat{\alpha})$

$$\begin{array}{ccccccc} P_{r+2} & \xrightarrow{\nabla} & P_1 \wedge P_r & \xrightarrow{1\wedge\hat{\alpha}} & P_1 \wedge T_n & \xrightarrow{\mu} & T_{n+1} \\ \big\updownarrow & & \big\updownarrow & & \big\uparrow & & \big\uparrow \\ P_3 \wedge S^{r-1} & \longrightarrow & P_1 \wedge P_1 \wedge S^{r-1} & \xrightarrow{1\wedge\alpha^*} & P_1 \wedge P_1 \wedge \Omega T_n & & \big| \\ & \overset{*}{\searrow} & \big\downarrow{\scriptstyle\mu\wedge 1} & & \big\downarrow{\scriptstyle\mu\wedge 1} & & \big| \\ & & P_2 \wedge S^{r-1} & \xrightarrow{1\wedge\alpha^*} & P_2 \wedge \Omega T_n & \longrightarrow & P_1 \wedge T_n\,. \end{array}$$ □

By including $S^1 \wedge P_1 \wedge \Omega T_n$ into $P_1 \wedge P_1 \wedge \Omega T_n$, we see that the strong conjecture implies the

Weak Conjecture 4.8. *There is a choice of ϵ so that the diagram:*

$$\begin{array}{ccc} S^1 \wedge P_1 \wedge \Omega T_n & \xrightarrow{1\wedge\epsilon} & S^1 \wedge T_n \\ {\scriptstyle ev}\big\downarrow & & \big\downarrow{\scriptstyle\sigma} \\ P_1 \wedge T_n & \xrightarrow{\mu} & T_{n+1} \end{array}$$

commutes up to homotopy.

Proposition 4.9. *The weak conjecture implies that $\widehat{\alpha}$ is modular for each $k \geq 2$.*

Proof. We factor $d_2(\widehat{\alpha})$:

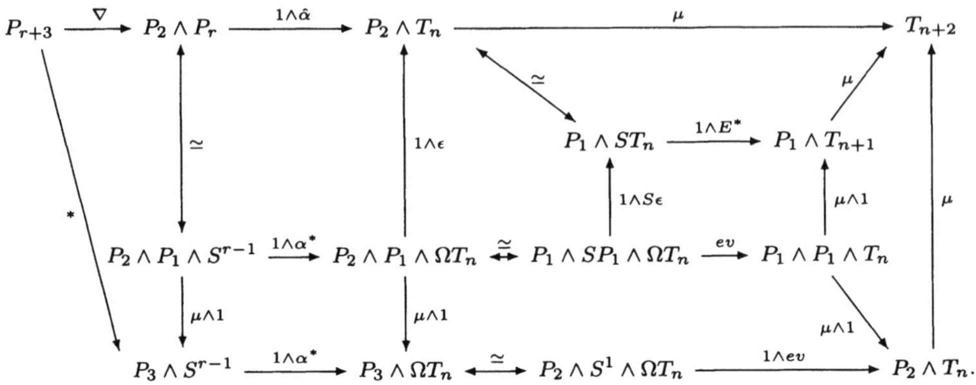

□

We call ϕ 0-modular if the composition $d_0(\phi)$:

$$P_{r+1} \xrightarrow{\nabla} P_1 \wedge P_{r-1} \xrightarrow{1\wedge\phi^*} P_1 \wedge \Omega T_n \xrightarrow{\epsilon} T_n$$

is null homotopic. It is easy to see that the weak conjecture implies that if ϕ is 0-modular, it is 1-modular and that if ϕ is 0-modular and $G(\phi) = *$ then $\phi \sim *$. If ϵ can be chosen so that $\widehat{\alpha}$ is 0-modular, there is then an isomorphism between the 0-modular H maps $T_r \longrightarrow T_n$ and $\pi_r(T_n)$. An even more complicated conjecture about ϵ will imply that $\widehat{\alpha}$ is always 0-modular. We will not pursue this line.

Proposition 4.10. *The weak conjecture holds when n is even.*

Proof. In this case we have an H-fibration sequence:

$$T_{2n} \longrightarrow \Omega T_{2n+1} \longrightarrow BW_n;$$

we will prove that such an ϵ exists by showing that the composition

$$P_1 \wedge \Omega T_{2n} \xrightarrow{e} \Omega(P_1 \wedge T_{2n}) \xrightarrow{\Omega\mu} \Omega T_{2n+1} \longrightarrow BW_{n+1}$$

is null homotopic. We begin by considering the diagram:

$$\begin{array}{ccc} P_1 \wedge T_{2n} & \xrightarrow{\qquad\mu\qquad} & T_{2n+1} \\ \big\downarrow & & \big\downarrow \\ P_1 \wedge \Omega^2 T_{2n+2} \simeq P_1 \wedge (S^{2n+3})^{P_2} & \xrightarrow{ev} & \Omega S^{2n+3} \end{array}$$

since both compositions are H maps of the second variable, they are homotopic iff they are homotopic when restricted to $P_1 \wedge P_{2n}$. This is clear. We now form a diagram which contains the loops on this diagram:

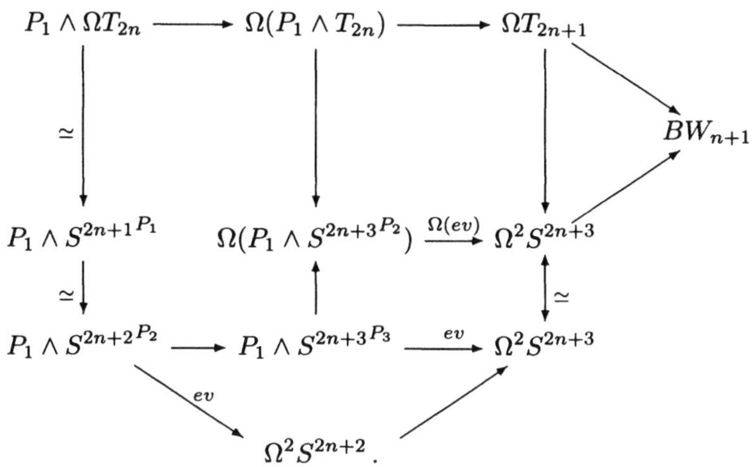

However the composition $\Omega S^{2n+2} \longrightarrow \Omega^2 S^{2n+3} \longrightarrow BW_{n+1}$ is null homotopic since $\Omega S^{2n+2} \longrightarrow \Omega^2 S^{2n+3} \longrightarrow BW_{n+1} \times S^{4n+3}$ is a fibration sequence [8]. This completes the proof. □

Theorem 4.11 (Barratt-Hilton Theorem). *Suppose that* $\alpha\colon T_r \longrightarrow T_n$ *and* $\beta\colon T_s \longrightarrow T_m$ *are* k *modular where* $n, m \geq k$. *Then*

$$\sigma^m(\alpha) \circ \sigma^r(\beta) = (-1)^{(n-r)(m-s)} \sigma^n(\beta)\sigma^s(\alpha)\colon T_{r+s} \longrightarrow T_{m+n}.$$

Proof. Exactly as in the case of spheres, we have $(\alpha \wedge 1) \circ (1 \wedge \beta) = \alpha \wedge \beta = (1 \wedge \beta) \circ (\alpha \wedge 1)$.

Now we also have:

$$\begin{array}{ccc} T_k \wedge T_s & \xrightarrow{1\wedge\beta} & T_k \wedge T_m \\ \downarrow{\scriptstyle\mu} & & \downarrow{\scriptstyle\mu} \\ T_{k+s} & \xrightarrow{\sigma^n(\beta)} & T_{k+m} \end{array}$$

for each k, while:

$$\begin{array}{ccc} T_r \wedge T_k & \xrightarrow{\alpha\wedge 1} & T_n \wedge T_k \\ \downarrow & & \downarrow \\ T_{r+k} & \xrightarrow{(-1)^{k(n-r)}\sigma(\alpha)} & T_{n+k}\,. \end{array}$$

It follows that the indicated equation holds when preceded by $\mu\colon T_r \wedge T_s \longrightarrow T_{r+s}$. However, the inclusion of P_{r+s} into T_{r+s} factors through μ, so the equation holds when restricted to P_{r+s}. Since both composites are H maps, they are homotopic by the universal property. □

Appendix A.

So far we have discussed composition behavior which mimics the behavior of the sphere spectrum localized at 2. Here we will discuss compositions in the sphere spectrum localized at a prime $p > 2$. Recall that we have:

$$S_\ell = \begin{cases} S^{2k+1}, & \text{if } \ell = 2k+1 \\ \widehat{S}^{2k} = J_{p-1}(S^{2k}), & \text{if } \ell = 2k. \end{cases}$$

The fact that half of the spaces in this spectrum are not spheres presents a difficulty in forming compositions. The question we deal with first is this: is there a sensible way of forming a "composite" of maps $f\colon S^r \longrightarrow \widehat{S}^{2n}$ and $g\colon S^{2n} \longrightarrow S_\ell$ to obtain a map $g * f\colon S^r \longrightarrow S_\ell$? If we allow ourselves to suspend once this can be done:

$$S^{r+1} \xrightarrow{\sigma f} S^{2n+1} \xrightarrow{\sigma g} S_{\ell+1}\,.$$

It is remarkable that the case $p = 3$ is easier than that of larger primes.

Proposition A.1. *Suppose $p = 3$ and X is a space such that ΩX is homotopy commutative in the loop space structure. Then:*

a) *ΩS_ℓ is homotopy commutative in the loop space structure.*
b) *Each map $g\colon S^{2n} \longrightarrow X$ extends to a map*

$$\widehat{g}\colon \widehat{S}^{2n} \longrightarrow X\,.$$

c) *If $\widehat{g}, \tilde{g}\colon \widehat{S}^{2n} \longrightarrow X$ are any two extensions of g, $\Omega\widehat{g} \sim \Omega\tilde{g}$.*
d) *The diagram:*

$$\begin{array}{ccc} \Omega\widehat{S}^{2k} & \xrightarrow{\Omega\widehat{g}} & \Omega X \\ \downarrow & & \downarrow \\ \Omega^2 S^{2k+1} & \xrightarrow{\Omega^2\Sigma g} & \Omega^2\Sigma X \end{array}$$

commutes up to homotopy.

Corollary A.2. *Suppose $p = 3$. If $f\colon S^r \longrightarrow \widehat{S}^{2k}$ and $g\colon S^{2k} \longrightarrow S_\ell$ we can define a composition*

$$g * f\colon S^r \longrightarrow S_\ell$$

by $g * f = \widehat{g} \circ f$. This is well defined and

$$\sigma(g * f) = \sigma(g) \circ \sigma(f)\colon S^{r+1} \longrightarrow S_{\ell+1}.$$

Proof. a) is well known if ℓ is odd and is proven in [7] in case ℓ is even. b) is due to Hua Feng [5]. It follows since when $p = 3$, $\widehat{S}^{2k} = S^{2k} \cup e^{4k}$ where the attaching map is the Whitehead product. The obstruction to defining $\widehat{g}$ is thus the composition $S^{4k-1} \longrightarrow S^{2k} \longrightarrow X$. The adjoint of this is the composition $S^{4k-2} \longrightarrow \Omega S^{2k} \longrightarrow \Omega X$ where the first map is a commutator. To prove c), note that two extensions $\widehat{g}$ and $\tilde{g}$ agree on S^{2n}. Let $\delta\colon S^{4k} \longrightarrow X$ be the difference element so that $\widehat{g}$ is homotopic to the composition:

$$\widehat{S}^{2n} \xrightarrow{\Delta} \widehat{S}^{2n} \vee S^{4n} \xrightarrow{\tilde{g}\vee\delta} X\,.$$

Consider the diagram of fibrations:

$$\begin{array}{ccccc}
\Omega SK * \Omega S^{4n} & \xrightarrow{\Omega\gamma * 1} & \Omega\widehat{S}^{2n} * \Omega S^{4n} & & \\
\downarrow & & \downarrow & & \\
SK_n \vee S^{4n} & \xrightarrow{\gamma\vee 1} & \widehat{S}^{2n} \vee S^{4n} & \xrightarrow{\tilde{g}\vee\delta} & X \\
\downarrow & & \downarrow & & \\
SK_n \times S^{4n} & \xrightarrow{\gamma\times 1} & \widehat{S}^{2n} \times S^{4n} & &
\end{array}$$

where $K_n = S^{2n-1} \cup e^{2np-2}$ is the $2np-2$ skeleton of $\Omega\widehat{S}^{2n}$ and $\gamma\colon SK_n \longrightarrow \widehat{S}^{2n}$ is the adjoint of the inclusion. Since ΩX is homotopy commutative, $(\tilde{g} \vee \delta)(\gamma \vee 1)$ extends over $SK_n \times S^{4n}$, as this space is the mapping cone of a Whitehead product. It follows that the composite along the top is null-homotopic. But since $\Omega\gamma$ has a right homotopy inverse, the composite:

$$\Omega\widehat{S}^{2n} * \Omega S^{4n} \longrightarrow \widehat{S}^{2n} \vee S^{4n} \longrightarrow X$$

is null-homotopic. Since the right hand sequence is a fibration, we get an extension:

$$\begin{array}{ccc}
\Omega(\widehat{S}^{2n} \vee S^{4n}) & \xrightarrow{\Omega(\tilde{g}\vee\delta)} & \Omega X \\
\downarrow & \nearrow_{\Gamma} & \\
\Omega(\widehat{S}^{2n} \times S^{4n}) & &
\end{array}$$

and Γ must be $\Omega\tilde{g} \times \Omega\delta$. Fitting these diagrams together, we get

$$\begin{array}{ccccc}
 & & \Omega(\widehat{S}^{2n} \vee S^{4n}) & & \\
 & \nearrow^{\Omega\Delta} & & \searrow^{\Omega(\tilde{g}\vee\delta)} & \\
\Omega\widehat{S}^{2n} & & \downarrow & & \Omega X \\
 & \searrow_{1\times\Omega\pi} & & \nearrow_{\Omega(\tilde{g})+\Omega\delta} & \\
 & & \Omega\widehat{S}^{2n} \times \Omega S^{4n}\,. & &
\end{array}$$

Hence $\Omega\widehat{g} = \Omega\tilde{g} + \Omega(\delta)\circ\Omega\pi$. But by [8, Proposition 7] $\Omega\pi \sim *$ so $\Omega\widehat{g} \sim \Omega\tilde{g}$. Finally, to prove d) observe both composites:

$$\widehat{S}^{2n} \xrightarrow{\widehat{g}} X \xrightarrow{\iota} \Omega\Sigma X$$

$$\widehat{S}^{2n} \longrightarrow \Omega S^{2n+1} \xrightarrow{\Omega\Sigma g} \Omega\Sigma X$$

extend $S^{2n} \longrightarrow X \longrightarrow \Omega\Sigma X$, so we may apply c) replacing X by $\Omega\Sigma X$. □

Theorem A.3. *For any p and any map* $g\colon S^{2m} \longrightarrow S_\ell$. *There is an extension* $\widehat{g}\colon \widehat{S}^{2m} \longrightarrow S_\ell$ *such that the diagram:*

$$\begin{array}{ccc} \widehat{S}^{2m} & \xrightarrow{\widehat{g}} & S_\ell \\ \downarrow{\scriptstyle E} & & \downarrow{\scriptstyle E} \\ \Omega S^{2m+1} & \xrightarrow{\Omega(\sigma g)^*} & \Omega S_{\ell+1} \end{array}$$

commutes up to homotopy.

Note A.4. *We make no statement about the uniqueness of* $\Omega\widehat{g}$ *in the general case.*

Proof. In case ℓ is odd, S_ℓ is an H space and g can then be extended to a map $g_\infty\colon S^{2m}_\infty \longrightarrow S_\ell$. The diagram clearly commutes in this case. Suppose now that $\ell = 2n$. We first lift g to a map $g'\colon S^{2m} \longrightarrow SK_n$ so that $g' \sim g$. Then consider the diagram:

$$\begin{array}{ccccc} S^{2m} & \xrightarrow{g'} & SK_n & \xrightarrow{\gamma} & \widehat{S}^{2n} \\ \downarrow & & \downarrow & & \downarrow \\ \widehat{S}^{2m} = J_{p-1}(S^{2m}) & \xrightarrow{J_{p-1}(g')} & J_{p-1}(SK_n) & \xrightarrow{\phi} & \widehat{S}^{2n} \\ \downarrow & & \downarrow & & \downarrow \\ J_\infty(S^{2m}) & \longrightarrow & J_\infty(SK_n) & \longrightarrow & S^{2n}_\infty . \end{array}$$

The proof of A3 is complete when we construct ϕ so that the two diagrams on the right homotopy commute. First we must prove

Proposition A.5. *Suppose* $f\colon SX \longrightarrow SY$ *and the composite:*

$$X \xrightarrow{f^*} \Omega SY \xrightarrow{H_k} \Omega SY^k$$

is null homotopic for each $k > 1$. *Then there is a commutative diagram:*

$$\begin{array}{ccc} \Omega SX & \xrightarrow{\Omega f} & \Omega SY \\ \downarrow{\scriptstyle H_k} & & \downarrow{\scriptstyle H_k} \\ \Omega SX^{(k)} & & \Omega SY^{(k)} \\ \downarrow & & \downarrow \\ \Omega^k S^k X^{(k)} & \xrightarrow{\Omega(f\wedge\cdots\wedge f)} & \Omega^k S^k Y^{(k)} . \end{array}$$

Proof. Of course, if $f = Sf'$, this is clear by naturality of the Hopf invariant. We will prove this using the results of Boardman-Steer [4]. They describe Hopf invariants $\lambda_n\colon [SA, SB] \longrightarrow [S^nA, S^nB^{(n)}]$ for each $n \geq 1$. Let $ev\colon S\Omega SX \longrightarrow SX$ be the evaluation map. Then [4, 3.15] $\lambda_n(ev)\colon S^n(\Omega SX) \longrightarrow S^nX^{(n)}$ is the adjoint of the composite $\Omega SX \xrightarrow{H_n} \Omega SX^{(n)} \longrightarrow \Omega^n S^n X^{(n)}$. Consequently, the diagram in question is equivalent to the diagram:

$$\begin{array}{ccc} S^k\Omega SX & \xrightarrow{S^k\Omega f} & S^k\Omega SY \\ \Big\downarrow{\scriptstyle\lambda_k(ev)} & & \Big\downarrow{\scriptstyle\lambda_k(ev)} \\ S^kX^{(k)} & \xrightarrow{f\wedge\cdots\wedge f} & S^kY^{(k)}. \end{array}$$

To establish this, we apply the composition formula [4, 3.16] to the composites in the square:

$$\begin{array}{ccc} S\Omega SX & \xrightarrow{ev} & SX \\ \Big\downarrow{\scriptstyle S\Omega f} & & \Big\downarrow{\scriptstyle f} \\ S\Omega SY & \xrightarrow{ev} & SY. \end{array}$$

Since $\lambda_q(f) \sim *$ for each $q > 1$, $\lambda_k(f \circ ev) = (f \wedge \cdots \wedge f) \circ \lambda_k(ev)$. However $\lambda_q(S\Omega f) \sim *$ for each $q > 1$, so $\lambda_k(ev \circ S\Omega f) = \lambda_k(ev) \circ S^k\Omega f$. This establishes the result. □

We now apply this to the map $\sigma\gamma\colon S^2K_n \longrightarrow S^{2n+1}$:

$$\begin{array}{ccc} \Omega S^2K_m & \xrightarrow{\Omega(\sigma g)} & \Omega S^{2m+1} \\ \Big\downarrow{\scriptstyle\lambda_p} & & \Big\downarrow{\scriptstyle\lambda_p} \\ \Omega^p S^{2p}K_m^{(p)} & \longrightarrow & \Omega^p S^{(2m+1)p}. \end{array}$$

$\Omega S^2K_m \simeq (SK_m)_\infty$ and $J_{p-1}(SK_m)$ maps trivially under H_p and hence under λ_p. Thus the composition:

$$J_{p-1}(SK_m) \longrightarrow (SK_m)_\infty \longrightarrow S^{2m}_\infty \xrightarrow{H_p} S^{2mp}_\infty \longrightarrow \Omega^p S^{(2m+1)p}$$

is null-homotopic. Since $\dim\big(J_{p-1}(SK)\big) = (2mp-1)(p-1) < 2(mp+1)p-3$, the composite of the first three maps:

$$J_{p-1}(SK_m) \longrightarrow (SK_m)_\infty \longrightarrow S^{2m}_\infty \xrightarrow{H_p} S^{2mp}_\infty$$

is null-homotopic. We have proven the first part of

Corollary A.6. *There is a map $\phi\colon J_{p-1}(SK_m) \longrightarrow \widehat{S}^{2m}$ such that the diagram:*

$$\begin{array}{ccc} J_{p-1}(SK_m) & \xrightarrow{\phi} & \widehat{S}^{2m} \\ \Big\downarrow & & \Big\downarrow \\ (SK_m)_\infty & \xrightarrow{(\sigma\gamma)_\infty} & S^{2m}_\infty \end{array}$$

homotopy commutes.

Furthermore, $\phi\big|_{SK_m} \sim \gamma$.

Proof. For the second part, we observe that the composites:

$$SK_m \longrightarrow J_{p-1}(SK_m) \xrightarrow{\phi} \widehat{S}^{2m} \xrightarrow{E} S^{2m}_\infty$$
$$SK_m \xrightarrow{\gamma} \widehat{S}^{2m} \xrightarrow{E} S^{2m}_\infty$$

are homotopic, so the difference $\phi|_{SK_m} - \gamma$ factors through the fiber of E. We will alter ϕ so that this difference vanishes. To do this, note that the choice of ϕ can be modified by any element of $[J_{p-1}(SK_m), \Omega^2 S^{2mp+1}]$. But the restriction:

$$[J_{p-1}(SK_m), \Omega^2 S^{2mp+1}] \longrightarrow [SK_m, \Omega^2 S^{2mp+1}]$$

is onto since $SJ_{p-1}(SK_m)$ splits. Therefore an appropriate choice of ϕ yields $\phi|_{SK_m} \sim \gamma$. □

□

References

[1] D. Anick, *Differential algebras in topology,* Research notes in Mathematics, A.K. Peters, Ltd. (1993).

[2] D. Anick and B. Gray, *Small H spaces related to Moore Spaces,* Topology **34** (1995), 859–881.

[3] A.K. Bousfield and D.M. Kan, *The homotopy spectral sequence of a space with coefficients in a ring,* Topology **11** (1977), 79–106.

[4] M. Boardman and B. Steer, *On Hopf Invariants,* Math. Helv. **42** (1967), 180–221.

[5] H. Feng, *On Reduced join and Barratt-Hilton formula, on delooping some spaces, a note on closed model categories,* Thesis, University of Illinois at Chicago (1993).

[6] B. Gray, *The odd components of the unstable homotopy groups of spheres,* Mimeographed notes (1967).

[7] B. Gray, *On Toda's fibrations,* Math. Proc. Cambridge Philos. Soc. **97** (1985), 289–298.

[8] B. Gray, *On the iterated suspension,* Topology **27** (1988), 301–310.

[9] B. Gray, *EHP spectra and periodicity, I: Geometric Constructions,* Trans. Amer. Math. Soc. **340** (1993), 565–586.

[10] J.C. Moore, *Course Lectures at Princeton University,* (Fall 1976).

[11] H. Toda, *Composition Methods in homotopy groups of spheres,* Princeton University Press, (1962).

[12] S. Theriault, *A Reconstruction of Anick's fibration,* Thesis, University of Toronto, (1997).

Department of Mathematics, Statistics, and Computer Science (M/C 249), University of Illinois at Chicago, 851 South Morgan Street, Chicago, IL 60607-7045
E-mail address: `brayton@uic.edu`

Progress in Mathematics, Vol. 196

Tate cohomology in axiomatic stable homotopy theory

J. P. C. Greenlees

CONTENTS

1.	Introduction	149
2.	Axiomatic Tate cohomology in a stable homotopy category	151
3.	Finite localizations	153
4.	The category of G-spectra	154
5.	The derived category of a commutative ring	156
6.	The category of modules over a highly structured ring	160
7.	The derived category of kG	160
8.	Chromatic categories	161
9.	Splittings of the Tate construction	162
10.	Calculation by comparison	163
11.	Calculations by associative algebra	166
12.	Calculations by commutative algebra	170
13.	Gorenstein localizations	173
References		174

1. Introduction

The purpose of the present note is to show how the axiomatic approach to Tate cohomology of [18, Appendix B] can be implemented in the axiomatic stable homotopy theory of Hovey–Palmieri–Strickland [32]. Much of the work consists of collecting known results in a single language and a single framework. The very effortlessness of the process is an effective advertisement for the language, and a call for further investigation of other instances. The main point is to recognize and compare incarnations of the same phenomenon in different contexts: the splitting and duality phenomena described in Sections 9 and 13 are particularly notable. More practically, Theorem 11.1 is new, and Theorem 12.1 extends results of [19].

A *stable homotopy category* [32, 1.1.4] is a triangulated category $\mathcal{C}$ with arbitrary coproducts, and so that all cohomology theories are representable. It is also required to have a compatible symmetric monoidal structure with unit S and a set $\mathcal{G}$ of strongly dualizable objects generating all of $\mathcal{C}$ using triangles, coproducts and retracts. If in addition the objects of $\mathcal{G}$ are small, the stable homotopy category is said to be *algebraic*.

We shall illustrate our constructions in several contexts specified in greater detail later. The following list gives the context followed by an associated stable homotopy category. Each of these admits a number of variations.

- Equivariant topology: the homotopy category of G-spectra (Section 4).
- Commutative algebra: the derived category of a commutative ring R (Section 5).
- Brave new commutative algebra: the homotopy category of modules over a highly structured commutative ring spectrum $\mathbf{R}$ (Section 6).
- Representation theory: the derived category of the group ring kG of a finite group G (Section 7).
- The bordism approach to stable homotopy theory: various chromatic categories (Section 8).

The paper is in three parts.

Part I: General formalities (Sections 2 and 3). In Section 2 we summarize necessary definitions and give the Tate construction in a stable homotopy category associated to a smashing localization. We establish the fundamental formal properties that make it reasonable to call this a Tate construction. In Section 3 we recall from [39] that finite localizations are smashing and hence give rise to Tate theories: the minor novelty is to emphasize the view that this is an Adams projective resolution in the sense of [1].

Part II: Examples (Sections 4 to 8). We describe the above contexts in more detail, and consider the construction in each one, identifying it in more familiar terms.

Part III: Special properties (Sections 9 to 13). The final sections give some more subtle results about the construction which require additional hypotheses. In Section 9, we discuss dichotomy results stating that the Tate construction is either periodic or split. We then turn to methods of calculation. The first is the familiar calculation using associative algebra, generalizing the use of group cohomology in descent spectral sequences (one uses homological algebra over the endomorphism ring of the basic building block). We describe this in Section 11, and give a new example in the case of equivariant topology with a compact Lie group of equivariance. This method applies fairly generally, provided the stable homotopy category arises from an underlying Quillen model category. Less familiar is the calculation in terms of commutative algebra. This arises when the (commutative) endomorphism ring of the unit object has a certain duality property (it is 'homotopically Gorenstein'). This is quite exceptional, but it applies in a surprisingly large number of familiar examples: in the cohomology of groups [19, 9, 8], in equivariant cohomology theories [17, 25], and in chromatic stable homotopy theory (Gross–Hopkins duality). Its occurrence in commutative algebra is investigated in [21], and shown to be very special.

2. Axiomatic Tate cohomology in a stable homotopy category

In this section we describe the Tate construction. Since it depends on a suitable Bousfield localization, we briefly recall the terminology in a suitable form (see [32, Section 3] for more detail). We consider a functor $L\colon \mathcal{C} \longrightarrow \mathcal{C}$ on the stable homotopy category $\mathcal{C}$. The *acyclics* of L are the objects X so that $LX \simeq *$. The functor L is a Bousfield localization if it is exact, equipped with a natural tranformation $X \longrightarrow LX$, idempotent, and its class of acyclics is an ideal.

A Bousfield localization L is determined by its class $\mathcal{D}$ of acyclics as follows: Y is *L-local* if and only if $[D,Y]_* = 0$ for all D in $\mathcal{D}$, and a map $X \longrightarrow Y$ is the Bousfield localization if and only if Y is local and the fibre lies in $\mathcal{D}$. The usual notation for the localization triangle is $CX \longrightarrow X \longrightarrow LX$. Furthermore, any such class of acyclics is a *localizing ideal* (i.e., it is closed under completing triangles, sums and smashing with an arbitrary object). A localization is said to be *smashing* if the natural map $X \wedge LS \longrightarrow LX$ is an equivalence for all X. It is equivalent to require either that L commutes with arbitrary sums, or that the class $L\mathcal{C}$ of L-local objects is a localizing ideal [32, 3.3.2].

We shall define a Tate construction associated to any smashing localization.

Notation 2.1. (*General context*)

- $\mathcal{C}$: a stable homotopy category
- $\mathcal{G}$: a set of generators for $\mathcal{C}$
- $\mathcal{D}$: the localizing ideal of acyclics for a smashing Bousfield localization $(\cdot)[\mathcal{D}^{-1}]$
- $\mathcal{C}[\mathcal{D}^{-1}]$: the localizing ideal of $[\mathcal{D}^{-1}]$-local objects.

The notation $L_{\mathcal{D}}$ is often used for $(\cdot)[\mathcal{D}^{-1}]$; the present notation better reflects the character of a smashing localization, and corresponds to that in [24]. The idea is that we should think of $X[\mathcal{D}^{-1}]$ as a localization *away from* $\mathcal{D}$. More precisely the archetype is localization away from a closed subset in algebraic geometry. The notation comes from the case when the closed subset is defined by the vanishing of a single function f. In this very special case, the localization is realized by inverting the multiplicatively closed set $\{1, f, f^2, \dots\}$ in the sense of commutative algebra. We therefore use the corresponding 'sections with support' notation for the fibre of this localization:

$$\Gamma_{\mathcal{D}}(X) \longrightarrow X \longrightarrow X[\mathcal{D}^{-1}].$$

We can use this to define an associated completion.

Lemma 2.2. *The natural transformation $X \longrightarrow F(\Gamma_{\mathcal{D}}(S), X)$ is Bousfield completion whose class of acyclics is the class of $[\mathcal{D}^{-1}]$-local objects.*

Proof. First we must show that if E is $[\mathcal{D}^{-1}]$-local, then $[E, F(\Gamma_{\mathcal{D}}(S), X)]_* = 0$. By [32, 3.1.8], $S[\mathcal{D}^{-1}]$ is a ring object in $\mathcal{C}$ and $E = E[\mathcal{D}^{-1}]$ is a $S[\mathcal{D}^{-1}]$-module. Hence $E \wedge \Gamma_{\mathcal{D}}(S)$ is a retract of $E \wedge S[\mathcal{D}^{-1}] \wedge \Gamma_{\mathcal{D}}(S)$; since $[\mathcal{D}^{-1}]$ is idempotent and smashing, $S[\mathcal{D}^{-1}] \wedge \Gamma_{\mathcal{D}}(S) \simeq *$.

Secondly we must show that the fibre $F(S[\mathcal{D}^{-1}], X)$ is $[\mathcal{D}^{-1}]$-local. However if D lies in $\mathcal{D}$ then $D \wedge S[\mathcal{D}^{-1}] \simeq D[\mathcal{D}^{-1}] \simeq *$. □

We write $X^\wedge_{\mathcal{D}} = F(\Gamma_{\mathcal{D}}(S), X)$ for this Bousfield completion, and also introduce the following notation for its fibre:

$$\Delta_{\mathcal{D}}(X) \longrightarrow X \longrightarrow X^\wedge_{\mathcal{D}}.$$

We now define the $\mathcal{D}$-Tate construction by

$$t_{\mathcal{D}}(X) = X^\wedge_{\mathcal{D}}[\mathcal{D}^{-1}].$$

This gives the diagram

$$\begin{array}{ccccc} \Gamma_{\mathcal{D}}(X) & \longrightarrow & X & \longrightarrow & X[\mathcal{D}^{-1}] \\ \downarrow & & \downarrow & & \downarrow \\ \Gamma_{\mathcal{D}}(X^\wedge_{\mathcal{D}}) & \longrightarrow & X^\wedge_{\mathcal{D}} & \longrightarrow & X^\wedge_{\mathcal{D}}[\mathcal{D}^{-1}] = t_{\mathcal{D}}(X). \end{array}$$

Lemma 2.3. *The map $\Gamma_{\mathcal{D}}(X) \longrightarrow \Gamma_{\mathcal{D}}(X^\wedge_{\mathcal{D}})$ is an equivalence.*

Proof. We need only remark that $\Gamma_{\mathcal{D}}(\Delta_{\mathcal{D}}(X)) \simeq *$; however by definition $\Delta_{\mathcal{D}}(X)$ lies in the class of $[\mathcal{D}^{-1}]$-local objects. □

Corollary 2.4. (*Hasse Principle*) *The diagram*

$$\begin{array}{ccc} X & \longrightarrow & X[\mathcal{D}^{-1}] \\ \downarrow & & \downarrow \\ X^\wedge_{\mathcal{D}} & \longrightarrow & t_{\mathcal{D}}(X) \end{array}$$

is a homotopy pullback square. □

Corollary 2.5. (*Warwick Duality* [18]) *There is an equivalence*

$$t_{\mathcal{D}}(X) = X^\wedge_{\mathcal{D}}[\mathcal{D}^{-1}] \simeq \Delta_{\mathcal{D}}(\Sigma\Gamma_{\mathcal{D}}(X)).$$

Proof. This is a composite of three equivalences,

$$X^\wedge_{\mathcal{D}}[\mathcal{D}^{-1}] \longleftarrow \Delta_{\mathcal{D}}(X^\wedge_{\mathcal{D}}[\mathcal{D}^{-1}]) \longrightarrow \Delta_{\mathcal{D}}(\Sigma\Gamma_{\mathcal{D}}(X^\wedge_{\mathcal{D}})) \longrightarrow \Delta_{\mathcal{D}}(\Sigma\Gamma_{\mathcal{D}}(X)).$$

The first is an equivalence since $(\cdot)[\mathcal{D}^{-1}]^\wedge_{\mathcal{D}} \simeq *$ (the class $\mathcal{E}$ of acyclics for $(\cdot)^\wedge_{A}$ consists of $[\mathcal{D}^{-1}]$-local objects) so that $X^\wedge_{\mathcal{D}}[\mathcal{D}^{-1}]^\wedge_{\mathcal{D}} \simeq *$. The second is an equivalence since $\Delta_{\mathcal{D}}(X^\wedge_{\mathcal{D}}) \simeq *$ (defining property of $\Delta_{\mathcal{D}}((\cdot))$ together with idempotence of $(\cdot)^\wedge_{\mathcal{D}}$). The third is an equivalence since $\Gamma_{\mathcal{D}}(\Delta_{\mathcal{D}}((\cdot))) \simeq *$ by 2.3 so that $\Delta_{\mathcal{D}}(\Sigma\Gamma_{\mathcal{D}}(\Delta_{\mathcal{D}}(X))) \simeq *$. □

This shows that the cohomology as well as the homology only depends on the localization away from $\mathcal{D}$. More precisely, the definition

$$t_{\mathcal{D}}(X) = F(\Gamma_{\mathcal{D}}(S), X)[\mathcal{D}^{-1}]$$

shows that $T \wedge t_{\mathcal{D}}(X)$ only depends on the localization $T[\mathcal{D}^{-1}]$. The second avatar $t_{\mathcal{D}}(X) \simeq F(S[\mathcal{D}^{-1}], \Sigma\Gamma_{\mathcal{D}}(X))$ gives

$$[T, t_{\mathcal{D}}(X)]_* = [T \wedge S[\mathcal{D}^{-1}], \Sigma\Gamma_{\mathcal{D}}(X)]_* = [T[\mathcal{D}^{-1}], \Sigma\Gamma_{\mathcal{D}}(X)]_*,$$

which again only depends on $T[\mathcal{D}^{-1}]$.

Remark 2.6. The definition of the Tate construction we have given is at a natural level of generality. One might be tempted to consider $L_{\mathcal{D}}L_{\mathcal{E}}X$ for arbitrary $\mathcal{D}$ and $\mathcal{E}$. However, if one wants Warwick duality, one requires (i) $L_{\mathcal{E}}L_{\mathcal{D}}X \simeq *$, so that $\mathcal{E} \supseteq L_{\mathcal{D}}\mathcal{C}$ and (ii) $C_{\mathcal{D}}C_{\mathcal{E}}X \simeq *$, so that $C_{\mathcal{E}}X$ is $L_{\mathcal{D}}$-local, and $\mathcal{E} \subseteq L_{\mathcal{D}}\mathcal{C}$. Thus we require $\mathcal{E} = L_{\mathcal{D}}\mathcal{C}$, and this must be a localizing ideal. Thus $L_{\mathcal{D}}$ must be smashing, and determines $\mathcal{E}$.

3. Finite localizations

In this section we describe one very fruitful source of smashing localizations. This is explicit in Section 3.3 of [32], and especially Theorem 3.3.5. It generalizes the finite localization of Mahowald–Sadofsky and Miller [36, 39]. We recall the construction for future reference, and emphasize the connection with Adams projective resolutions.

Recall that a full subcategory is *thick* if it is closed under completing triangles and taking retracts. The piece of data we need is a $\mathcal{G}$-ideal $\mathcal{A}$ of small objects (i.e., a thick subcategory of small objects, closed under smashing with elements of $\mathcal{G}$). If $\mathcal{C}$ is not algebraic, we must suppose in addition that $\mathcal{A}$ is essentially small, consists of strongly dualizable objects and is closed under Spanier–Whitehead duality; if $\mathcal{C}$ is algebraic these conditions are automatic. In practice we will specify $\mathcal{A}$ by giving a set $\mathcal{T}$ of small generators: $\mathcal{A} = \mathcal{G}\text{-ideal}(\mathcal{T})$. We then need to form the localizing ideal $\mathcal{D} = \text{locid}(\mathcal{A})$ generated by $\mathcal{A}$: this is the smallest thick subcategory containing $\mathcal{A}$ which is closed under arbitrary sums and smashing with arbitrary elements of $\mathcal{C}$.

Context 3.1. (For a finite localization)

- $\mathcal{C}$: a stable homotopy category
- $\mathcal{G}$: a set of generators for $\mathcal{C}$
- $\mathcal{T}$: a set of small objects of $\mathcal{C}$
- $\mathcal{A} = \mathcal{G}\text{-ideal}(\mathcal{T})$
- $\mathcal{D} = \text{locid}(\mathcal{T}) = \text{locid}(\mathcal{A})$.

In these circumstances, we write write $\mathcal{A}$ or $\mathcal{T}$ in place of $\mathcal{D}$ in the notation, so that $t_{\mathcal{A}}(X) = t_{\mathcal{T}}(X) = t_{\mathcal{D}}(X)$ and so forth.

Miller has shown that there is a smashing localization functor $(\cdot)[\mathcal{A}^{-1}]$ whose acyclics are precisely $\mathcal{D}$, and whose small acyclics are precisely $\mathcal{A}$; this is known as a *finite localization* and the notation $L^f_{\mathcal{A}}$ is used in [32]. The construction is described in 3.3 below. The associated functor $(\cdot)^{\wedge}_{\mathcal{A}}$ whose acyclics are the objects $X[\mathcal{A}^{-1}]$ is denoted by $L_{\mathcal{A}}$ in [32].

There is a convenient lemma for showing a set of elements in a localizing subcategory is a generating set. It would be more traditional to view it as a convergence theorem for a projective resolution in the sense of Adams [1].

Proposition 3.2. *If $\mathcal{T} \subseteq \mathcal{D}$ is a set of objects then $\mathcal{D} = \mathrm{locid}(\mathcal{T})$ provided one of the two following conditions holds.*

(i) *$\mathcal{T}$ is a set of small objects and detects triviality in $\mathcal{D}$, in the sense that if X is in $\mathcal{D}$ then $[T, X]_* = 0$ for all T in $\mathcal{T}$ implies $X \simeq *$.*

(ii) *The objects of $\mathcal{G}$ are small, and for any $X \in \mathcal{D}, S' \in \mathcal{G}$ and any $x \in [S', X]_*$ there is a map $t \colon \Sigma^n T \longrightarrow X$ with $x \in \mathrm{im}(t_* \colon [S', \Sigma^n T]_* \longrightarrow [S', X]_*)$ and T in $\mathcal{T}$.*

Proof. We need to prove that if X is an arbitrary object of $\mathcal{D}$, we may form X from copies of elements of $\mathcal{T}$ using sums and completion of triangles. We give the proof assuming that Condition (ii) holds; the proof when Condition (i) holds is similar except that $[S', \cdot]_*$ for $S' \in \mathcal{G}$ is replaced by $[T, \cdot]_*$ for $T \in \mathcal{T}$.

By hypothesis we may form a projective resolution in the sense of Adams:

$$\begin{array}{ccccccc}
X = & X_0 & \xrightarrow{i_0} & X_1 & \xrightarrow{i_1} & X_2 & \xrightarrow{i_2} \cdots \\
 & \uparrow{\scriptstyle t_0} & & \uparrow{\scriptstyle t_1} & & \uparrow{\scriptstyle t_2} & \\
 & T_0 & & T_1 & & T_2 &
\end{array}$$

Thus each T_i is a sum of suspensions of elements of $\mathcal{T}$, each t_i is surjective in $[S', \cdot]_*$ for all $S' \in \mathcal{G}$ and X_{i+1} is formed as the cofibre of $t_i \colon T_i \longrightarrow X_i$. Note that $X_\infty = \mathrm{Tcl}_n X_n$ has trivial $[S', \cdot]_*$ for all $S' \in \mathcal{G}$ since i_s is zero in $[S', \cdot]_*$ by construction; thus $X_\infty \simeq *$ since $\mathcal{G}$ gives a set of generators. Defining X^i as the fibre of $X \longrightarrow X_i$ we find that X^i is constructed from sums of suspensions of elements of $\mathcal{T}$ by a finite number of cofibre sequences. Passing to direct limits, we obtain a cofibre sequence $X^\infty \longrightarrow X \longrightarrow X_\infty$, so that $X^\infty \simeq X$. □

Remark 3.3. Note that the argument essentially gives the construction of a finite localization. Take a set $\mathcal{T}$ of small generators of the $\mathcal{G}$-ideal $\mathcal{A}$ and the localizing ideal $\mathcal{D}$, and ensure it is closed under duality. We now form a projective resolution as in the proof of 3.2, but without assuming that X lies in $\mathcal{D}$. Ensure t_i is surjective in $[T, \cdot]_*$ for each i. Then the triangle $X^\infty \longrightarrow X \longrightarrow X_\infty$ has X^∞ in $\mathcal{D}$ by construction, and $[T, X^\infty]_* = 0$ for all T in $\mathcal{T}$.

This completes the discussion of formalities. In the rest of the paper we want to discuss a number of examples from this point of view, and show how comparisons between the examples give rise to means of calculation.

4. The category of G-spectra

In this section we consider the category $\mathcal{C} = G$-**spectra** of G-spectra for a compact Lie group G, and localizations associated to a family $\mathcal{F}$ of subgroups. We recover the constructions of [23]; indeed these constructions motivated investigation of its other manifestations.

Thus we suppose $\mathcal{F}$ is closed under conjugation and passage to subgroups, and we let $\mathcal{T} = \{G/H_+ \mid H \in \mathcal{F}\}$. Thus $\mathcal{A}$ is the class of retracts of finite $\mathcal{F}$-spectra, and $\mathcal{D}$ is the class of all $\mathcal{F}$-spectra. We recall that in the homotopy category of G-spectra, the class of $\mathcal{F}$-spectra can be described in three ways, as is implicit in [34].

Lemma 4.1. *The following three classes of G-spectra are equal, and they are called $\mathcal{F}$-spectra.*

(i) *G-spectra formed from spheres $G/H_+ \wedge S^n$ with $H \in \mathcal{F}$,*
(ii) *G-spectra X so that the natural map $E\mathcal{F}_+ \wedge X \longrightarrow X$ is an equivalence, and*
(iii) *G-spectra X so that the geometric fixed point spectra $\Phi^H X$ are non-equivariantly contractible for $H \in \mathcal{F}$.*

Proof. The equality of Classes (i) and (ii) is straightforward.

Since Φ^H commutes with smash products [34, II.9.12], and it agrees with H-fixed point spaces on suspension spectra [34, II.9.9], it follows that $E\mathcal{F}_+ \wedge X$ lies in the third class, so Class (ii) is contained in Class (iii). Suppose then that X is in Class (iii); we must show it is also in Class (ii). By hypothesis, the map $E\mathcal{F}_+ \wedge X \longrightarrow X$ has the property that applying Φ^H gives a non-equivariant equivalence for all H. It remains to observe that geometric fixed points detect weak equivalences. This is well known, but I do not know a good reference: it follows from the fact that Lewis–May fixed points tautologically detect weak equivalences, by an induction on isotropy groups. The basis is the relation between geometric and Lewis–May fixed points [34, II.9.8]: for any H-spectrum X, $\Phi^H X \simeq (\tilde{E}\mathcal{P} \wedge X)^H$ where $\mathcal{P}$ is the family of proper subgroups of H. □

From the equality of Classes (i) and (ii) $E\mathcal{F}_+ \wedge X$ lies in $\mathcal{D}$, and from the fact that $\tilde{E}\mathcal{F}$ is $\mathcal{F}$-contractible we see that $X \longrightarrow \tilde{E}\mathcal{F} \wedge X$ is localization away from $\mathcal{D}$. Hence $\Gamma_{\mathcal{F}} S = E\mathcal{F}_+$ and $S[\mathcal{F}^{-1}] = \tilde{E}\mathcal{F}$. Now the equality of Classes (i) and (ii) can be recognized as the statement that localization away from the class of $\mathcal{F}$-spectra is smashing. It follows that in this case Diagram 2 is the diagram

$$\begin{array}{ccccc} E\mathcal{F}_+ \wedge X & \longrightarrow & X & \longrightarrow & \tilde{E}\mathcal{F} \wedge X \\ \simeq\downarrow & & \downarrow & & \downarrow \\ F(E\mathcal{F}_+, E\mathcal{F}_+ \wedge X) & \longrightarrow & F(E\mathcal{F}_+, X) & \longrightarrow & t_{\mathcal{F}}(X), \end{array}$$

which is Diagram C of [23].

The skeletal filtration gives rise to spectral sequences for calculating the homotopy groups of these spectra based on group cohomology [23], and we discuss this in more abstract terms in Section 11. More interesting is that for well behaved cohomology theories (such as those which are Noetherian, complex orientable and highly structured), one may prove a local cohomology theorem in which case the homotopy groups may be calculated by commutative algebra [18]. We discuss this in more abstract terms in Section 12. The formal framework for such spectral sequences is described in Section 10.

5. The derived category of a commutative ring

In this section we consider the category $\mathcal{C} = D(R)$ for a commutative ring R, and localizations associated to an ideal I of R. In particular, we obtain a new approach to the results of [18] and an improved perspective on the role of finiteness conditions.

We wish to consider the class of acyclics for a localization, and there are several candidates for this. The most natural is the class

$$\mathcal{D} = \{M \mid \mathrm{supp}(H_*(M)) \subseteq V(I)\},$$

but we should also consider

$$\mathcal{D}' = \{M \mid \text{every element of } H_*(M) \text{ is } I\text{-power torsion}\}.$$

It is straightforward to check they are both candidates.

Lemma 5.1. *The classes $\mathcal{D}$ and $\mathcal{D}'$ are localizing ideals.* □

It is also easy to see that $\mathcal{D}' \subseteq \mathcal{D}$.

Lemma 5.2. *If I is finitely generated then $\mathcal{D}' = \mathcal{D}$, but this is not true in general.*

Proof. Suppose M is a module with support in $V(I)$, and $x \in M$ has annihilator J. Since R/J has support $V(J)$, we see that $V(J) \subseteq V(I)$ so that $\sqrt{J} \supseteq \sqrt{I} \supseteq I$. If I is finitely generated, some power of I lies in J.

To give an example where equality fails we need only display an ideal J so that no power of $\sqrt{J}$ lies in J, since then we may take $I = \sqrt{J}$ and $M = R/J$. For instance if R is polynomial on a countably infinite number of generators, $x_1, x_2, x_3, \ldots$ over a field and $J = (x_1, x_2^2, x_3^3, \ldots)$ we find that $\sqrt{J}$ is the maximal ideal $(x_1, x_2, x_3, \ldots)$ no power of which lies in J. □

It is useful to have a specific generator for $\mathcal{D}$ as a localizing ideal. Perhaps the most natural candidate for a generator of $\mathcal{D}$ is R/I, but this can only generate $\mathcal{D}'$. For the rest of the section we assume that I is finitely generated, say $I = (x_1, \ldots, x_n)$, and thus $\mathcal{D} = \mathcal{D}'$. We show that R/I does give a generator, but there are other candidates which are usually convenient.

Warning 5.3. If R/I does not have a finite resolution by finitely generated projectives, it need not be small.

We may define the unstable Koszul complex for the sequence $x_1^d, x_2^d, \ldots, x_n^d$ by

$$UK_d^\bullet(\mathbf{x}) = (R \xrightarrow{x_1^d} R) \otimes \cdots \otimes (R \xrightarrow{x_n^d} R).$$

We also write $UK^\bullet(\mathbf{x}) = UK_1^\bullet(\mathbf{x})$. The unstable Koszul complexes have the advantage of being small, and explicitly constructed from free modules.

We may also define the stable Koszul complex

$$K^\bullet(I) = (R \longrightarrow R[1/x_1]) \otimes \cdots \otimes (R \longrightarrow R[1/x_n])$$

and define $\check{C}^\bullet(I)$ by the existence of a fibre sequence $K^\bullet(I) \longrightarrow R \longrightarrow \check{C}^\bullet(I)$. It is not hard to check [17] that both $K^\bullet(I)$ and $\check{C}^\bullet(I)$ are independent of the generators of the ideal, up to quasi-isomorphism. Since $K^\bullet(I)$ and $\check{C}^\bullet(I)$ are only complexes of flat modules and not projective modules, it is necessary to replace them by complexes $PK^\bullet(I)$ and $P\check{C}^\bullet(I)$ of projectives when calculating maps out of them in the derived category.

We start by showing what can be constructed from $UK^\bullet(\mathbf{x})$.

Lemma 5.4. (i) *Provided $d_1, d_2, \ldots, d_n \geq 1$, the unstable Koszul complex $UK^\bullet(x_1^{d_1}, x_2^{d_2}, \ldots, x_n^{d_n})$ lies in the thick subcategory generated by $UK^\bullet(\mathbf{x})$.*
(ii) *The stable Koszul complex $K^\bullet(I)$ lies in the localizing subcategory generated by the unstable Koszul complex $UK^\bullet(\mathbf{x})$.*

Proof. (i) First we deal with the case $n = 1$. We proceed by induction on d using the square

$$\begin{array}{ccc} R & \xrightarrow{x^{d-1}} & R \\ {\scriptstyle 1}\big\downarrow & & \big\downarrow{\scriptstyle x} \\ R & \xrightarrow{x^{d}} & R \end{array}$$

to construct a cofibre sequence $UK^\bullet_{d-1}(x) \longrightarrow UK^\bullet_d(x) \longrightarrow UK^\bullet(x)$. The general case follows since the argument remains valid after tensoring with any free object.

(ii) The map $R \longrightarrow R[1/x]$ is the direct limit of the maps $R \xrightarrow{x^d} R$, and hence $K^\bullet(x)$ is equivalent to the homotopy direct limit of the terms $UK^\bullet(x^d)$. Tensoring these together and using the fact that $\underrightarrow{\mathrm{holim}}_d$ commutes with tensor products, we find

$$K^\bullet(I) \simeq \underrightarrow{\mathrm{holim}}_d UK^\bullet_d(\mathbf{x}).$$

□

We also need a related result in the other direction.

Lemma 5.5. *The unstable Koszul complex $UK^\bullet(\mathbf{x})$ lies in the thick subcategory generated by the stable Koszul complex $K^\bullet(I)$.*

Proof. Consider the self-map of the cofibre sequence $K^\bullet(x) \longrightarrow R \longrightarrow R[1/x]$ given by multiplication by x. Since x is an equivalence of $R[1/x]$, the octahedral axiom shows there is a fibre sequence $UK^\bullet(x) \longrightarrow K^\bullet(x) \xrightarrow{x} K^\bullet(x)$. We may tensor this argument with any object X, so that we find a fibre sequence

$$K^\bullet(x_1, \ldots, x_{n-1}) \otimes UK^\bullet(x_n) \longrightarrow K^\bullet(I) \xrightarrow{x_n} K^\bullet(I).$$

Repeating this, we see that $UK^\bullet(\mathbf{x})$ lies in the thick subcategory generated by $K^\bullet(I)$. □

Proposition 5.6. *If $I = (x_1, x_2, \ldots, x_n)$ is finitely generated, the class $\mathcal{D}$ is generated as a localizing ideal by R/I, by $K^\bullet(I)$ and by $UK^\bullet(\mathbf{x})$.*

Proof. We start by showing that $\mathcal{D}$ is generated by $UK^\bullet(\mathbf{x})$. Since $UK^\bullet(\mathbf{x})$ is small, we may apply Proposition 3.2 (i). It suffices to check that $UK^\bullet(\mathbf{x})$ detects triviality of objects D of $\mathcal{D}$. Suppose then that $H_*(X)$ is I-power torsion and $t \in H_*(X)$. It suffices by 5.4 to show that the corresponding map $t\colon R \longrightarrow X$ extends over $R \longrightarrow UK^\bullet(x_1^{d_1}, x_2^{d_2}, \ldots, x_n^{d_n})$ for some $d_1, d_2, \ldots, d_n \geq 1$.

Suppose by induction on m that t has been extended to

$$t'\colon UK^\bullet(x_1^{d_1}, x_2^{d_2}, \ldots, x_m^{d_m}) \longrightarrow X.$$

This is clear for $m = 0$, so the induction starts, and we suppose $0 < m < n$. Now note that t' is I-power torsion, since $[T, X]$ is I-power torsion for any finite complex T of free modules. Choose d_{m+1} so that $x_{m+1}^{d_{m+1}} t' = 0$. Construct a cofibre sequence by tensoring

$$R \stackrel{x_{m+1}^{d_{m+1}}}{\longrightarrow} R \longrightarrow UK^\bullet(x_{m+1}^{d_{m+1}})$$

with $UK^\bullet(x_1^{d_1}, x_2^{d_2}, \ldots, x_n^{d_m})$. Exactness of $[\cdot, X]$ shows that t' extends along

$$UK^\bullet(x_1^{d_1}, x_2^{d_2}, \ldots, x_m^{d_m}) \longrightarrow UK^\bullet(x_1^{d_1}, x_2^{d_2}, \ldots, x_m^{d_m}, x_{m+1}^{d_{m+1}}),$$

completing the inductive step. This completes the proof that $UK^\bullet(\mathbf{x})$ generates $\mathcal{D}$.

By 5.5 it follows that $K^\bullet(I)$ also generates $\mathcal{D}$, and the fact that R/I generates $\mathcal{D}$ follows if we can show $UK^\bullet(\mathbf{x})$ lies in the localizing ideal generated by R/I.

Lemma 5.7. *The localizing ideal containing R/I contains any complex X so that $H_*(X)$ is bounded in both directions and I-power torsion.*

Proof. First, we prove by induction on k that a module M (regarded as an object of the derived category in degree 0 with zero differential) lies in $\mathrm{locid}(R/I)$ provided $I^k M = 0$. If $k = 0$ this means $M = 0$, so we suppose $k \geq 1$. First, the short exact sequence $I^k/I^{k+1} \longrightarrow R/I^{k+1} \longrightarrow R/I^k$ gives a triangle, with the first and third term already known to be in the ideal, so R/I^{k+1} lies in the ideal. Now suppose $I^{k+1}M = 0$. There is a surjective map $T_0 \longrightarrow M_0 = M$ of modules where T_0 is a sum of copies of R/I^{k+1}, and the kernel K_0 also satisfies $I^{k+1}K_0 = 0$. We may thus iterate the construction and apply 3.2 (ii) to deduce M lies in the localizing ideal generated by R/I^{k+1}.

The modules M are Eilenberg–Mac Lane objects, and we show that if X is bounded, it has a finite Postnikov tower. After suspension we may suppose $H_i(X) = 0$ for $i < 0$. Since X is equivalent to the subcomplex X' zero in negative degrees, with X'_0 the 0-cycles, and agreeing with X in positive degrees, we may suppose X is zero in negative degrees. There is then a canonical map $X = X^0 \longrightarrow M_0$ which is an isomorphism in degree 0 where $M_0 = H_0(X)$. The fibre X^1 then has $H_i(X^1) = 0$ for $i < 1$, and $H_i(X^1) \cong H_i(X)$ for $i \geq 1$, and we may iterate the construction. Defining X_k by the triangle $X^k \longrightarrow X \longrightarrow X_k$ we see that $X_0 \simeq 0$, and by the octahedral axiom there is a cofibre seqence

$$\Sigma^k M_k \longrightarrow X_{k+1} \longrightarrow X_k.$$

Since M_k lies in the localizing ideal generated by R/I, so does X_k for all k. By the boundedness hypothesis, $X^N \simeq 0$ for N sufficiently large, and so $X_N \simeq X$. □

Since $UK^\bullet(\mathbf{x})$ satisfies the conditions of the lemma, this completes the proof of 5.6. □

It is not hard to construct the relevant localizations and completions.

Lemma 5.8. *If I is finitely generated,*
(i) $M[\mathcal{D}^{-1}] = M \otimes \check{C}^\bullet(I)$,
(ii) $\Gamma_{\mathcal{D}}(M) = M \otimes K^\bullet(I)$,
(iii) $M^\wedge_{\mathcal{D}} = \mathrm{Hom}(PK^\bullet(I), M)$.

Proof. (i) To see that $M \otimes \check{C}^\bullet(I)$ is local, we need only check it admits no morphism from $UK^\bullet(\mathbf{x})$ except zero. However $\check{C}^\bullet(I)$ admits a finite filtration with subquotients $R[1/x]$ for $x \in I$ so it suffices to show $[UK^\bullet(\mathbf{x}), R[1/x]]_* = 0$. This follows since x is nilpotent on $UK^\bullet(\mathbf{x})$ and an isomorphism on $R[1/x]$. To see that $M \longrightarrow M \otimes \check{C}^\bullet(I)$ is a $\mathcal{D}$-equivalence we need only verify that the $M \otimes K^\bullet(I)$ can be built from $UK^\bullet(\mathbf{x})$. However M can be built from R, and we saw in 5.4 that $K^\bullet(I)$ can be built from $UK^\bullet(\mathbf{x})$.

Part (ii) follows from the defining fibre sequence of $\check{C}^\bullet(I)$, and Part (iii) follows from 2.2. □

We write

$$H^*_I(M) = H^*(K^\bullet(I) \otimes M) = H_*(\Gamma_{\mathcal{D}}(M));$$

this is the local cohomology of M, and if R is Noetherian it calculates the right derived functors of

$$\Gamma_I(M) = \{x \in M \mid I^n x = 0 \text{ for } n >> 0\}$$

for modules M [29]. We write

$$H^I_*(M) = H_*(\mathrm{Hom}(PK^\bullet(I), M)) = H_*(M^\wedge_{\mathcal{D}});$$

this is the local homology of M [22]. If, in addition, R is Noetherian or good in the sense of [22], then this local homology gives the left derived functors of completion. In particular, if M is of finite type, $M^\wedge_{\mathcal{D}} = M^\wedge_I$. Furthermore, the Tate cohomology coincides with that of [18]. As pointed out in [18], Warwick duality is a generalization of the isomorphism $\mathbb{Z}^\wedge_p[1/p] = \lim\limits_{\leftarrow}(\mathbb{Z}/p^\infty, p)$.

Remark 5.9. If I is finitely generated, we have described both a construction and a method of calculation for useful localizations. It would be interesting to have analogues when I is not finitely generated.

6. The category of modules over a highly structured ring

In this section we suppose that $\mathbf{R}$ is a commutative S-algebra in the sense of [14], and we allow the equivariant case. Such objects are essentially equivalent to E_∞ ring spectra, so there is a good supply: in particular, any commutative ring R gives rise to the Eilenberg–Mac Lane S-algebra HR. We then let $\mathcal{C}$ denote the homotopy category of highly structured module spectra over $\mathbf{R}$ and consider localizations and completions associated to a finitely generated ideal I of the coefficient ring $\mathbf{R}_*$.

Much of the discussion of the previous section applies in the present case, and was presented in [24], so we shall be brief. Thus we may form the stable and unstable Koszul modules by using cofibre sequences and smash products. Thus for example, $UK^\bullet(x)$ is the fibre of $\mathbf{R} \xrightarrow{x} \mathbf{R}$; we avoid the common notation $\Sigma^{-1}\mathbf{R}/x$ for fear of confusion. Now $UK^\bullet(\mathbf{x}) = UK^\bullet(x_1) \wedge_{\mathbf{R}} UK^\bullet(x_2) \wedge_{\mathbf{R}} \dots \wedge_{\mathbf{R}} UK^\bullet(x_n)$; similarly $K^\bullet(x)$ is the fibre of $\mathbf{R} \longrightarrow \mathbf{R}[1/x]$, and

$$K^\bullet(I) = K^\bullet(x_1) \wedge_{\mathbf{R}} K^\bullet(x_2) \wedge_{\mathbf{R}} \dots \wedge_{\mathbf{R}} K^\bullet(x_n)$$

where $I = (x_1, x_2, \dots, x_n)$. We take $\mathcal{A}$ to be the class of retracts of finite $\mathbf{R}$-modules $\mathbf{M}$ so that $\mathbf{M}_*$ is I-power torsion This is generated by $\mathcal{T} = \{UK^\bullet(\mathbf{x})\}$, and generates the localizing ideal of all $\mathbf{M}$ so that each element of $\mathbf{M}_*$ is I-power torsion (i.e., $\mathbf{M}_*$ is in the class $\mathcal{D}(\mathbf{R}_*, I)$ in the sense of Section 5). We write $\Gamma_I(\mathbf{M}) = \Gamma_{\mathcal{D}}(\mathbf{M})$, $\mathbf{M}[I^{-1}] = \mathbf{M}[\mathcal{D}^{-1}]$ and $t_I(\mathbf{M}) = t_{\mathcal{D}}(\mathbf{M})$.

The statement and proof of Lemma 5.8 apply without change. Because the construction comes with an evident filtration we may obtain spectral sequences by taking homotopy, and the E_1-term is a chain complex representing the corresponding constructions of Section 5. This gives spectral sequences

$$\begin{aligned}
\check{H}^*_I(\mathbf{R}_*; \mathbf{M}_*) &\Longrightarrow \mathbf{M}[I^{-1}]_* \\
\hat{H}^*_I(\mathbf{R}_*; \mathbf{M}_*) &\Longrightarrow t_I(\mathbf{M})_* \\
H^*_I(\mathbf{R}_*; \mathbf{M}_*) &\Longrightarrow \Gamma_I(\mathbf{M})_* \\
H^I_*(\mathbf{R}_*; \mathbf{M}_*) &\Longrightarrow (\mathbf{M}^\wedge_I)_*
\end{aligned}$$

for calculating their homotopy.

7. The derived category of kG

For a finite group G and a field k we consider the derived category $\mathcal{C} = D(kG)$, and take $\mathcal{A}$ to be the category of finite complexes of projectives. This is generated by $\mathcal{T} = \{kG\}$, and the generation is so systematic algebraically that it leads to the usual method for calculating Tate cohomology using projective resolutions and their duals. The relationship of the derived category $D(kG)$ to the category of G-spectra is analogous to the relationship of $D(\mathbf{R}_*)$ to the category of highly structured modules over $\mathbf{R}$.

It is proved in [32, 9.6] that the localization $M \longrightarrow M[\mathcal{A}^{-1}]$ is obtained by tensoring with a Tate resolution. Since any Tate resolution admits a finite filtration with subquotients $R[1/x]$ as in [19], it follows that every object of $\mathcal{C}$ with bounded cohomology is already complete. Thus we find that if M has bounded cohomology, $t_{\mathcal{A}}(M) = M[\mathcal{A}^{-1}] = M \otimes t_{\mathcal{A}}(k)$ and so the Tate construction defined by localization agrees with Tate homology in the classical sense.

There are at least three other examples to consider here, but some work is needed to give them substance. Recall that an indecomposable module M has vertex H if it is a summand in a module induced from H but not from any proper subgroup of H.

Variation 7.1. Consider a family $\mathcal{F}$ of subgroups, and the category $\mathcal{A}_{\mathcal{F}}$ of finite complexes of modules with vertex in $\mathcal{F}$. The case $\mathcal{F} = \{1\}$ is that given above. The $\mathcal{G}$-ideal $\mathcal{A}_{\mathcal{F}}$ is generated by $\mathcal{T}_{\mathcal{F}} = \{k[G/H] \mid H \in \mathcal{F}\}$. It is then appropriate to use Amitsur–Dress $\mathcal{F}$-cohomology [13]. Perhaps there is again a local cohomology theorem in the sense of Section 12 below, using the ideal of positive degree elements, but the appropriate theory of varieties has not been developed. It would also be interesting to know the relationship to ordinary group cohomology and the ideal $I_{\mathcal{F}}$ of cohomology elements restricting to zero in the cohomology of H for all $H \in \mathcal{F}$.

Variation 7.2. We choose a block β of kG and take $\mathcal{A}_{\beta}$ to be the category of finite complexes of projectives in β.

Variation 7.3. We may consider the stable module category $\mathcal{C} = StMod(kG)$, which is proved in [32, 9.6.4] to be a localization of $D(kG)$. It would then be interesting to investigate complexity quotients in the sense of Carlson–Donovan–Wheeler [10, 11, 5, 6] from the present point of view.

8. Chromatic categories

Another important class of examples arises in the approach to stable homotopy theory through bordism. For background and further information see [40]. Thus we work in the stable homotopy category of spectra in the sense of algebraic topology, and we choose a prime $p > 0$. For $0 \leq n \leq \infty$ we shall need the spectrum $E(n)$ representing Johnson–Wilson cohomology theory and the Morava K-theory spectrum $K(n)$. For $0 < n < \infty$ these have coefficient rings

$$E(n)_* = \mathbb{Z}_{(p)}[v_1, v_2, \ldots, v_{n-1}, v_n, v_n^{-1}],$$

and $K(n)_* = \mathbb{Z}/p[v_n, v_n^{-1}]$. The cases $n = 0, \infty$ are somewhat exceptional: by convention, for $n = 0$ we have $E(0) = K(0) = H\mathbb{Q}$ and for $n = \infty$ we have $E(\infty) = BP$ and $K(\infty) = H\mathbb{Z}/p$. Recall that a spectrum is said to be of type n if $K(i)_*(X) = 0$ for $i < n$ and $K(n)_*(X) \neq 0$.

Bousfield localization L_n with respect to $E(n)$ is the localization whose acyclics are the spectra X with $E(n) \wedge X \simeq *$. A well known theorem of Hopkins–Ravenel states that L_n is smashing. The usual notation is $C_n X \longrightarrow X \longrightarrow L_n X$. The completion $X^{\wedge}_{\mathcal{D}} = F(C_n S, X)$ is more mysterious, but when $n = 0$ it is profinite completion $F(S^{-1}\mathbb{Q}/\mathbb{Z}, X)$.

n	$E(n)$	$K(n)$	$F(n)$	L_{n-1}	$L_{K(n)}$
0	$H\mathbb{Q}$	$H\mathbb{Q}$	$S^0_{(p)}$	*	rationalization
1	$K_{(p)}$	K/p	S^0/p^k	invert p	$L_{K(1)}$
2	Ell	Ell$/(p, v_1)$	$S^0/(p^k, v_1^l)$	$L_1 =$ "$(\cdot)[(p, v_1)^{-1}]$"	$L_{K(2)}$
$\vdots$	$\vdots$	$\vdots$	$\vdots$	$\vdots$	$\vdots$
∞	BP	$H\mathbb{F}_p$	*	p-localization	p-adic completion

Following [33], let us consider a slightly simpler example. Let $\mathcal{C}$ be the $E(n)$-local category, and $\mathcal{A}$ the thick subcategory generated by $L_nF(n)$ for a finite type n spectrum $F(n)$. In this case $X[\mathcal{A}^{-1}] = L_{n-1}X$ [33, 6.10] and $X^\wedge_{\mathcal{A}} = L_{K(n)}X$ [33, 7.10]. The fibre of $X \longrightarrow L_{n-1}X$ is usually known as the nth monochromatic piece when X is $E(n)$-local, so we have $\Gamma_{\mathcal{A}}(X) = M_nX$. The fibre of $K(n)$ completion is sometimes known as $C_{K(n)}$, but we simply write $\Delta_{\mathcal{A}}(X) = \Delta_{K(n)}(X)$.

Corollary 8.1. (*Warwick Duality*) *If X is $E(n)$-local then*

$$L_{n-1}L_{K(n)}X \simeq \Sigma\Delta_{K(n)}(M_nX).$$

We note that if $n = 0$ this states M_0X is rational, and if $n = 1$ it states that the cofibre of $M_1X \longrightarrow L_{K(1)}M_1X$ is the rationalization of $L_{K(1)}X$.

If we take $\mathcal{C}$ to be the entire category of p-local spectra there are two related examples. Indeed, we may still consider the smashing localization $L_{n-1} = L_{E(n-1)}$, but it does not seem so easy to describe the associated completion. In particular it is not equal to $L_{K(n)}$ (indeed, although $L_{n-1}S^0$ is $K(n)$-acyclic, there are many spectra, such as $F(n+1)$, which are $K(n)$-acyclic but not $E(n-1)$-local). We may also consider spectra $\mathrm{Tel}(n) = F(n)[1/v_n]$, and the smashing localization L^f_{n-1} which is Bousfield localization with respect to $\mathrm{Tel}(0) \vee \mathrm{Tel}(1) \vee \ldots \vee \mathrm{Tel}(n-1)$; this is finite localization with respect to $F(n)$ [39] and it is therefore smashing, and we may again consider the associated completion, which is again different from $L_{K(n)}$ for similar reasons. There is a natural transformation $L^f_n \longrightarrow L_n$, which is believed not to be an isomorphism for $n \geq 2$ [35].

9. Splittings of the Tate construction

We describe two different classes of splittings of the Tate construction. Each requires special properties of the localization.

First, continuing with the notation of the previous section, note that $t_{\mathcal{A}}(X) = L_{n-1}L_{K(n)}X$ is the subject of Hopkins's chromatic splitting conjecture [31, 4.2]. When $X = S^0$ (albeit not in the $E(n)$-local category $\mathcal{C}$) this is conjectured to split into 2^n pieces. More precisely there is a cofibre sequence

$$L_{n-1}S^0_p \longrightarrow L_{n-1}L_{K(n)}S^0 \longrightarrow \Sigma F(L_{n-1}S^0, L_nS^0_p),$$

which is conjectured to split, and furthermore, $F(L_{n-1}S^0, L_nS^0_p)$ is also conjectured to split as a wedge of $2^n - 1$ suitable localizations of spheres. To obtain the cofibre sequence, apply $F(\cdot, X)[\mathfrak{D}^{-1}]$ to the cofibre sequence

$$\Sigma^{-1}S[\mathfrak{D}^{-1}] \longrightarrow \Gamma_{\mathfrak{D}}(S) \longrightarrow S$$

to obtain

$$X[\mathfrak{D}^{-1}] \longrightarrow t_{\mathfrak{D}}(X) \longrightarrow \Sigma F(S[\mathfrak{D}^{-1}], X)$$

since $F(S[\mathfrak{D}^{-1}], X)$ is already $[\mathfrak{D}^{-1}]$-local.

Secondly, there is a dichotomy between the periodic and split behaviour of the Tate construction, typified by the cohomology of finite groups. Although Tate cohomology is often associated with periodic behaviour, it is the split case that is generic. On the one hand, when G has periodic cohomology there is a 'periodicity element' x in $H^*(G)$ and the Tate cohomology $\hat{H}^*(G) = H^*(G)[1/x]$ is periodic under multiplication by x. By contrast, when group cohomology $H^*(G)$ has a regular sequence of length 2, Benson–Carlson [4] and Benson–Greenlees [7] have shown that the mod p Tate cohomology $\hat{H}^*(G)$ of a finite group splits

$$\hat{H}^*(G) = H^*(G) \oplus \Sigma^1 H_*(G)$$

(where the suspension is homological) both as a module over $H^*(G)$ and as a module over the Steenrod algebra. Even this context does not provide a true dichotomy, since there are groups with depth 1 which are not periodic, but this mixed behaviour is exceptional.

The analogous statement concerns the standard cofibre sequence

$$X_{\mathfrak{D}}^{\wedge} \longrightarrow t_{\mathfrak{D}}(X) \longrightarrow \Sigma\Gamma_{\mathfrak{D}}(X)$$

when $X = S$. The dichotomy principle would suggest that in most cases, either $t_{\mathfrak{D}}(S)$ is obtained from $S_{\mathfrak{D}}^{\wedge}$ by inverting some multiplicatively closed subset of $\pi_*(S_{\mathfrak{D}}^{\wedge})$, or else the cofibre sequence splits, and that the split case is generic. The hypotheses for a splitting must include the requirement that the norm map $\Sigma^{-1}\Gamma_{\mathfrak{D}}(X) \longrightarrow X_{\mathfrak{D}}^{\wedge}$ is zero in homotopy, and probably also that $\pi_*(X_{\mathfrak{D}}^{\wedge})$ is of depth at least 2. However the proofs from the case of group cohomology do not extend in any simple way since they use the fact that homology and cohomology are identified in the Tate cohomology by their occurrence in positive and negative degrees.

A second case where the dichotomy holds is in commutative algebra [18]. When the ring is Noetherian and of Krull dimension 1, the rationality theorem [18, 7.1] holds: $\hat{H}_I^*(R) = S^{-1}(R_I^{\wedge})$ where S is the set of regular elements of R. This is the periodic case. It is immediate that if the ring is of I-depth two or more the Tate cohomology splits since the local homology is in degree 0 whilst the local cohomology is only non-zero at or above the depth.

10. Calculation by comparison

We discuss two quite different methods of calculation. To introduce the discussion, we explain the two methods as they apply to calculating the homology $H_*(G; M)$

of a finite group with coefficients in a chain complex M of kG-modules. The first is quite familiar, and states there is a spectral sequence

$$H^*(G; H_*(M)) \Longrightarrow H_*(G; M).$$

The second method is the local cohomology theorem, stating that there is a spectral sequence

$$H^*_I(H^*(G; M)) \Longrightarrow H_*(G; M)$$

where I is the ideal of positive codegree elements of $H^*(G)$ [19], and $H^*_I(\cdot)$ denotes local cohomology in the sense of Grothendieck [29] (the definition was recalled in Section 5).

The generalization we have in mind concerns finite localizations in the case that $\mathcal{A}$ is generated by a single object A. We require that A is a commutative comonoid in the sense that it has a commutative and associative coproduct $A \longrightarrow A \wedge A$ and a counit $A \longrightarrow S$. We require in addition that A is strongly dualizable and self-dual up to an invertible element, in the sense that $DA \simeq A \wedge S^{-\tau}$ for some object $S^{-\tau}$ admitting a smash inverse $S^{-\tau} \wedge S^{\tau} \simeq S$.

Example 10.1. (i) The motivating example has $\mathcal{C} = \mathcal{D}(kG)$ for a finite group G and $A = kG$. Note that we have an augmentation $kG \longrightarrow k$, and a diagonal map $kG \longrightarrow kG \otimes kG$. Furthermore kG is self-dual.

(ii) Alternatively, for a compact Lie group G, we may take $\mathcal{C}$ to be a category of G-spectra (or of module G-spectra over a ring G-spectrum $\mathbf{R}$) and $A = G_+$ (or $\mathbf{R} \wedge G_+$). Again we have an augmentation $G_+ \longrightarrow S^0$, and a diagonal map $G_+ \longrightarrow G_+ \wedge G_+$. We also have the duality statement $DG_+ \simeq \Sigma^{-d} G_+$ where $d = \dim(G)$. This helps explain the notation $S^{-\tau}$, which is chosen since, in the geometric context, Atiyah duality shows τ corresponds to the tangent bundle.

(iii) Rather differently, we may take $\mathcal{C}$ to be the category of p-local spectra, (or of p-local $\mathbf{R}$-module spectra over a ring spectrum $\mathbf{R}$) and $A = \Sigma^{-d} F(n)$ (or $A = \mathbf{R} \wedge \Sigma^{-d} F(n)$) where

$$F(n) = S^0/(p^{i_0}, v_1^{i_1}, v_2^{i_2}, \ldots, v_{n-1}^{i_{n-1}})$$

for suitable $i_0, i_1, \ldots, i_{n-1}$ and $d = \dim(F(n))$. Collapse onto the top cell gives an augmentation $\Sigma^{-d} F(n) \longrightarrow S^0$. In favourable cases we have the duality statement $DF(n) \simeq \Sigma^{-d} F(n)$, and $F(n)$ may be taken to be a commutative ring spectrum [12], and the dual to the product gives a coproduct map

$$\Sigma^{-d} F(n) \longrightarrow \Sigma^{-d} F(n) \wedge \Sigma^{-d} F(n).$$

We need to consider the graded commutative ring $k_* = [S, S]_*$, where S is the unit in $\mathcal{C}$, and two k_*-algebras. Firstly, since A is a commutative comonoid, $l_* = [A, S]_*$ is a commutative k_*-algebra, and $[A, Z]_*$ is a module over l_* for any Z. Secondly, we consider the k_*-algebra $\mathcal{E}_* = \mathrm{End}(A)_*$, which need not be commutative.

Context 10.2. (For calculation)

- A a commutative comonoid object
- $\mathcal{A}$ generated by A
- $DA \simeq S^{-\tau} \wedge A$
- $k_* = [S, S]_*$
- $l_* = [A, S]_*$
- $I = \ker(k_* \longrightarrow l_*)$
- $\mathcal{E}_* = [A, A]_*$.

In our examples these are as follows.

Example 10.3. (i) When $A = kG$ we have $k_* = l_* = k$ and $\mathrm{End}(kG)_* = kG$.

(ii) When $A = \mathbf{R} \wedge G_+$ we have $k_* = \mathbf{R}_*^G$, $l_* = \mathbf{R}_*$ and $\mathrm{End}(\mathbf{R} \wedge G_+)_* = \mathbf{R}_*(G_+)$.

(iii) When $A = \Sigma^{-d} F(n)$ we have $k_* = \mathbf{R}_*$, $l_* = \mathbf{R}_*(F(n))$ and

$$\mathrm{End}(\mathbf{R} \wedge \Sigma^{-d} F(n))_* = \mathbf{R}_*(F(F(n), F(n))).$$

Given these data, there are two functors we can apply:

$$[A, \cdot]_* \colon \mathcal{C} \longrightarrow \mathrm{End}(A)_*\text{-mod}$$

(corresponding to non-equivariant homotopy in Example (ii)), and

$$[S, \cdot]_* \colon \mathcal{C} \longrightarrow k_*\text{-mod}$$

(corresponding to equivariant homotopy in Example (ii)).

It is then natural to seek spectral sequences reversing these two functors. In the first case we may hope they take the form

10.4.

$$H_*(\mathrm{End}(A)_*; [A, S^\tau \wedge X]_*) \Longrightarrow (\Gamma_{\mathcal{A}}(X))_*$$
$$H^*(\mathrm{End}(A)_*; [A, X]_*) \Longrightarrow (X_{\mathcal{A}}^{\wedge})_*$$
$$\text{and} \qquad \hat{H}^*(\mathrm{End}(A)_*; [A, X]_*) \Longrightarrow t_{\mathcal{A}}(X)_*.$$

A construction in some cases is given in Section 11, and the twisting S^τ in the first spectral sequence will be explained.

In the second case we let

$$I = \ker(k_* \colon [S, S]_* \longrightarrow [A, S]_*)$$

be the augmentation ideal, and apply local cohomology, local homology and local Tate cohomology as appropriate and hope the spectral sequences take the form

10.5.

$$H_I^*(X_*) \Longrightarrow (\Gamma_{\mathcal{A}}(X))_*$$
$$H_*^I(X_*) \Longrightarrow (X_{\mathcal{A}}^{\wedge})_*$$
$$\text{and} \qquad \hat{H}_*^I(X_*) \Longrightarrow t_{\mathcal{A}}(X)_*.$$

A construction in some cases is given in Section 12. When the first spectral sequence exists we say that *the local cohomology theorem holds*. Provided this happens for good enough reasons, the other two spectral sequences exist as a consequence.

The content should be clearer when we give some examples. It is not surprising that to prove the existence of either set of spectral sequences we have to assume the existence of additional structure beyond that present in the stable homotopy category.

11. Calculations by associative algebra

The point of this section is to generalize the Atiyah–Hirzebruch Tate spectral sequence of [16]:

$$\hat{H}^*(G; E^*(X)) \Longrightarrow t(E)^*_G(X)$$

for finite groups G, or in other words to prove the expectations suggested in 10.4 hold under suitable circumstances. The construction does not work entirely in a stable homotopy category, but rather relies on the existence of a suitable Quillen model category from which the stable homotopy category is formed by inverting weak equivalences.

We thus suppose given a stable homotopy category, and consider the $\mathcal{G}$-ideal $\mathcal{A}$ generated by a single object A. The aim is to find ways to calculate $\Gamma_A(X)_*$, $(X^\wedge_A)_*$ and $t_A(X)_*$ in terms of the $[A,A]_*$-module $[A,X]_*$. In view of the notational conflict we remind the reader that in the context of G-spectra, where $A = G_+$ the group $[S,\cdot]_*$ is *equivariant* homotopy and $[A,\cdot]_*$ is *non-equivariant* homotopy. The present discussion covers a number of new examples: the generalization is cruder than that of [23], but more general. The discussion of convergence in [23, Appendix B] applies without change.

To avoid the appearance of empty generalization, we state an unequivocal theorem in the equivariant homotopy context of Section 4 (with $\mathcal{A}$ generated by $\mathbf{R} \wedge G_+$).

Theorem 11.1. *Suppose G is a compact Lie group of dimension d, $\mathbf{R}$ is an equivariant S-algebra, and $\mathbf{M}$ an $\mathbf{R}$-module. Provided we have the Künneth theorem*

(KT1) $$\mathbf{M}_*(G_+ \wedge T \wedge Y) = \mathbf{R}_*(G_+) \otimes_{\mathbf{R}_*} \mathbf{M}_*(T \wedge Y)$$

and the universal coefficient theorem

(UCT) $$[G_+ \wedge T, \mathbf{M} \wedge Y]^G_* = [T, \mathbf{M} \wedge Y]_* = \mathrm{Hom}_{\mathbf{R}_*}(\mathbf{R}_*(T), \mathbf{M}_*(Y))$$

when $T = G_+^{\wedge s}$ for $s \geq 0$, there are spectral sequences

$$H_*(\mathbf{R}_*(G_+); \mathbf{M}_*(S^d \wedge Y)) \Longrightarrow \mathbf{M}^G_*(EG_+ \wedge Y),$$
$$H_*(\mathbf{R}_*(G_+); \mathbf{M}^*(Y)) \Longrightarrow \mathbf{M}^*_G(EG_+ \wedge Y),$$
and $$\hat{H}_*(\mathbf{R}_*(G_+), \mathbf{M}_*(Y)) \Longrightarrow t(\mathbf{M})^G_*(Y),$$

where the homology and cohomology on the left is that of the Frobenius algebra $\mathbf{R}_*(G_+)$.

We return to this particular case at the end of the section. The rest of the discussion is conducted in general terms.

We want to view the construction of $\Gamma_A S$ as a "resolution" for $X = S$ using sums of objects of $\mathcal{A}$. More precisely, we use the method of 3.2 (i) without assuming X is in $\mathfrak{D}$. The dual resolution is thus

$$\begin{array}{ccccccccc} * & = & (S)^0 & \xrightarrow{j^0} & (S)^1 & \xrightarrow{j^1} & (S)^2 & \xrightarrow{j^2} & \cdots \\ & & \downarrow & & \downarrow{\scriptstyle q_0} & & \downarrow{\scriptstyle q_1} & & \\ & & * & & T_0 & & T_1 & & \end{array}$$

where each T_i is a sum of suspensions of objects of $\mathcal{A}$. This is associated with the sequence

$$T_0 \xleftarrow{\delta_1} \Sigma^{-1}T_1 \xleftarrow{\delta_2} \Sigma^{-2}T_2 \xleftarrow{\delta_3} \Sigma^{-3}T_3 \longleftarrow \cdots .$$

We want to apply simplicial methods, so we suppose there is an underlying model category, from which the stable homotopy category is formed by passage to homotopy. Furthermore, we require a compatible symmetric monoidal structure and that A is a strict comonoid object.

Example 11.2. The example relevant to the theorem is the homotopy category of modules over the equivariant S-algebra $\mathbf{R}$ for a compact Lie group G, and $A = \mathbf{R} \wedge G_+$. This is the homotopy category of the model category of equivariant $\mathbf{R}$-modules [14]. However, in this case it is more elementary to make the construction described below at the space level, apply the suspension spectrum functor and take the extended $\mathbf{R}$-module: this strategy will give parts of the theorem under weaker hypotheses.

We form the homogeneous bar construction [38] as a simplicial object, and take its geometric realization

$$\Gamma_A S = S^\infty = B(A, A, S).$$

This ensures $T_i = \Sigma^i A^{\wedge(i+1)}$. By smashing with X we obtain a resolution for arbitrary X. Thus we may define

$$t_A(X) = F(B(A, A, S), X) \wedge \tilde{B}(A, A, S),$$

where $\tilde{B}(A, A, S)$ is the mapping cone of $B(A, A, S) \longrightarrow B(S, S, S) = S$.

To relate the resolution to an algebraic one, we apply a homology theory to obtain

$$[A, T_0]_* \xleftarrow{(\delta_1)_*} [A, \Sigma^{-1}T_1]_* \xleftarrow{(\delta_2)_*} [A, \Sigma^{-2}T_2]_* \xleftarrow{(\delta_3)_*} [A, \Sigma^{-3}T_3]_* \longleftarrow \cdots .$$

In the equivariant context we have $[A, \Sigma^{-i}T_i]_* = \mathbf{R}_*(G_+^{\wedge i+1})$. To ensure it is a resolution, we assume there is a Künneth theorem

(KT1) $$[A, A \wedge Z] = [A, A]_* \otimes_{[A,S]_*} [A, Z]_*$$

for relevant Z (namely $Z = A^{\wedge i}$). In the equivariant context this is a Künneth theorem for the (non-equivariant) homology theory $\mathbf{R}_*(\cdot)$. This ensures that the simplicial contraction in geometry is converted to one in algebra and the bar filtration spectral sequence for calculating $[A, B(A, A, S)]_*$ has its E^1-term given by the algebraic bar construction $B([A, A]_*, [A, A]_*, [A, X]_*)$. To calculate $\Gamma_A(X)_*$, we need the second Künneth theorem

$$\text{(KT2)} \qquad [S, Z]_* = [A, S]_* \otimes_{[A,A]_*} [A, S^\tau \wedge Z]_*$$

for relevant Z (namely $Z = A^{\wedge(i+1)} \wedge X$). In the equivariant context, this states that the change of groups isomorphisms $[S, G_+ \wedge T]^G_* = [S, S^d \wedge T]_* = [G_+, T]^G_*$ are reflected in algebra.

Lemma 11.3. *The Künneth theorem* (KT2) *for* $Z = A \wedge T$ *follows from the Künneth theorem* (KT1) *for* $Z = S^\tau \wedge T$.

Proof. Assuming (KT1) for $Z = S^\tau \wedge T$, we calculate

$$\begin{aligned} [A, S]_* \otimes_{[A,A]_*} [A, A \wedge S^\tau \wedge T]_* &= [A, S]_* \otimes_{[A,A]_*} [A, A]_* \otimes_{[A,S]_*} [A, S^\tau \wedge T]_* \\ &= [A, S]_* \otimes_{[A,S]_*} [A, S^\tau \wedge T]_* \\ &= [A, S^\tau \wedge T]_* \\ &= [S, S^\tau \wedge DA \wedge T]_* \\ &= [S, A \wedge T]_* \end{aligned}$$

as required. □

This is enough to give a spectral sequence with

$$E^1 = [A, S]_* \otimes_{[A,A]_*} B([A, A]_*, [A, A]_*, [A, S^\tau \wedge X]_*);$$

it therefore takes the form

$$E^2_{*,*} = H_*(\mathrm{End}(A)_*; [A, S^\tau \wedge X]_*) \Longrightarrow \Gamma_A(X)_*.$$

It is easy to see this spectral sequence is conditionally convergent in the sense of Boardman [2]. The homology in the E_2-term is defined to be the homology of the bar construction, but in favourable cases it can be calculated in various other ways. For example in the case of G-spectra this spectral sequence takes the form

$$H_*(\mathbf{R}_*(G_+); (S^d \wedge X)_*) \Longrightarrow X^G_*(EG_+).$$

Note that we have two possible definitions of the $\mathbf{R}_*(G_+)$-module structure on X_*. A diagram chase verifies they agree.

Lemma 11.4. *The action of* $\mathbf{R}_*(G_+)$ *on* $X_* = [G_+, X]^G_* = [\mathbf{R} \wedge G_+, X]^{\mathbf{R},G}_*$ *implied by the Künneth theorem and the action of* G *on* X *agrees with the action of* $[\mathbf{R} \wedge G_+, \mathbf{R} \wedge G_+]^{\mathbf{R},G}_*$ *by composition.* □

For cohomology we want to have the universal coefficient theorem (UCT)

$$[A \wedge Z, X]_* = \mathrm{Hom}_{[A,A]_*}([A, A \wedge Z]_*, [A, X]_*) = \mathrm{Hom}_{[A,S]_*}([A, Z]_*, [A, X]_*),$$

where the second equality is (KT1) and a change of rings isomorphism. This is enough to get a spectral sequence with

$$E_1 = \mathrm{Hom}_{[A,A]_*}(B([A,A]_*,[A,A]_*,[A,S]_*),[A,X]_*);$$

it therefore takes the form

$$E_2^{*,*} = H^*(\mathrm{End}(A)_*;[A,X]_*) \Longrightarrow [\Gamma_A(S),X]_* = (X_A^\wedge)_*.$$

Convergence is again conditional in the sense of Boardman. In the equivariant case this spectral sequence becomes

$$H^*(\mathbf{R}_*(G_+);X_*) \Longrightarrow X_G^*(EG_+).$$

When it comes to Tate cohomology we need to ask about splicing, both in topology and algebra. In topology we have

$$\ldots \longleftarrow DA^2 \longleftarrow DA \longleftarrow A \longleftarrow A^2 \longleftarrow \ldots$$

where the splicing is via

$$DA \overset{Dt_0}{\longleftarrow} DS = S \overset{t_0}{\longleftarrow} A.$$

To obtain a spectral sequence we may either apply $[A, \cdot \wedge X]_*$ and use the first avatar $t_A(X) = F(B(A,A,S),X) \wedge \tilde{B}(A,A,S)$, or apply $[A \wedge \cdot, X]_*$ and use the second avatar $t_A(X) = F(\tilde{B}(A,A,S),\Sigma X \wedge B(A,A,S))$. The first will make the relation to homology clearer and the second will make the relation to cohomology clearer, but since the resolution is self-dual, the two are essentially equivalent, and we only discuss the first. Convergence is again covered by the relevant argument (10.8) from [23].

In view of the equality $[A \wedge A, S]_* = [A, DA]_*$, we conclude that the E^2-term agrees with the homological one in positive filtration degrees, and with the cohomological one (shifted by one degree) in filtration degrees ≤ -2. More precisely, if $\mathcal{E}_* = \mathrm{End}(A)_* = [A,A]_*$, and $\tilde{\mathcal{E}}_* = [A \wedge A, S]^* = [A, DA]_*$, we have the algebraic resolution

$$\ldots \longleftarrow \tilde{\mathcal{E}}_*^{\otimes 2} \longleftarrow \tilde{\mathcal{E}}_* \longleftarrow \mathcal{E}_* \longleftarrow \mathcal{E}_*^{\otimes 2} \longleftarrow \ldots .$$

Using this particular resolution to define the E_2-term we have a spectral sequence

$$\hat{H}_*([A,A]_*,[A,X]_*) \Longrightarrow t_A(X)_*.$$

This is again conditionally convergent in the sense of Boardman. In the equivariant case this spectral sequence becomes

$$\hat{H}_*(\mathbf{R}_*(G_+),X_*) \Longrightarrow t^{\mathbf{R}}(X)_*^G.$$

For a more satisfactory account of the algebra, we assume $\mathcal{E}_*$ is projective as an l_*-module. Next, we express this in terms of a single type of resolution. Thus, by (KT1),

$$\tilde{\mathcal{E}}_* = [A,DA]_* = [A, A \wedge S^{-\tau}]_* = [A,A]_* \otimes_{[A,S]_*} [A,S^{-\tau}]_* = \mathcal{E}_* \otimes_{l_*} \lambda_*$$

where $\lambda_* = [A, S^{-\tau}]_*$. On the other hand, by (UCT),

$$\tilde{\mathcal{E}}_* = [A \wedge A, S]_* = \mathrm{Hom}([A,A]_*,[A,S]_*) = \mathrm{Hom}(\mathcal{E}_*, l_*),$$

so we conclude

$$\mathrm{Hom}(\mathcal{E}_*, l_*) = \mathcal{E}_* \otimes \lambda_*.$$

Next, we assume that the first Künneth theorem (KT1) applies also to $S^\tau \wedge S^{-\tau}$, so that λ_* is invertible and hence projective. Then we can specify a projective complete resolution by taking a resolution of l_*, dualizing and splicing. This is essentially the Tate cohomology of a Frobenius algebra, but with the twisting module inserted.

Proof of 11.1. We work in the category of **R**-modules and take $X = \mathbf{M} \wedge Y$ in the first and third case, and $X = F(Y, \mathbf{M})$ in the second. □

12. Calculations by commutative algebra

In this section we discuss the more subtle question of when the local cohomology theorem holds for $\mathcal{A}$ so that there is a calculation by commutative algebra in the sense of 10.5. This requires better behaviour of the cohomology theory concerned, and considerably more substance to the proof. We discuss two somewhat different methods for proving a local cohomology theorem. In a sense, the second method is a partial unravelling of the first: cellular constructions are replaced by multiple complexes. Both methods apply to the local cohomology theorem for finite groups, but beyond this they have different domains of relevance.

We discuss the more sophisticated example first [17, 24], because the formal machinery highlights the structure of the proof whilst hiding the technical difficulties.

Indeed if **R** is a highly structured commutative ring G-spectrum we have seen in Section 6 that, by construction, for any finitely generated ideal I in $\mathbf{R}^G_*$ we have spectral sequences

$$\hat{H}^*_I(\mathbf{R}^G_*; \mathbf{M}^G_*) \Longrightarrow t_I(\mathbf{M})^G_*$$
$$H^*_I(\mathbf{R}^G_*; \mathbf{M}^G_*) \Longrightarrow \Gamma_I(\mathbf{M})^G_*$$
$$H^I_*(\mathbf{R}^G_*; \mathbf{M}^G_*) \Longrightarrow (\mathbf{M}^\wedge_I)^G_*.$$

What we really want is to obtain similar spectral sequences for calculating $t_{\mathcal{A}}(\mathbf{M})^G_*$, $\Gamma_{\mathcal{A}}(\mathbf{M})^G_*$ and $(\mathbf{M}^\wedge_{\mathcal{A}})^G_*$ for the class $\mathcal{A}$ generated by G_+ using the ideal

$$I = \ker(\mathbf{R}^*_G \longrightarrow \mathbf{R}^*).$$

We assume here that $\mathbf{R}^*_G$ is Noetherian, so that I is finitely generated, but see [25] for an example where this is not true. To obtain the desired spectral sequences we need to check that each of the constructions with $\mathcal{A}$ is equivalent to the corresponding construction on **R**-modules for the ideal I. In fact, we need only check that

$$\Gamma_I(\mathbf{R}) \simeq \Gamma_I(\mathbf{R} \wedge EG_+) \simeq \Gamma_{\{G_+\}}\mathbf{R} = \mathbf{R} \wedge EG_+.$$

The second equivalence is a formal consequence of the fact that I restricts to zero non-equivariantly. The first equivalence contains the real work: it is equivalent to the statement that $\Gamma_I(\mathbf{R}\wedge\tilde{E}G)\simeq *$, where $\tilde{E}G$ is the unreduced suspension of EG. If G acts freely on a product of spheres (for example if it is a p-group) this follows from the existence of Euler classes (obviously elements of I) and the construction of $\tilde{E}G$ in terms of representation spheres [17]. To extend this to other groups some sort of transfer argument is necessary (see [20, 27] for examples).

This construction will give means of calculation whenever we have two suitably related smashing localizations. For example we may consider the localization $(\cdot)[\mathfrak{D}^{-1}]$ with acyclics $\mathfrak{D}$ and the localization $(\cdot)[I^{-1}]$ for an ideal I in the coefficient ring S_*. The requirements are then

- $\Gamma_I(S)\wedge S[\mathfrak{D}^{-1}]\simeq *$ and
- $S[I^{-1}]\wedge\Gamma_{\mathfrak{D}}(S)\simeq *$.

Together, these give the equivalence

$$\Gamma_I(S)\simeq\Gamma_{\mathfrak{D}}(S),$$

and hence the corresponding equivalences of other localization and colocalization functors. If we suppose $\mathfrak{D}$ is generated by the single augmented object A as before, and define $I=\ker([S,S]_*\longrightarrow[A,S]_*)$, then the second requirement is again a formal consequence of the fact that elements of I restrict to zero. One expects the first requirement to use special properties of the context, as it did in the equivariant case.

We now turn to the second method for proving a local cohomology theorem, and work with the group cohomology of a finite group in the derived category $D(kG)$ as in Section 7. We are considering the relationship with the derived category of the graded ring $R=H^*(G;k)$ and the ideal I of positive dimensional elements as in [19]. We may view these results as relating various completions and Tate cohomologies in the two categories by spectral sequences. We take this opportunity to extend the results of [19] to unbounded complexes. Since $H_*(G;M)$ is already I-complete if M is bounded below, the second spectral sequence is only of interest in the unbounded case.

Theorem 12.1. *Suppose G is a finite group, and M is a complex of kG-modules, and let I denote the ideal of positive codegree elements of the graded ring $H^*(G)$. There are spectral sequences*

$$H_I^*(H^*(G;M))\Longrightarrow H_*(G;M),$$
$$H_*^I(H^*(G);H_*(G;M))\Longrightarrow H^*(G;M^\wedge_{\{kG\}})$$
and
$$\hat{H}_*^I(H^*(G);H_*(G;M))\Longrightarrow H^*(G;t_{\{kG\}}(M)).$$

We explain the changes that need to be made to the arguments of [19] to cover the unbounded case. The idea is to use the algebraic spheres of Benson–Carlson [3] to construct algebraic analogues of tori B on which G acts freely. Suppose first that G has periodic cohomology with periodicity element $\zeta\in H^r(G)$. We may then form the algebraic sphere B as a complex of projectives representing ζ viewed as an r-extension of k by k. The complex B is concentrated in degrees between 0 and $r-1$, and we may form a projective resolution of k by concatenating copies of B to fill all degrees ≥ 0. Now, return to the general case. If k is of characteristic $p > 0$ and G is of p-rank r, then B may be formed as the tensor product of r algebraic spheres representing cohomology classes of a system of parameters. This is a complex graded over $\mathbb{Z}^r$ concentrated in a box with the lowest corner at the origin, and it is a complex of projectives by the theory of varieties. From B we may construct a multigraded projective resolution T of k by stacking boxes in the region with all coordinates ≥ 0. More generally, if $\sigma \subseteq \{1, 2, \ldots, r\}$ we may form $T[\sigma]$ by stacking boxes to fill the region defined by requiring $n_i \geq 0$ if $i \neq \sigma$. Thus $T[\emptyset] = T$, and $T[\{1, 2, \ldots, r\}]$ fills all of $\mathbb{Z}^r$. We then form a dual Koszul complex $L_\bullet$ of multigraded chain complexes:

$$L_\bullet = \left(\bigoplus_{|\sigma|=r} T[\sigma] \longrightarrow \bigoplus_{|\sigma|=r-1} T[\sigma] \longrightarrow \cdots \longrightarrow \bigoplus_{|\sigma|=0} T[\sigma] \right).$$

The idea of the proof is to consider the double complex

$$\mathrm{Hom}(L_\bullet, M)^G.$$

If one takes homology in the Koszul direction first one obtains $\mathrm{Hom}(T^!, M)^G$, where $T^!$ is the complex concentrated in negative multidegrees; provided M is bounded below this is isomorphic to the rth suspension of $T \otimes_G M$, and this has homology $H_*(G; M)$ by definition. If M is not bounded below, the first complex has infinite products where the second has infinite sums.

Now

$$\mathrm{Hom}(T[\sigma], M) = \mathrm{Hom}(\underleftarrow{\lim}_k \Sigma^{-k|\sigma|} T, M)$$

and, provided M is bounded below, this is equal to $\underrightarrow{\lim}_k \mathrm{Hom}(\Sigma^{-k|\sigma|} T, M)$ because the limit is achieved in each total degree. Thus, if one takes homology in the kG-resolution direction first, one obtains the stable Koszul complex of $H^*(G; M)$. To avoid the requirement of boundedness we simply use the double complex

$$\underrightarrow{\lim}_s \mathrm{Hom}(L[\geq s], M)$$

from the start, where $L[\geq s]$ is the quotient of L by the subcomplex of boxes which are at least s boxes below zero in some coordinate.

As is familiar from the case of commutative algebra, to construct the second spectral sequence we should consider the double complex

$$\underleftarrow{\mathrm{holim}}_s L[\geq s] \otimes_G M.$$

If we take homology in the Koszul direction first we obtain

$$\begin{aligned} \underleftarrow{\mathrm{holim}}_s T^![\geq s] \otimes_G M &= \underleftarrow{\mathrm{holim}}_s \mathrm{Hom}((T^![\geq s])^*, M)^G \\ &= \mathrm{Hom}(\underrightarrow{\mathrm{holim}}_s (T^![\geq s])^*, M)^G \\ &\simeq \mathrm{Hom}(\Sigma^r T, M)^G. \end{aligned}$$

On the other hand, if we take homology in the kG-resolution degree we obtain a homotopy inverse limit of complexes, each term of which is a suspension of

$H_*(G;M)$, and so that the differentials are products of the chosen generators of I. By definition this is $\operatorname{holim}_{\leftarrow s} UK_s(\mathbf{x}) \otimes_{H^*(G)} H_*(G;M)$, and by definition, its homology is the local homology in the statement.

For the Tate spectral sequence we combine these methods to form the double complex

$$\operatorname{holim}_{\rightarrow t} \operatorname{Hom}(L[\leq t], \operatorname{holim}_{\leftarrow s} T^![\geq s] \otimes_G M).$$

13. Gorenstein localizations

In this final section we point out that a local cohomology theorem in the sense of Section 12 implies a strong duality theorem in certain cases. The idea is that the local cohomology theorem gives a covariant equivalence of two objects that are quite generally contravariantly equivalent using a universal coefficient theorem. The composite contravariant self-equivalence is the duality.

To motivate the name, we recall that under mild hypotheses, a commutative complete local k-algebra (R, I, k) of dimension d is Gorenstein if $H_I^*(R) = H_I^d(R)$ (i.e., R is Cohen–Macaulay) and in addition

$$R = \operatorname{Hom}_R(H_I^d(R), R^\vee) = \operatorname{Hom}_{R/I}(H_I^d(R), R/I)$$

where $M^\vee = \operatorname{Hom}_{R/I}(M, R/I)$. We want to consider a homotopy level version of the Gorenstein condition on the unit object S in a stable homotopy category $\mathcal{C}$. To make sense of this we need (i) a second stable homotopy category $\overline{\mathcal{C}}$ with unit object $\bar{S}$, (ii) a 'restriction' functor $r\colon \mathcal{C} \longrightarrow \overline{\mathcal{C}}$, thought of as a forgetful map, and required to be lax monoidal, and (iii) an 'inflation' functor $i\colon \overline{\mathcal{C}} \longrightarrow \mathcal{C}$, splitting the forgetful map, and also required to be lax monoidal. This gives sense to the statement that S is an $\bar{S}$-algebra. Now take $I = \ker(S_* \longrightarrow \bar{S}_*)$, and say that S is *homotopically I-Gorenstein* if it is complete and there is an equivalence

$$S \simeq F(\Gamma_I(S), S^\vee) = F_{\bar{S}}(\Gamma_I(S), \bar{S})$$

where $X^\vee = F_{\bar{S}}(X, \bar{S})$, and where the $\bar{S}$-function object is an additional piece of structure.

To see that the homotopical Gorenstein statement has force, suppose $\bar{S}_*$ is a field. We then remark that if S is homotopically Gorenstein and S_* is Cohen–Macaulay then S_* is Gorenstein. Indeed, if S_* is Cohen–Macaulay of dimension d, then $\pi_*(\Gamma_I(S)) = H_I^d(S_*)$ from the spectral sequence of Section 6, and in the presence of a universal coefficient theorem we find a spectral sequence

$$\operatorname{Ext}_{\bar{S}_*}^{*,*}(H_I^d(S_*), \bar{S}_*) \Rightarrow S_*.$$

If in addition $\bar{S}_*$ is a field, this states that S_* is the dual of $H_I^d(S_*)$ and so S_* is Gorenstein. See [21] for further investigation.

The principal example of the present formal setup is when $\mathcal{C}$ is the category of equivariant $\mathbf{R}$-modules for a highly structured split ring spectrum $\mathbf{R}$ and $\overline{\mathcal{C}}$

is the category of non-equivariant $\mathbf{R}$-modules. The relevant functors have been constructed by Elmendorf and May [15, 37].

In this case the augmentation is right adjoint to product with $A = G_+$, and there is additional structure since the completion $X^\wedge_{\mathcal{A}} = F(EG_+, X)$ and the torsion $\Gamma_{\mathcal{A}}(X) = EG_+ \wedge X$ both have homotopy described in nonequivariant terms. It is pointed out in the appendix to [21] that when there is a local cohomology theorem, $\mathbf{R}^\wedge_{\mathcal{A}} = F(EG_+, \mathbf{R})$ is homotopically Gorenstein. Recalling that $S = \mathbf{R}$ in the equivariant category and $\bar{S} = \mathbf{R}$ in the non-equivariant category, we may summarize the proof as follows

$$F_{\bar{S}}(\Gamma_I(S^\wedge_{\mathcal{A}}), \bar{S}) \simeq F_{\bar{S}}(\Gamma_{\mathcal{A}}(S^\wedge_{\mathcal{A}}), \bar{S}) \simeq F_{\bar{S}}(\Gamma_{\mathcal{A}}(S), \bar{S}) \simeq F_S(\Gamma_{\mathcal{A}}(S), S) = S^\wedge_{\mathcal{A}}.$$

The first equivalence is the local cohomology theorem, the second is 2.3 and the third is the split condition.

We remark that one expects a twisting in the application of the universal coefficient theorem when G is not a finite group. For example with a compact Lie group G, the twisting is given by the adjoint bundle in the Adams isomorphism. Similarly the twisting is given by the dualizing module for a virtual Poincaré duality group as in [8]. The twisting is built from the invertible object S^τ in the sense that it is essentially S^τ on each copy of A used to build $\Gamma_{\mathcal{A}}(S)$. Thus, when G is a compact Lie group of dimension d, the adjoint bundle is a trivial d-dimensional bundle over any cell $S^n \wedge G_+$.

The existence and implications of the homotopy Gorentstein duality statement has been investigated for the cohomology of groups [19, 9, 8, 21], and for coefficients of equivariant cohomology theories in [17, 25, 26, 27]. We remark here that there is a precise formal similarity with Gross–Hopkins duality [28, 30, 41], which states that the Brown–Comenetz dual IM_nX of the monochromatic section M_nX is a twisted suspension of $L_{K(n)}DX$ for suitable finite spectra X, where M_nX is the fibre of $L_nX \longrightarrow L_{n-1}X$. Hopkins and Ravenel have proved there are spectral sequences for calculating the homotopy of M_nX and $L_{K(n)}X$ whose E_2-terms are the cohomology of the profinite group $\Gamma = S_n \rtimes \mathrm{Gal}(\mathbb{F}_{p^n}/\mathbb{F}_p)$ with suitable coefficients, where S_n is the Morava stabilizer group. Furthermore, Γ is a p-adic Lie group; if it is p-torsion free it is a Poincaré duality group, and in general its cohomology has a local cohomology theorem as in [8] (the proof in the discrete case carries over to the profinite case in the category of Symonds–Weigel [42]). The local cohomology theorem at the E_2 level is the precise counterpart of the Gross–Hopkins duality between the spectra.

References

[1] J. F. Adams, Lectures on generalized cohomology, Lecture Notes in Math. **99**, Springer-Verlag (1969), 1–138.

[2] J. M. Boardman, Conditionally convergent spectral sequences, Contemporary Math. **239**, Amer. Math. Soc. (1999), 49–84.

[3] D. J. Benson and J. F. Carlson, Complexity and multiple complexes, *Math. Z.* **195** (1987), 221–238.

[4] D. J. Benson and J. F. Carlson, Products in negative cohomology, *J. Pure Appl. Algebra* **82** (1992), 107–129.

[5] D. J. Benson, J. F. Carlson and J. Rickard, Complexity and varieties for infinitely generated modules I, *Math. Proc. Cambridge Philos. Soc.* **118** (1995), 223–243.

[6] D. J. Benson, J. F. Carlson and J. Rickard, Complexity and varieties for infinitely generated modules II, *Math. Proc. Cambridge Philos. Soc.* **120** (1996), 597–615.

[7] D. J. Benson and J. P. C. Greenlees, The action of the Steenrod algebra on Tate cohomology, *J. Pure Appl. Algebra* **85** (1993), 21–26.

[8] D. J. Benson and J. P. C. Greenlees, Commutative algebra for cohomology rings of classifying spaces of virtual duality groups, *J. Algebra* **192** (1997), 678–700.

[9] D. J. Benson and J. P. C. Greenlees, Commutative algebra for cohomology rings of classifying spaces of compact Lie groups, *J. Pure Appl. Algebra* **122** (1997), 41–53.

[10] J. F. Carlson, P. W. Donovan and W. W. Wheeler, Complexity and quotient categories for group algebras, *J. Pure Appl. Algebra* **93** (1994), 147–167.

[11] J. F. Carlson and W. W. Wheeler, Varieties and localizations of module categories, *J. Pure Appl. Algebra* **102** (1995), 137–153.

[12] E. S. Devinatz, Small ring spectra, *J. Pure Appl. Algebra* **81** (1992), 11–16.

[13] A. W. M. Dress, Contributions to the theory of induced representation, Lecture Notes in Math. **342**, Springer-Verlag (1973), 183–240.

[14] A. D. Elmendorf, I. Kriz, M. A. Mandell and J. P. May, Rings, modules and algebras in stable homotopy theory, American Mathematical Society Surveys and Monographs, American Mathematical Society (1997).

[15] A. D. Elmendorf and J. P. May, Algebras over equivariant sphere spectra, *J. Pure Appl. Algebra* **116** (1997), 139–149.

[16] J. P. C. Greenlees, Representing Tate cohomology of G-spaces, *Proc. Edinburgh Math. Soc.* **30** (1987), 435–443.

[17] J. P. C. Greenlees, K-homology of universal spaces and local cohomology of the representation ring, *Topology* **32** (1993), 295–308.

[18] J. P. C. Greenlees, Tate cohomology in commutative algebra, *J. Pure Appl. Algebra* **94** (1994), 59–83.

[19] J. P. C. Greenlees, Commutative algebra in group cohomology, *J. Pure Appl. Algebra* **98** (1995), 151–162.

[20] J. P. C. Greenlees, Augmentation ideals of equivariant cohomology rings, *Topology* **37** (1998), 1313–1323.

[21] J. P. C. Greenlees and G. Lyubeznik, Rings with a local cohomology theorem and applications to cohomology rings of groups, *J. Pure Appl. Algebra* **149** (2000), 267–285.

[22] J. P. C. Greenlees and J. P. May, Derived functors of completion and local homology, *J. Algebra* **149** (1992), 438–453.

[23] J. P. C. Greenlees and J. P. May, Generalized Tate cohomology, Mem. Amer. Math. Soc. **543** (1995).

[24] J. P. C. Greenlees and J. P. May, Completions in algebra and topology, Handbook of Algebraic Topology (ed. I. M. James), North-Holland (1995), 255–276.

[25] J. P. C. Greenlees and J. P. May, Localization and completion theorems for MU-module spectra, *Ann. of Math.* **146** (1997), 509–544.

[26] J. P. C. Greenlees and H. Sadofsky, The Tate spectrum of v_n-periodic complex oriented theories, *Math. Z.* **222** (1996), 391–405.

[27] J. P. C. Greenlees and H. Sadosfsky, Tate cohomology of theories with one dimensional coefficient ring, *Topology* **37** (1997), 279–292.

[28] B. H. Gross and M. J. Hopkins, Equivariant bundles on the Lubin–Tate moduli space, Contemporary Math. **158**, Amer. Math. Soc. (1994), 23–88.

[29] A. Grothendieck (notes by R. Hartshorne), Local cohomology, Lecture Notes in Math. **41**, Springer-Verlag (1967).

[30] M. J. Hopkins and B. H. Gross, The rigid analytic period mapping, Lubin–Tate space and stable homotopy theory, *Bull. Amer. Math Soc.* **30** (1994), 76–86.

[31] M. Hovey, Bousfield localization functors and Hopkins' chromatic splitting conjecture, Contemporary Math. **181**, Amer. Math. Soc. (1995), 225–250.

[32] M. Hovey, J. H. Palmieri and N. P. Strickland, Axiomatic stable homotopy theory, Mem. Amer. Math. Soc. **610** (1997).

[33] M. Hovey and N. P. Strickland, Morava K-theories and localization, Mem. Amer. Math. Soc. **666** (1999).

[34] L. G. Lewis, J. P. May and M. Steinberger (with contributions by J. E. McClure), Equivariant stable homotopy theory, Lecture Notes in Math. **1213**, Springer-Verlag (1986).

[35] M. E. Mahowald, D. C. Ravenel and P. Shick, The triple loop space approach to the telescope conjecture, Contemporary Math. **271**, Amer. Math. Soc. (to appear).

[36] M. E. Mahowald and H. Sadofsky, v_n-telescopes and the Adams spectral sequence, *Duke Math. J.* **78** (1995), 101–129.

[37] J. P. May, Equivariant and non-equivariant module spectra, *J. Pure Appl. Algebra* **127** (1998), 83–97.

[38] J. P. May, The geometry of iterated loop spaces, Lecture Notes in Math. **271**, Springer-Verlag (1972).

[39] H. R. Miller, Finite localization, *Bol. Soc. Mat. Mexicana* **37** (1992), 383–389.

[40] D. C. Ravenel, Nilpotence and periodicity in stable homotopy theory, Princeton University Press (1992).

[41] N. P. Strickland, Gross–Hopkins duality, *Topology* **39** (2000), 1021–1033.

[42] P. A. Symonds and T. Weigel, Cohomology for p-adic analytic groups, Progress in Math. **184**, Birkhäuser (2000), 349–410.

School of Mathematics and Statistics,
Hicks Building,
Sheffield S3 7RH, UK
E-mail address: `j.greenlees@sheffield.ac.uk`

Progress in Mathematics, Vol. 196

Serre's theorem and the $\mathcal{N}il_l$ filtration of Lionel Schwartz

Jesper Grodal

Abstract. We give three different cohomological characterizations of classifying spaces of p-compact toral groups amongst finite Postnikov systems satisfying mild conditions. This leads to a unifying generalization of previous versions of Serre's theorem on the homotopy groups of a finite complex.

1. Introduction

In the heart of all generalizations of Serre's theorem on the homotopy groups of a finite complex (cf. e.g. [17, 12, 10, 11, 4, 6]) there have implicitly or explicitly been statements about the 'non-finiteness' of the mod p cohomology of a finite Postnikov system. The negations of these statements say that a space with 'finite' mod p cohomology has infinitely many nontrivial mod p homotopy groups (that is, infinitely many homotopy groups are not uniquely p-divisible). Such statements can then in some cases (but not all, cf. Remark 4.5) afterwards be improved to showing that the space has infinitely many homotopy groups containing p-torsion, by employing a method of McGibbon and Neisendorfer [12]. We put these generalizations into a common framework by offering the following cohomological characterizations of classifying spaces of p-compact toral groups amongst finite Postnikov systems.

Theorem 1.1. *Let X be a connected finite Postnikov system with $\pi_1 X$ a finite p-group and $H^*(X;\mathbf{F}_p)$ of finite type. The following conditions are equivalent.*

1. $QH^*(X;\mathbf{F}_p) \in \overline{\mathcal{N}il}_2$.
2. $H^*(X;\mathbf{F}_p)$ *is of finite transcendence degree.*
3. $H^*(X;\mathbf{F}_p)$ *is noetherian.*
4. X *is* $\mathbf{F}_p$*-equivalent to the classifying space of a p-compact toral group.*

Here Q denotes the indecomposables, and $\overline{\mathcal{N}il}_2$ refers to the $\mathcal{N}il$ filtration of the category $\mathcal{U}$ of unstable modules over the Steenrod algebra of Schwartz [15]. (Note that locally finite modules over the Steenrod algebra lie in $\overline{\mathcal{N}il}_2$.) Recall that two spaces are said to be $\mathbf{F}_p$-equivalent if they become homotopy equivalent after

1991 *Mathematics Subject Classification.* Primary: 55Q52; Secondary: 55S45, 55S10.
Key words and phrases. Postnikov system, unstable module, $\mathcal{N}il$ filtration, Serre's theorem, p-toral group, p-compact group.

Bousfield $H_*(-;\mathbf{F}_p)$ localization [1]. Furthermore, recall that a classifying space of a p-compact toral group is a space which fits as the total space in a fibration sequence over $K(P,1)$ with fiber $K(\hat{\mathbf{Z}}_p^n,2)$ for some $n<\infty$, where $\hat{\mathbf{Z}}_p$ denotes the p-adic integers and P is a finite p-group (cf. [5]).

The theorem extends results in [6], and also the generalization of Serre's theorem of [4] can be easily recovered and generalized (Theorem 4.4).

To prove the results we study the $\mathcal{N}il_l$ filtration and its relation to the Eilenberg–Moore spectral sequence, using the methods of Schwartz [15], and use this to reduce the theorem to results of [6].

I am grateful to W. Dwyer and B. Shipley for helpful discussions and suggestions, and to B. Shipley for making an early thesis proposal [18] on a similar topic available to me.

2. Notation and preliminaries

By a space we mean, for simplicity, an object in the homotopy category of finite pointed CW-complexes. A connected finite Postnikov system is a connected space such that $\pi_i X = 0$ for i large. Throughout we abbreviate $H^*(X;\mathbf{F}_p)$ to H^*X where p is a fixed but arbitrary prime.

We now state some basic facts about the Eilenberg–Moore spectral sequence and the $\mathcal{N}il_l$ filtration.

2.1. The Eilenberg–Moore spectral sequence

The cohomological Eilenberg–Moore spectral sequence of a fibration $D\to E\to B$ over a connected space B is a second quadrant spectral sequence with E_2-term given by $E_2^{s,*}=\mathrm{Tor}^s_{H^*B}(\mathbf{F}_p,H^*E)$ as an unstable module over the Steenrod algebra $\mathcal{A}$ [14]. The columns $E_r^{s,*}$, $s\le 0, r\ge 2$ are likewise unstable $\mathcal{A}$-modules, and the differentials $d_r\colon E_r^{s,*}\to E_r^{s+r,*}$ are $\mathcal{A}$-linear of degree $-(r-1)$. The spectral sequence converges strongly to H^*D when H^*E and H^*B are of finite type and $\pi_1 B$ acts nilpotently on H^*D [3]. Specifically, there is a cocomplete filtration of $\mathcal{A}$-modules of H^*D,

$$H^*D\supset\cdots F_s\supset F_{s+1}\supset\cdots\supset F_0\supset F_1=0$$

such that $\Sigma^{-s}F_s/F_{s+1}\simeq E_\infty^{s,*}$.

2.2. The $\mathcal{N}il_l$ filtration

An unstable module M is said to be l-nilpotent if it is the colimit of unstable modules each having a finite filtration whose filtration quotients are l-suspensions. (For some equivalent definitions of l-nilpotency see for example [16], which is also a general reference for other facts concerning unstable modules over the Steenrod algebra.) Let $\mathcal{N}il_l$ denote the full subcategory of $\mathcal{U}$ of l-nilpotent modules. We hence get a decreasing filtration

$$\cdots\subset\mathcal{N}il_2\subset\mathcal{N}il_1=\mathcal{N}il\subset\mathcal{N}il_0=\mathcal{U}$$

of $\mathcal{U}$ with the property that $\mathcal{N}il_1$ equals the usual subcategory of nilpotent modules $\mathcal{N}il$ and $\cap_l \mathcal{N}il_l = 0$. It is often more convenient to work with the slightly larger subcategory $\overline{\mathcal{N}il_l}$ which is the smallest Serre class in $\mathcal{U}$ that contains $\mathcal{N}il_l$ and $\mathcal{B}$, the category of locally finite modules, and is closed under colimits. A more concrete description of $\overline{\mathcal{N}il_l}$ is given by observing that $M \in \overline{\mathcal{N}il_l}$ if and only if $M^{\geq l} \in \mathcal{N}il_l$.

3. Relations between QH^*X and $H^*\Omega X$

In this section, we study the relation between the $\mathcal{N}il_l$ filtration and the Eilenberg–Moore spectral sequence using the methods of Schwartz [15]. Our Theorem 3.1 can fairly easily be derived directly from [15, 16], but since we need the results under slightly weaker assumptions than the ones used in [15, 16], and think that insight is gained from a direct proof, we give one.

Define the nilpotency degree of an unstable module M as the largest l, such that $M \in \overline{\mathcal{N}il_l}$. If M is locally finite we say that the nilpotency degree of M is infinite. We start by giving a proof of the key result about the nilpotency degree.

Theorem 3.1. *Let X be a connected space with $\pi_1 X$ a finite p-group and H^*X of finite type. For any $l \geq 0$, $QH^*X \in \overline{\mathcal{N}il_{l+1}}$ if and only if $H^*\Omega X \in \overline{\mathcal{N}il_l}$.*

Proof. Since we assume that $\pi_1 X$ is a finite p-group, it acts nilpotently on the $\mathbf{F}_p$-vector space $H^*\Omega X$, since the group-ring $\mathbf{F}_p\pi_1 X$ has nilpotent augmentation ideal. Hence the Eilenberg–Moore spectral sequence for the path-loop fibration over X converges strongly to $H^*\Omega X$.

For an unstable algebra K we have by [16, Lemma 8.7.6] that

$$QK \in \overline{\mathcal{N}il_n} \quad \text{implies that} \quad \mathrm{Tor}^s_K(\mathbf{F}_p, \mathbf{F}_p) \in \overline{\mathcal{N}il_{n-s-1}}. \tag{3.1}$$

Assume that $QH^*X \in \overline{\mathcal{N}il_{l+1}}$. This implies that $\mathrm{Tor}^s_{H^*X}(\mathbf{F}_p, \mathbf{F}_p) \in \overline{\mathcal{N}il_{l-s}}$, so $\Sigma^{-s} F_s/F_{s+1} \in \overline{\mathcal{N}il_{l-s}}$. Thus, $F_s/F_{s+1} \in \overline{\mathcal{N}il_l}$ for all s which gives $H^*\Omega X \in \overline{\mathcal{N}il_l}$.

To prove the other direction assume that $H^*\Omega X \in \overline{\mathcal{N}il_l}$ so $\Sigma \bar{H}^*\Omega X \in \overline{\mathcal{N}il_{l+1}}$. If $QH^*X \in \mathcal{B}$ we are done; else let n be the nilpotency degree of QH^*X. Consider the canonical map

$$QH^*X = E_2^{-1,*} \to E_\infty^{-1,*} \hookrightarrow \Sigma \bar{H}^*\Omega X.$$

We claim that this map has kernel in $\overline{\mathcal{N}il_{n+1}}$. To see this, first note that since $QH^*X \in \overline{\mathcal{N}il_n}$ we have that $E_r^{-r-1} \in \overline{\mathcal{N}il_{n+r}}$ by (3.1). Therefore the image of $d_r\colon E_r^{-r-1,*} \to E_r^{-1,*}$ is in $\overline{\mathcal{N}il_{n+1}}$ by the fact that d_r is $\mathcal{A}$-linear of degree $-(r-1)$, and this establishes the claim. Since $\overline{\mathcal{N}il_l}$ is a Serre class we get that $n \geq \min\{n+1, l+1\}$, so $n \geq l+1$ as wanted. □

Corollary 3.2. *Let X be a connected space with $\pi_1 X$ a finite p-group and H^*X of finite type. We have that QH^*X is locally finite if and only if $H^*\Omega X$ is locally finite.* □

Remark 3.3. Note that one of the main results in [5] can be formulated as saying that if $H^*\Omega X$ is finite over $\mathbf{F}_p$ then QH^*X is finite over $\mathbf{F}_p$ —here the converse is however far from being true.

Theorem 3.1 makes it desirable to understand the relationship between the nilpotency degree of K and QK for an unstable algebra K. In general the nilpotency degree can differ radically if K is not in $\overline{\mathcal{N}il}_1$, for example for any finite group G whose order is divisible by p, $QH^*BG \in \mathcal{B}$ but $H^*BG \notin \overline{\mathcal{N}il}_1$. However the next proposition shows that this is the only thing that can go wrong.

Proposition 3.4. *Let K be a connected unstable algebra and assume that $K \in \overline{\mathcal{N}il}_1$. Then $K \in \overline{\mathcal{N}il}_l$ if and only if $QK \in \overline{\mathcal{N}il}_l$.*

Proof. It is clear that $QK \in \overline{\mathcal{N}il}_l$ if $K \in \overline{\mathcal{N}il}_l$. To see the converse suppose that $QK \in \overline{\mathcal{N}il}_l$. The statement is trivially true for $l = 0, 1$, so suppose that $l \geq 2$ and assume by induction that the statement is true for $l - 1$.

It is straight forward to see that $\bar{K}^0 = 0$ and $\bar{K} \in \mathcal{N}il_{l-1}$ ensures that $\bar{K} \otimes \bar{K} \in \overline{\mathcal{N}il}_l$ (cf. [15, Cor. 1.9]). (Here $\bar{K}$ denotes the augmentation ideal of K.) Hence both QK and $\bar{K} \otimes \bar{K}$ lie in $\overline{\mathcal{N}il}_l$, so $\bar{K} \in \overline{\mathcal{N}il}_l$ as well, by the defining sequence of QK and the fact that $\overline{\mathcal{N}il}_l$ is a Serre class. This shows that $K \in \overline{\mathcal{N}il}_l$ as wanted. □

4. Generalizations of Serre's theorem

In this section we use the results of the preceding section to prove the promised generalizations of Serre's theorem on the homotopy groups of a finite complex.

Before doing this, however, we prove a general result which says that the transcendence degree of the cohomology ring decreases when passing to covers. To get the result in its best form we use the Sullivan p-adic completion (see [13]), which coincides with the Bousfield–Kan $\mathbf{F}_p$-completion on spaces with mod p cohomology of finite type [13, 3.4]. (We will actually only need the result for spaces with mod p cohomology of finite type, where references to [13] can be replaced by references to [9].)

Theorem 4.1. *Let X be a Sullivan p-adically complete space. Then the transcendence degree of $H^*X\langle 1\rangle$ is less than or equal to the transcendence degree of H^*X.*

Proof. First note that we may assume that X is connected. Recall that, by the work of Morel [13], for any Sullivan p-adically complete space X and any elementary abelian p-group V, $[BV, X] = \mathrm{Hom}_{\mathcal{K}}(H^*X, H^*BV)$, where $[-,-]$ denotes free homotopy classes of maps and $\mathrm{Hom}_{\mathcal{K}}$ denotes Hom in the category of unstable algebras over the Steenrod algebra. By [8] the transcendence degree of H^*X is equal to the transcendence degree of the functor $[B-, X] = \mathrm{Hom}_{\mathcal{K}}(H^*X, H^*B-)$ from elementary abelian p-groups to profinite sets. One definition of the transcendence degree of the functor $[B-, X]$ is the rank of the largest elementary abelian p-group V such that there exists $s \in [BV, X]$ which cannot be written as $s = s' \circ B\varphi$, where $\varphi \in \operatorname{End} V$ is singular (see [8], [6, Prop. 5.8]). Since $[BV, X]$ is obtained from $[BV, X]_{\mathrm{pt}}$ by taking the quotient under the action of $\pi_1 X$ we especially see that $[B-, X]$ and $[B-, X]_{\mathrm{pt}}$ have the same transcendence degree.

The principal fibration $\pi_1 X \to X\langle 1\rangle \to X$ shows that the map

$$[BV, X\langle 1\rangle]_{\mathrm{pt}} \to [BV, X]_{\mathrm{pt}}$$

is injective, so the transcendence degree of $[B-, X\langle 1\rangle]$ is less than or equal to that of $[B-, X]$. Since $X\langle 1\rangle$ is again Sullivan p-adically complete (see [13, 1.3, 1.4]) we conclude that the transcendence degree of $H^*X\langle 1\rangle$ is less than or equal to that of H^*X. □

Theorem 4.2. *Assume that X is a connected finite Postnikov system, with $\pi_1 X$ a finite p-group and H^*X of finite type. Then $QH^*X \in \overline{\mathcal{N}il}_2$ if and only if X is $\mathbf{F}_p$-equivalent to the classifying space of a p-compact toral group.*

Proof. If X is $\mathbf{F}_p$-equivalent to the classifying space of a p-compact toral group, then ΩX is $\mathbf{F}_p$-equivalent to a disjoint union of circles, so especially $H^*\Omega X \in \overline{\mathcal{N}il}_1$, so $QH^*X \in \overline{\mathcal{N}il}_2$ by Theorem 3.1.

To see the converse, assume that $QH^*X \in \overline{\mathcal{N}il}_2$. Since $\pi_1 X$ is a finite p-group the Bousfield–Kan $\mathbf{F}_p$-completion of X is $\mathbf{F}_p$-complete and is again a finite Postnikov system by [2, II.5.1], so we may assume that X is $\mathbf{F}_p$-complete. Hence ΩX is also $\mathbf{F}_p$-complete [2] [5, 11.9]. By for example the Eilenberg–Moore spectral sequence, $H^*\Omega X$ is also of finite type so both X and ΩX are Sullivan p-adically complete. By Theorem 3.1 $H^*(\Omega X) \in \overline{\mathcal{N}il}_1$, so Theorem 4.1 implies that also $H^*((\Omega X)\langle 1\rangle) \in \overline{\mathcal{N}il}_1$. Likewise, as above, $H^*((\Omega X)\langle 1\rangle)$ is of finite type.

We therefore have that $(\Omega X)\langle 1\rangle$ is a one-connected finite Postnikov system with cohomology in $\overline{\mathcal{N}il}_1$ and of finite type, which by [6, Thm. 1.1] implies that $\tilde{H}^*((\Omega X)\langle 1\rangle) = 0$. So X is homotopy equivalent to its second Postnikov stage $P_2 X$, since both spaces are $\mathbf{F}_p$-complete. Since $H^*\Omega P_2 X \in \overline{\mathcal{N}il}_1$, the abelian group $\pi_2 X$ cannot have p-torsion, as this would imply the existence of an element of infinite height in the mod p reduced cohomology ring of $\Omega_0 P_2 X = K(\pi_2 X, 1)$ by standard group cohomology (or [9]). (Ω_0 denotes the zero component of the loop space.) Hence X is an $\mathbf{F}_p$-complete space with homotopy only in dimensions 1 and 2, and $\pi_2 X$ p-torsion free. But this means that $\pi_2 X$ is a torsion free Ext-p-complete abelian group and hence isomorphic to $\hat{\mathbf{Z}}_p^n$ for some n (cf. [2, p. 181], [7]). Since H^*X is of finite type we have $n < \infty$. This completes the proof. □

Combining Theorem 4.2 with the results in [6] now enables us to give a proof of the main Theorem 1.1.

Proof of Theorem 1.1. By [5, Prop. 6.9, 12.1], 4 implies 3. Condition 3 obviously implies 1 and 2, and the implication 1 implies 4 follows from Theorem 4.2. The remaining 2 implies 4 follows easily from [6]. Namely, assume that the transcendence degree of H^*X is finite. We can assume that X is $\mathbf{F}_p$-complete. By Theorem 4.1, $H^*X\langle 1\rangle$ likewise has finite transcendence degree. Now, $X\langle 1\rangle$ is an $\mathbf{F}_p$-complete one-connected finite Postnikov system with $H^*X\langle 1\rangle$ of finite type, which by [6, Thm. 1.2] means that $X\langle 1\rangle$ is homotopy equivalent to $K(\hat{\mathbf{Z}}_p^n, 2)$ for some $n < \infty$. Hence X is $\mathbf{F}_p$-equivalent to the classifying space of a p-compact toral group. □

Remark 4.3. Note how 1 and 2 are complementary, in the sense that 2 is a statement about the ring structure of H^*X, whereas 1 is a statement about the Steenrod algebra action on what is left when you kill all products.

Using the 'McGibbon–Neisendorfer trick', which basically says that an arbitrary space cannot have locally finite cohomology *and* have infinitely many nontrivial homotopy groups which are all p-torsion free, we can now easily strengthen the result of Dwyer and Wilkerson [4]. We show that the 2-connected assumption on their result was only necessary to exclude classifying spaces of p-compact toral groups.

Theorem 4.4. *Assume that X is a connected space, with $\pi_1 X$ a finite p-group and H^*X of finite type. If QH^*X is locally finite as a module over the Steenrod algebra then X is either $\mathbf{F}_p$-equivalent to the classifying space of a p-compact toral group, or it contains p-torsion in infinitely many of its homotopy groups.*

Proof. Assume that X is not $\mathbf{F}_p$-equivalent to the classifying space of a p-compact toral group. By Theorem 4.2, X cannot be $\mathbf{F}_p$-equivalent to a finite Postnikov system, in other words $\pi_i(X; \mathbf{Z}/p) \neq 0$ for infinitely many i. But $H^*\Omega X$ is also locally finite by Corollary 3.2 which by [12, p. 255] implies that $\pi_i X$ actually has p-torsion for infinitely many i. □

Remark 4.5. The above theorem is not a true generalization of Theorem 4.2 for a good reason. The assumptions in Theorem 4.4 cannot be weakened to the assumption $QH^*X \in \overline{\mathcal{N}il}_2$, as is demonstrated by setting $X = BSU$, the classifying space of the infinite special unitary group. Indeed, $H^*SU \in \overline{\mathcal{N}il}_1$, so $QH^*BSU \in \overline{\mathcal{N}il}_2$ by Theorem 3.1. But BSU is obviously neither the classifying space of a p-compact toral group nor does it contain any p-torsion in its homotopy groups by Bott periodicity. (See also [11].)

References

[1] A. K. Bousfield. The localization of spaces with respect to homology. *Topology*, 14:133–150, 1975.

[2] A. K. Bousfield and D. M. Kan. *Homotopy Limits, Completions and Localizations.* Lecture Notes in Math. vol. 304. Springer, 1972.

[3] W. G. Dwyer. Strong convergence of the Eilenberg–Moore spectral sequence. *Topology*, 13:255–265, 1974.

[4] W. G. Dwyer and C. W. Wilkerson. Spaces of null homotopic maps. *Astérisque*, 191:97–108, 1990. International Conference on Homotopy Theory (Marseille-Luminy, 1988).

[5] W. G. Dwyer and C. W. Wilkerson. Homotopy fixed point methods for Lie groups and finite loop spaces. *Ann. of Math.*, 139:395–442, 1994.

[6] J. Grodal. The transcendence degree of the mod p cohomology of finite Postnikov systems. In *Stable and unstable homotopy (Toronto, ON, 1996)*, pages 111–130. Amer. Math. Soc., Providence, RI, 1998.

[7] D. K. Harrison. Infinite abelian groups and homological methods. *Ann. of Math.*, 69:366–391, 1959.

[8] H.-W. Henn, J. Lannes, and L. Schwartz. The categories of unstable modules and unstable algebras over the Steenrod algebra modulo nilpotent objects. *Amer. J. Math.*, 115:1053–1106, 1993.

[9] J. Lannes. Sur les espaces fonctionnels dont la source est le classifiant d'un p-groupe abélien élémentaire. *Inst. Hautes Études Sci. Publ. Math.*, 75:135–244, 1992.

[10] J. Lannes and L. Schwartz. A propos de conjectures de Serre et Sullivan. *Invent. Math.*, 83:593–603, 1986.

[11] J. Lannes and L. Schwartz. Sur les groupes d'homotopie des espaces dont la cohomologie modulo 2 est nilpotente. *Israel J. Math.*, 66:260–273, 1989.

[12] C. A. McGibbon and J. Neisendorfer. On the homotopy groups of a finite dimensional space. *Comment. Math. Helv.*, 59:253–257, 1984.

[13] F. Morel. Quelques remarques sur la cohomologie modulo p continue des pro-p-espaces et les résultats de J. Lannes concernant les espaces fonctionnels hom(BV, X). *Ann. Sci. École Norm. Sup. (4)*, 26(3):309–360, 1993.

[14] D. Rector. Steenrod operations in the Eilenberg–Moore spectral sequence. *Comment. Math. Helv.*, 45:540–552, 1970.

[15] L. Schwartz. La filtration nilpotente de la categorie $\mathcal{U}$ et la cohomologie des espaces de lacets. In *Algebraic topology —rational homotopy (Louvain-la-Neuve, 1986)*, Lecture Notes in Math. vol. 1318, pages 208–218. Springer, 1988.

[16] L. Schwartz. *Unstable Modules over the Steenrod Algebra and Sullivan's Fixed Point Set Conjecture.* The University of Chicago Press, 1994.

[17] J.-P. Serre. Cohomologie modulo 2 des complexes d'Eilenberg–Mac Lane. *Comment. Math. Helv.*, 27:198–232, 1953.

[18] B. Shipley. On Serre's conjecture. 1992 MIT Thesis proposal.

Department of Mathematics
Massachusetts Institute of Technology
Cambridge, MA 02139, USA
E-mail address: `jg@math.mit.edu`

Progress in Mathematics, Vol. 196

New relationships among loopspaces, symmetric products, and Eilenberg MacLane spaces

Nicholas J. Kuhn

Abstract. Let $T(j)$ be the dual of the jth stable summand of Ω^2S^3 (at the prime 2) with top class in dimension j. Then it is known that $T(j)$ is a retract of a suspension spectrum, and that the homotopy colimit of a certain sequence $T(j) \to T(2j) \to \dots$ is an infinite wedge of stable summands of $K(V,1)$'s, where V denotes an elementary abelian 2 group. In particular, when one starts with $T(1)$, one gets $K(Z/2,1) = RP^\infty$ as one of the summands.

I discuss a generalization of this picture using higher iterated loopspaces and Eilenberg MacLane spaces. I consider certain finite spectra $T(n,j)$ for $n,j \geq 0$ (with $T(1,j) = T(j)$), dual to summands of $\Omega^{n+1}S^N$, conjecture generalizations of the above, and prove that these conjectures are correct in cohomology. So, for example, $T(n,j)$ has unstable cohomology, and the cohomology of the hocolimit of a certain sequence $T(n,j) \to T(n,2j) \to \dots$ agrees with the cohomology of the wedge of stable summands of $K(V,n)$'s corresponding to the wedge occurring in the $n = 1$ case above.

One can also map the $T(n,j)$ to each other as n varies, and here the cohomological calculations imply a homotopical conclusion: the hocolimits that are nonzero, $T(\infty, 2^k)$, for $k \geq 0$, map to each other, giving rise to a fitration of $H\mathbf{Z}/2$ which is equivalent to the mod 2 symmetric powers of spheres filtration.

Our homotopical constructions use Hopf invariant methods and loopspace technology. These are quite general and should be of independent interest. To study the action of the Steenrod operations on the cohomology of our spectra, we derive a Nishida formula for how $\chi(Sq^i)$ acts on Dyer–Lashof operations. This should be of use in other settings. In an appendix, we explain connections with recent work by Greg Arone and Mark Mahowald on the Goodwillie tower of the identity.

1. Introduction

With all spaces and spectra localized at 2, let $T(j)$ be the $(2j)^{th}$ dual of the jth stable summand of Ω^2S^3. These finite complexes were explored in the 1970's and

Partially supported by the N.S.F. and the C.N.R.S..

1980's in work by M. Mahowald, E. Brown, S. Gitler, F. Peterson, R. Cohen, G. Carlsson, H. Miller, J. Lannes, and P. Goerss, among others. (Entries into the extensive literature include [33, 5, 8, 37, 28, 12, 15].) They played an essential role in a number of the major achievements in homotopy theory during this time: Mahowald's construction [33] of an infinite family of 2-primary elements in $\pi_*^S(S^0)$ having Adams filtration 2; Goerss, Lannes, and F. Morel's work [12] on representing mod 2 homology by maps from (desuspensions of) the $T(j)$'s; and Miller's proof of the Sullivan conjecture [37].

All of this work is a reflection of unexpected "unstable" properties of the $T(j)$'s. [33] is based on two facts: that as modules over the Steenrod algebra, the cohomology of the $T(j)$ are dual Brown–Gitler modules, and that one can construct maps $T(j) \to T(2j)$ realizing on cohomology certain canonical maps between these. [8, 37] are then based on the connection, just on the level of cohomology, between the classifying spaces BV of elementary abelian 2 groups V, and the homotopy colimits of the sequences

$$T(j) \to T(2j) \to T(4j) \to \cdots .$$

It is not hard to show that this cohomological connection can be realized homotopically: these hocolimits are always infinite wedges of stable wedge summands of BV's. In particular, if one starts with $T(1)$, one gets $B(\mathbf{Z}/2)$ as a summand. Finally, that $T(j)$ has unstable cohomology is explained by the fact that $T(j)$ is homotopic to a dual Brown–Gitler spectrum, which can be shown to be a wedge summand of suspension spectrum [27, 11, 15]. ([12] shows much more.)

In this paper, we first show that, at least on the level of cohomology, certain finite complexes $T(n,j)$ arising from $\Omega^{n+1}S^N$ appear to be unstable, and to be related to the Eilenberg-MacLane spaces $K(V,n)$ in the same way that the $T(j)$ are related to the spaces BV. Second, we let "n go to ∞", and obtain homotopical connections between these finite complexes and symmetric powers of spheres.

What I prove involves, first of all, some new observations about loopspace machinery and the Nishida relations which should be of independent interest. For Theorem 1.6 (which describes $T(n,j)$ as n goes to ∞), the author's old work on the Whitehead conjecture [19, 20] is needed. The proof of Theorem 1.9 (which describes how the $T(n,j)$ are cohomologically related to $K(V,n)$'s) uses much of what the author knows about the relationship between the category of unstable modules over the Steenrod algebra and the "generic representation" category of [23, 24, 25].

What I can't yet prove, but only conjecture, seems to suggest that there is a remarkable "naturally occurring" infinite loopspace (or perhaps E_∞–ring spectrum) waiting to be discovered.

To explain our main results, we need to introduce our cast of characters. Recall that [34], if X is path connected, there is a stable decomposition

$$\Sigma^\infty \Omega^n \Sigma^n X \simeq \bigvee_{j \geq 1} \Sigma^\infty D_{n,j} X,$$

where $D_{n,j}X = \mathcal{C}(n,j)_+ \wedge_{\Sigma_j} X^{[j]}$. Here $\mathcal{C}(n,j)$ is the configuration space of j tuples of distinct 'little cubes' in I^n, a space acted on freely by the jth symmetric group Σ_j, and $X^{[j]}$ denotes the j-fold smash product of X with itself.

For a given n and j, there is a natural number d and a natural equivalence

$$D_{n,j}(\Sigma^d X) \simeq \Sigma^{dj} D_{n,j} X,$$

thus allowing $D_{n,j}X$ to be defined for a finite *spectrum*[1].

Definition 1.1. For $n \geq 0, j \geq 0$, let $T(n,j)$ be the S-dual of $D_{n+1,j}(S^{-n})$.

$T(n,j)$ is a finite spectrum with top cell in dimension nj, and with bottom mod 2 homology in dimension $n\alpha(j)$, where $\alpha(j)$ denotes the number of 1's in the 2-adic expansion of j. As examples, we note that, for all j and n, $T(0,j) = S^0 = T(n,0)$, $T(n,1) = S^n$, $T(1,j) = T(j)$ as above, and $T(n,2) =$ cofiber $\{\Sigma^n RP_+^{n-1} \to S^n\}$.

This bigraded family of finite spectra has some extra structure we will need. The H-space structure on loopspaces induces copairings

$$\Psi : T(n,k) \to \bigvee_{i+j=k} T(n,i) \wedge T(n,j).$$

Evaluation on loopspaces induces maps

$$\delta : T(n,j) \to \Sigma^{-1} T(n+1,j).$$

Finally, looping Hopf invariants, together with the above periodicity, induces "Frobenious" maps

$$\Phi : T(n,j) \to T(n,2j).$$

These three families of maps will be shown to be compatible in the expected ways. In particular, δ and Φ commute up to homotopy.

Our first result is a description of $H^*(T(n,j);\mathbf{Z}/2)$ as a module over the mod 2 Steenrod algebra $\mathcal{A}$. Following the lead of others in the $n = 1$ case [8, 37, 30], we describe the bigraded object $H^*(T(n,*);\mathbf{Z}/2)$, with the extra structure afforded by Ψ^* and Φ^*. We need first to define variants on the category $\mathcal{U}$ of unstable $\mathcal{A}$ modules, and the category $\mathcal{K}$ of unstable $\mathcal{A}$ algebras.

Let $\mathcal{U}_\rho$ be the category whose objects are pairs (M,ρ): $M = M_{*,*}$ is an $\mathbf{N} \times \mathbf{N}[\frac{1}{2}]$ graded $\mathbf{Z}/2$ vector space[2] whose columns $M_{*,j}$ are unstable $\mathcal{A}$ modules, and $\rho : M \to M$ is a collection of $\mathcal{A}$ linear maps $\rho : M_{*,2j} \to M_{*,j}$. Morphisms in $\mathcal{U}_\rho$ are just maps $f : M \to N$ preserving all structure.

Let $\mathcal{K}_\rho$ be the category of "restricted algebras in $\mathcal{U}_\rho$", i.e. commutative, unital algebras K in $\mathcal{U}_\rho$ (a category with a tensor product) satisfying the "restriction axiom": $Sq^{|x|}x = (\rho(x))^2$ for all $x \in K$.

Let $U_\rho : \mathcal{U}_\rho \to \mathcal{K}_\rho$ be the free functor, left adjoint to the forgetful functor. Explicitly, $U_\rho(M,\rho) = S^*(M)/(Sq^{|x|}x - (\rho(x))^2)$.

[1] There are more sophisticated ways to do this. See §2.

[2] Often $\mathbf{N} \times \mathbf{N}$ graded vector spaces will be considered $\mathbf{N} \times \mathbf{N}[\frac{1}{2}]$ graded by setting $M_{*,j} = \{0\}$ for $j \notin \mathbf{N}$.

If $I = (i_1, \dots, i_l)$, we set $Sq^I = Sq^{i_1} \dots Sq^{i_l}, l(I) = l$, and $e(I) = (i_1 - 2i_2) + \cdots + (i_{l-1} - 2i_l) + i_l$. I is called *admissible* if $i_s \geq 2i_{s+1}$ for all s. Define $E(n), L(k) \subset \mathcal{A}$ by

$$E(n) = \langle Sq^I \mid I \text{ is admissible and } e(I) > n \rangle$$

$$L(k) = \langle Sq^I \mid I \text{ is admissible and } l(I) > k \rangle.$$

Both of these are known to be left $\mathcal{A}$ modules [40, Prop.1.6.2], [36] . Now let $F(n,k)$ be the unstable $\mathcal{A}$ module $\Sigma^n(A/(E(n) + L(k)))$, and then let $F_\rho(n) \in \mathcal{U}_\rho$ be the pair $(\bigoplus_{k\geq 0} F(n,k), \rho)$, where $F(n,k)$ has second grading 2^k, and $\rho : F(n,k+1) \to F(n,k)$ is the projection.

Theorem 1.2. *Let $n \geq 1$. With multiplication and restriction given by Ψ^* and Φ^*,*

$$H^*(T(n,*); \mathbf{Z}/2) \simeq U_\rho(F_\rho(n))$$

as objects in $\mathcal{K}_\rho$. In particular, $H^*(T(n,j); \mathbf{Z}/2)$ is an unstable $\mathcal{A}$ module.

This theorem suggests

Conjecture 1.3. $T(n,j)$ is a stable wedge summand of a suspension spectrum.

This is known to be true when $n = 1$ [27, 11, 15].

To discuss stablizing $T(n,j)$ with respect to δ, we make the following definition.

Definition 1.4. $T(\infty, j) = \text{hocolim } \{ \; T(0,j) \xrightarrow{\delta} \cdots \xrightarrow{\delta} \Sigma^{-n}T(n,j) \xrightarrow{\delta} \cdots \}$.

Theorem 1.5.

(1) *$T(\infty, j) \simeq *$ unless j is a power of 2.*

(2) *$H^*((T(\infty, 2^k); \mathbf{Z}/2) \simeq \mathcal{A}/L(k)$ as $\mathcal{A}$ modules.*

The $\mathcal{A}$ module $\mathcal{A}/L(k)$ is already known to arise as the cohomology of a spectrum: it is the cohomology of $SP^{2^k}_\Delta(S^0)$, the cofiber of the diagonal map $\Delta : SP^{2^{k-1}}(S^0) \to SP^{2^k}(S^0)$ between symmetric products of the sphere spectrum S^0 [38].

Theorem 1.6. *The sequence $T(\infty,1) \to T(\infty,2) \to T(\infty,4) \to \dots$ is equivalent to the sequence $SP^1_\Delta(S^0) \to SP^2_\Delta(S^0) \to SP^4_\Delta(S^0) \to \dots$. In particular, $T(\infty, 2^k) \simeq SP^{2^k}_\Delta(S^0)$.*

Thus the maps $T(\infty, 2^k) \to T(\infty, 2^{k+1})$ have the striking properties proved in [19], e.g. they induce the zero map in homotopy groups in positive degrees.

Corollary 1.7. $\underset{n,k\to\infty}{\text{hocolim}} \; \Sigma^{-n}T(n, 2^k) \simeq H\mathbf{Z}/2$.

We now turn our discussion to how $T(n,j)$ stablizes with respect to Φ.

Definition 1.8. $\Phi^{-1}T(n,j) = \text{hocolim } \{ \; T(n,j) \xrightarrow{\Phi} T(n,2j) \xrightarrow{\Phi} T(n,4j) \xrightarrow{\Phi} \cdots \}$

Our last theorem identifies $H^*(\Phi^{-1}T(n,j);\mathbf{Z}/2)$ as the cohomology of an infinite wedge of certain stable summands of the Eilenberg MacLane spaces $K(V,n)$, in a manner that is independent of n. In particular, just as $H^*(K(\mathbf{Z}/2,1);\mathbf{Z}/2)$ was shown in [8] to be an $\mathcal{A}$ module direct summand of $H^*(\Phi^{-1}T(1,1);\mathbf{Z}/2)$, so is $H^*(K(\mathbf{Z}/2,n);\mathbf{Z}/2)$ an $\mathcal{A}$ module summand of $H^*(\Phi^{-1}T(n,1);\mathbf{Z}/2)$.

To be more precise, we need yet more notation. As in [23, 24, 25], let $\mathcal{F}$ be the category with objects the functors

$$F : \text{finite dimensional } \mathbf{Z}/2 \text{ vector spaces} \rightarrow \mathbf{Z}/2 \text{ vector spaces},$$

and with morphisms the natural transformations. For example, S^j and S_j, defined by $S^j(V) = V^{\otimes j}/\Sigma_j$ and $S_j(V) = (V^{\otimes j})^{\Sigma_j}$, are objects in $\mathcal{F}$.

Let Λ be an indexing set for the simple objects in this abelian category: algebraic group considerations suggest a number of Λ's, e.g. the set of 2-regular partitions [24, Sections 5 and 6]. Given $\lambda \in \Lambda$, let $F_\lambda \in \mathcal{F}$ be the corresponding simple object, V_λ a vector space large enough so that $F_\lambda(V_\lambda) \neq 0$, $e_\lambda \in \mathbf{Z}_2[End(V_\lambda)]$ an idempotent chosen so that $\mathbf{Z}/2[End(V_\lambda)]e_\lambda$ is the projective cover of the $\mathbf{Z}/2[End(V_\lambda)]$ module $F_\lambda(V_\lambda)$, and $K(\lambda,n) = e_\lambda\Sigma^\infty K(V_\lambda,n)$ the corresponding stable summand of $K(V_\lambda,n)$. Finally, given $\lambda \in \Lambda$ and $j = 0,1,\ldots,$ define $a(\lambda,j) \in \mathbf{N}$ by

$$a(\lambda,j) = \dim_{\mathbf{Z}/2} \operatorname{Hom}_{\mathcal{F}}(F_\lambda, S^{2^k j}), \text{ for } k >> 0.$$

Theorem 1.9. $H^*(\Phi^{-1}T(n,j);\mathbf{Z}/2) \simeq H^*(\bigvee_{\lambda\in\Lambda} a(\lambda,j)K(\lambda,n);\mathbf{Z}/2)$ *as $\mathcal{A}$ modules.*

(Here $\bigvee_i b_iY_i$ means that each Y_i occurs in the wedge sum with multiplicity b_i.)

We remark that these large $\mathcal{A}$ modules are nevertheless of finite type.

Conjecture 1.10. $\Phi^{-1}T(n,j) \simeq \bigvee_{\lambda\in\Lambda} a(\lambda,j)K(\lambda,n)$.

Some form of the following has been known to the experts[3] since the late 1980's.

Proposition 1.11. *This conjecture is true when $n = 1$. In particular, $\Phi^{-1}T(1,1)$ has $B(\mathbf{Z}/2)$ as a stable summand.*

The organization of the rest of the paper is as follows.

§2, §3, and §4 are devoted to the geometric constructions used to define the three families of maps Ψ,Φ,δ on the $T(n,j)$. In hopes that these will be useful in other settings, we develop this material with perhaps more care than is traditional (at one point, proving a lemma using ideas from "Goodwillie calculus"). Theorem 2.4 summarizes our main geometric results. In §5, properties of these constructions are combined with standard formula [9] for the homology of iterated loopspaces to give descriptions of $H^*(T(n,j);\mathbf{Z}/2), \Psi^*, \Phi^*$, and δ^* in terms of

[3] By "experts" here I mean at least the authors of [29], as well as myself.

Dyer-Lashof-like operations. The standard Nishida relations then yield recursive formulae for how $\chi(Sq^i)$ acts on $H^*(T(n,j);\mathbf{Z}/2)$; we deduce more useful formulae for how Sq^i acts in §6. These should be of some independent interest. Theorem 1.2 and Theorem 1.5 are then deduced in §7. Using the author's proof of the Whitehead conjecture, Theorem 1.6 is quickly deduced from Theorem 1.5 in §8.

The proof of Theorem 1.9 is rather different. Recall [23] that there are adjoint functors

$$\mathcal{U} \underset{r}{\overset{l}{\rightleftarrows}} \mathcal{F},$$

where $r(F) = \mathrm{Hom}_{\mathcal{F}}(S_*, F)$, with the Steenrod operations acting on the right of the S_j in the obvious way. Let $I_\lambda \in \mathcal{F}$ be the injective envelope of the simple functor F_λ, and let $\Phi^{-1}S^j \in \mathcal{F}$ be defined by

$$\Phi^{-1}S^j = \mathrm{colim}\left\{S^j \xrightarrow{\Phi} S^{2j} \xrightarrow{\Phi} S^{4j} \xrightarrow{\Phi} S^{8j} \cdots\right\},$$

where $\Phi : S^j \to S^{2j}$ is the squaring map.

The "Vanishing Theorem" of [24] says that $\Phi^{-1}S^j$ is an injective object in the category $\mathcal{F}_\omega \subset \mathcal{F}$ of locally finite functors. It follows formally that there is a decomposition in $\mathcal{F}$

$$\Phi^{-1}S^j \simeq \bigoplus_{\lambda\in\Lambda} a(\lambda,j)I_\lambda.$$

Precomposing this with the functor S_n, and then applying the functor r, yields a decomposition in $\mathcal{U}$

$$\Phi^{-1}r(S^j \circ S_n) \simeq \bigoplus_{\lambda\in\Lambda} a(\lambda,j)r(I_\lambda \circ S_n).$$

The classical description of $H^*(K(V,n);\mathbf{Z}/2)$ reveals that

$$r(I_\lambda \circ S_n) = H^*(K(\lambda,n);\mathbf{Z}/2),$$

so the righthand side of this last decomposition agrees with the righthand side of the the isomorphism in Theorem 1.9. Meanwhile, the lefthand side of the isomorphism of Theorem 1.9 is known by Theorem 1.2; this is then shown to agree with $\Phi^{-1}r(S^j \circ S_n)$ by using a new result of ours [26] that calculates $r(S^j \circ F)$ as a functor of $r(F)$.

§9 contains the details of this outline of the proof of Theorem 1.9. Finally in §10, we prove Proposition 1.11, as well as discussing approaches to the conjectures. In the appendix, we relate our spectra $T(\infty, 2^k)$ to work of Arone and Mahowald [3]. Theorem 1.6 thus gives new information about their constructions.

We wish to give hearty thanks to Doug Ravenel. This project had its origins in a question that he was asking in late 1994: our Conjecture 1.10 amounts to a refinement and extension of this. Most of our results were presented in Gargnano, Italy in 1995 and Toronto, Canada in 1996. An earlier version of this preprint was circulated in the summer of 1996. In the two years that have followed, we have noticed that Theorem 1.6 follows from Theorem 1.5, and managed to connect our constructions to those of Arone and Mahowald. We hope the reader familiar with our older version will appreciate these improvements.

2. Geometric Constructions

We begin by being a bit more specific about some notation introduced in the introduction. A point $\mathbf{c} \in \mathcal{C}(n,j)$ is a j tuple $\mathbf{c} = (c_1, \dots c_j)$ in which each $c_i : I^n \to I^n$ is a product of n linear embeddings from the unit interval I to itself, and the interiors of the images of the c_i are disjoint. Then the book of Gaunce Lewis, et. al. [32] shows that the functor

$$D_{n,j}X = \mathcal{C}(n,j)_+ \wedge_{\Sigma_j} X^{[j]}$$

is well defined in the category of spectra.

Standard properties of equivariant homotopy then allow us to write

$$\begin{aligned} T(n,j) &= F(D_{n+1,j}S^{-n}, S^0) \\ &= F(\mathcal{C}(n+1,j)_+ \wedge_{\Sigma_j} S^{-nj}, S^0) \\ &= F(\mathcal{C}(n+1,j)_+, S^{nj})^{\Sigma_j}. \end{aligned}$$

This gives an interesting alternative (and technically simpler) definition of the spectra $T(n,j)$, reminiscent of some of the constructions recently occurring in the "Goodwillie Calculus" literature [3]. (See the Appendix.)

Definition 2.1. Let $\tilde{D}_{n,j}X = F(\mathcal{C}(n,j)_+, X^{[j]})^{\Sigma_j}$.

With this definition, we have $T(n,j) = \tilde{D}_{n+1,j}S^n$, and, more generally, if X is a finite spectrum, then $\tilde{D}_{n,j}X =$ S-dual $(D_{n,j}(\text{S-dual }(X)))$.

In the usual way, the little cubes operad structure on the spaces $\mathcal{C}(n,j)$ induces natural maps

$$\mu : D_{n,i}X \wedge D_{n,j}X \to D_{n,i+j}X,$$

$$\Theta : D_{n,i}D_{n,j}X \to D_{n,ij}X,$$

and dually, natural maps

$$\Psi : \tilde{D}_{n,i+j}X \to \tilde{D}_{n,i}X \wedge \tilde{D}_{n,j}X,$$

and

$$\Gamma : \tilde{D}_{n,ij}X \to \tilde{D}_{n,i}\tilde{D}_{n,j}X.$$

In particular, we obtain maps

$$\Psi : T(n,i+j) \to T(n,i) \wedge T(n,j),$$

and

$$\Gamma : T(n,2j) \to \tilde{D}_{n+1,2}T(n,j).$$

These two families of maps provide sufficient structure for the purposes of computing the mod 2 cohomology of the $T(n,j)$.

We turn our attention to constructing the maps

$$\delta : T(n,j) \to \Sigma^{-1}T(n+1,j).$$

In [18] we noted that the evaluation map

$$\epsilon : \Sigma\Omega^{n+1}\Sigma^{n+1}X \to \Omega^n\Sigma^{n+1}X$$

induces maps

$$\epsilon : \Sigma D_{n+1,j}X \to D_{n,j}\Sigma X.$$

We note that the same geometric construction also yields natural maps

$$\delta : \tilde{D}_{n,j}X \to \Sigma^{-1}\tilde{D}_{n+1,j}\Sigma X.$$

Both of these families are induced by explicit Σ_j equivariant maps

$$\beta : \mathcal{C}(n+1,j)_+ \wedge S^1 \to \mathcal{C}(n,j)_+ \wedge S^j,$$

defined as follows.

Given a linear embedding $c : I \to I$, let $c^* : I \to I$ be the associated "Thom-Pontryagin collapse" map. Explicitly,

$$c^*(t) = \begin{cases} 0 & \text{if } t \leq Im(c) \\ s & \text{if } c(s) = t \\ 1 & \text{if } t \geq Im(c). \end{cases}$$

Note that $(c \circ d)^* = d^* \circ c^*$.

Given a little $n+1$ cube $c : I^{n+1} \to I^{n+1}$, we write $c = c' \times c''$, where $c' : I^n \to I^n$, and $c'' : I \to I$. Regarding S^1 as $I/\partial I$, and S^j as $(I/\partial I)^{[j]}$, we have the following definition.

Definition 2.2. (Compare with [34, page 47].)

$$\beta(c_1, \ldots, c_j, t) = (c_1', \ldots, c_j', c_1''^*(t), \ldots, c_j''^*(t)).$$

A straightforward check of definitions yields the next proposition, which shows how δ is related to the maps Ψ and Γ.

Proposition 2.3.

(1) *The composite* $\Sigma\tilde{D}_{n,i+j}X \xrightarrow{\delta} \tilde{D}_{n+1,i+j}\Sigma X \xrightarrow{\Psi} \tilde{D}_{n+1,i}\Sigma X \wedge \tilde{D}_{n+1,j}\Sigma X$ *is null if* $i > 0$ *and* $j > 0$.

(2) *There are commutative diagrams:*

$$\begin{array}{ccccc} \Sigma\tilde{D}_{n,ij}X & & \xrightarrow{\quad\quad\delta\quad\quad} & & \tilde{D}_{n+1,ij}X \\ \downarrow{\scriptstyle \Sigma\Gamma} & & & & \downarrow{\scriptstyle \Gamma} \\ \Sigma\tilde{D}_{n,i}\tilde{D}_{n,j}X & \xrightarrow{\delta} & \tilde{D}_{n+1,i}\Sigma\tilde{D}_{n,j}X & \xrightarrow{\tilde{D}_{n+1,i}\delta} & \tilde{D}_{n+1,i}\tilde{D}_{n+1,j}\Sigma X \end{array}$$

Our last and most delicate construction is of the family

$$\Phi : T(n, j) \to T(n, 2j).$$

The next theorem summarizes the properties we need to know.

Theorem 2.4. *There exist maps $\Phi_{n,j} : T(n, j) \to T(n, 2j)$ such that the following five properties hold.*

(1) $\Phi_{0,j} : T(0, j) = S^0 \to T(0, 2j) = S^0$ *is multiplication by* $(2j)!/j!2^j$.

(2) *There are commutative diagrams:*

$$\begin{array}{ccc} \Sigma T(n,j) & \xrightarrow{\Sigma\Phi_{n,j}} & \Sigma T(n,2j) \\ \Big\downarrow{\scriptstyle\delta} & & \Big\downarrow{\scriptstyle\delta} \\ T(n+1,j) & \xrightarrow{\Phi_{n+1,j}} & T(n+1,2j) \end{array}$$

(3) *For $n \geq 1$, there are commutative diagrams:*

$$\begin{array}{ccc} T(n,i+j) & \xrightarrow{\Phi_{n,i+j}} & T(n,2(i+j)) \\ \Big\downarrow{\scriptstyle\Psi} & & \Big\downarrow{\scriptstyle\Psi} \\ T(n,i)\wedge T(n,j) & \xrightarrow{\Phi_{n,i}\wedge\Phi_{n,j}} & T(n,2i)\wedge T(n,2j) \end{array}$$

(4) *If $n \geq 1$, i and j are odd, and $i + j = 2k$, the composite*

$$T(n,k) \xrightarrow{\Phi_{n,k}} T(n,2k) \xrightarrow{\Psi} T(n,i) \wedge T(n,j)$$

is null.

(5) *For $n \geq 1$, there are commutative diagrams:*

$$\begin{array}{ccc} T(n,2j) & \xrightarrow{\Phi_{n,2j}} & T(n,4j) \\ \Big\downarrow{\scriptstyle\Gamma} & & \Big\downarrow{\scriptstyle\Gamma} \\ \tilde{D}_{n,2}T(n,j) & \xrightarrow{\tilde{D}_{n,2}\Phi_{n,j}} & \tilde{D}_{n,2}T(n,2j) \end{array}$$

Proof. Fix $N \geq 0, J \geq 0$. Let $\mathcal{S}(N, J)$ be the collection of sets of maps $S = \{ \Phi_{n,j} \mid n \leq N, j \leq J \}$ such that properties (1)–(5) are true whenever the maps $\Phi_{n,j}$ appearing in those statements are chosen from S. (In other words, $S \in \mathcal{S}(N, J)$ makes true a finite number of the infinite lists of statements in (1)–(5).)

There are restriction maps $\mathcal{S}(N,J) \to \mathcal{S}(N-1,J)$ and $\mathcal{S}(N,J) \to \mathcal{S}(N,J-1)$. The theorem amounts to saying that the inverse limit, $\lim \mathcal{S}(N,J)$, taken over all N and J, is nonempty.

Since (1) and (2) determine $\Phi_{0,j}$ and $\Phi_{n,0}$, $\mathcal{S}(N,J)$ can be regarded as a subset of $\prod_{n=1}^{N}\prod_{j=1}^{J}\{T(n,j), T(n,2j)\}$, which is finite, as each $T(n,j)$ is a finite complex, and each $T(n,j)$ with $n \geq 1, j \geq 2$ is torsion. Since the inverse limit of nonempty finite sets is nonempty[4], the next theorem completes the proof of the theorem. □

Theorem 2.5. *$\mathcal{S}(N,J)$ is nonempty.*

There are two ingredients in our construction of a set $\{\Phi_{n,j}\} \in \mathcal{S}(N,J)$. The first is the use of vector bundle trivializations to construct natural equivalences

$$\omega_{n,j} : D_{n,j}(\Sigma^d X) \simeq \Sigma^{dj} D_{n,j} X,$$

for n and j in any *finite* range, compatible with the structure maps (ϵ, μ, Θ). The second is the use of Hopf invariants to construct maps, for $d > n$,

$$h^d_{n,j} : D_{n+1,2j} S^{d-n} \to D_{n+1,j} S^{2d-n}$$

with appropriate properties.

The next two theorems, whose proofs occupy the next two sections, more precisely describe what we need.

Theorem 2.6. *Fix N and J. Then there exists $d > 0$, and natural equivalences*

$$\omega_{n,j} : D_{n,j}(\Sigma^d X) \simeq \Sigma^{dj} D_{n,j} X,$$

defined for $1 \leq n \leq N, 1 \leq j \leq J$, such that the following diagrams commute:

(1) *for all $1 \leq n \leq N-1, 1 \leq j \leq J$,*

$$\begin{array}{ccc} \Sigma D_{n+1,j}(\Sigma^d X) & \xrightarrow{\omega_{n+1,j}} & \Sigma^{1+dj} D_{n+1,j}(X) \\ \downarrow \epsilon & & \downarrow (-1)^{d(j-1)}\epsilon \\ D_{n,j}(\Sigma^{d+1} X) & \xrightarrow{\omega_{n,j}} & \Sigma^{dj} D_{n,j}(\Sigma X) \end{array}$$

(2) *for all $1 \leq n \leq N, i+j \leq J$,*

$$\begin{array}{ccc} D_{n,i}(\Sigma^d X) \wedge D_{n,j}(\Sigma^d X) & \xrightarrow{\omega_{n,i}\wedge\omega_{n,j}} & \Sigma^{di} D_{n,i}(X) \wedge \Sigma^{dj} D_{n,j}(X) \\ \downarrow \mu & & \downarrow \mu \\ D_{n,i+j}(\Sigma^d X) & \xrightarrow{\omega_{n,i+j}} & \Sigma^{d(i+j)} D_{n,i+j}(X) \end{array}$$

[4]A standard application of the Tychonoff Theorem.

(3) *for all* $1 \le n \le N, ij \le J$,

$$\begin{array}{ccccc} D_{n,i}D_{n,j}(\Sigma^d X) & \xrightarrow{D_{n,i}\omega_{n,j}} & D_{n,i}\Sigma^{dj}D_{n,j}(X) & \xrightarrow{\omega^j_{n,i}} & \Sigma^{dij}D_{n,i}D_{n,j}(X) \\ \downarrow{\scriptstyle \Sigma\Gamma} & & & & \downarrow{\scriptstyle \Gamma} \\ \Sigma\tilde{D}_{n,ij}X & & \xrightarrow{\quad\delta\quad} & & \tilde{D}_{n+1,ij}X \end{array}$$

Theorem 2.7. *For all* $0 \le n < d$ *and for all* j, *there exist maps*

$$h^d_{n,j} : D_{n+1,2j}S^{d-n} \to D_{n+1,j}S^{2d-n}$$

with the following properties.

(1) *If* d *is even,* $h^d_{0,j} : D_{1,2j}S^d = S^{2jd} \to D_{1,j}S^{2d} = S^{2jd}$ *is multiplication by* $(2j)!/j!2^j$.

(2) *There are commutative diagrams:*

$$\begin{array}{ccc} \Sigma D_{n+1,2j}S^{d-n} & \xrightarrow{\Sigma h^d_{n,j}} & \Sigma D_{n+1,j}S^{2d-n} \\ \downarrow{\scriptstyle \epsilon} & & \downarrow{\scriptstyle \epsilon} \\ D_{n,2j}S^{d-n+1} & \xrightarrow{h^d_{n-1,j}} & D_{n,j}S^{2d-n+1}. \end{array}$$

(3) *There are commutative diagrams:*

$$\begin{array}{ccc} D_{n+1,2i}S^{d-n} \wedge D_{n+1,2j}S^{d-n} & \xrightarrow{h^d_{n,i}\wedge h^d_{n,j}} & D_{n+1,i}S^{2d-n} \wedge D_{n+1,j}S^{2d-n} \\ \downarrow{\scriptstyle \mu} & & \downarrow{\scriptstyle \mu} \\ D_{n+1,2(i+j)}S^{d-n} & \xrightarrow{h^d_{n,i+j}} & D_{n+1,i+j}S^{2d-n}. \end{array}$$

(4) *If* i *and* j *are odd, and* $i + j = 2k$, *the composite*

$$D_{n+1,i}S^{d-n} \wedge D_{n+1,j}S^{d-n} \xrightarrow{\mu} D_{n+1,2k}S^{d-n} \xrightarrow{h^d_{n,k}} D_{n+1,k}S^{2d-n}$$

is null.

(5) *There are commutative diagrams:*

$$\begin{array}{ccc} D_{n,2}D_{n+1,2j}S^{d-n} & \xrightarrow{D_{n,2}h^d_{n,j}} & D_{n,2}D_{n+1,j}S^{2d-n} \\ \downarrow \Theta & & \downarrow \Theta \\ D_{n+1,4j}S^{d-n} & \xrightarrow{h^d_{n,2j}} & D_{n+1,2j}S^{2d-n}. \end{array}$$

Assuming these two theorems, we note that Theorem 2.5 follows easily. First choose d as in Theorem 2.6 (but with J replaced by $2J$). We can also assume d is even. Then, with $h^d_{n,j}$ as in Theorem 2.7, we define $\Phi_{n,j} : T(n,j) \to T(n,2j)$ to be the S-dual of the composite

$$D_{n+1,2j}S^{-n} \xrightarrow{\omega^{-1}_{n,2j}} \Sigma^{-2dj} D_{n+1,2j}S^{d-n} \xrightarrow{h^d_{n,j}} \Sigma^{-2dj} D_{n+1,j}S^{2d-n} \xrightarrow{\omega^2_{n,j}} D_{n+1,j}S^{-n}.$$

Courtesy of Theorem 2.6, each statement in Theorem 2.7 translates immediately into the corresponding statement in Theorem 2.4, proving Theorem 2.5.

3. Quasiperiodicity of the Sphere Spectrum

In this section we prove Theorem 2.6, which asserts that given N and J, there exists $d > 0$ and natural equivalences

$$\omega_{n,j} : D_{n,j}(\Sigma^d X) \simeq \Sigma^{dj} D_{n,j}X,$$

defined for $1 \le n \le N, 1 \le j \le J$ which are appropriately compatible with the three families of structure maps

$$\epsilon : \Sigma D_{n+1,j}X \to D_{n,j}\Sigma X,$$

$$\mu : D_{n,i}X \wedge D_{n,j}X \to D_{n,i+j}X, \text{ and}$$

$$\Theta : D_{n,i}D_{n,j}X \to D_{n,ij}X.$$

To put this theorem in context, recall that as an aid to constructing power operations and studying Thom isomorphisms, the authors of [6] defined the notion of an H^d_∞–ring spectrum. For the sphere spectrum S^0 to admit an H^d_∞ structure would be roughly equivalent to natural equivalences $\omega_{n,j}$ as in the theorem for all $n < \infty, j < \infty$. Though it is easy to see that this cannot be done, our theorem says that it partially *can* be. If one defines the notion of an H^d_n structure in the obvious way, we know of no reason why the following conjecture might not be true.

Conjecture 3.1. Localized at a prime p, for each n, S^0 admits the structure of an H^d_n–ring spectrum for some $d > 0$.

The origin of the natural equivalences is as follows.

Suppose ξ and ζ are two r dimensional vector bundle over a space B, respectively classified by maps $f_\xi, f_\zeta : B \to BO$. Then a homotopy $H : B \times I \to BO$ between f_ξ and f_ζ induces an bundle isomorphism $\omega_H : \xi \to \zeta$ and thus a homeomorphism $\omega_H : M(\xi) \to M(\zeta)$ of Thom spaces. In particular, given a map $i : B \to C$ to a contractible space C, and an extension $F : C \to BO$ of f_ξ, there is an induced homeomorphism of spaces

$$\omega_F : M(\xi) \to \Sigma^r(B_+).$$

Furthermore, given a second extension $F' : C' \to BO$, ω_F and $\omega_{F'}$ will be homotopic if the map

$$F \cup_{f_\xi} F' : C \cup_B C' \to BO$$

is null. This last map can be regarded an obstruction $o(F, F') : \Sigma B \to BO$.

We apply these general remarks to the case of interest. Let $\xi_{n,j}$ be the vector bundle

$$\mathcal{C}(n,j) \times_{\Sigma_j} \mathbf{R}^j \to B(n,j) = \mathcal{C}(n,j)/\Sigma_j,$$

with classifying map $f_{n,j} : B(n,j) \to BO$. This is easily seen to be a bundle of finite order, and an extension $F : CB(n,j) \to BO$ of $df_{n,j}$ to the cone on $B(n,j)$ induces a homeomorphism

$$\omega_F : \mathcal{C}(n,j)_+ \wedge_{\Sigma_j} S^{dj} \to \Sigma^{ds}(B(n,j)_+),$$

and thus a Σ_j–equivariant homeomorphism

$$\omega_F : \mathcal{C}(n,j)_+ \wedge S^{dj} \to \Sigma^{ds}(\mathcal{C}(n,j)_+),$$

and finally a natural equivalence

$$\omega_F : D_{n,j}(\Sigma^d X) \simeq \Sigma^{dj} D_{n,j} X.$$

A straightforward check of definitions shows

Lemma 3.2. *In this situation, if $F : CB(n,j) \to BO$ is the restriction of a map $F' : CB(n+1,j) \to BO$ extending $df_{n+1,j}$ then the following diagram commutes:*

$$\begin{array}{ccc}
\Sigma D_{n+1,j}(\Sigma^d X) & \xrightarrow{\omega_{F'}} & \Sigma^{1+dj} D_{n+1,j}(X) \\
\Big\downarrow \epsilon & & \Big\downarrow (-1)^{d(j-1)}\epsilon \\
D_{n,j}(\Sigma^{d+1} X) & \xrightarrow{\omega_F} & \Sigma^{dj} D_{n,j}(\Sigma X).
\end{array}$$

Now fix N and J as in Theorem 2.6. Let $d > 0$ and let $\mathcal{F} = \{F_j : CB(N,j) \to BO \mid j = 1, \ldots, J\}$ be a collection of extensions of the maps $df_{N,j}$. We define the obstruction set $o(\mathcal{F})$ to be the following set of maps:

$$o^\mu_{i,j}(\mathcal{F}) : \Sigma(B(N,i) \times B(N,j)) \to BO,$$

for $i + j = J$, and

$$o^{\Theta}_{i,j}(\mathcal{F}) : \Sigma(\mathcal{C}(N,i) \times_{\Sigma_i} B(N,j)^i) \to BO,$$

for $ij = J$, where these maps are defined as follows.

For $o^{\mu}_{i,j}(\mathcal{F})$, we regard $\Sigma(B(N,i) \times B(N,j))$ as

$$C(B(N,i) \times B(N,j)) \cup_{B(N,i)\times B(N,j)} CB(N,i) \times CB(N,j),$$

and we let

$$o^{\mu}_{i,j}(\mathcal{F}) = \begin{cases} F_{i+j} \circ \mu & \text{on } C(B(N,i) \times B(N,j)), \\ \mu_{BO} \circ (F_i \times F_j) & \text{on } CB(N,i) \times CB(N,j). \end{cases}$$

Here $\mu_{BO} : BO \times BO \to BO$ is the H-space structure map.

For $o^{\Theta}_{i,j}(\mathcal{F})$, we regard $\Sigma(\mathcal{C}(N,i) \times_{\Sigma_i} B(N,j)^i)$ as

$$C(\mathcal{C}(N,i) \times_{\Sigma_i} B(N,j)^i) \cup_{\mathcal{C}(N,i)\times_{\Sigma_i} B(N,j)^i} \mathcal{C}(N,i) \times_{\Sigma_i} CB(N,j)^i,$$

and we let

$$o^{\Theta}_{i,j}(\mathcal{F}) = \begin{cases} F_{ij} \circ \Theta & \text{on } C(\mathcal{C}(N,i) \times_{\Sigma_i} B(N,j)^i), \\ \Theta_{BO} \circ (Id \times_{\Sigma_i} (F_j)^i) & \text{on } \mathcal{C}(N,i) \times_{\Sigma_i} CB(N,j)^i. \end{cases}$$

Here $\Theta_{BO} : \mathcal{C}(n,i) \times_{\Sigma_i} BO^i \to BO$ is the infinite loopspace structure map.

Theorem 2.6 will follow if we can show that there is a choice of d and $\mathcal{F}$ for which $o(\mathcal{F})$ is a set of null maps. Firstly, we note that there *do* exist collections $\mathcal{F}$ as above: we just need to choose d equal to a common multiple of the orders of the bundles $\xi_{N,1}, \dots, \xi_{N,J}$. By making d possibly bigger, we can even ensure that $\mathcal{F}$ is the restriction of a similar family $\tilde{\mathcal{F}}$ defined for the pair $(N+1, J)$, and the obstruction set $o(\mathcal{F})$ is the restriction of $o(\tilde{\mathcal{F}})$.

Given a family $\mathcal{F}$, let $r\mathcal{F}$ be the family with j^{th} function equal to rF_j. Note that if F_j extends $df_{N,j}$, then rF_j extends $(rd)f_{N,j}$. It is easy to check

Lemma 3.3.

(1) $o^{\mu}_{i,j}(r\mathcal{F}) = r o^{\mu}_{i,j}(\mathcal{F}) \in K^1(B(N,i) \times B(N,j))$.

(2) $o^{\Theta}_{i,j}(r\mathcal{F}) = r o^{\Theta}_{i,j}(\mathcal{F}) \in K^1(\mathcal{C}(N,i) \times_{\Sigma_i} B(N,j)^i)$.

Proposition 3.4. *Let $X(N)$ be one of the spaces $B(N,j)$, $B(N,i) \times B(N,j)$, or $\mathcal{C}(n,i) \times_{\Sigma_i} B(N,j)^i$. If $x \in K^*(X(N))$ is in the image of the restriction from $K^*(X(N+1))$, then x is torsion.*

Postponing the proof of this proposition for the moment, we show that there is a choice of d and $\mathcal{F}$ for which $o(\mathcal{F})$ is a set of null maps. Start with any family $\mathcal{F}$ (and associated d) as above. Let r be a common multiple of the orders of the obstructions $o^{\mu}_{i,j}(\mathcal{F})$ and $o^{\Theta}_{i,j}(\mathcal{F})$. (Proposition 3.4 tells us that these elements *do* have finite order.) Then the family $r\mathcal{F}$ has an obstruction set consisting only of null maps, as needed.

It remains to prove Proposition 3.4. This will follow from three lemmas.

Lemma 3.5. *Let $f : X \to Y$ be a map between finite complexes. If $H_*(f;\mathbf{Q}) = 0$, then $Im\{E^*(f) : E^*(Y) \to E^*(X)\}$ is torsion for all generalized cohomology theories E^*.*

Proof. For finite complexes Z, $E^*(Z_{\mathbf{Q}}) \simeq E^*(Z) \otimes \mathbf{Q}$. $H_*(f;\mathbf{Q}) = 0$ implies that $f_{\mathbf{Q}} \simeq *$, and thus that $E^*(f) \otimes \mathbf{Q} = 0$. □

Lemma 3.6. *If $X(N)$ is as in Proposition 3.4, $X(N)$ has the homotopy type of a finite complex.*

Proof. There are many ways to see this. The author's favorite is to note that the explicit cell decomposition for $B(2,j)$ given by Fox and Neuwirth in [10] generalizes to $B(n,j)$: $B(n,j)$ has the homotopy type of an $(n-1)(j-1)$ dimensional cell complex with exactly n^{j-1} cells. □

Lemma 3.7. *With $X(N)$ as in Proposition 3.4,*

$$H_*(X(N);\mathbf{Q}) \to H_*(X(N+1);\mathbf{Q})$$

is 0.

Proof. This follows from standard homology calculations [9]. □

4. Hopf Invariants

In this section we use Hopf invariants to define maps

$$h^d_{n,j} : D_{n+1,2j}S^{d-n} \to D_{n+1,j}S^{2d-n},$$

for $0 \leq n < d$, and then show that they have the properties listed in Theorem 2.7.

The maps are not hard to define. Let

$$H_Y : \Omega\Sigma Y \to \Omega\Sigma(Y \wedge Y)$$

be the classic Hopf invariant. Replacing Y by $\Sigma^n X$, and looping n times, defines an unstable natural map

$$\Omega^n H_{\Sigma^n X} : \Omega^{n+1}\Sigma^{n+1}X \to \Omega^{n+1}\Sigma^{n+1}(\Sigma^n X \wedge X).$$

Now let $D_n X$ denote $\bigvee_{j=1}^{\infty} D_{n,j}X$, and, for connected X, let

$$s_n : D_n X \simeq \Omega^n \Sigma^n X$$

be the natural *stable* Snaith equivalence as studied in [32, Chapter VII].

Finally,

$$H_n(X) : D_{n+1}X \to D_{n+1}(\Sigma^n X \wedge X)$$

will be the stable map given by the composite $s_{n+1}^{-1} \circ (\Omega^n H_{\Sigma^n X}) \circ s_{n+1}$.

Definition 4.1. For all $0 \leq n < d$, and for all j,

$$h^d_{n,j} : D_{n+1,2j}S^{d-n} \to D_{n+1,j}S^{2d-n}$$

is defined to be the $(2j,j)^{th}$ component of $H_n(S^{d-n})$.

The first of the properties in Theorem 2.7 is easily checked. If d is even, $h^d_{0,j} : S^{2jd} \to S^{2jd}$ is multiplication by $(2j)!/j!2^j$, as cup product considerations easily show that $H : \Omega S^{d+1} \to S^{2d+1}$ induces multiplication by this number in cohomology in dimension $2dj$ [16, p.294].

Property (2) of Theorem 2.7, the compatibility of $h^d_{n,j}$ with the maps ϵ, follows from the main result of [18]: under the Snaith equivalence, the evaluation

$$\epsilon : \Sigma\Omega^{n+1}\Sigma^{n+1}X \to \Omega^n\Sigma^{n+1}X$$

is carried to

$$\bigvee_{j=1}^{\infty} \epsilon : \bigvee_{j=1}^{\infty} \Sigma D_{n+1,j}X \to \bigvee_{j=1}^{\infty} D_{n,j}\Sigma X.$$

The remaining three properties follow from the next two propositions.

Proposition 4.2. *There is a commutative diagram:*

$$\begin{array}{ccc} D_nD_{n+1}X & \xrightarrow{D_nH_n(X)} & D_nD_{n+1}(\Sigma^nX \wedge X) \\ \downarrow \Theta & & \downarrow \Theta \\ D_{n+1}X & \xrightarrow{H_n(X)} & D_{n+1}(\Sigma^nX \wedge X). \end{array}$$

Here $\Theta : D_nD_{n+1}X \to D_{n+1}X$ is the restriction of the structure map $\Theta : D_{n+1}D_{n+1}X \to D_{n+1}X$.

Proposition 4.3. *The $(i,j)^{th}$ component of $H_n(X)$ is null unless $i \leq 2j$.*

This tells us that $H_n(X)$ can be regarded as an "upper triangular matrix" of maps. With this information fed into Proposition 4.2, the three last properties of Theorem 2.7 can be read off immediately.

Proof of Proposition 4.3. The $(i,j)^{th}$ component of $H_n(X)$ is a natural transformation

$$D_{n+1,i}X \to D_{n+1,j}(\Sigma^nX \wedge X).$$

In the terminology of [13], the domain is a homogeneous functor of degree i, while the range is a functor of degree $2j$. Thus there are *no* nontrivial natural transformations from the former to the latter if $i > 2j$. □

Remark 4.4. This proposition presumably has a direct proof, along the lines of the proofs of similar results in [22].

Proof of Proposition 4.2. This is a consequence of the fact that $H_n(X)$ corresponds to an n fold loop map. Let C_nX denote the usual approximation to $\Omega^n\Sigma^nX$, with monad structure map $\Theta : C_nC_n \to C_n$, and let Y denote $\Sigma^nX \wedge X$.

With this notation, we assert that there is a commutative diagram:

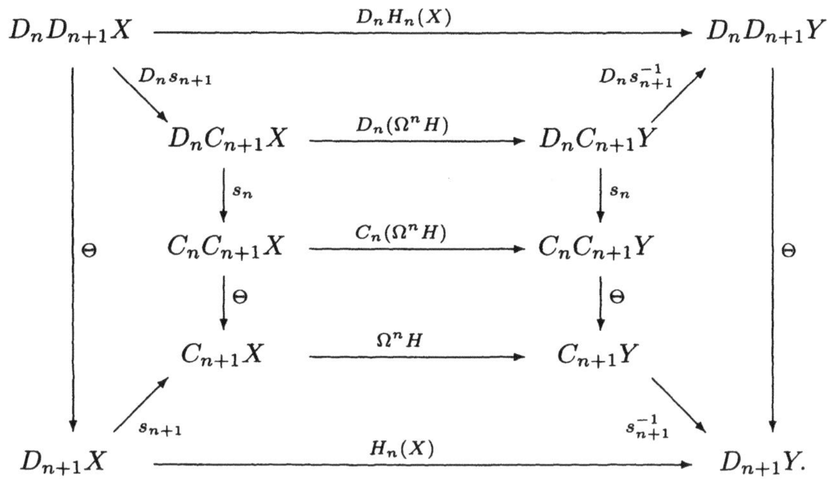

The lower central square commutes since $\Omega^n H$ is a C_n–map. The upper square commutes by naturality. Finally the argument in [21, §4] shows that the two side trapezoids commute. □

5. Cohomology Calculations

We use the following notational conventions in the next three sections. $H_*(X)$ and $H^*(X)$ will denote homology and cohomology with $\mathbf{Z}/2$ coefficients. The binomial coefficient $\binom{b}{a}$ is defined, for *all* integers a and b, as the ath Taylor coefficient of $(x+1)^b$ if $a \geq 0$, and 0 otherwise. We will use, without further comment, that $\binom{b}{a} = \binom{a-b-1}{a}$.

In this section we describe $H^*(T(n,*))$, and the maps Ψ^*, Φ^*, and δ^*, in terms of "dual" Dyer-Lashof operations. We begin by remarking that since $T(n,j)$ is the S–dual of $D_{n+1,j}S^{-n}$, and $H_*(D_{n+1,j}S^{-n})$ embeds in $H_*(D_{\infty,j}S^{-n})$, we will not need to confront the Browder operations, and the "top" Dyer-Lashof operation will be additive (as are the others).

As part of the general theory [9], the product maps μ induce a bigraded product on $H_*(D_{n+1,*}S^{-n})$, and associated to the structure maps Θ, there are Dyer-Lashof operations

$$Q^s : H_q(D_{n+1,j}S^{-n}) \to H_{q+s}(D_{n+1,2j}S^{-n}).$$

These are defined for $s \leq q+n$, and are 0 for $s < q$. Furthermore, these satisfy the Cartan formula, Adem relations, and restriction axiom: $Q^{|x|}x = x^2$. $H_*(D_{n+1,*}S^{-n})$ is the free object with all this structure, generated by a class in degree $-n$.

There is a canonical isomorphism $H^q(T(n,j)) = H_{-q}(D_{n+1,j}S^{-n})$. Under this isomorphism, Ψ^* will correspond to μ_*, and will induce a bigraded product (occasionally denoted " $*$ ") on $H^*(T(n,*))$. We define operations

$$\tilde{Q}^s : H^q(T(n,j)) \to H^{q+s}(T(n,2j))$$

to correspond to

$$Q^{-s} : H_{-q}(D_{n+1,j}S^{-n}) \to H_{-q-s}(D_{n+1,2j}S^{-n}).$$

These are defined for $s \geq q-n$, and are 0 for $s > q$. These satisfy the Cartan formula,

$$\tilde{Q}^t(x * y) = \sum_{r+s=t} \tilde{Q}^r x * \tilde{Q}^s y,$$

Adem relations,

$$\tilde{Q}^r\tilde{Q}^s x = \sum_i \binom{s-i-1}{r-2i} \tilde{Q}^{r+s-i}\tilde{Q}^i x,$$

and restriction axiom,

$$\tilde{Q}^{|x|}x = x^2.$$

(We note that in the Adem relations, whenever the iterated operation on the left is defined, so are those appearing with nonzero coefficient on the right, though not conversely[5].)

Theorem 5.1. *$H^*(T(n,*))$ is the free object with all this structure, generated by a class x_n in degree n. Explicitly, if*

$$\tilde{R}_n = \langle \tilde{Q}^I x_n \mid I \text{ is admissible } \rangle / \langle \tilde{Q}^I x_n \mid I \text{ is admissible and } e(I) > n \rangle,$$

$H^*(T(n,*)) = S^*(\tilde{R}_n)/(\tilde{Q}^{|x|}x - x^2)$.

Thus, as a bigraded algebra, $H^*(T(n,*))$ is a polynomial algebra on the set $\{\tilde{Q}^I x_n \mid I \text{ is admissible and } e(I) < n\}$, with $\tilde{Q}^I x_n \in H^*(T(n, 2^{l(I)}))$.

Here, if $I = (i_1, \dots, i_l)$, $\tilde{Q}^I = \tilde{Q}^{i_1} \dots \tilde{Q}^{i_l}$, and $e(I)$, $l(I)$, and *admissible* mean what they did in §1. There is a little wrinkle here however: as $\tilde{Q}^0$ is not the identity, an admissible sequence can end with 0's.

The geometric results of §2 allow us to quickly deduce the behavior of δ^* and Φ^*.

Proposition 5.2. *$\delta^* : H^{*+1}(T(n+1,j)) \to H^*(T(n,*))$ is determined by*
(1) $\delta^*(\tilde{Q}^I x_{n+1}) = \tilde{Q}^I x_n$, and
(2) δ^* is 0 on decomposables.

Proof. This follows from Proposition 2.3, and the fact that Dyer–Lashof operations commute with the evaluation [9, p.6, p.218]. □

[5] The relation $\tilde{Q}^1\tilde{Q}^2 = \tilde{Q}^3\tilde{Q}^0$ illustrates this.

Proposition 5.3. $\Phi^* : H^*(T(n,*)) \to H^*(T(n,*))$ *is determined by*

(1) *When* $n = 0$, $\Phi^*(x_0^{2j}) = x_0^j$.

(2) $\Phi^*(\tilde{Q}^s x) = \tilde{Q}^s(\Phi^* x)$ *if* $s > |x| - n$.

(3) *Whenever the iterated operation* $\tilde{Q}^I x_n$ *is defined,* $\Phi^*(\tilde{Q}^I x_n) = \tilde{Q}^{I'} x_n$ *if* $I = (I', 0)$, *and is 0 otherwise.*

(4) *When* $n \geq 1$, Φ^* *is an algebra map (with the second grading in the domain of* Φ^* *doubled).*

Proof. This follows from Theorem 2.4 and the last proposition. As $(2j)!/j!2^j$ is always odd, statement (1) of Theorem 2.4 implies that statement (1) here is true. Statement (2) here is implied by statement (5) of Theorem 2.4. To see that statement (3) is true, we first prove this in the special case when I consists only of 0's. Note that (1) includes the $n = 0$ subcase of this special case, and then the statement for general n follows by combining the last proposition with statement (2) of Theorem 2.4 (which implies that Φ^* and δ^* commute). Now use (2) to deduce (3) for general I from the special case already established. Finally, (4) follows from statements (3) and (4) of Theorem 2.4. □

Note that as a corollary of Proposition 5.2, we have partially proved Theorem 1.5.

Corollary 5.4.

(1) $T(\infty, j) \simeq *$ *unless* j *is a power of 2.*

(2) $H^*((T(\infty, 2^k)) = \tilde{R}[k]$, *where* $\tilde{R}[k] = \langle \tilde{Q}^I x_0 \mid I$ *is admissible and* $l(I) = k\rangle$.

6. New Nishida Relations

In the last section, we determined $H^*(T(n,*))$ in terms of dual Dyer–Lashof operations. Here we describe the Steenrod algebra action.

The standard Nishida relations [9, p.6, p.214] tell us how $(Sq^r)_*$ commutes with Q^s in $H_*(D_{n+1,*}S^{-n})$. Since $\chi(Sq^r)$[6] acting on $H^*(T(n,*))$ corresponds to $(Sq^r)_*$ acting on $H_{-*}(D_{n+1,*}S^{-n})$, we immediately have the following formula.

Lemma 6.1.

$$\chi(Sq^r)\tilde{Q}^s x = \sum_i \binom{-r-s}{r-2i} \tilde{Q}^{r+s-i}\chi(Sq^i)x.$$

Though this does completely specify the $\mathcal{A}$ module structure on $H^*(T(n,*))$, it is in a form completely unsuitable for proving theorems like those in the introduction. The point of this section is to prove

Theorem 6.2.

$$Sq^r\tilde{Q}^s x = \sum_i \binom{s-i-1}{r-2i} \tilde{Q}^{r+s-i} Sq^i x.$$

[6] χ is the antiautomorphism of the connected Hopf algebra $\mathcal{A}$.

The reader may find it amusing to compare this formula to the Adem relation of the last section,

$$\tilde{Q}^r\tilde{Q}^s x = \sum_i \binom{s-i-1}{r-2i} \tilde{Q}^{r+s-i}\tilde{Q}^i x,$$

the Adem relations in $\mathcal{A}$,

$$Sq^r Sq^s x = \sum_i \binom{s-i-1}{r-2i} Sq^{r+s-i} Sq^i x,$$

and the formula defining the "Singer construction" [41]

$$Sq^r(t^{s-1}\otimes x) = \sum_i \binom{s-i-1}{r-2i} t^{r+s-i-1}\otimes Sq^i x.$$

Proof of Theorem 6.2. With Sq denoting the total square $1 + Sq^1 + Sq^2 + \dots$, to verify the formula, it suffices to check that it is consistent with the identity $Sq(\chi(Sq)) = 1$ and Lemma 6.1 above. Fixing n and s, we compute

$$\sum_r Sq^{n-r}\chi(Sq^r)\tilde{Q}^s x$$

$$= \sum_r Sq^{n-r}\left[\sum_i \binom{-r-s}{r-2i} \tilde{Q}^{r+s-i}\chi(Sq^i)x\right]$$

$$= \sum_{i,j}\left[\sum_r \binom{r+s-i-j-1}{n-r-2j}\binom{-r-s}{r-2i}\right] \tilde{Q}^{n+s-i-j} Sq^j\chi(Sq^i)x$$

$$= \sum_{i,j}\left[\sum_p \binom{i+s-j-1+p}{n-2i-2j-p}\binom{-2i-s-p}{p}\right] \tilde{Q}^{n+s-i-j} Sq^j\chi(Sq^i)x$$

(letting $p = r - 2i$)

$$= \sum_{i,j}\binom{-(i+j)}{n-2(i+j)} \tilde{Q}^{n+s-(i+j)} Sq^j\chi(Sq^i)x$$

(using J. Adem's formula [1, (25.3)]: $\sum_p \binom{b+p}{c-p}\binom{a-p}{p} \equiv \binom{a+b+1}{c} \mod 2$)

$$= \sum_k \binom{-k}{n-2k} \tilde{Q}^{n+s-k}\left[\sum_i Sq^{k-i}\chi(Sq^i)x\right]$$

$$= \binom{0}{n}\tilde{Q}^{n+s}x = \begin{cases} \tilde{Q}^s x & \text{if } n = 0 \\ 0 & \text{otherwise.}\end{cases}$$

□

Remark 6.3. Our method of proof also shows that the analogues of the formula in Lemma 6.1,

$$\chi(Sq^r)Sq^s x = \sum_i \binom{-r-s}{r-2i} Sq^{r+s-i}\chi(Sq^i)x,$$

and

$$\chi(Sq^r)(t^{s-1}\otimes x) = \sum_i \binom{-r-s}{r-2i} t^{r+s-i-1}\otimes Sq^i x,$$

respectively hold in the Steenrod algebra and Singer construction. The formula in $\mathcal{A}$ already appears in the literature as [4, (4.4)], where it is given a proof in the style of Bullett and MacDonald [7].

7. The Proofs of Theorem 1.2 and Theorem 1.5

To prove Theorem 1.2, first recall the description of $H^*(T(n,*))$ given in Theorem 5.1:

$$H^*(T(n,*)) = S^*(\tilde{R}_n)/(\tilde{Q}^{|x|}x - x^2),$$

where

$$\tilde{R}_n = \langle \tilde{Q}^I x_n \mid I \text{ is admissible } \rangle / \langle \tilde{Q}^I x_n \mid I \text{ is admissible and } e(I) > n \rangle.$$

Note that $\tilde{R}_n$ is closed under both the action of $\mathcal{A}$ and Φ^*, thanks to our Nishida relations and Proposition 5.3, i.e. $(\tilde{R}_n, \Phi^*)$ is an object in $\mathcal{U}_\rho$. Thus Theorem 1.2 will follow from the next two proposition.

Proposition 7.1. $(\tilde{R}_n, \Phi^*) \simeq F_\rho(n)$ *as objects in* $\mathcal{U}_\rho$.

Proposition 7.2. *Let* $n \geq 1$. *In* $S^*(\tilde{R}_n)$, *the ideal generated by elements of the form* $\tilde{Q}^{|x|}x - x^2$ *equals the ideal generated by elements of the form* $Sq^{|y|}y - (\Phi^* y)^2$.

Both propositions will follow from the next result.

Theorem 7.3. $Sq^I\tilde{Q}^J x_n = (\Phi^*)^{l(I)}(\tilde{Q}^I\tilde{Q}^J x_n)$, *whenever the iterated operation* $\tilde{Q}^I\tilde{Q}^J x_n$ *is defined.*

Proposition 7.1 then follows from

Corollary 7.4. *If* I *is admissible,* $Sq^I(\tilde{Q}^0)^k x_n = \begin{cases} \tilde{Q}^I(\tilde{Q}^0)^{k-l(I)} & \text{if } l(I) \leq k, \\ 0 & \text{if } l(I) > k. \end{cases}$

This same corollary, together with Corollary 5.4 proves Theorem 1.5.

Proof of Proposition 7.2. Let $F(x) = \tilde{Q}^{|x|}x - x^2$ and $G(x) = Sq^{|x|}x - (\Phi^* x)^2$. Using the fact that $\tilde{R}_n$ is unstable, it is easy to deduce that the two ideals in question are generated by elements of the form $F(x)$ and $G(x)$ respectively, where $x \in \tilde{R}_n$. We claim that the sets of such elements are the same; more precisely, $F(\tilde{Q}^I x_n) = G(\tilde{Q}^I\tilde{Q}^0 x_n)$ and $G(\tilde{Q}^I x_n) = F(\Phi^*(\tilde{Q}^I x_n))$.

To see that these hold, we let $d = |I| + n$ and compute:

$$F(\tilde{Q}^I x_n) = \tilde{Q}^d \tilde{Q}^I x_n - (\tilde{Q}^I x_n)^2$$
$$= Sq^d \tilde{Q}^I \tilde{Q}^0 x_n - (\Phi^*(\tilde{Q}^I \tilde{Q}^0 x_n))^2,$$

using Theorem 7.3 and Proposition 5.3,

$$= G(\tilde{Q}^I \tilde{Q}^0 x_n).$$

Similarly,

$$G(\tilde{Q}^I x_n) = Sq^d \tilde{Q}^I x_n - (\Phi^*(\tilde{Q}^I x_n))^2$$
$$= \Phi^*(\tilde{Q}^d \tilde{Q}^I x_n) - (\Phi^*(\tilde{Q}^I \tilde{Q}^0 x_n))^2,$$

using Theorem 7.3 and Proposition 5.3,

$$= \tilde{Q}^d \Phi^*(\tilde{Q}^I x_n) - (\Phi^*(\tilde{Q}^I \tilde{Q}^0 x_n))^2,$$

using part (2) of Proposition 5.3 (since $n \geq 1$),

$$= F(\Phi^*(\tilde{Q}^I x_n)).$$

□

It remains to prove Theorem 7.3. This will follow from a couple of lemmas.

Lemma 7.5. *$Sq^r \tilde{Q}^J \tilde{Q}^0 x_n = \tilde{Q}^r \tilde{Q}^J x_n$, whenever the iterated operation $\tilde{Q}^r \tilde{Q}^J x_n$ is defined.*

Proof. This is proved by induction on $l(J)$. The induction is started by using the Nishida relations to verify that $Sq^r \tilde{Q}^0 x_n = \tilde{Q}^r x_n$.

For the inductive step, suppose $J = (j, J')$. Then

$$\begin{aligned} Sq^r \tilde{Q}^J \tilde{Q}^0 x_n &= Sq^r \tilde{Q}^j \tilde{Q}^{J'} \tilde{Q}^0 x_n \\ &= \sum_i \binom{r-j-1}{r-2i} \tilde{Q}^{r+j-i} Sq^i \tilde{Q}^{J'} \tilde{Q}^0 x_n \quad \text{(using the Nishida relations)} \\ &= \sum_i \binom{r-j-1}{r-2i} \tilde{Q}^{r+j-i} \tilde{Q}^i \tilde{Q}^{J'} x_n \quad \text{(by induction)} \\ &= \tilde{Q}^r \tilde{Q}^j \tilde{Q}^{J'} x_n \quad \text{(using the Adem relations)} \\ &= \tilde{Q}^r \tilde{Q}^J x_n. \end{aligned}$$

□

Lemma 7.6. *$Sq^I \tilde{Q}^J (\tilde{Q}^0)^{l(I)} x_n = \tilde{Q}^I \tilde{Q}^J x_n$, whenever the iterated operation $\tilde{Q}^I \tilde{Q}^J x_n$ is defined.*

Proof. This is proved by induction on $l(I)$, and the last lemma is the case $l(I) = 1$. Let $I = (I', i)$. Then

$$\begin{aligned} Sq^I \tilde{Q}^J (\tilde{Q}^0)^{l(I)} x_n &= Sq^{I'} Sq^i \tilde{Q}^J (\tilde{Q}^0)^{l(I)} x_n \\ &= Sq^{I'} \tilde{Q}^i \tilde{Q}^J (\tilde{Q}^0)^{l(I)-1} x_n \quad \text{(by the case } l(I) = 1) \\ &= \tilde{Q}^{I'} \tilde{Q}^i \tilde{Q}^J x_n \qquad \text{(by induction)} \\ &= \tilde{Q}^I \tilde{Q}^J x_n. \end{aligned}$$

□

Proof of Theorem 7.3. Applying $(\Phi^*)^{l(I)}$ to the formula in the previous lemma yields

$$(\Phi^*)^{l(I)} (Sq^I \tilde{Q}^J (\tilde{Q}^0)^{l(I)} x_n) = (\Phi^*)^{l(I)} (\tilde{Q}^I \tilde{Q}^J x_n).$$

As it has a topological origin, $(\Phi^*)^{l(I)}$ commutes with Steenrod operations. By Proposition 5.3, $(\Phi^*)^{l(I)} (\tilde{Q}^J (\tilde{Q}^0)^{l(I)} x_n) = \tilde{Q}^J x_n$. The theorem follows. □

8. The Whitehead Conjecture Resolution and Theorem 1.6

In this section, we note that the homotopical equivalence of Theorem 1.6 can be deduced from the homological isomorphism of Theorem 1.5, using work of Lannes and Zarati [31] to improve previous work of the author [19, 20].

Letting $Z_k = T(\infty, 2^k)$ in the next theorem, Theorem 1.6 follows from Theorem 1.5.

Theorem 8.1. *Any sequence of 2 complete, connective spectra*

$$Z_0 \xrightarrow{\Phi} Z_1 \xrightarrow{\Phi} Z_2 \xrightarrow{\Phi} \cdots$$

that realizes the length filtration of $\mathcal{A}$ in cohomology is equivalent to the sequence

$$SP^1_\Delta(S^0) \to SP^2_\Delta(S^0) \to SP^4_\Delta(S^0) \to \cdots.$$

This is proved in [19, 20], assuming the extra geometric condition:

$$\Sigma^{-k}(Z_k/Z_{k-1}) \text{ is a wedge summand of a suspension spectrum.}$$

We note that this geometric condition is *automatically* satisfied! Under our cohomological hypothesis, $H^*(\Sigma^{-k}(Z_k/Z_{k-1}))$ is isomorphic to $H^*(M(k))$, where $M(k)$ is the stable wedge summand of $B(\mathbf{Z}/2)^k$ associated to the Steinberg module.

Now consider the Adams spectral sequence for computing maps from $M(k)$ to $\Sigma^{-k}(Z_k/Z_{k-1})$. An $\mathcal{A}$–module isomorphism $H^*(\Sigma^{-k}(Z_k/Z_{k-1})) \simeq H^*(M(k))$ can be regarded as an element in $E_2^{0,0}$. The following proposition implies that such an element is a permanent cycle, i.e. one can topologically realize this isomorphism.

Proposition 8.2. [31, Proposition 5.4.7.1] *If M is an unstable $\mathcal{A}$–module, and N is a summand of $H^*(B(\mathbf{Z}/2)^k)$, then $\mathrm{Ext}^{s,t}_{\mathcal{A}}(M, N) = 0$ for all $t - s < 0$.*

Lannes and Zarati prove this using ideas of W.Singer. As explained in [15], this proposition can also be deduced from [5, Lemma 2.3(i)] (slightly modified) in the spirit of Carlsson's work [8].

9. The Proof of Theorem 1.9

This sections contains the details of the proof of Theorem 1.9, which was outlined at the end of §1.

As in [23], $F \in \mathcal{F}$ is said to be *finite* if it has a finite length composition series with simple subquotients, and is said to be *locally finite* (written $F \in \mathcal{F}_\omega$) if it is the union of its finite subfunctors. Recall that $I_\lambda \in \mathcal{F}$ is the injective envelope of the simple functor F_λ. The I_λ are locally finite [23]. Then the general theory of locally Noetherian abelian categories [40, p.92] [39, Theorem 5.8.11] implies that, if $J \in \mathcal{F}_\omega$ is any injective, then there is a decomposition in $\mathcal{F}$

$$J \simeq \bigoplus_{\lambda \in \Lambda} a(\lambda, J) I_\lambda,$$

where $a(\lambda, J) = \dim_{\mathbf{Z}/2} \operatorname{Hom}_{\mathcal{F}}(F_\lambda, J)$.

Applying this to the case $J = \Phi^{-1} S^j$, and noting [17] that

$$\dim_{\mathbf{Z}/2} \operatorname{Hom}_{\mathcal{F}}(F_\lambda, \Phi^{-1} S^j) = \dim_{\mathbf{Z}/2} \operatorname{Hom}_{\mathcal{F}}(F_\lambda, S^{2^k j}), \text{ for } k >> 0,$$

we deduce that

$$\Phi^{-1} S^j \simeq \bigoplus_{\lambda \in \Lambda} a(\lambda, j) I_\lambda,$$

with $a(\lambda, j)$ as in the introduction.

Recall that $r : \mathcal{F} \to \mathcal{U}$ is defined by letting $r(F)_j = \operatorname{Hom}_{\mathcal{F}}(S_j, F)$. The fact that S_j is finite implies that r will commute with filtered direct limits. In particular, we can deduce the decomposition in $\mathcal{U}$

$$\Phi^{-1} r(S^j \circ S_n) \simeq \bigoplus_{\lambda \in \Lambda} a(\lambda, j) r(I_\lambda \circ S_n).$$

Proposition 9.1. *$r(I_\lambda \circ S_n) \simeq H^*(K(\lambda, n); \mathbf{Z}/2)$ as $\mathcal{A}$ modules.*

Momentarily postponing the proof of this, to prove Theorem 1.9, we need to show

$$H^*(\Phi^{-1} T(n, j); \mathbf{Z}/2) \simeq \Phi^{-1} r(S^j \circ S_n) \text{ as } \mathcal{A} \text{ modules.}$$

Note that this asserts that a certain inverse limit of finite dimensional modules is isomorphic to a certain direct limit of *nilclosed* modules (i.e. modules of the form $r(F)$).

To show this, observe that $\Phi^{-1} r(S^* \circ S_n)$ is $\mathbf{N} \times \mathbf{N}[\frac{1}{2}]$ graded. It is even an object in $\mathcal{K}_\rho$, using $\Phi^{-1} : \Phi^{-1} r(S^{2j} \circ S_n) \to \Phi^{-1} r(S^j \circ S_n)$ as the restriction.

Theorem 9.2. *$H^*(\Phi^{-1} T(n, *); \mathbf{Z}/2) \simeq \Phi^{-1} r(S^* \circ S_n)$ as objects in $\mathcal{K}_\rho$.*

Returning to the proof of Proposition 9.1, we first note that $H^*(K(\lambda,n);\mathbf{Z}/2) = H^*(K(V_\lambda,n);\mathbf{Z}/2)e_\lambda$ and $r(I_\lambda \circ S_n) = r(I_{V_\lambda} \circ S_n)e_\lambda$, where $I_W \in \mathcal{F}$ is the injective defined by $I_W(V) = (\mathbf{Z}/2)^{\mathrm{Hom}(V,W)}$. Thus we need just show that

$$r(I_W \circ S_n) = H^*(K(W,n);\mathbf{Z}/2).$$

Now one has the classic calculation [40, p. 184]

$$H^*(K(\mathbf{Z}/2,n);\mathbf{Z}/2) = U(F(n)),$$

where $F(n) = \mathcal{A}/E(n)$ is the free unstable module on an n dimensional class, and where $U : \mathcal{U} \to \mathcal{K}$ is the free functor, left adjoint to the forgetful functor. Explicitly, $U(M) = S^*(M)/(Sq^{|x|}x - x^2)$. Similarly, $H^*(K(W,n);\mathbf{Z}/2) = U_W(F(n))$ where $U_W : \mathcal{U} \to \mathcal{K}$ is given by $U_W(M) = U(M \otimes W^*)$.

A simple calculation reveals that $F(n) = r(S_n)$[7] (see e.g. [25, Prop.8.1]), so the proof of Proposition 9.1 is completed with

Lemma 9.3. [26] *There are natural isomorphisms* $U_W(r(F)) \simeq r(I_W \circ F)$, *for all* $F \in \mathcal{F}_\omega$.

Sketch proof. It is easy to reduce to the case when $W = \mathbf{Z}/2$. Let $I = I_{\mathbf{Z}/2}$. By filtering $U(M)$ one then verifies that if M is nilclosed, so is $U(M)$. Thus to identify $U(r(F))$ with $r(I \circ F)$, it suffices to check that $l(U(r(F))) = I \circ F$, where $l : \mathcal{U} \to \mathcal{F}$ is left adjoint to r. The functor l is exact, preserves tensor products, and can be regarded as localization away from nilpotent modules [14, 23]. Thus it carries

$$S^*(r(F))/(Sq^{|x|}x - x^2)$$

to the functor that sends V to

$$S^*(l(r(F))(V))/(x - x^2).$$

Since $l(r(F)) = F$, and $I(V) = S^*(V)/(x - x^2)$ [23], this functor is just $I \circ F$. □

To prove Theorem 9.2, we need to use the main result of [26].

As in [25], let $\mathcal{U}^2$ be the category of $\mathbf{N} \times \mathbf{N}$ graded modules over the bigraded algebra $A \otimes A$, unstable in each grading. For $M \in \mathcal{U}^2$, there are natural maps $\Phi_1 : M_{m,*} \to M_{2m,*}$ and $\Phi_2 : M_{*,n} \to M_{*,2n}$[8], and we let $\mathcal{K}^2$ denote the category of commutative algebras M in $\mathcal{U}^2$ satisfying the "restriction" axiom: for all $x \in M$, $(\Phi_1 \otimes \Phi_2)(x) = x^2$.

Let $U_2 : \mathcal{U}^2 \to \mathcal{K}^2$ be left adjoint to the forgetful functor: explicitly, $U_2(M) = S^*(M)/((\Phi_1 \otimes \Phi_2)(x) - x^2)$.

Given $M \in \mathcal{U}$, $M \otimes F(1)$ is an object in $\mathcal{U}^2$. $F(1)$ can be regarded as the module $\langle x_1, \ldots, x^{2^k}, \ldots \rangle$, with x^{2^k} having bidegree $(1, 2^k)$. Now define

$$\mathrm{Hom}_{\mathcal{F}}(S_*, F) \otimes F(1) \to \mathrm{Hom}_{\mathcal{F}}(S_*, S^* \circ F)$$

[7] This is false at odd primes: $F(n)$ is not nilclosed in the odd prime case.

[8] These are the Steenrod squares in the right degree

by sending $(S_i \xrightarrow{\alpha} F) \otimes x^{2^k}$ to the composite $S_i \xrightarrow{\alpha} F \to S^{2^k} \circ F$. Since $\mathrm{Hom}_{\mathcal{F}}(S_*, S^* \circ F)$ is easily checked to be in $\mathcal{K}^2$, this map extends to a natural map in $\mathcal{K}^2$:

$$\Theta_F : U_2(\mathrm{Hom}_{\mathcal{F}}(S_*, F) \otimes F(1)) \to \mathrm{Hom}_{\mathcal{F}}(S_*, S^* \circ F).$$

Theorem 9.4. [26] *For all $F \in \mathcal{F}_\omega$, Θ_F is an isomorphism.*

This is proved in a manner similar to the way Lemma 9.3 is proved.

Corollary 9.5. $r(S^* \circ S_n) \simeq U_2(F(n) \otimes F(1))$, *as objects in* $\mathcal{K}^2$.

Corollary 9.6. $\Phi^{-1} r(S^* \circ S_n) \simeq U_\rho(F(n) \otimes \Phi^{-1} F(1))$, *as objects in* $\mathcal{K}_\rho$.

Here $\Phi^{-1}F(1) = \langle x^{2^k} \mid k \in \mathbf{Z}\rangle$, with the restriction map (part of the $\mathcal{K}_\rho$ structure), taking x^{2^k} to $x^{2^{k-1}}$.

By Theorem 1.2, $H^*(\Phi^{-1}T(n,*); \mathbf{Z}/2) \simeq U_\rho(F_\rho(n)_{\hat{\Phi}})$ as objects in $\mathcal{K}_\rho$, where $F_\rho(n)_{\hat{\Phi}}$ denotes the inverse limit

$$F_\rho(n) \xleftarrow{\rho} F_\rho(n) \xleftarrow{\rho} F_\rho(n) \xleftarrow{\rho} \cdots.$$

The following observation completes the proof of Theorem 9.2, and thus the proof of Theorem 1.9.

Lemma 9.7. $F_\rho(n)_{\hat{\Phi}} = F(n) \otimes \Phi^{-1}F(1)$, *as objects in* $\mathcal{U}_\rho$.

10. Towards the Conjectures

In this section we outline some possible approaches to the conjectures of the introduction.

We start with a rigorous proof of Proposition 1.11.

Proof of Proposition 1.11. Let $X(j) = \bigvee_{\lambda \in \Lambda} a(\lambda, j) K(\lambda, 1)$, and recall that we wish to topologically realize an $\mathcal{A}$ module isomorphism:

$$H^*(\Phi^{-1}T(1,j); \mathbf{Z}/2) \simeq H^*(X(j); \mathbf{Z}/2).$$

But Proposition 8.2 tells us that any such $\mathcal{A}$–module map *can* be realized: in the Adams spectral sequence for computing maps from $X(j)$ to $\Phi^{-1}T(1,j)$, $E_2^{s,t} = 0$ for $t - s < 0$. □

Thus far, we have been unable to find any way in which this proof, or the related proofs of Conjecture 1.3 [27, 11, 15] in the $n = 1$ case, generalize to prove the $n > 1$ cases of the conjectures. These Adams spectral sequence based proofs rely on magical properties of the spectra $T(j)$ and $K(V, 1)$, which, in turn, are (partly) due to the fact that $H^*(T(j); \mathbf{Z}/2)$ and $H^*(K(V, 1))$ are injective in $\mathcal{U}$. A search for similar proofs of the conjectures leads to the following questions.

Question 10.1. For $n > 1$, do $H^*(T(n,j); \mathbf{Z}/2)$ and $H^*(K(\mathbf{Z}/2, n); \mathbf{Z}/2)$ have any sort of injectivity properties in some well chosen subcategory of $\mathcal{U}$?

Question 10.2. Is $\mathrm{Ext}_{\mathcal{A}}^{s,t}(H^*(K(W,n)), H^*(K(V,n))) = 0$ if $t - s < 0$?

Question 10.3. Is $\mathbf{Z}/2[\mathrm{Hom}(V,W)] \to \mathrm{Hom}_{\mathcal{A}}(H^*(K(W,n)), H^*(K(V,n)))$ an isomorphism?

Note that an affirmative answer to the second question would allow us to prove Conjecture 1.10 along the lines of the above proof of Proposition 1.11. It is not hard to show that, if the last question has an affirmative answer, then Conjecture 1.10 would follow if one could just construct a family of stable maps

$$T(n, 2^k) \to K(\mathbf{Z}/2, n) \tag{10.1}$$

nonzero in cohomology in dimension n.

Related to the [15] proof of Conjecture 1.3 in the $n = 1$ case, we note that, by [15, Proposition 1.6], if Conjecture 1.10 were true, then one could conclude that $\epsilon : \Sigma^\infty \Omega^\infty T(n,j) \to T(n,j)$ is onto in mod 2 homology, which is a weak form of Conjecture 1.3.

Now we discuss a rather intriguing "conceptual" approach to Conjectures 1.3 and 1.10. The idea would be to start with the $n = 0$ case (!) of the conjectures, using the concept of S–algebras (a.k.a. E_∞–ring spectra).

Question 10.4. Let Λ denote a divided power algebra over $\mathbf{Z}_2$. Does there exist an $\mathbf{N}$–graded commutative augmented S–algebra structure on

$$T = \bigvee_{j \geq 0} T(0,j) = \bigvee_{j \geq 0} S^0$$

such that

(1) $\pi_0(T) = \Lambda$,

(2) $\Phi : T \to T$ is a map of S–algebras, and

(3) T admits a nonzero S-algebra map to $\Sigma^\infty(\mathbf{Z}/2)_+$?

An affirmative answer to parts (1) and (3) would presumably yield a construction of maps as in (10.1) upon applying the "bar construction" n times to map (3).

We can refine this question, motivated by work in [25].

The key is to rearrange the untidy right side of the isomorphism

$$H^*(\Phi^{-1}T(n,j); \mathbf{Z}/2) \simeq H^*(\bigvee_{\lambda \in \Lambda} a(\lambda, j) K(\lambda, n); \mathbf{Z}/2).$$

We know that this module corresponds to the functor $(\Phi^{-1}S^j) \circ S_n \in \mathcal{F}$. The proof in [25] that $\Phi^{-1}S^j$ is injective in $\mathcal{F}_\omega$ reveals that

$$\Phi^{-1}S^j \simeq \lim_{s \to \infty} I_{(\mathbf{F}_{2^s})^*}[j],$$

where $(\mathbf{F}_{2^s})^*$ is the $\mathbf{F}_2$ linear dual of the finite field $\mathbf{F}_{2^s}$, and $I_{(\mathbf{F}_{2^s})^*}[j]$ is the jth eigenspace of $I_{(\mathbf{F}_{2^s})^*}$ under the action of $\mathbf{F}_{2^s}^\times$. Furthermore, if we extend the scalars to the algebraic closure $\bar{\mathbf{F}}_2$, this isomorphism is well behaved with respect to pairings (between various j's).

It follows that

$$H^*(\Phi^{-1}T(n,*);\bar{\mathbf{F}}_2) \simeq H^*_{\text{cont}}(K(\bar{\mathbf{F}}_2,n);\bar{\mathbf{F}}_2)[*]$$

as $\mathbf{N}[\frac{1}{2}]$ graded algebras in $\mathcal{U}$, where we write

$$H^*_{\text{cont}}(K(\bar{\mathbf{F}}_2,n);\bar{\mathbf{F}}_2) = \lim_{s\to\infty} H^*(K(\mathbf{F}_{2^s},n);\bar{\mathbf{F}}_2).$$

Just as one can discuss S–algebras, one can discuss $SW(\bar{\mathbf{F}}_2)$–algebras, where $W(\bar{\mathbf{F}}_2)$ are the Witt vectors of $\bar{\mathbf{F}}_2$.

Question 10.5. With T the S-algebra as in Question 10.4 above, does there exist an equivalence of $\mathbf{N}[\frac{1}{2}]$ graded $SW(\bar{\mathbf{F}}_2)$–algebras

$$\Phi^{-1}T \wedge_S SW(\bar{\mathbf{F}}_2) \simeq \Sigma^\infty((\bar{\mathbf{F}}_2)^*)_+ \wedge_S SW(\bar{\mathbf{F}}_2)?$$

As before, an affirmative answer to this formidable question would presumably yield a proof of Conjecture 1.10 upon applying the bar construction to the equivalence n times.

We end with a question about the most straightforward way to try to get at these sorts of things.

Question 10.6. Does there exist a "naturally occurring" spectrum E, with a group action, such that the group action can be used to establish a splitting

$$\Sigma E_n \simeq \bigvee_j T_1(n,j),$$

where E_n is the nth infinite loop space of the spectrum E, and $T_1(n,j)$ is a desuspension of $\Sigma T(n,j)$?

When $n=1$, this would be consistent with [12]. However, anyone searching for such a spectrum should make sure their search is compatible with results in [35].

Appendix A. Connections with Work of Arone and Mahowald

In this appendix, we explain how our constructions are related to those appearing in [3] in their work on the Goodwillie tower of the identity. (Our arguments are a bit sketchy as we plan to elaborate on these ideas elsewhere.)

Recall our definition: $\tilde{D}_{n,j}(X) = F(\mathcal{C}(n,j)_+, X^{[j]})^{\Sigma_j}$. We begin by rewriting this in a useful way.

Let $\Delta(n,j) \subset S^{nj}$ be the singular part of the Σ_j–space S^{nj}. Then $\mathcal{C}(n,j)$ is equivariantly homotopy equivalent to $S^{nj} - \Delta(n,j)$ (the configuration space). Thus, by equivariant Alexander duality [32, Theorem III.4.1],

$$F(\mathcal{C}(n,j)_+, (\Sigma^n X)^{[j]}) \simeq S^{nj}/\Delta(n,j) \wedge X^{[j]}$$

as Σ_j spectra. Now note that this latter spectrum is clearly Σ_j–free, as $S^{nj}/\Delta(n,j)$ is, thus its fixed point spectrum is naturally equivalent to its orbit spectrum [32, Theorem II.7.1]. We have proved

Proposition A.1. *$\tilde{D}_{n,j}(\Sigma^n X)$ is naturally equivalent to $((S^{nj}/\Delta(n,j)) \wedge X^{[j]})_{\Sigma_j}$.*

Checking definitions reveals

Lemma A.2. *$\beta : \mathcal{C}(n+1,j)_+ \wedge S^1 \to \mathcal{C}(n,j)_+ \wedge S^j$ is equivariantly S-dual to the evident diagonal map $S^1 \wedge (S^{nj}/\Delta(n,j)) \to S^{(n+1)j}/\Delta(n+1,j)$.*

Definition A.3. Let $\tilde{D}_j(X) = \operatorname{hocolim}_n \Sigma^{-n}\tilde{D}_{n,j}(\Sigma^n X)$, with the colimit induced by either of the maps in the last lemma.

Note that, with this notation, $T(\infty,j) = \Sigma\tilde{D}_j(S^{-1})$.

Now let K_j be the Σ_j-space introduced in [3]: K_j is the unreduced suspension of $\tilde{K}_j$, the classifying space of the poset of the nontrivial partitions of a set with cardinality j. (By nontrivial, we mean to exclude the partitions (j) and $(1,1,\ldots,1)$.)

Proposition A.4. [3, early versions] *and* [2, §6] *There is a Σ_j equivariant map*

$$\operatorname{hocolim}_n \Sigma^{-n}(S^{nj}/\Delta(n,j)) \to \Sigma K_j$$

that is a nonequivariant equivalence.

Corollary A.5. $\tilde{D}_j(X) = (\Sigma K_j \wedge X^{[j]})_{h\Sigma_j}$.

Combining this corollary with Theorem 1.6 yields

Theorem A.6. $(K_{2^k} \wedge S^{-2^k})_{h\Sigma_{2^k}} \simeq \Sigma^{-2} SP^{2^k}_{\Delta}(S^0)$.

In work in progress, we have established the following.

Proposition A.7. *Localized at 2, there are cofibration sequences*

$$\Sigma\tilde{D}_{2^{k-1}}(S^{2n-1}) \to \Sigma\tilde{D}_{2^k}(S^{n-1}) \to \tilde{D}_{2^k}(S^n)$$

which are short exact in cohomology.

The first map here is constructed with Hopf invariant techniques, and is the generalization of $\Phi : T(\infty, 2^{k-1}) \to T(\infty, 2^k)$.

Using these sequences when $n = 0$ and $n = 1$, one can deduce

Corollary A.8. *Localized at 2, there are equivalences*

(1) $(K_{2^k})_{h\Sigma_{2^k}} \simeq \Sigma^{-1} SP^{2^k}_{\Delta}(S^0)/SP^{2^{k-1}}_{\Delta}(S^0)$,

(2) $(K_{2^k} \wedge S^{2^k})_{h\Sigma_{2^k}} \simeq SP^{2^k}(S^0)/SP^{2^{k-1}}(S^0)$.

Part (2) of this corollary is due to Arone and Mahowald who sketch the following elegant and direct short proof in their early versions of [3]. (See also [2].)

Lemma A.9. *The space S^{nj}/Σ_j is homeomorphic to $SP^j(S^n)/SP^{j-1}(S^n)$.*

Lemma A.10. *$(\Delta(n,j) \wedge S^j)_{\Sigma_j}$ is contractible.*

Sketch proof. The partition filtration of $\Delta(n,j)$ induces a filtration of $(\Delta(n,j) \wedge S^j)_{\Sigma_j}$ in which each subquotient has the form $SP^i(S^1)/SP^{i-1}(S^1)$, and so is contractible. □

Corollary A.11. *There are homotopy equivalences of spaces*

$$SP^j(S^{n+1})/SP^{j-1}(S^{n+1}) \simeq (S^{nj}/\Delta(n,j) \wedge S^j)_{\Sigma_j}.$$

Proof. $SP^j(S^{n+1})/SP^{j-1}(S^{n+1}) \simeq (S^{nj} \wedge S^j)_{\Sigma_j} \simeq (S^{nj}/\Delta(n,j) \wedge S^j)_{\Sigma_j}$. □

Now Corollary A.8(2) follows by letting n go to infinity, and using Proposition A.4.

We finish with one last observation. Let $\mathcal{D}_j(X) = F(\Sigma K_j, X^{[j]})_{h\Sigma_j}$. Arone and Mahowald [3] show that $\Omega^\infty \mathcal{D}_j(X)$ is the jth fiber of the Goodwillie tower of the identity applied to a space X. Arone and Dwyer [2] show that, if X is an odd dimensional sphere, then $\Sigma^{2k}\mathcal{D}_{2^k}(X) \simeq (\Sigma K_{2^k} \wedge X^{[2^k]})_{h\Sigma_{2^k}}$. Thus we have

Corollary A.12. *If X is an odd dimensional sphere, then $\mathcal{D}_{2^k}(X) \simeq \Sigma^{-2k}\tilde{D}_{2^k}(X)$.*

Corollary A.13. $\mathcal{D}_{2^k}(S^{-1}) \simeq \Sigma^{-(2k+1)}SP^{2^k}_{\Delta}(S^0)$.

References

[1] J. Adem, *The relations on Steenrod powers of cohomology classes*, Algebraic Geometry and Topology (R. H. Fox, et. al., editors), Princeton University Press, 1957, 191–242.

[2] G. Z. Arone and W. G. Dwyer, *Partition complexes, Tits buildings, and symmetric product spectra*, preliminary preprint, 1997.

[3] G. Arone and M. Mahowald, *The Goodwillie tower of the identity functor and the unstable periodic homotopy of spheres*, Invent. Math. **135**(1999), 743–788.

[4] M. G. Barratt and H. R. Miller, *On the antiautomorphism of the Steenrod algebra*, A. M. S. Cont. Math. **12**(1982), 47–52.

[5] E. H. Brown and R. L. Cohen, *The Adams spectral sequence of $\Omega^2 S^3$ and Brown-Gitler spectra*, Annals of Math. Study **113**(1987), 101–125.

[6] R. R. Bruner, J. P. May, J. McClure, and M. Steinberger *H_∞ Ring Spectra and their Applications*, Springer L. N. Math. **1176**, 1986.

[7] S. R. Bullett and I. G. MacDonald, *On the Adem relations*, Topology **21**(1982), 329–332.

[8] G. Carlsson, *G. B. Segal's Burnside ring conjecture for* $(\mathbf{Z}/2)^k$, Topology **22**(1983), 83–103.

[9] F. R. Cohen, T. Lada, and J. P. May, *The Homology of Iterated Loop Spaces*, Springer L. N. Math. **533**, 1976.

[10] R. Fox and L. Neuwirth, *The braid groups*, Math. Scand. **10**(1962), 119–126.

[11] P. Goerss, *A direct construction for the duals of Brown-Gitler spectra*, Ind. Math. J. **34**(1985), 733–751.

[12] P. Goerss, J. Lannes, and F. Morel, *Vecteurs de Witt non-commutatifs et representabilité de l'homologie modulo p*, Invent. Math. **108**(1992), 163–227.

[13] T. G. Goodwillie, *Calculus III: the Taylor series of a homotopy functor*, preprint, c.1994.

[14] H.-W. Henn, J. Lannes, and L. Schwartz *The categories of unstable modules and unstable algebras modulo nilpotent objects*, Amer. J. Math. **115**(1993), 1053–1106.

[15] D. J. Hunter and N. J. Kuhn, *Characterizations of spectra with $\mathcal{U}$-injective cohomology which satisfy the Brown–Gitler property*, to appear in Trans. A.M.S.

[16] D. Husemoller, *Fibre Bundles*, Springer Graduate Texts in Math. **20**, 1966.

[17] P. Krason and N. J. Kuhn, *On embedding polynomial functors in symmetric powers*, J. Algebra. **99**(1994), 281–294.

[18] N. J. Kuhn, *The geometry of the James Hopf maps*, Pac. J. Math. **102**(1982), 397–412.

[19] N. J. Kuhn, *A Kahn-Priddy sequence and a conjecture of G. W. Whitehead*, Math. Proc. Camb. Phil. Soc. **92**(1982), 467–483.

[20] N. J. Kuhn, *Spacelike resolutions of spectra*, A.M.S. Cont. Math. Series 19(1983), 153–165.

[21] N. J. Kuhn, *Suspension spectra and homology equivalences*, Trans. A. M. S. **283**(1984), 303–313.

[22] N. J. Kuhn, *The transfer and the James-Hopf invariants*, Math. Zeit. **196**(1987), 391–405.

[23] N. J. Kuhn, *Generic representation theory of the finite general linear groups and the Steenrod algebra: I*, Amer. J. Math. **116**(1994), 327–360.

[24] N. J. Kuhn, *Generic representation theory of the finite general linear groups and the Steenrod algebra: II*, K-Theory J. **8**(1994), 395–428.

[25] N. J. Kuhn, *Generic representation theory of the finite general linear groups and the Steenrod algebra: III*, K-Theory J. **9**(1995), 273–303.

[26] N. J. Kuhn, *Computations in generic representation theory: maps from symmetric powers to composite functors*, Trans. A.M.S. **350**(1998), 4221–4233.

[27] J. Lannes, *Sur le n dual du n-ème spectre de Brown-Gitler*, Math. Zeit. **159**(1988), 29–42.

[28] J.Lannes, *Sur les espaces fonctionnels dont la source est le classifiant d'un p-groupe abélien élémentaire*, Pub. I.H.E.S. **75**(1992), 135–244.

[29] J. Lannes and L. Schwartz, *Sur la structure des A-modules instables injectifs*, Topology **28**(1989), 153–169.

[30] J. Lannes and S. Zarati, *Sur les $\mathcal{U}$-injectifs*, Ann. Scient. Ec. Norm. Sup **19**(1986), 1–31.

[31] J. Lannes and S. Zarati, *Sur les foncteurs dérivés de la déstablisation*, Math. Zeit. **194**(1987), 25–59.

[32] L. G. Lewis, J. P. May, J. McClure, and M. Steinberger *Equivariant Stable Homotopy Theory*, Springer L. N. Math. **1213**, 1986.

[33] M. Mahowald, *A new infinite family in ${}_2\pi_*^S$*, Topology **16**(1977), 249–256.

[34] J. P. May, *The Geometry of Iterated Loop Spaces*, Springer L. N. Math. **271**, 1972.

[35] C. A. McGibbon, *Infinite loop spaces and Neisendorfer localization*, Proc. A.M.S. **125**(1997), 309–313.

[36] H. R. Miller, *An algebraic analogue of a conjecture of G. W. Whitehead*, Proc. A.M.S. **84**(1982), 131–137.

[37] H. R. Miller, *The Sullivan conjecture on maps from classifying spaces,* Ann. Math. **120**(1984), 39–87.

[38] S. A. Mitchell and S. B. Priddy, *Stable splittings derived from the Steinberg module,* Topology **22**(1983), 285–298.

[39] N. Popescu, *Abelian Categories with Applications to Rings and Modules*, Academic Press, London, 1973.

[40] L. Schwartz, *Unstable modules over the Steenrod algebra and Sullivan's fixed point conjecture,* Chicago Lectures in Math., Unversity of Chicago Press, 1994.

[41] W. M. Singer, *A new chain complex for the homology of the Steenrod algebra,* Math. Proc. Camb. Phil. Soc. **90**(1981), 279–292.

Department of Mathematics
University of Virginia
Charlottesville, VA 22903
E-mail address: `njk4x@adams.math.virginia.edu`

Progress in Mathematics, Vol. 196

Chern characters for the equivariant K-theory of proper G-CW-complexes

Wolfgang Lück and Bob Oliver

Abstract. We first construct a classifying space for defining equivariant K-theory for proper actions of discrete groups. This is then applied to construct equivariant Chern characters with values in Bredon cohomology with coefficients in the representation ring functor $R(-)$ (tensored by the rationals). And this in turn is applied to prove some versions of the Atiyah-Segal completion theorem for real and complex K-theory in this setting.

In an earlier paper [8], we showed that for any discrete group G, equivariant K-theory for finite proper G-CW-complexes can be defined using equivariant vector bundles. This was then used to prove a version of the Atiyah-Segal completion theorem in this situation. In this paper, we continue to restrict attention to actions of discrete groups, and begin by constructing an appropriate classifying space which allows us to define $K_G^*(X)$ for an arbitrary proper G-complex X. We then construct rational-valued equivariant Chern characters for such spaces, and use them to prove some more general versions of completion theorems.

In fact, we construct two different types of equivariant Chern character, both of which involve Bredon cohomology with coefficients in the system $\big(G/H \mapsto R(H)\big)$. The first,

$$\mathrm{ch}_X^* : K_G^*(X) \longrightarrow H_G^*(X; \mathbb{Q} \otimes R(-)),$$

is defined for arbitrary proper G-complexes. The second, a refinement of the first, is a homomorphism

$$\widetilde{\mathrm{ch}}_X^* : K_G^*(X) \longrightarrow \mathbb{Q} \otimes H_G^*(X; R(-)),$$

but defined only for finite dimensional proper G-complexes for which the isotropy subgroups on X have bounded order. When X is a *finite* proper G-complex (i.e., X/G is a finite CW-complex), then $H_G^*(X; R(-))$ is finitely generated, and these two target groups are isomorphic. The second Chern character is important when proving the completion theorems. The idea for defining equivariant Chern characters with values in Bredon cohomology $H_G^*(X; \mathbb{Q} \otimes R(-))$ was first due to Słomińska [12]. A complex-valued Chern character was constructed earlier by Baum and Connes [4], using very different methods.

The second author is partly supported by UMR 7539 of the CNRS.

The completion theorem of [8] is generalized in two ways. First, we prove it for real as well as complex K-theory. In addition, we prove it for families of subgroups in the sense of Jackowski [7]. This means that for each finite proper G-complex X and each family $\mathcal{F}$ of subgroups of G, $K_G^*(E_{\mathcal{F}}(G) \times X)$ is shown to be isomorphic to a certain completion of $K_G^*(X)$. In particular, when $\mathcal{F} = \{1\}$, then $E_{\mathcal{F}}(G) = EG$, and this becomes the usual completion theorem.

The classifying spaces for equivariant K-theory are constructed here using Segal's Γ-spaces. This seems to be the most convenient form of topological group completion in our situation. However, although Γ-spaces do produce spectra, as described in [10], the spectra they produce are connective, and hence not what is needed to define equivariant K-theory directly. So instead, we define $K_G^{-n}(-)$ and $KO_G^{-n}(-)$ for all $n \geq 0$ using classifying *spaces* constructed from a Γ-space, then prove Bott periodicity, and use that to define the groups in positive degrees. One could, of course, construct an equivariant spectrum (or an $\mathrm{Or}(G)$-spectrum in the sense of [6]) by combining our classifying space K_G with the Bott map $\Sigma^2 K_G \to K_G$; but the approach we use here seems the simplest way to do it.

By comparison, in [6], equivariant K-*homology* groups $K_*^G(X)$ were defined by using certain covariant functors $\mathbf{K}^{\mathrm{top}}$ from the orbit category $\mathrm{Or}(G)$ to spectra. This construction played an important role in [6] in reformulating the Baum-Connes conjecture. In general, one expects an equivariant homology theory to be classified by a covariant functor from the orbit category to spaces or spectra, and an equivariant cohomology theory to be classified by a contravariant functor. But in fact, when defining equivariant K-theory here, it turned out to be simplest to do so via a classifying G-space, rather than a classifying functor from $\mathrm{Or}(G)$ to spaces.

We would like in particular to thank Chuck Weibel for suggesting Segal's paper and the use of Γ-spaces, as a way to avoid certain problems we encountered when first trying to define the multiplicative structure on $K_G(X)$.

The paper is organized as follows. The classifying spaces for $K_G^{-n}(-)$ and $KO_G^{-n}(-)$ are constructed in Section 1; and the connection with G-vector bundles is described. Products are then constructed in Section 2, and are used to define Bott homomorphisms and ring structures on $K_G^*(X)$; and thus to complete the construction of equivariant K-theory as a multiplicative equivariant cohomology theory. Homomorphisms in equivariant K-theory involving changes of groups are then constructed in Section 3. Finally, the equivariant Chern characters are constructed in Section 5, and the completion theorems are formulated and proved in Section 6. Section 4 contains some technical results about rational characters.

1. A classifying space for equivariant K-theory

Our classifying space for equivariant K-theory for proper actions of an infinite discrete group is constructed using Γ-spaces in the sense of Segal. So we begin by summarizing the basic definitions in [10].

Let Γ be the category whose objects are finite sets, and where a morphism $\theta : S \to T$ sends each element $s \in S$ to a subset $\theta(s) \subseteq T$ such that $s \neq s'$ implies $\theta(s) \cap \theta(s') = \emptyset$. Equivalently, if $\mathcal{P}(S)$ denotes the set of subsets of S, one can regard a morphism in Γ as a map $\mathcal{P}(S) \to \mathcal{P}(T)$ which sends disjoint unions to disjoint unions. For all $n \geq 0$, $\mathbf{n}$ denotes the object $\{1, \ldots, n\}$. (In particular, $\mathbf{0}$ is the empty set.) There is an obvious functor from the simplicial category Δ to Γ, which sends each object $[n] = \{0, 1, \ldots, n\}$ in Δ to $\mathbf{n}$, and where a morphism in Δ — an order preserving map $\varphi : [m] \to [n]$ — is sent to the morphism $\theta_\varphi : \mathbf{m} \to \mathbf{n}$ in Γ which sends i to $\{j \,|\, \varphi(i-1) < j \leq \varphi(i)\}$.

A Γ-space is a functor $\underline{A} : \Gamma^{\mathrm{op}} \to \mathcal{S}paces$ which satisfies the following two conditions:

(i) $\underline{A}(\mathbf{0})$ is a point; and

(ii) for each $n > 1$, the map $\underline{A}(\mathbf{n}) \longrightarrow \prod_{i=1}^n \underline{A}(\mathbf{1})$, induced by the inclusions $\kappa_i : \mathbf{1} \to \mathbf{n}$ ($\kappa_i(1) = \{i\}$), is a homotopy equivalence.

(In fact, Segal only requires that $\underline{A}(\mathbf{0})$ be contractible; but for our purposes it is simpler to assume it is always a point.) Note that each $A(S)$ has a basepoint: the image of $A(\mathbf{0})$ induced by the unique morphism $S \to \mathbf{0}$. We write $A = \underline{A}(\mathbf{1})$, thought of as the "underlying space" of the Γ-space $\underline{A}$. A Γ-space $\underline{A} : \Gamma^{\mathrm{op}} \to \mathcal{S}paces$ can be regarded as a simplicial space via restriction to Δ, and $|\underline{A}|$ denotes its topological realization (nerve) as a simplicial space.

If $\underline{A}$ is a Γ-space, then $\underline{BA}$ denotes the Γ-space $\underline{BA}(S) = |\underline{A}(S \times -)|$; and this is iterated to define $\underline{B^nA}$ for all n. Thus, $B^nA = \underline{B^nA}(\mathbf{1})$ is the realization of the n-simplicial space which sends $(S_1, \ldots, S_n)$ to $\underline{A}(S_1 \times \cdots \times S_n)$. Since $\underline{A}(\mathbf{0})$ is a point, we can identify ΣA $(= \Sigma(\underline{A}(\mathbf{1})))$ as a subspace of $BA \cong |\underline{A}|$; and this induces by adjointness a map $A \to \Omega BA$. Upon iterating this, we get maps $\Sigma(B^nA) \to B^{n+1}A$ for all n; and these make the sequence $A, BA, B^2A, \ldots$ into a spectrum. This is "almost" an Ω-spectrum, in that $B^nA \simeq \Omega B^{n+1}A$ for all $n \geq 1$ [10, Proposition 1.4].

Note that for any Γ-space $\underline{A}$, the underlying space $A = \underline{A}(\mathbf{1})$ is an H-space: multiplication is defined to be the composite of a homotopy inverse of the equivalence $\underline{A}(\mathbf{2}) \xrightarrow{\simeq} \underline{A}(\mathbf{1}) \times \underline{A}(\mathbf{1})$ with the map $\underline{A}(\mathbf{2}) \to \underline{A}(\mathbf{1})$ induced by $m_2 : \mathbf{1} \to \mathbf{2}$ ($m_2(1) = \{1, 2\}$). Then $A \simeq \Omega BA$ if $\pi_0(A)$ is a group; and ΩBA is the topological group completion of A otherwise. All of this is shown in [10, §1].

We work here with equivariant Γ-spaces; i.e., with functors

$$\underline{A} : \Gamma^{\mathrm{op}} \to G\text{-}\mathcal{S}paces$$

for which $\underline{A}(\mathbf{0})$ is a point, and for which $\big(\underline{A}(\mathbf{n})\big)^H \to \prod_{i=1}^n \big(\underline{A}(\mathbf{1})\big)^H$ is a homotopy equivalence for all $H \subseteq G$. In other words, restriction to fixed point sets of any $H \subseteq G$ defines a Γ-space $\underline{A}^H$; and the properties of equivariant Γ-spaces follow immediately from those of nonequivariant ones. For example, Segal's [10, Proposition 1.4] implies immediately that for any equivariant Γ-space $\underline{A}$, $B^nA \to \Omega B^{n+1}A$ is a weak equivalence for all $n \geq 1$ in the sense that it restricts to an equivalence $(B^nA)^H \simeq (\Omega B^{n+1}A)^H$ for all $H \subseteq G$. This motivates the following definitions.

If $\mathcal{F}$ is any family of subgroups of G, then a *weak $\mathcal{F}$-equivalence* of G-spaces is a G-map whose restriction to fixed point sets of any subgroup in $\mathcal{F}$ is a weak homotopy equivalence in the usual sense. The following lemma about maps to weak equivalences is well known; we note it here for later reference.

Lemma 1.1. *Fix a family $\mathcal{F}$ of subgroups of G, and let $f : Y \to Y'$ be any weak $\mathcal{F}$-equivalence. Then for any G-complex X all of whose isotropy subgroups are in $\mathcal{F}$, the map*

$$f_* : [X,Y]_G \xrightarrow{\cong} [X,Y']_G$$

is a bijection. More generally, if $A \subseteq X$ is any G-invariant subcomplex, and all isotropy subgroups of $X \smallsetminus A$ are in $\mathcal{F}$, then for any commutative diagram

$$\begin{array}{ccc} A & \xrightarrow{\alpha_0} & Y \\ \downarrow & & \downarrow{\scriptstyle f} \\ X & \xrightarrow{\alpha} & Y' \end{array}$$

of G-maps, there is an extension of α_0 to a G-map $\widetilde{\alpha} : X \to Y$ such that $f \circ \widetilde{\alpha} \simeq \alpha$ (equivariantly homotopic), and $\widetilde{\alpha}$ is unique up to equivariant homotopy.

Proof. The idea is the following. Fix a G-orbit of cells $\bigl(G/H \times D^n \to X\bigr)$ in X whose boundary is in A. Then, since $Y^H \to (Y')^H$ is a weak homotopy equivalence, the map $eH \times D^n \to X^H \to (Y')^H$ can be lifted to Y^H (up to homotopy), and this extends equivariantly to a G-map $G/H \times D^n \to Y$. Upon continuing this procedure, we obtain a lifting of α to a G-map $\widetilde{\alpha} : X \to Y$ which extends α_0. This proves the existence of a lifting in the above square (and the surjectivity of f_* in the special case); and the uniqueness of the lifting follows upon applying the same procedure to the pair $X \times I \supseteq (X \times \{0,1\}) \cup (A \times I)$. □

Now fix a discrete group G. Let $\mathcal{E}(G)$ be the category whose objects are the elements of G, and with exactly one morphism between each pair of objects. Let $\mathcal{B}(G)$ be the category with one object, and one morphism for each element of G. (Note that $|\mathcal{E}(G)| = EG$ and $|\mathcal{B}(G)| = BG$; hence the notation.) When necessary to be precise, g_a will denote the morphism $a \to ga$ in $\mathcal{E}(G)$. We let G act on $\mathcal{E}(G)$ via right multiplication: $x \in G$ acts on objects by sending a to ax and on morphisms by sending g_a to g_{ax}. Thus, for any $H \subseteq G$, the orbit category $\mathcal{E}(G)/H$ is the groupoid whose objects are the cosets in G/H, and with one morphism $g_{aH} : aH \to gaH$ for each $g \in G$: a category which is equivalent to $\mathcal{B}(H)$. Note in particular that $\mathcal{B}(G) \cong \mathcal{E}(G)/G$.

In order to deal simultaneously with real and complex K-theory, we let F denote one of the fields $\mathbb{C}$ or $\mathbb{R}$. Set $F^\infty = \bigcup_{n=1}^\infty F^n$: the space of all infinite sequences in F with finitely many nonzero terms. Let F-mod be the category whose objects are the finite dimensional vector subspaces of F^∞, and whose morphisms are F-linear *isomorphisms.* The set of objects of F-mod is given the discrete topology, and the space of morphisms between any two objects has the usual topology.

For any finite set S, an *S-partitioned* vector space is an object V of F-mod, together with a direct sum decomposition $V = \bigoplus_{s \in S} V_s$. Let $F\langle S\rangle$-mod denote the category of S-partitioned vector spaces in F-mod, where morphisms are isomorphisms which respect the decomposition. In particular, $F\langle\mathbf{0}\rangle$-mod has just one object $0 \subseteq F^\infty$ and one morphism. A morphism $\theta : S \to T$ induces a functor $F\langle\theta\rangle$ from $F\langle T\rangle$-mod to $F\langle S\rangle$-mod, by sending $V = \bigoplus_{t\in T} V_t$ to $W = \bigoplus_{s\in S} W_s$ where $W_s = \bigoplus_{t\in\theta(s)} V_t$.

Let $\underline{\mathrm{Vec}}_G^F$ be the Γ-space defined by setting

$$\underline{\mathrm{Vec}}_G^F(S) \stackrel{\text{def}}{=} \left|\mathrm{func}(\mathcal{E}(G), F\langle S\rangle\text{-mod})\right|$$

for each finite set S. Here, $\mathrm{func}(\mathcal{C},\mathcal{D})$ denotes the category of functors from $\mathcal{C}$ to $\mathcal{D}$. We give this the G-action induced by the action on $\mathcal{E}(G)$ described above. This is made into a functor on Γ via composition with the functors $F\langle\theta\rangle$.

By definition, $\underline{\mathrm{Vec}}_G^F(\mathbf{0})$ is a point. To see that $\underline{\mathrm{Vec}}_G^F$ is an equivariant Γ-space, it remains to show for each n and H that the map $\left(\underline{\mathrm{Vec}}_G^F(\mathbf{n})\right)^H \to \prod_{i=1}^n \left(\underline{\mathrm{Vec}}_G^F(\mathbf{1})\right)^H$ is a homotopy equivalence. The target is the nerve of the category of functors from $\mathcal{E}(G)/H$ to n-tuples of objects in F-mod, while the source can be thought of as the nerve of the full subcategory of functors from $\mathcal{E}(G)/H$ to n-tuples of vector subspaces which are independant in F^∞. And these two categories are equivalent, since every object in the larger one is isomorphic to an object in the smaller (and the set of objects is discrete).

For all finite $H \subseteq G$, $\left(\mathrm{Vec}_G^F\right)^H$ is the disjoint union, taken over isomorphism classes of finite dimensional H-representations, of the classifying spaces of their automorphism groups. We will see later that Vec_G^F classifies G-vector bundles over proper G-complexes. So it is natural to define equivariant K-theory using the its group completion $KF_G \stackrel{\text{def}}{=} \Omega B\mathrm{Vec}_G^F$, regarded as a pointed G-space.

In the following definition, $[-,-]_G$ and $[-,-]_G^{\cdot}$ denote sets of homotopy classes of G-maps, and of pointed G-maps, respectively.

Definition 1.2. *For each proper G-complex X, set*

$$K_G(X) = [X, K\mathbb{C}_G]_G \quad \textit{and} \quad KO_G(X) = [X, K\mathbb{R}_G]_G.$$

For each proper G-CW-pair (X,A) and each $n \geq 0$, set

$$K_G^{-n}(X,A) = [\Sigma^n(X/A), K\mathbb{C}_G]_G^{\cdot} \quad \textit{and} \quad KO_G^{-n}(X,A) = [\Sigma^n(X/A), K\mathbb{R}_G]_G^{\cdot}.$$

The usual cohomological properties of the $KF_G^{-n}(-)$ follow directly from the definition. Homotopy invariance and excision are immediate; and the exact sequence of a pair and the Mayer-Vietoris sequence of a pushout square are shown using Puppe sequences to hold in degrees ≤ 0. Note in particular the relations

$$\begin{aligned} KF_G^{-n}(X) &\cong \mathrm{Ker}\big[KF_G(S^n \times X) \longrightarrow KF_G(X)\big] \\ KF_G^{-n}(X,A) &\cong \mathrm{Ker}\big[KF_G^{-n}(X \cup_A X) \longrightarrow KF_G^{-n}(X)\big], \end{aligned} \tag{1.3}$$

for any proper G-CW-pair (X,A) and any $n \geq 0$.

The following lemma will be needed in the next section. It is a special case of the fact that $\underline{\mathrm{Vec}}_G^F$ and KF_G (at least up to homotopy) are independent of our choice of category of F-vector spaces.

Lemma 1.4. *For any monomorphism $\alpha : F^\infty \to F^\infty$, the induced map $\alpha_* : \underline{\mathrm{Vec}}_G^F \to \underline{\mathrm{Vec}}_G^F$, defined by composition with F-$\mathtt{mod} \xrightarrow{\alpha(-)} F$-$\mathtt{mod}$, is G-homotopic to the identity. In particular, α_* induces the identity on $K_G(X)$.*

Proof. The functor $\big(V \mapsto \alpha(V)\big)$ is naturally isomorphic to the identity. □

In [8], we defined $\mathbb{K}_G(X)$, for any proper G-complex X, to be the Grothendieck group of the monoid of vector bundles over X. We next construct natural homomorphisms $\mathbb{K}_G(X) \to K_G(X)$, for all proper G-complexes X, which are isomorphisms if X/G is a finite complex (this is the situation where the $\mathbb{K}_G^*(X)$ form an equivariant cohomology theory).

For each $n \geq 0$, let F^n-$\mathtt{mod} \subseteq F$-$\mathtt{mod}$ be the full subcategory of n-dimensional vector subspaces in F^∞. Let F^n-$\mathtt{frame}$ denote the category whose objects are the pairs (V, b), where V is an object of F^n-$\mathtt{mod}$ and b is an ordered basis of V; and whose morphisms are the isomorphisms which send ordered basis to ordered basis. The set of objects is given the topology of a disjoint union of copies of $GL_n(F)$ (one for each V in F^n-$\mathtt{mod}$). Note that there is a unique morphism between any pair of objects in F^n-$\mathtt{frame}$. Set

$$\mathrm{Vec}_G^{F,n} = \big|\mathrm{func}(\mathcal{E}(G), F^n\text{-}\mathtt{mod})\big| \qquad \text{and} \qquad \widetilde{\mathrm{Vec}}_G^{F,n} = \big|\mathrm{func}(\mathcal{E}(G), F^n\text{-}\mathtt{frame})\big|,$$

with the action of $G \times GL_n(F)$ on $\widetilde{\mathrm{Vec}}_G^{F,n}$ induced by the G-action on $\mathcal{E}(G)$ and the $GL_n(F)$-action on the set of ordered bases of each n-dimensional V. Let $\tau_n : \widetilde{\mathrm{Vec}}_G^{F,n} \to \mathrm{Vec}_G^{F,n}$ be the G-map induced by the forgetful functor F^n-$\mathtt{frame} \to F^n$-$\mathtt{mod}$. Then $GL_n(F)$ acts freely and properly on $\widetilde{\mathrm{Vec}}_G^{F,n}$. And τ_n induces a G-homeomorphism $\widetilde{\mathrm{Vec}}_G^{F,n}/GL_n(F) \cong \mathrm{Vec}_G^{F,n}$, since for any $\varphi : V \to V'$ in F-$\mathtt{mod}$, a lifting of V or V' to F^n-$\mathtt{frame}$ determines a unique lifting of the morphism.

Let $H \subseteq G \times GL_n(F)$ be any subgroup. If $H \cap (1 \times GL_n(F)) \neq 1$, then $(\widetilde{\mathrm{Vec}}_G^{F,n})^H = \emptyset$, since $GL_n(F)$ acts freely on $\widetilde{\mathrm{Vec}}_G^{F,n}$. So assume $H \cap (1 \times GL_n(F)) = 1$. Then H is the graph of some homomorphism $\varphi : H' \to GL_n(F)$ $(H' \subseteq G)$, and $(\widetilde{\mathrm{Vec}}_G^{F,n})^H$ is the nerve of the (nonempty) category of φ-equivariant functors $\mathcal{E}(G) \to F^n$-$\mathtt{frame}$, with a unique morphism between any pair of objects (since there is a unique morphism between any pair of objects in F^n-$\mathtt{frame}$). In particular, this shows that $(\widetilde{\mathrm{Vec}}_G^{F,n})^H$ is contractible.

Thus, $\widetilde{\mathrm{Vec}}_G^{F,n}$ is a universal space for those $(G \times GL_n(F))$-complexes upon which $GL_n(F)$ acts freely (cf. [8, §2]). The frame bundle of any n-dimensional G-F-vector bundle over a G-complex X is such a complex, and hence n-dimensional G-F-vector bundles over X are classified by maps to $\mathrm{Vec}_G^{F,n} = \widetilde{\mathrm{Vec}}_G^{F,n}/GL_n(F)$. It follows that

$$\mathrm{EVec}_G^{F,n} = \widetilde{\mathrm{Vec}}_G^{F,n} \times_{GL_n(F)} F^n \longrightarrow \mathrm{Vec}_G^{F,n}$$

is a universal n-dimensional G-F-vector bundle. And $[X, \mathrm{Vec}_G^{F,n}]_G \cong \mathbb{V}\mathrm{ect}_G^{F,n}(X)$: the set of isomorphism classes of n-dimensional G-F-vector bundles over X.

If E is any G-F-vector bundle over X, we let $[\![E]\!] \in KF_G(X) = [X, KF_G]_G$ be the composite of the classifying map $X \to \mathrm{Vec}_G^F$ for E with the group completion map $\mathrm{Vec}_G^F \to \Omega B\mathrm{Vec}_G^F = KF_G$. Any pair E, E' of vector bundles over X is induced by a G-map

$$X \longrightarrow \mathrm{Vec}_G^F \times \mathrm{Vec}_G^F = |\operatorname{func}(\mathcal{E}(G), F\text{-}\mathtt{mod} \times F\text{-}\mathtt{mod})| \simeq |\operatorname{func}(\mathcal{E}(G), F\langle \mathbf{2}\rangle\text{-}\mathtt{mod})|;$$

and upon composing with the forgetful functor $F\langle\mathbf{2}\rangle\text{-}\mathtt{mod} \to F\text{-}\mathtt{mod}$ we get the classifying map for $E \oplus E'$. The direct sum operation on $\mathbb{V}\mathrm{ect}_G^F(X)$ is thus induced by the H-space structure on Vec_G^F, and $[\![E \oplus E']\!] = [\![E]\!] + [\![E']\!]$ for all E, E'.

Proposition 1.5. *The assignment* $\big([E] \mapsto [\![E]\!]\big)$ *defines a homomorphism*

$$\gamma_X : \mathbb{K}F_G(X) \longrightarrow KF_G(X),$$

for any proper G-complex X. This extends to natural homomorphisms $\gamma_{X,A}^{-n}$: $\mathbb{K}F_G^{-n}(X,A) \to KF_G^{-n}(X,A)$, *for all proper G-CW-pairs (X,A) and all $n \geq 0$; which are isomorphisms when restricted to the category of finite proper G-CW-pairs.*

Proof. By the above remarks, $\big([E] \mapsto [\![E]\!]\big)$ defines a homomorphism of monoids from $\mathbb{V}\mathrm{ect}_G^F(X)$ to $KF_G(X)$, and hence a homomorphism of groups

$$\gamma_X : \mathbb{K}F_G(X) \longrightarrow KF_G(X).$$

Homomorphisms $\gamma_{X,A}^{-n}$ (for all proper G-CW-pairs (X,A)) are then constructed via the definitions

$$\mathbb{K}F_G^{-n}(X) \stackrel{\text{def}}{=} \operatorname{Ker}[\mathbb{K}F_G(S^n \times X) \to \mathbb{K}F_G(X)]$$

and $\mathbb{K}F_G^{-n}(X,A) \stackrel{\text{def}}{=} \operatorname{Ker}[\mathbb{K}F_G^{-n}(X \cup_A X) \to \mathbb{K}F_G^{-n}(X)]$ used in [8], together with the analogous relations (1.3) for $K_G^*(-)$. These homomorphisms clearly commute with boundary maps.

It remains to check that γ_X^{-n} is an isomorphism whenever X is a finite proper G-complex. Since $\mathbb{K}F_G(-)$ and $KF_G(-)$ are both cohomology theories in this situation, it suffices, using the Mayer-Vietoris sequences for pushout squares

$$\begin{array}{ccc} G/H \times S^{m-1} & \longrightarrow & G/H \times D^m \\ {\scriptstyle\varphi}\big\downarrow & & \big\downarrow \\ X & \longrightarrow & (G/H \times D^m) \cup_\varphi X, \end{array}$$

to do this when $X = G/H \times S^m$ for finite $H \subseteq G$ and any $m \geq 0$. Using (1.3) again, it suffices to show that $\gamma_X = \gamma_X^0$ is an isomorphism whenever $X = G/H \times Y$ for any finite complex Y with trivial G-action. By definition,

$$KF_G(G/H \times Y) = \big[G/H \times Y, KF_G\big]_G \cong [Y, (KF_G)^H];$$

while $\mathbb{K}F_G(G/H \times Y)$ is the Grothendieck group of the monoid

$$\mathbb{V}\mathrm{ect}_G^F(G/H \times Y) \cong \big[G/H \times Y, \mathrm{Vec}_G^F\big]_G \cong \big[Y, (\mathrm{Vec}_G^F)^H\big].$$

Since $\pi_0((\mathrm{Vec}_G^F)^H)$ is a free abelian monoid (the monoid of isomorphism classes of H-representations), [10, Proposition 4.1] applies to show that $[-, (KF_G)^H]$ is universal among representable functors from compact spaces to abelian groups which extend $\mathrm{Vect}_G^F(G/H\times -) \cong \mathrm{Vect}_H^F(-)$. And since $\mathbb{K}_H$ is representable as a functor on compact spaces with trivial action (H is finite), it is the universal functor, and so $[Y,(KF_G)^H] \cong \mathbb{K}_H(Y) \cong \mathbb{K}_G(G/H\times Y)$. □

2. Products and Bott periodicity

We now want to construct Bott periodicity isomorphisms, and define the multiplicative structures on $K_G^*(X)$ and $KO_G^*(X)$. Both of these require defining pairings of classifying spaces; thus pairings of Γ-spaces. A general procedure for doing this was described by Segal [10, §5], but a simpler construction is possible in our situation.

Fix an isomorphism $\mu : F^\infty \otimes F^\infty \to F^\infty$ ($F = \mathbb{C}$ or $\mathbb{R}$), induced by some bijection between the canonical bases. This induces a functor

$$\mu_* : F\langle S\rangle\text{-mod} \times F\langle T\rangle\text{-mod} \longrightarrow F\langle S\times T\rangle\text{-mod},$$

and hence (for any discrete groups H and G)

$$\mu_* : \underline{\mathrm{Vec}}_H^F(S) \wedge \underline{\mathrm{Vec}}_G^F(T) \longrightarrow \underline{\mathrm{Vec}}_{H\times G}^F(S\times T). \tag{2.1}$$

This is an $(H\times G)$-equivariant map of bi-Γ-spaces, and after taking their nerves (and loop spaces) we get maps

$$\begin{array}{lll} \Omega B\mathrm{Vec}_H^F \wedge \Omega B\mathrm{Vec}_G^F \longrightarrow \Omega^2(B\mathrm{Vec}_H^F \wedge B\mathrm{Vec}_G^F) \xrightarrow{\Omega^2|\mu_*|} & & \Omega^2 B^2\mathrm{Vec}_{H\times G}^F \\ \quad = KF_H \wedge KF_G & & \simeq \Omega B\mathrm{Vec}_{H\times G}^F = KF_{H\times G} \end{array} \tag{2.2}$$

By Lemma 1.4, these maps are all independent (up to homotopy) of the choice of $\mu : F^\infty \otimes F^\infty \to F^\infty$.

Lemma 2.3. *For any discrete groups H and G, any H-space X, and any G-space Y, the following square commutes:*

$$\begin{array}{ccc} \mathbb{K}F_H(X)\otimes \mathbb{K}F_G(Y) & \xrightarrow{\gamma_X\times\gamma_Y} & KF_H(X)\otimes KF_G(Y) \\ \otimes\downarrow & & \mu_*\downarrow \\ \mathbb{K}F_{H\times G}(X\times Y) & \xrightarrow{\gamma_{X\times Y}} & KF_{H\times G}(X\times Y) \end{array}$$

where μ_* is the homomorphism induced by (2.2).

Proof. The pullback of the universal bundle $E\mathrm{Vec}_{H\times G}^F$, via the pairing $\mathrm{Vec}_H^F \wedge \mathrm{Vec}_G^F \to \mathrm{Vec}_{H\times G}^F$ of (2.1), is isomorphic to the tensor product of the universal bundles $E\mathrm{Vec}_H^F$ and $E\mathrm{Vec}_G^F$. This is clear if we identify $E\mathrm{Vec}_G^F \cong |\mathrm{func}(\mathcal{E}(G), F\text{-Bdl})|$ (and similarly for the other two bundles), where F-$\mathtt{Bdl}$ is the category of pairs (V,x) for V in F-mod and $x\in V$. □

We now consider case where $H = 1$, and hence where $KF_H = \mathbb{Z} \times BU$ or $\mathbb{Z} \times BO$. The product map (2.2), after composition with the Bott elements in $\pi_2(BU)$ or $\pi_8(BO)$, induces Bott maps

$$\beta_*^{\mathbb{C}} : \Sigma^2 K_G \longrightarrow K_G \qquad \text{and} \qquad \beta_*^{\mathbb{R}} : \Sigma^8 KO_G \longrightarrow KO_G. \tag{2.4}$$

Proposition 2.5. *For any proper CW-pair (X, A), the Bott homomorphisms*

$$b_{X,A}^{\mathbb{C}} : K_G^{-n}(X, A) \longrightarrow K_G^{-n-2}(X, A)$$
$$b_{X,A}^{\mathbb{R}} : KO_G^{-n}(X, A) \longrightarrow KO_G^{-n-8}(X, A)$$

are isomorphisms; and commute with the homomorphisms

$$\gamma_{X,A}^{-n} : \mathbb{K}F_G^{-n}(X, A) \to KF_G^{-n}(X, A).$$

Proof. The last statement follows immediately from Lemma 2.3.

By Lemma 1.1, it suffices to prove that the adjoint maps

$$K_G \longrightarrow \Omega^2 K_G \qquad \text{and} \qquad KO_G \longrightarrow \Omega^8 KO_G$$

to the pairings in (2.4) are weak homotopy equivalences after restricting to fixed point sets of finite subgroups of G. In other words, it suffices to prove that $b_X^{\mathbb{C}} : K_G(X) \to K_G^{-2}(X)$ and $b_X^{\mathbb{R}} : KO_G(X) \to KO_G^{-8}(X)$ are isomorphisms when $X = G/H \times S^n$ for any $n \geq 0$ and any finite $H \subseteq G$. And this follows since the Bott maps for $\mathbb{K}_G$ and $\mathbb{KO}_G$ are isomorphisms [8, Theorems 3.12 & 3.15], since $\mathbb{K}F_G^{-n}(X) \cong KF_G^{-n}(X)$ (Proposition 1.5), and since these isomorphisms commute with the Bott maps. □

The $K_G^{-n}(X)$ and $KO_G^{-n}(X)$ can now be extended to (additive) equivariant cohomology theories in the usual way. But before stating this explicitly, we first consider the ring structure on $K_G(X)$. This is defined to be the composite

$$[X, KF_G]_G \times [X, KF_G]_G \longrightarrow [X, KF_{G\times G}]_G \xrightarrow{\Delta^*} [X, KF_G]_G,$$

where the first map is induced by the pairing in (2.2), and the second by restriction to the diagonal subcategory $\mathcal{E}(G) \subseteq \mathcal{E}(G\times G)$.

Before we can prove the ring properties of this multiplication, we must look more closely at the homotopy equivalence $\Omega B\mathrm{Vec}_G^F \xrightarrow{\simeq} \Omega^2 B^2 \mathrm{Vec}_G^F$ which appears in the definition of the product. In fact, there is more than one natural map from $\Omega^n B^n \mathrm{Vec}_G^F$ to $\Omega^{n+1} B^{n+1} \mathrm{Vec}_G^F$. For each $n \geq 0$ and each $k = 0, \ldots, n$, let $\iota_n^k : \Omega^n B^n \mathrm{Vec}_G^F \to \Omega^{n+1} B^{n+1} \mathrm{Vec}_G^F$ denote the map induced as $\Omega^n(f)$, where f is adjoint to the map $\Sigma B^n \mathrm{Vec}_G^F \to B^{n+1} \mathrm{Vec}_G^F$, induced by identifying $B^n \mathrm{Vec}_G^F(S_1, \ldots, S_n)$ with $B^{n+1} \mathrm{Vec}_G^F(\ldots, S_{k-1}, \mathbf{1}, S_k, \ldots)$.

By a weak G-equivalence $f : X \to Y$ is meant a map of G-spaces which restricts to a weak equivalence $f^H : X^H \to Y^H$ for all $H \subseteq G$; i.e., a weak $\mathcal{F}$-equivalence in the notation of Lemma 1.1 when $\mathcal{F}$ is the family of all subgroups of G. Since we are interested equivariant Γ-spaces only as targets of maps from G-complexes, it suffices by Lemma 1.1 to work in a category where weak G-equivalences are inverted.

Lemma 2.6. *Let $\underline{A}$ be any G-equivariant Γ-space. Then for any $n \geq 1$, the maps $\iota_n^k : \Omega^n B^n A \to \Omega^{n+1} B^{n+1} A$ (for $0 \leq k \leq n$) are all equal in the homotopy category of G-spaces where weak G-equivalences are inverted.*

Proof. For any $\sigma \in \Sigma_n$, let $\sigma_* : \Omega^n B^n A \to \Omega^n B^n A$ be the map induced by permuting the coordinates of $B^n A$ as an n-simplicial set, and by switching the order of looping. Then any two of the ι_{n-1}^k differ by composition with some appropriate σ_*, and so it suffices to show that the σ_* are all homotopic to the identity.

Consider the following commutative diagram

$$\begin{array}{ccccc} \Omega BA & \xrightarrow{\varphi} & \Omega^{n+1} B^{n+1} A & \xleftarrow{\iota_n^n} & \Omega^n B^n A \\ {\scriptstyle \mathrm{Id}}\downarrow & & {\scriptstyle (1\times\sigma)_*}\downarrow & & {\scriptstyle \sigma_*}\downarrow \\ \Omega BA & \xrightarrow{\varphi} & \Omega^{n+1} B^{n+1} A & \xleftarrow{\iota_n^n} & \Omega^n B^n A, \end{array}$$

for any $\sigma \in \Sigma_n \subseteq \Sigma_{n+1}$, where $\varphi = \iota_n^0 \circ \cdots \circ \iota_1^0$ is induced by identifying $\underline{A}(S)$ with $\underline{A}(S, \mathbf{1}, \ldots, \mathbf{1})$. The diagram commutes, and all maps in it are weak G-equivalences by [10, Proposition 1.4]. So $(1 \times \sigma)_*$ and σ_* are both homotopic to the identity after inverting weak G-equivalences. □

We are now ready to show:

Theorem 2.7. *For any discrete group and any proper G-complex X, the pairings μ_X define a structure of graded ring on $K_G^*(X)$ and on $KO_G^*(X)$, which make $K_G^*(-)$ and $KO_G^*(-)$ into multiplicative cohomology theories. The Bott isomorphisms*

$$b_X^{\mathbb{C}} : K_G^{-n}(X) \to K_G^{-n-2}(X) \qquad \text{and} \qquad b_X^{\mathbb{R}} : KO_G^{-n}(X) \to KO_G^{-n-8}(X)$$

are $K_G(X)$- or $KO_G(X)$-linear. And the canonical homomorphisms

$$\gamma_X^{\mathbb{C}} : \mathbb{K}_G^*(X) \to K_G^*(X) \qquad \text{and} \qquad \gamma_X^{\mathbb{R}} : \mathbb{KO}_G^*(X) \to KO_G^*(X)$$

are homomorphisms of rings.

Proof. As usual, set $F = \mathbb{C}$ or $\mathbb{R}$. We first check that μ_X makes $KF_G(X)$ into a commutative ring — i.e., that it is associative and commutative and has a unit.

To see that there is a unit, let $[F^1] \in \mathrm{Vec}_G^F$ denote the vertex for the constant functor $\mathcal{E}(G) \mapsto F^1 \in F\langle \mathbf{1} \rangle\text{-}\mathrm{mod}$, and set $[F^1]_\Omega = \iota_0^0([F^1]) \in \Omega B\mathrm{Vec}_G^F$. The following diagram commutes:

$$\begin{array}{ccccc} \Omega B\mathrm{Vec}_G^F & \xrightarrow{[F^1]\wedge -} & \mathrm{Vec}_G^F \wedge \Omega B\mathrm{Vec}_G^F & \xrightarrow{\mu_*} & \Omega B\mathrm{Vec}_G^F \\ {\scriptstyle \mathrm{Id}}\downarrow & & {\scriptstyle \iota_0^0\wedge\mathrm{Id}}\downarrow\simeq & & {\scriptstyle \iota_1^0}\downarrow\simeq \\ \Omega B\mathrm{Vec}_G^F & \xrightarrow{[F^1]_\Omega\wedge -} & \Omega B\mathrm{Vec}_G^F \wedge \Omega B\mathrm{Vec}_G^F & \xrightarrow{\mu_*} & \Omega^2 B^2\mathrm{Vec}_G^F; \end{array}$$

and the composite in the top row is homotopic to the identity by Lemma 1.4. So the element $1 \in KF_G(X)$, represented by the constant map $X \mapsto [F^1]_\Omega \in KF_G$, is an identity for multiplication in $KF_G(X)$.

The commutativity of $KF_G(X)$ follows from Lemma 2.6 (the uniqueness of the map $\Omega BA \to \Omega^2 B^2 A$ after inverting weak G-equivalences); together with the fact that the pairing

$$\mu_* : B\mathrm{Vec}_G^F \wedge B\mathrm{Vec}_G^F \longrightarrow B^2\mathrm{Vec}_G^F$$

commutes up to homotopy using Lemma 1.4. And associativity follows since the triple products are induced by maps

$$(\Omega B\mathrm{Vec}_G^F)^{\wedge 3} \longrightarrow \Omega^3\big((B\mathrm{Vec}_G^F)^{\wedge 3}\big) \underset{\Omega^3|\mu_*\circ(\mathrm{Id}\wedge\mu_*)|}{\overset{\Omega^3|\mu_*\circ(\mu_*\wedge\mathrm{Id})|}{\rightrightarrows}} \Omega^3 B^3\mathrm{Vec}_G^F \xleftarrow{\ \simeq\ } \Omega B\mathrm{Vec}_G^F;$$

where the two maps in the middle are homotopic by Lemma 1.4, and the last could be any of the three possible maps by Lemma 2.6.

The extension of the product to negative gradings is straightforward, via the identifications of (1.3). For any $n, m \geq 0$, the composite

$$KF_G(S^n \times X) \otimes KF_G(S^m \times X) \xrightarrow{\mathrm{proj}^*} KF_G(S^n \times S^m \times X) \otimes KF_G(S^n \times S^m \times X) \xrightarrow{\ \mu\ } KF_G(S^n \times S^m \times X)$$

restricts to a product map $KF_G^{-n}(X) \otimes KF_G^{-m}(X) \to KF_G^{-n-m}(X)$. To see that the product has image in $KF_G^{-n-m}(X)$, just note that

$$\begin{aligned} KF_G^{-n-m}(X) &\cong \mathrm{Ker}\big[KF_G(S^{n+m} \times X) \longrightarrow KF_G(X)\big] \\ &= \mathrm{Ker}\big[KF_G(S^n \times S^m \times X) \longrightarrow KF_G(S^n \times X) \oplus KF_G(S^m \times X)\big]. \end{aligned}$$

This product is clearly associative, and graded commutative (where the change in sign comes from the degree of the switching map $S^{n+m} \to S^{m+n}$).

We next check that this product commutes with the Bott maps in the obvious way, so that it can be extended to $K_G^i(X)$ for all i. This means showing that the two maps

$$KF(S^n) \otimes KF_G(X) \otimes KF_G(X) \rightrightarrows KF_G(S^n \times X)$$

induced by the products constructed above are equal. And this follows from the same argument as that used to prove associativity of the internal product on $KF_G(X)$.

Finally, $\gamma : \mathbb{K}F_G^*(X) \to KF_G^*(X)$ is a ring homomorphism by Lemma 2.3. □

3. Induction, restriction, and inflation

In this section we explain how the natural maps defined on $\mathbb{K}_G(X)$ and $\mathbb{KO}_G(X)$ by induction and restriction carry over to $K_G(X)$ and $KO_G(X)$. Namely, we want to construct for any pair $H \subseteq G$ of discrete groups, any $F = \mathbb{C}$ or $\mathbb{R}$, any G-complex X, and any H-complex Y, natural induction and restriction maps

$$\mathrm{Ind}_H^G : KF_H^*(Y) \xrightarrow{\cong} KF_G^*(G\times_H Y)$$

and

$$\mathrm{Res}_H^G : KF_G^*(X) \longrightarrow KF_H^*(X|_H).$$

Furthermore, when $H \lhd G$ is a normal subgroup, we construct an inflation homomorphism

$$\mathrm{Infl}_{G/H}^G : KF_{G/H}^*(X/H) \longrightarrow KF_G^*(X),$$

which is an isomorphism whenever H acts freely on X. These maps correspond under the natural homomorphism $\mathbb{K}F_G^*(X) \to KF_G^*(X)$ to the obvious homomorphisms induced by induction, restriction, and pullback of vector bundles. They are all induced using the following maps between classifying spaces for equivariant K-theory.

Lemma 3.1. *Let $f : G' \to G$ be any homomorphism of discrete groups. Then composition with the induced functor $\mathcal{E}(f) : \mathcal{E}(G') \to \mathcal{E}(G)$ induces an G'-equivariant map $f^* : \underline{\mathrm{Vec}}_G^F \to \underline{\mathrm{Vec}}_{G'}^F$ of Γ-spaces, and hence a G'-equivariant map $f^* : KF_G \to KF_{G'}$ of classifying spaces. And for any subgroup $L \subseteq G'$ such that $L \cap \mathrm{Ker}(f) = 1$, f^* restricts to a homotopy equivalence $(KF_G)^{f(L)} \simeq (KF_{G'})^L$.*

Proof. This is immediate, except for the last statement. And if $L \subseteq G'$ is such that $L \cap \mathrm{Ker}(f) = 1$, then $L \cong f(L)$, the categories $\mathcal{E}(G')/L$ and $\mathcal{E}(G)/f(L)$ are both equivalent to the category $\mathcal{B}(L)$ with one object and endomorphism group L; and thus $(\underline{\mathrm{Vec}}_G^F)^{f(L)}(S) = |\,\mathrm{func}(\mathcal{E}(G)/f(L), F\langle S\rangle\text{-mod})|$ is homotopy equivalent to $(\underline{\mathrm{Vec}}_{G'}^F)^L(S) = |\,\mathrm{func}(\mathcal{E}(G')/L, F\langle S\rangle\text{-mod})|$ for each S in Γ. □

We first consider the restriction and induction homomorphisms.

Proposition 3.2. *Fix $F = \mathbb{C}$ or $\mathbb{R}$, and let $H \subseteq G$ be any pair of discrete groups. Let $i^* : KF_G \to KF_H$ be the map of Lemma 3.1.*

(a) *For any proper G-CW-pair (X, A), i^* induces a homomorphism of rings*

$$\mathrm{Res}_H^G : KF_G^*(X, A) \longrightarrow KF_H^*(X, A).$$

(b) *For any proper H-CW-pair (Y, B), i^* induces an isomorphism*

$$\mathrm{Ind}_H^G : KF_H^*(Y, B) \xrightarrow{\cong} KF_G^*(G\times_H Y, G\times_H B),$$

which is natural in (Y, B), and also natural with respect to inclusions of subgroups.

The restriction and induction maps both commute with the maps between $\mathbb{K}F_G(-)$ and $\mathbb{K}F_H(-)$ induced by induction and restriction of equivariant vector bundles.

Proof. It suffices to prove this when $A = \emptyset = B$ and $* = 0$. The fact that $i^* : KF_G \to KF_H$ commutes with the Bott homomorphisms and the products follows directly from the definitions. So part (a) is clear.

The inverse of the homomorphism in (b) is defined to be the composite

$$[G \times_H Y, KF_G]_G \cong [Y, KF_G]_H \xrightarrow{i^* \circ -} [Y, KF_H]_H.$$

And since i^* restricts to a homotopy equivalence $(KF_G)^L \to (KF_H)^L$ for each finite $L \subseteq H$ (Lemma 3.1), this map is an isomorphism by Lemma 1.1.

The last statement is clear from the construction and the definition of $\gamma : \mathbb{K}F_G(-) \to KF_G(-)$. □

We next consider the inflation homomorphism.

Proposition 3.3. *Fix $F = \mathbb{C}$ or $\mathbb{R}$. Let G be any discrete group, and let $N \lhd G$ be a normal subgroup. Then for each proper G-CW-pair (X, A), there is an inflation map*

$$\mathrm{Infl}_{G/N}^G : KF_{G/N}^*(X/N, A/N) \longrightarrow KF_G^*(X, A),$$

which is natural in (X, A), which is a homomorphism of rings (if $A = \emptyset$), and which commutes with the homomorphism $\mathbb{K}F_{G/N}(X/N, A/N) \to \mathbb{K}F_G(X, A)$ induced considering G/N-vector bundles as G-vector bundles. And if N acts freely on X, then $\mathrm{Infl}_{G/N}^G$ is an isomorphism.

Proof. Let $f : G \to G/N$ denote the natural homomorphism, and let $f^* : KF_{G/N} \to KF_G$ be the induced map of Lemma 3.1. Define $\mathrm{Infl}_{G/N}^G$ to be the composite

$$[X/N, KF_{G/N}]_{G/N} \cong [X, KF_{G/N}]_G \xrightarrow{f^* \circ -} [X, KF_G]_G.$$

If N acts freely on X, then for each isotropy subgroup L of X, $L \cap N = 1$, so $(f^*)^L : (KF_{G/N})^L \to (KF_G)^L$ is a homotopy equivalence by Lemma 3.1, and the inflation map is an isomorphism by Lemma 1.1. The other statements are clear. □

Another type of natural map will be needed when constructing the equivariant Chern character. Fix a discrete group G and a finite normal subgroup $N \lhd G$, and let $\mathrm{Irr}(N)$ be the set of isomorphism classes of irreducible *complex* N-representations. Let X be any proper G/N-complex. For any $V \in \mathrm{Irr}(N)$ and any G-vector bundle $E \to X$, let $\mathrm{Hom}_N(V, E)$ denote the vector bundle over X whose fiber over $x \in X$ is $\mathrm{Hom}_N(V, E_x)$ (each fiber of E is an N-representation). If $H \subseteq G$ is any subgroup which centralizes N, then we can regard $\mathrm{Hom}_N(V, E)$ as an H-vector bundle by setting $(hf)(x) = h \cdot f(x)$ for any $h \in H$ and any $f \in \mathrm{Hom}_N(V, E)$. We thus get a homomorphism

$$\Psi : \mathbb{K}_G(X) \longrightarrow \mathbb{K}_H(X) \otimes R(N),$$

where $\Psi([E]) = \sum_{V \in \mathrm{Irr}(N)} [\mathrm{Hom}_N(V, E)] \otimes [V]$. We need a similar homomorphism defined on $K_G^*(X)$.

Proposition 3.4. *Let G be a discrete group, let $N \lhd G$ be any finite normal subgroup, and let $H \subseteq G$ be any subgroup such that $[H, N] = 1$. Then for any proper G/N-complex X, there is a homomorphism of rings*

$$\Psi = \Psi_{G;N,H}^X : K_G^*(X) \longrightarrow K_H^*(X) \otimes R(N),$$

which is natural in X and is natural with respect to the degree-shifting maps $K_G^*(X) \to K_G^{*+n}(S^n \times X)$, and which has the following properties:

(a) *For any (complex) G-vector bundle $E \to X$,*

$$\Psi([\![E]\!]) = \sum_{V \in \mathrm{Irr}(N)} [\![\mathrm{Hom}_N(V,E)]\!] \otimes [V].$$

(b) *For any $G' \subseteq G$, $N' \subseteq N \cap G'$, and $H' \subseteq H \cap G'$, the following diagram commutes:*

$$\begin{array}{ccc} K_G^*(X) & \xrightarrow{\Psi^X_{G;N,H}} & K_H^*(X) \otimes R(N) \\ {\scriptstyle \mathrm{Res}^G_{G'}} \downarrow & & \downarrow {\scriptstyle \mathrm{Res}^H_{H'} \otimes \mathrm{Res}^N_{N'}} \\ K_{G'}^*(X) & \xrightarrow{\Psi^X_{G';N',H'}} & K_{H'}^*(X) \otimes R(N'). \end{array}$$

Proof. Fix G, H, and N. For any irreducible N-representation V and any surjective homomorphism $p : \mathbb{C}[N] \twoheadrightarrow V$, composition with p defines a monomorphism

$$\mathrm{Hom}_N(V,W) \xrightarrow{-\circ p} \mathrm{Hom}_N(\mathbb{C}[N],W) = W$$

for any N-representation W; and thus allows us to identify $\mathrm{Hom}_N(V,W)$ as a subspace of W. In particular, there is a functor

$$p^* : \mathrm{func}(\mathsf{Or}(G)/N, \mathbb{C}\langle S\rangle\text{-}\mathsf{mod}) \longrightarrow \mathrm{func}(\mathsf{Or}(H), \mathbb{C}\langle S\rangle\text{-}\mathsf{mod})$$

which sends any α to the functor $h \mapsto \mathrm{Hom}_N(V, \alpha(hN)) \subseteq \alpha(hN)$. If $p' : \mathbb{C}[N] \twoheadrightarrow V'$ is another surjection of N-representations, where $V \cong V'$, then any isomorphism $V \xrightarrow{\cong} V'$ defines a natural isomorphism between p^* and $(p')^*$. We thus get a map of Γ-spaces

$$\psi_p : \underline{\mathrm{Vec}}_G^{\mathbb{C}} \longrightarrow \underline{\mathrm{Vec}}_H^{\mathbb{C}}$$

which is unique (independant of the projection p) up to H-equivariant homotopy. So this induces homomorphisms $\psi_V : K_G^{-n}(X) \to K_H^{-n}(X)$, for all proper G/N-complex X (and all $n \geq 0$), which depend only on V and not on p. The ψ_V clearly commute with the Bott maps, and thus extend to homomorphisms $\psi_V : K_G^*(X) \to K_H^*(X)$. So we can define Ψ by setting $\Psi(x) = \sum_{V \in \mathrm{Irr}(N)} \psi_V(x) \otimes [V]$. Point (a) is immediate; as is naturality in X and naturality for restriction to $G' \subseteq G$ or $H' \subseteq H$. Naturality with respect to the degree-shifting maps holds by construction.

We next show that Ψ is natural in N; i.e., that point (b) holds when $G' = G$ and $H' = H$. Let ψ_V be the homomorphisms defined above, for each irreducible N-representation V; and let $\psi'_W : K_G^*(X) \to K_H^*(X)$ be the corresponding homomorphism for each irreducible N'-representation W. For each $V \in \mathrm{Irr}(N)$ and each $W \in \mathrm{Irr}(N')$, set

$$n_W^V = \dim_{\mathbb{C}}\big(\mathrm{Hom}_{N'}(W,V)\big) = \dim_{\mathbb{C}}\big(\mathrm{Hom}_N(\mathrm{Ind}_{N'}^N(W),V)\big).$$

Thus, n_W^V is the multiplicity of W in the decomposition of $V|_{N'}$, as well as the multiplicity of V in the decomposition of $\mathrm{Ind}_{N'}^N(W)$. So for any $x \in K_G^*(X)$,

$$\begin{aligned}(\mathrm{Id} \otimes \mathrm{Res}_{N'}^N)(\Psi_{G;N,H}(x)) &= \sum_{V \in \mathrm{Irr}(N)} \psi_V(x) \otimes [V|_{N'}] \\ &= \sum_{W \in \mathrm{Irr}(N')} \Big(\sum_{V \in \mathrm{Irr}(N)} n_W^V \cdot \psi_V(x) \Big) \otimes [W];\end{aligned}$$

and we will be done upon showing that $\psi'_W = \sum_V n_W^V \cdot \psi_V$ for each $W \in \mathrm{Irr}(N')$. Fix a surjection $p_0 : \mathbb{C}[N'] \twoheadrightarrow W$, and a decomposition $\mathrm{Ind}_{N'}^N(W) = \sum_{i=1}^k V_i$ (where the V_i are irreducible and $k = \sum_V n_W^V$). For each $1 \le i \le k$, let $p_i : \mathbb{C}[N] \twoheadrightarrow V_i$ be the composite of $\mathrm{Ind}_{N'}^N(p_0)$ followed by projection to V_i. Then

$$\psi_{p_0} = \bigoplus_{i=1}^k \psi_{p_i} : \big(\underline{\mathrm{Vec}}_G^{\mathbb{C}}\big)^N \longrightarrow \underline{\mathrm{Vec}}_H^{\mathbb{C}}$$

as maps of Γ-spaces, and so $\psi_W \simeq \sum_{i=1}^k \psi_{V_i}$ as maps $K_G^*(X) \to K_H^*(X)$.

It remains to show that Ψ is a homomorphism of rings. Since it is natural in N, and since $R(N)$ is detected by characters, it suffices to prove this when N is cyclic. For any $x, y \in K_G(X)$,

$$\Psi(x) \cdot \Psi(y) = \sum_{V,W \in \mathrm{Irr}(N)} (\psi_V(x) \cdot \psi_W(y)) \otimes [V \otimes W]$$

and

$$\Psi(xy) = \sum_{U \in \mathrm{Irr}(N)} \psi_U(xy) \otimes [U].$$

And thus $\Psi(x) \cdot \Psi(y) = \Psi(xy)$ since

$$\psi_U \circ \mu_* = \bigoplus_{\substack{V,W \in \mathrm{Irr}(G) \\ V \otimes W \cong U}} \mu_* \circ (\psi_V \wedge \psi_W) : \big(\underline{\mathrm{Vec}}_G^{\mathbb{C}}\big)^N \wedge \big(\underline{\mathrm{Vec}}_G^{\mathbb{C}}\big)^N \longrightarrow \underline{\mathrm{Vec}}_H^{\mathbb{C}},$$

as maps of Γ-spaces, for each $U \in \mathrm{Irr}(N)$. □

4. Characters and class functions

Throughout this section, G will be a *finite* group. We prove here some results showing that certain class functions are characters; results which will be needed in the next two sections.

For any field K of characteristic zero, a *K-character* of G means a class function $G \to K$ which is the character of some (virtual) K-representation of G. Two elements $g, h \in G$ are called *K-conjugate* if g is conjugate to h^a for some a prime to $n = |g| = |h|$ such that $(\zeta \mapsto \zeta^a) \in \mathrm{Gal}(K\zeta/K)$, where $\zeta = \exp(2\pi i/n)$. For example, g and h are $\mathbb{Q}$-conjugate if $\langle g \rangle$ and $\langle h \rangle$ are conjugate as subgroups, and are $\mathbb{R}$-conjugate if g is conjugate to h or h^{-1}.

Proposition 4.1. *Fix a finite extension K of $\mathbb{Q}$, and let $A \subseteq K$ be its ring of integers. Let $f : G \to A$ be any function which is constant on K-conjugacy classes. Then $|G|{\cdot}f$ is an A-linear combination of K-characters of G.*

Proof. Set $n = |G|$, for short. Let $V_1, \ldots, V_k$ be the distinct irreducible $K[G]$-representations, let χ_i be the character of V_i, set $D_i = \mathrm{End}_{K[G]}(V_i)$ (a division algebra over K), and set $d_i = \dim_K(D_i)$. Then by [11, Theorem 25, Cor. 2],

$$|G|{\cdot}f = \sum_{i=1}^{k} r_i\chi_i \qquad \text{where} \qquad r_i = \frac{1}{d_i}\sum_{g\in G} f(g)\chi_i(g^{-1});$$

and we must show that $r_i \in A$ for all i. This means showing, for each $i = 1, \ldots, k$, and each $g \in G$ with K-conjugacy class $\mathrm{conj}_K(g)$, that $|\mathrm{conj}_K(g)|{\cdot}\chi_i(g) \in d_iA$.

Fix i and g; and set $C = \langle g\rangle$, $m = |g| = |C|$, and $\zeta = \exp(2\pi i/m)$. Then $\mathrm{Gal}(K(\zeta)/K)$ acts freely on the set $\mathrm{conj}_K(g)$: the element $(\zeta \mapsto \zeta^a)$ acts by sending h to h^a. So $[K(\zeta){:}K]\big|\,|\mathrm{conj}_K(g)|$.

Let $V_i|_C = W_1^{a_1} \oplus \cdots \oplus W_t^{a_t}$ be the decomposition as a sum of irreducible $K[C]$-modules. For each j, $K_j \overset{\text{def}}{=} \mathrm{End}_{K[C]}(W_j)$ is the field generated by K and the r-th roots of unity for some $r|m$ ($m = |C|$), and $\dim_{K_j}(W_j) = 1$. So

$$\dim_K(W_j)\big|[K(\zeta){:}K].$$

Also, $d_i\big|\dim_K(W_j^{a_j})$, since $W_j^{a_j}$ is a D_i-module; and thus $d_i\big|a_j{\cdot}|\mathrm{conj}_K(g)|$. So if we set $\xi_j = \chi_{W_j}(g) \in A$, then

$$|\mathrm{conj}_K(g)|{\cdot}\chi_i(g) = |\mathrm{conj}_K(g)|{\cdot}\sum_{j=1}^{t} a_j\xi_j \in d_iA,$$

and this finishes the proof. □

For each prime p and each element $g \in G$, there are unique elements g_r of order prime to p and g_u of p-power order, such that $g = g_rg_u = g_ug_r$. As in [11, §10.1], we refer to g_r as the p'-component of g. We say that a class function $f : G \to \mathbb{C}$ is *p-constant* if $f(g) = f(g_r)$ for each $g \in G$. Equivalently, f is p-constant if and only if $f(g) = f(g')$ for all $g, g' \in G$ such that $[g, g'] = 1$ and $g^{-1}g'$ has p-power order.

Lemma 4.2. *Fix a finite group G, a prime p, and a field K of characteristic zero. Then a p-constant class function $\varphi : G \to K$ is a K-character of G if and only if $\varphi|_H$ is a K-character of H for all subgroups $H \subseteq G$ of order prime to p.*

Proof. Recall first that G is called K-elementary if for some prime q, $G = C_m \rtimes Q$, where C_m is cyclic of order m, $q\nmid m$, Q is a q-group, and the conjugation action of Q on $K[C_m]$ leaves invariant each of its field components. By [11, §12.6, Prop. 36], a K-valued class function of G is a K-character if and only if its restriction to any K-elementary subgroup of G is a K-character. Thus, it suffices to prove the lemma when G is K-elementary.

Assume first that G is q-K-elementary for some prime $q \neq p$. Fix a subgroup $H \subseteq G$ of p-power index and order prime to p, and let $\alpha : G \twoheadrightarrow H$ be the surjection with $\alpha|_H = \mathrm{Id}$. Set $p^a = |\operatorname{Ker}(\alpha)|$. Then

$$\operatorname{Aut}(\operatorname{Ker}(\alpha)) \cong (\mathbb{Z}/p^a)^* \cong (1 + p\mathbb{Z}/p^a) \times (\mathbb{Z}/p)^*,$$

where the first factor is a p-group. Hence for any $g \in H$ and $x \in \operatorname{Ker}(\alpha)$, either $[g, x] = 1$ and hence $g = (gx)_r$; or $gxg^{-1} = x^i$ for some $i \not\equiv 1 \pmod p$ and hence g is conjugate to gx. In either case, $\varphi(gx) = \varphi(g)$. Thus, $\varphi = (\varphi|_H)\circ\alpha$, and this is a K-character of G since $\varphi|_H$ is by assumption a K-character of H.

Now assume G is p-K-elementary. Write $G = C_m \rtimes P$, where $p\nmid m$ and P is a p-group. Let S be the set of primes which divide m. For each $I \subseteq S$, let $C_I \subseteq C_m$ be the product of the Sylow p-subgroups for $p \in I$, set $G_I = C_I \rtimes P$, and let $\alpha_I : G \twoheadrightarrow G_I$ be the homomorphism which is the identity on G_I.

For each $I \subseteq S$, we can consider $K[C_I]$ as a G-representation via the conjugation action of P; and each C_I-irreducible summand of $K[C_I]$ is P-invariant and hence G-invariant. Thus, each irreducible $K[C_I]$-representation can be extended to a $K[G_I]$-representation upon which $P \cap C_G(C_I)$ acts trivially. Hence, since $\varphi|_{C_I}$ is a K-character of C_I; there is a K-character χ_I of G_I such that $\chi_I(gx) = \chi_I(x) = \varphi(x)$ for all $x \in C_I$ and $g \in P$ such that $[g, C_I] = 1$.

Now set

$$\chi = \sum_{J \subseteq I \subseteq S} (-1)^{|I \smallsetminus J|}(\chi_I \circ \alpha_J),$$

a K-character of G. We claim that $\varphi = \chi$. Since both are class functions, it suffices to show that $\varphi(gx) = \chi(gx)$ for all commuting $g \in P$ and $x \in C_m = C_S$. Fix such g and x, and let $X \subseteq S$ be the set of all primes $p \big| |x|$. Then $[g, C_X] = 1$, and so

$$\begin{aligned}
\chi(gx) &= \sum_{J \subseteq I \subseteq S} (-1)^{|I \smallsetminus J|}\chi_I(\alpha_J(gx)) = \sum_{J \subseteq I \subseteq S} (-1)^{|I \smallsetminus J|}\chi_I(g\cdot\alpha_J(x)) \\
&= \sum_{J \subseteq I \subseteq X} (-1)^{|I \smallsetminus J|}\varphi(\alpha_J(x)) + \sum_{J \subseteq I \not\subseteq X} (-1)^{|I \smallsetminus J|}\chi_I(g\cdot\alpha_J(x)) \\
&= \varphi(x) = \varphi(gx).
\end{aligned}$$

Note, in the second line, that all terms in the second sum cancel since $\alpha_J(x) = \alpha_{J'}(x)$ if $J = J' \cap X$, and all terms in the first sum cancel except that where $J = I = X$. $\square$

When $A = \mathbb{Z}$ and $K = \mathbb{Q}$, Proposition 4.1 and Lemma 4.2 combine to give:

Corollary 4.3. *Fix a finite group G and a prime p. Let $f : G \to \mathbb{Z}$ be any function which is p-constant, and constant on $\mathbb{Q}$-conjugacy classes in G. Set $|G| = m\cdot p^r$ where $p\nmid m$. Then $m\cdot f$ is a $\mathbb{Q}$-character of G.*

5. The equivariant Chern character

We construct here two different equivariant Chern characters, both defined on the equivariant complex K-theory of proper G-complexes. The first is defined for arbitrary X (with proper G-action), and sends $K_G^*(X)$ to the Bredon cohomology group $H_G^*(X;\mathbb{Q}\otimes_{\mathbb{Z}}R(-))$. The second is defined only when X is finite dimensional and has bounded isotropy, and takes values in $\mathbb{Q}\otimes_{\mathbb{Z}}H_G^*(X;R(-))$.

We first fix our notation for dealing with Bredon cohomology [5]. Let $\mathsf{Or}(G)$ denote the orbit category: the category whose objects are the orbits G/H for $H\subseteq G$, and where $\mathrm{Mor}_{\mathsf{Or}(G)}(G/H,G/K)$ is the set of G-maps. A coefficient system for Bredon cohomology is a functor $F:\mathsf{Or}(G)^{\mathrm{op}}\to\mathsf{Ab}$. For any such functor F and any G-complex X, the Bredon cohomology $H_G^*(X;F)$ is the cohomology of a certain cochain complex $C_G^*(X;F)$, where $C_G^n(X;F)$ is the direct product over all orbits of n-cells of type G/H of the groups $F(G/H)$. This can be expressed functorially as a group of morphisms of functors on $\mathsf{Or}(G)$:

$$C_G^n(X,F)=\mathrm{Hom}_{\mathsf{Or}(G)}\big(\underline{C}_n(X),F\big),$$

where $\underline{C}_n(X):\mathsf{Or}(G)^{\mathrm{op}}\to\mathsf{Ab}$ is the functor $\underline{C}_n(X)(G/H)=C_n(X^H)$.

Clearly, the coefficient system F need only be defined on the subcategory of orbit types which occur in the G-complex X. In particular, since we work here only with proper actions, we restrict attention to the full subcategory $\mathsf{Or}_f(G)$ of orbits G/H for *finite* $H\subseteq G$. Let $R(-)$ denote the functor on $\mathsf{Or}_f(G)$ which sends G/H to $R(H)$: a functor on the orbit category via the identification $R(H)\cong K_G^0(G/H)$. More precisely, a morphism $G/H\to G/K$ in $\mathsf{Or}_f(G)$, where $gH\mapsto gaK$ for some $a\in G$ with $a^{-1}Ha\subseteq K$, is sent to the homomorphism $R(K)\to R(H)$ induced by restriction and conjugation by a.

Since $R(-)$ is a functor from the orbit category to rings, there is a pairing

$$C_G^*(X;R(-))\otimes C_G^*(X;R(-))\longrightarrow C_G^*(X\times X;R(-))$$

for any proper X, and hence a similar pairing in cohomology. Via restriction to the diagonal subspace $X\subseteq X\times X$ this defines a ring structure on $H_G^*(X;R(-))$.

The equivariant Chern character will be constructed here by first reinterpreting $H_G^*(X;\mathbb{Q}\otimes R(-))$ as a certain group of homomorphisms of functors, and then directly constructing a map from $K_G(X)$ to such homomorphisms. This will be done with the help of another category, $\mathsf{Sub}_f(G)$, which is closely related to $\mathsf{Or}_f(G)$. The objects of $\mathsf{Sub}_f(G)$ are the finite subgroups of G, and

$$\mathrm{Mor}_{\mathsf{Sub}_f(G)}(H,K)\subseteq\mathrm{Hom}(H,K)/\mathrm{Inn}(K)$$

is the subset consisting of those monomorphisms induced by conjugation and inclusion in G. There is a functor $\mathsf{Or}_f(G)\to\mathsf{Sub}_f(G)$ which sends an orbit G/H to the subgroup H, and which sends a morphism $(xH\mapsto xaK)$ in $\mathsf{Or}_f(G)$ to the homomorphism $(x\mapsto a^{-1}xa)$ from H to K. Via this functor, we can think of $\mathsf{Sub}_f(G)$ as a quotient category of $\mathsf{Or}_f(G)$.

Let $\underline{C}_*^{\mathrm{qt}}(X), \underline{H}_*^{\mathrm{qt}}(X) : \mathsf{Sub}_f(G)^{\mathrm{op}} \longrightarrow \mathsf{Ab}$ be the functors

$$\underline{C}_*^{\mathrm{qt}}(X)(H) = C_*(X^H/C_G(H)) \qquad \text{and} \qquad \underline{H}_*^{\mathrm{qt}}(X)(H) = H_*(X^H/C_G(H)). \tag{5.1}$$

For any functor $F : \mathsf{Sub}_f(G)^{\mathrm{op}} \to \mathsf{Ab}$, regarded also as a functor on $\mathsf{Or}_f(G)^{\mathrm{op}}$,

$$\mathrm{Hom}_{\mathsf{Or}_f(G)}(\underline{C}_*(X), F) \cong \mathrm{Hom}_{\mathsf{Sub}_f(G)}(\underline{C}_*^{\mathrm{qt}}(X), F), \tag{5.2}$$

since

$$\mathrm{Hom}_{C_G(H)}(C_*(X^H), F(H)) \cong \mathrm{Hom}(C_*(X^H/C_G(H)), F(H))$$

for each H (and $C_G(H)$ is the group of automorphisms of G/H in $\mathsf{Or}_f(G)$ sent to the identity in $\mathsf{Sub}_f(G)$). In particular, (5.2) will be applied when $F = R(-)$, regarded as a functor on $\mathsf{Sub}_f(G)$ as well as on $\mathsf{Or}_f(G)$.

As noted above, for any coefficient system F, the cochain complex $C_G^*(X; F)$ can be identified as a group of homomorphisms of functors on $\mathsf{Or}(G)$. The following lemma says that the Bredon cohomology groups $H_G^*(X; \mathbb{Q} \otimes R(-))$ have a similar description, but using functors on $\mathsf{Sub}_f(G)^{\mathrm{op}}$.

Lemma 5.3. *Fix a discrete group G and a proper G-complex X. Then (5.2) induces an isomorphism of rings*

$$\Phi_X : H_G^*(X; \mathbb{Q} \otimes R(-)) \xrightarrow{\ \cong\ } \mathrm{Hom}_{\mathsf{Sub}_f(G)}\big(\underline{H}_*^{\mathrm{qt}}(X), \mathbb{Q}\otimes R(-)\big).$$

Proof. Since

$$\begin{aligned} C_G^*(X; \mathbb{Q}\otimes R(-)) &\cong \mathrm{Hom}_{\mathsf{Or}_f(G)}(\underline{C}_*(X), \mathbb{Q}\otimes R(-)) \\ &\cong \mathrm{Hom}_{\mathsf{Sub}_f(G)}(\underline{C}_*^{\mathrm{qt}}(X), \mathbb{Q}\otimes R(-)), \end{aligned}$$

this will follow immediately once we show that $\mathbb{Q}\otimes R(-)$ is injective as a functor $\mathsf{Sub}_f(G)^{\mathrm{op}} \to \mathsf{Ab}$. It suffices to prove this after tensoring with $\mathbb{C}$; i.e., it suffices to prove that $\mathrm{Cl}(-)$ (complex valued class functions) is injective. And this holds since for any $F : \mathsf{Sub}_f(G)^{\mathrm{op}} \to \mathsf{Ab}$,

$$\mathrm{Hom}_{\mathsf{Sub}_f(G)}(F, \mathrm{Cl}(-)) \cong \prod_g \mathrm{Hom}_{\mathsf{Sub}_f(G)}(F, \mathrm{Cl}_g(-)) \cong \prod_g \mathrm{Hom}(F(\langle g\rangle), \mathbb{C});$$

where both products are taken over any set of conjugacy class representatives for elements of finite order in G, and where $\mathrm{Cl}_g(H)$ denotes the space of class functions on H which vanish on all elements not G-conjugate to g. □

We are now ready to define the Chern character

$$\mathrm{ch}_X^* : K_G^*(X) \longrightarrow H_G^*(X; \mathbb{Q}\otimes R(-))$$

for any proper G-complex X. Here and in the following theorem, we regard $K_G^*(-)$ as being $\mathbb{Z}/2$-graded; so that ch_X^* sends $K_G^0(X)$ to $H_G^{\mathrm{ev}}(X; \mathbb{Q} \otimes R(-))$ and sends $K_G^1(X)$ to $H_G^{\mathrm{odd}}(X; \mathbb{Q}\otimes R(-))$. By Lemma 5.3, it suffices to define homomorphisms

$$\mathrm{ch}_X^H : K_G^*(X) \longrightarrow \mathrm{Hom}\big(H_*(X^H/C_G(H)),\ \mathbb{Q}\otimes R(H)\big),$$

for each finite subgroup $H \subseteq G$, which are natural in H in the obvious way. We define ch_X^H to be the following composite:

$$K_G^*(X) \xrightarrow{\text{Res}} K_{N_G(H)}^*(X^H) \xrightarrow{\Psi} K_{C_G(H)}^*(X^H) \otimes R(H)$$
$$\xrightarrow{(\text{proj})^*} K_{C_G(H)}^*(EG \times X^H) \otimes R(H) \xrightarrow[\cong]{\text{Infl}^{-1}} K^*(EG \times_{C_G(H)} X^H) \otimes R(H)$$
$$\xrightarrow{\text{ch} \otimes \text{Id}} H^*\big(EG \times_{C_G(H)} X^H; \mathbb{Q}\big) \otimes R(H) \xleftarrow[\cong]{(\text{proj})^*} H^*\big(X^H/C_G(H); \mathbb{Q}\big) \otimes R(H)$$
$$\cong \text{Hom}\big(H_*(X^H/C_G(H)), \mathbb{Q} \otimes R(H)\big). \quad (5.4)$$

Here, Ψ is the homomorphism defined in Proposition 3.4, ch denotes the ordinary Chern character, and $(\text{proj})^*$ in the bottom line is an isomorphism since all fibers of the projection from $EG \times_{C_G(H)} X^H$ to $X^H/C_G(H)$ are $\mathbb{Q}$-acyclic (classifying spaces of finite groups). By the naturality properties of Ψ shown in Proposition 3.4, $\prod_H \mathrm{ch}_X^H$ takes values in $\text{Hom}_{\text{Sub}_f(G)}\big(\underline{\text{H}}_*^{\text{qt}}(X), \mathbb{Q} \otimes R(-)\big)$, and hence (via Lemma 5.3) defines an equivariant Chern character

$$\mathrm{ch}_X^* : K_G^*(X) \longrightarrow H_G^*(X; \mathbb{Q} \otimes R(-)) \cong \text{Hom}_{\text{Sub}_f(G)}\big(\underline{\text{H}}_*^{\text{qt}}(X), \mathbb{Q} \otimes R(-)\big).$$

All of the maps in (5.4) are homomorphisms of rings, and hence ch_X^* is also a homomorphism of rings. Also, the ch_X^* commute with degree-changing maps $K_G^*(X) \to K^{*+m}(S^m \times X)$ (i.e., product with the fundamental class of S^m) and similarly in cohomology, since all maps in (5.4) do so. They are thus natural with respect to boundary maps in Mayer-Vietoris sequences.

Theorem 5.5. *For any finite proper G-complex X, the Chern character ch_X^* extends to an isomorphism of rings*

$$\mathbb{Q} \otimes \mathrm{ch}_X^* : \mathbb{Q} \otimes K_G^*(X) \xrightarrow{\cong} H_G^*(X; \mathbb{Q} \otimes R(-)).$$

Proof. For any finite subgroup $H \subseteq G$,

$$K_G^0(G/H) \cong R(H) \cong H_G^0(G/H; R(-)),$$

and

$$K_G^1(G/H) = 0 = H_G^{\neq 0}(G/H; R(-)).$$

From the definition in (5.4) (and since the non-equivariant Chern character $K(\text{pt}) \to H^0(\text{pt})$ is the identity map), we see that $\mathbb{Q} \otimes \mathrm{ch}_{G/H}^*$ is the identity map under the above identifications. The Chern characters for $G/H \times D^n$ and $G/H \times S^{n-1}$) are thus isomorphisms for all n. The theorem now follows by induction on the number of orbits of cells in X, together with the Mayer-Vietoris sequences for pushouts $X = X' \cup_\varphi (G/H \times D^n)$ (and the 5-lemma). □

Theorem 5.5 means that the $\mathbb{Q}$-localization of the classifying space K_G splits as a product of equivariant Eilenberg-Maclane spaces. Hence for any proper G-complex X, there is an isomorphism $K_G^*(X; \mathbb{Q}) \xrightarrow{\cong} H_G^*(X; \mathbb{Q} \otimes R(-))$, where the first group is defined via the localized spectrum (and is *not* in general isomorphic to $\mathbb{Q} \otimes K_G^*(X, A)$).

The coefficient system $\mathbb{Q}\otimes R(-)$, and hence its cohomology, splits in a natural way as a product indexed over cyclic subgroups of G of finite order. For any cyclic group S of order $n<\infty$, we let $\mathbb{Z}[\zeta_S]\subseteq\mathbb{Q}(\zeta_S)$ denote the cyclotomic ring and field generated by the n-th roots of unity; but regarded as quotient rings of the group rings $\mathbb{Z}[S^*]\subseteq\mathbb{Q}[S^*]$ ($S^*=\mathrm{Hom}(S,\mathbb{C}^*)$). In other words, we fix an identification of the n-th roots of unity in $\mathbb{Q}(\zeta_S)$ with the irreducible characters of S. The kernel of the homomorphism $R(S)\cong\mathbb{Z}[S^*]\twoheadrightarrow\mathbb{Z}[\zeta_S]$ is precisely the ideal of elements whose characters vanish on all generators of S.

Lemma 5.6. *Fix a discrete group G, and let $\mathcal{S}(G)$ be a set of conjugacy class representatives for the cyclic subgroups $S\subseteq G$ of finite order. Then for any proper G-complex X, there is an isomorphism of rings*

$$H^*_G(X;\mathbb{Q}\otimes R(-))\cong\prod_{S\in\mathcal{S}(G)}\left(H^*(X^S/C_G(S);\mathbb{Q}(\zeta_S))\right)^{N(S)},$$

where $N(S)$ acts via the conjugation action on $\mathbb{Q}(\zeta_S)$ and via translation on $X^S/C_G(S)$. If, furthermore, the isotropy subgroups on X have bounded order, then the homomorphism of rings

$$H^*_G(X;R(-))\longrightarrow\prod_{S\in\mathcal{S}(G)}H\Big(\big(C^*(X^S/C_G(S);\mathbb{Z}[\zeta_S])\big)^{N(S)}\Big)$$
$$\longrightarrow\prod_{S\in\mathcal{S}(G)}\left(H^*(X^S/C_G(S);\mathbb{Z}[\zeta_S])\right)^{N(S)},\qquad(1)$$

induced by restriction to cyclic subgroups and by the projections $R(S)\twoheadrightarrow\mathbb{Z}[\zeta_S]$, has kernel and cokernel of finite exponent.

Proof. By (5.2),

$$C^*_G(X;R(-))\cong\mathrm{Hom}_{\mathrm{Or}_f(G)}(\underline{C}_*(X),R(-))\cong\mathrm{Hom}_{\mathrm{Sub}_f(G)}(\underline{C}^{\mathrm{qt}}_*(X),R(-)).$$

For each $S\in\mathcal{S}(G)$, let $\chi_S\in\mathrm{Cl}(G)$ be the idempotent class function: $\chi_S(g)=1$ if $\langle g\rangle$ is conjugate to S, and $\chi_S(g)=0$ otherwise. By Proposition 4.1, for each finite subgroup $H\subseteq G$, $(\chi_S)|_H$ is the character of an idempotent $e^H_S\in\mathbb{Q}\otimes R(H)$. Set $\mathbb{Q}R_S(H)=e^H_S\cdot(\mathbb{Q}\otimes R(H))$, and let $R_S(H)\subseteq\mathbb{Q}R_S(H)$ be the image of $R(H)$ under the projection. This defines a splitting $\mathbb{Q}\otimes R(-)=\prod_{S\in\mathcal{S}(G)}\mathbb{Q}R_S(-)$ of the coefficient system. For each S and H,

$$\mathbb{Q}R_S(S)=\mathbb{Q}(\zeta_S)\quad\text{and so}\quad\mathbb{Q}R_S(H)\cong\mathrm{map}_{N(S)}\Big(\mathrm{Mor}_{\mathrm{Sub}_f(G)}(S,H)\,,\,\mathbb{Q}(\zeta_S)\Big).$$

It follows that

$$\begin{aligned}C^*_G(X;\mathbb{Q}R_S(-))&\cong\mathrm{Hom}_{\mathrm{Sub}_f(G)}(\underline{C}^{\mathrm{qt}}_*(X),\mathbb{Q}R_S(-))\\&\cong\mathrm{Hom}_{\mathbb{Q}[N(S)]}(C_*(X^S/C_G(S)),\mathbb{Q}(\zeta_S));\end{aligned}$$

and hence $H^*_G(X;\mathbb{Q}R_S(-))\cong\left(H^*(X^S/C_G(S));\mathbb{Q}(\zeta_S)\right)^{N(S)}$.

Now assume there is a bound on the orders of isotropy subgroups on X, and let m be the least common multiple of the $|G_x|$. By Proposition 4.1 again,

$me_S^H \in R(H)$ for each $S \in \mathcal{S}(G)$ and each isotropy subgroup H. So there are homomorphisms of functors

$$R(-) \underset{j}{\overset{i}{\rightleftarrows}} \prod_{S\in\mathcal{S}(G)} R_S(-),$$

where i is induced by the projections $R(H) \twoheadrightarrow R_S(H)$ and j by the homomorphisms $R_S(H) \xrightarrow{me_S^H\cdot} R(H)$ (regarding $R_S(H)$ as a quotient of $R(H)$); and $i\circ j$ and $j\circ i$ are both multiplication by m. For each S, the monomorphism

$$C_G^*(X; R_S(-)) \cong \operatorname{Hom}_{\mathbb{Z}[N(S)]}(C_*(X^S/C_G(S)), \mathbb{Z}[\zeta_S]) \longrightarrow C^*(X^S/C_G(S); \mathbb{Z}[\zeta_S])$$

is split by the norm map for the action of $N(S)/C_G(S)$, and hence the kernel and cokernel of the induced homomorphism

$$H_G^*(X; R_S(-)) \longrightarrow \big(H^*(X^S/C_G(S); \mathbb{Z}[\zeta_S])\big)^{N(S)}$$

have exponent dividing $\varphi(m)$ (since $|N(S)/C_G(S)|\,\big|\,|\operatorname{Aut}(S)|\,\big|\,\varphi(m)$). The composite in (1) thus has kernel and cokernel of exponent $m\cdot\varphi(m)$. □

By the first part of Proposition 5.6, the equivariant Chern character can be regarded as a homomorphism

$$\mathrm{ch}_X^* : K_G^*(X) \longrightarrow \prod_{S\in\mathcal{S}(G)} \big(H^*(X^S/C_G(S); \mathbb{Q}(\zeta_S))\big)^{N(S)},$$

where $\mathcal{S}(G)$ is as above. This is by construction a product of ring homomorphisms.

We now apply the splitting of Lemma 5.6 to construct a second version of the equivariant rational Chern character: one which takes values in $\mathbb{Q}\otimes H_G^*(X; R(-))$ rather than in $H_G^*(X; \mathbb{Q}\otimes R(-))$. The following lemma handles the nonequivariant case.

Lemma 5.7. *There is a homomorphism* $n!\mathrm{ch} : K^*(X) \to H^{\le 2n}(X;\mathbb{Z})$, *natural on the category of CW-complexes, whose composite to* $H^*(X;\mathbb{Q})$ *is* $n!$ *times the usual Chern character truncated in degrees greater than* $2n$. *Furthermore,* $n!\mathrm{ch}$ *is natural with respect to suspension isomorphisms* $K^*(X) \cong \widetilde{K}^{*+m}(\Sigma^m(X_+))$, *and is multiplicative in the sense that* $\big(n!\mathrm{ch}(x)\big)\cdot\big(n!\mathrm{ch}(y)\big) = n!\cdot\big(n!\mathrm{ch}(xy)\big)$ *for all* $x, y \in K(X)$ *(in both cases after restricting to the appropriate degrees).*

Proof. Define $n!\mathrm{ch} : K^0(X) \to H^{\mathrm{ev},\le 2n}(X;\mathbb{Z})$ to be the following polynomial in the Chern classes:

$$n!\cdot\sum_{i=1}^{n}\Big(1 + x_i + \frac{x_i^2}{2!} + \cdots + \frac{x_i^n}{n!}\Big) \in \mathbb{Z}[c_1, \ldots, c_n] = \mathbb{Z}[x_1, x_2, \ldots, x_n]^{\Sigma_n}.$$

Here, as usual, c_k is the k-th elementary symmetric polynomial in the x_i. This is extended to $K^{-1}(X) \cong \widetilde{K}(\Sigma(X_+))$ in the obvious way. The relations all follow from the usual relations between Chern classes in the rings $H^*(BU(m))$. □

We are now ready to construct the integral Chern character. What this really means is that under certain restrictions on X, some multiple of the rational Chern character ch_X^* of Theorem 5.5 can be lifted to the integral Bredon cohomology group $H_G^*(X;R(-))$.

Proposition 5.8. *Let G be a discrete group, and let X be a finite dimensional proper G-complex whose isotropy subgroups have bounded order. Then there is a homomorphism*

$$\widetilde{\mathrm{ch}}_X^* : K_G^*(X) \longrightarrow \mathbb{Q}\otimes H_G^*(X;R(-)),$$

natural in such X, whose composite to $H_G^*(X;\mathbb{Q}\otimes R(-))$ is the map ch_X^* of Theorem 5.5. Furthermore, $\widetilde{\mathrm{ch}}_X^*$ induces an isomorphism of rings $\mathbb{Q}\otimes K_G^*(X) \xrightarrow{\cong} \mathbb{Q}\otimes H_G^*(X;R(-))$. And for any finite subgroup $K\subseteq G$, $\widetilde{\mathrm{ch}}_{G/K}^0$ is the identity map under the identifications $K_G(G/K)\cong R(G/K)\cong H_G^0(G/K;R(-))$.

Proof. Fix X, and choose any integer $n\geq \dim(X)/2$. Set $m=\mathrm{lcm}\{|G_x| \mid x\in X\}$ and $N=n!\cdot m^{4n}$. For each $S\in G$ of finite order, let $\widetilde{\mathrm{ch}}_X^S$ be the following composite:

$$\begin{aligned}
K_G^*(X) &\xrightarrow{\mathrm{Res}} K_{N_G(S)}^*(X^S) \xrightarrow{\Psi} K_{C_G(S)}^*(X^S)\otimes R(S)\\
&\xrightarrow{(\mathrm{proj})^*} K_{C_G(S)}^*(EG\times X^S)\otimes R(S) \xrightarrow[\cong]{\mathrm{Infl}^{-1}} K^*(EG\times_{C_G(S)}X^S)\otimes R(S)\\
&\xrightarrow{n!\mathrm{ch}} H^{\leq 2n}\big(EG\times_{C_G(S)}X^S\big)\otimes R(S) \xrightarrow{m^{4n}(\mathrm{proj}^*)^{-1}} H^*\big(X^S/C_G(S)\big)\otimes R(S)\\
&\longrightarrow H^*(X^S/C_G(S);\mathbb{Z}[\zeta_S]).
\end{aligned}$$

Here, Ψ is the homomorphism of Lemma 3.4, and Infl is the inflation isomorphism of Proposition 3.3. The first map in the bottom row is well defined since $H^*(X^S/C(S);\mathbb{Z}[\zeta_S]) \xrightarrow{\mathrm{proj}^*} H^*(EG\times_{C(S)}X^S;\mathbb{Z}[\zeta_S])$ has kernel and cokernel of exponent m^{2n} (this follows from the spectral sequence for the projection, all of whose fibers are of the form BG_x for $x\in X$). The last map is induced by the projection $R(S)\twoheadrightarrow \mathbb{Z}[\zeta_S]$. All of these maps are homomorphisms of rings (up to the obvious integer multiples).

Now let $\mathcal{S}(G)$ be any set of conjugacy class representatives for cyclic subgroups $S\subseteq G$ of finite order. Define $\widetilde{\mathrm{ch}}_X^*$ to be the composite

$$\begin{aligned}
\widetilde{\mathrm{ch}}_X^* : K_G^*(X) \xrightarrow{\frac{1}{N}\prod\widetilde{\mathrm{ch}}_X^S} &\ \mathbb{Q}\otimes\Big(\prod_{S\in\mathcal{S}(G)} \big(H^*(X^S/C_G(S);\mathbb{Z}[\zeta_S])\big)^{N(S)}\Big)\\
&\cong \mathbb{Q}\otimes H_G^*(X;R(-)),
\end{aligned}$$

where the isomorphism is that of Lemma 5.6. The naturality of $\widetilde{\mathrm{ch}}_X^*$, its independence of the choice of n, and its relation with ch_X^*, are immediate from the construction. Also, $\widetilde{\mathrm{ch}}_X^*$ is natural with respect to the degree-changing maps $K^*(X)\to K^{*+m}(S^m\times X)$ (and similarly in cohomology). In particular, this means that it commutes with all maps in Mayer-Vietoris sequences.

It remains to prove that $\widetilde{\mathrm{ch}}_X^*$ induces an isomorphism on $\mathbb{Q}\otimes K_G^*(X)$. This is done by induction on $\dim(X)$, using the obvious Mayer-Vietoris sequences. So it suffices to show it for (possibly infinite) disjoint unions $\coprod_{i\in I} G/H_i$ of orbits. Both groups are zero in odd degrees. And in even degrees,

$$\begin{array}{ccc} \mathbb{Q}\otimes K_G(\coprod_{i\in I} G/H_i) & \xrightarrow{\widetilde{\mathrm{ch}}^0_X} & \mathbb{Q}\otimes H^{\mathrm{ev}}_G(\coprod_{i\in I} G/H_i; R(-)) \\ \cong \mathbb{Q}\otimes (\prod_i R(H_i)) & & \cong \mathbb{Q}\otimes (\prod_i R(H_i)) \end{array}$$

is the identity map under these identifications. □

6. Completion theorems

Throughout this section, G is a discrete group. We want to prove completion theorems for finite proper G-complexes: theorems which show that $K^*_G(E\times X)$, when E is a "universal space" of a certain type, is isomorphic to a certain completion of $K^*_G(X)$. The key step will be to construct elements of $K^*_G(X)$ whose restrictions to orbits in X are sufficiently "interesting". And this requires a better understanding of the "edge homomorphism" for $K^*_G(X)$.

For any finite dimensional proper G-complex X, the skeletal filtration of $K^*_G(X)$ induces a spectral sequence

$$E_2^{p,2*} \cong H^p_G(X; R(-)) \Longrightarrow K^*_G(X).$$

If X also has bounded isotropy, the Chern character $\widetilde{\mathrm{ch}}_X$ of Proposition 5.8 is an isomorphism (after tensoring with $\mathbb{Q}$) from the limit of this spectral sequence to its E_2-term. It follows that the spectral sequence collapses rationally; i.e., that the images of all differentials in the spectral sequence consist of torsion elements.

Of particular interest is the edge homomorphism of the spectral sequence. This is a homomorphism

$$\epsilon_X : K^*_G(X) \longrightarrow H^0_G(X; R(-)),$$

which is induced by restriction to the 0-skeleton of X under the identification

$$\begin{aligned} H^0_G(X; R(-)) &= \mathrm{Ker}\big[K_G(X^{(0)}) \longrightarrow K^1_G(X^{(1)}, X^{(0)})\big] \\ &= \mathrm{Im}\big[K_G(X^{(1)}) \longrightarrow K_G(X^{(0)})\big]. \end{aligned}$$

Alternatively, $H^0_G(X; R(-))$ can be thought of as the inverse limit, taken over all isotropy subgroups H of X and all connected components of X^H, of the representation rings $R(H)$; and the edge homomorphism sends an element of $K^*_G(X)$ to the collection of its restrictions to elements of $K^*_G(Gx) \cong R(G_x)$ at all points $x\in X$.

As an application of the integral Chern character of Proposition 5.8, we get:

Proposition 6.1. *Let X be any finite dimensional proper G-complex whose isotropy subgroups have bounded order. Then for any $\xi \in H^0_G(X; R(-))$, there is $k>0$ such that $k\cdot\xi$ and ξ^k lie in the image of the edge homomorphism*

$$\epsilon_X : K_G(X) \longrightarrow H^0_G(X; R(-)).$$

Similarly, for any $\xi \in H^0_G(X; RO(-))$, *there is* $k > 0$ *such that* $k{\cdot}\xi$ *and* ξ^k *lie in the image of the edge homomorphism*

$$\epsilon_X : KO_G(X) \longrightarrow H^0_G(X; RO(-)).$$

Proof. The usual homomorphisms between $R(-)$ and $RO(-)$, and between $K^*_G(-)$ and $KO^*_G(-)$, induced by $(\mathbb{C}\otimes_{\mathbb{R}})$ and by forgetting the complex structure, show that up to 2-torsion, $KO^*_G(X)$ and $H^0_G(X; RO(-))$ are the fixed point sets under complex conjugation of the groups $K^*_G(X)$ and $H^0_G(X; R(-))$, respectively. So the edge homomorphism in the orthogonal case is also surjective modulo torsion. The rest of the argument is identical in the real and complex cases; we restrict to the complex case for simplicity.

By Proposition 5.8, the integral Chern character for $X^{(0)}$ is the identity under the usual identifications $K_G(G/K) \cong R(K) \cong H^0_G(G/K; R(-))$ for an orbit G/K (K finite). So by the naturality of $\widetilde{\mathrm{ch}}_X$, the composite

$$\mathbb{Q}\otimes K_G(X) \xrightarrow[\cong]{\widetilde{\mathrm{ch}}^0_X} \mathbb{Q}\otimes H^{\mathrm{ev}}_G(X; R(-)) \twoheadrightarrow \mathbb{Q}\otimes H^0_G(X; R(-)) \subseteq \mathbb{Q}\otimes K_G(X^{(0)}) \quad (1)$$

is just the map induced by restriction to $X^{(0)}$. So rationally, the edge homomorphism is just the projection of the integral Chern character $\widetilde{\mathrm{ch}}$ onto $H^0_G(X; R(-))$, and is in particular surjective. And hence, for any $\xi \in H^0_G(X; R(-))$, there is some $k > 0$ such that $k{\cdot}\xi \in \epsilon_X(K_G(X))$.

It remains to show that $\xi^k \in \mathrm{Im}(\epsilon_X)$ for some k. If we knew that the Atiyah-Hirzebruch spectral sequence

$$E_2^{p,2*} \cong H^p_G(X; R(-)) \Longrightarrow K^*_G(X)$$

were multiplicative (i.e., that the differentials were derivations), then the result would follow directly. As we have seen, all differentials in the spectral sequence have finite order. Hence, for each $r \geq 2$ and each $\eta \in E_r^{0,2*}$, there is some $k > 0$ such that

$$k{\cdot}d_r(\eta) = 0, \qquad \text{and hence} \qquad d_r(\eta^k) = k{\cdot}d_r(\eta)\eta^{k-1} = 0.$$

Upon iteration, this shows that for any $\xi \in H^0_G(X; R(-)) = E_2^{0,0}$, there exists $k > 0$ such that $k{\cdot}\xi$ and ξ^k both survive to $E_\infty^{0,0}$; and hence lie in the image of the edge homomorphism.

Rather than prove the multiplicativity of the spectral sequence, we give the following more direct argument. Identify

$$\xi \in H^0_G(X; R(-)) = \mathrm{Im}\big[K_G(X^{(1)}) \longrightarrow K_G(X^{(0)})\big].$$

Assume, for some $r \geq 2$, that ξ lies in the image of $K_G(X^{(r-1)})$; we prove that some power of ξ lies in the image of $K_G(X^{(r)})$.

Fix $\widetilde{\xi} \in K_G(X^{(r-1)})$ such that $\mathrm{res}_{X^{(0)}}(\widetilde{\xi}) = \xi$. Since $r \geq 2$,

$$\begin{aligned}&\mathrm{Im}\big[H^0_G(X^{(r-1)}; R(-)) \longrightarrow H^0_G(X^{(0)}; R(-))\big]\\ &= \mathrm{Im}\big[H^0_G(X^{(r)}; R(-)) \longrightarrow H^0_G(X^{(0)}; R(-))\big].\end{aligned}$$

Hence, since the Chern character is rationally an isomorphism, there exists k such that $k\cdot\xi$ lies in the image of $K_G(X^{(r)})$, or equivalently such that

$$k\cdot\widetilde{\xi} \in \mathrm{Ker}\Big[K_G(X^{(r-1)}) \longrightarrow K_G(X^{(0)}) \xrightarrow{d} K_G^1(X^{(r)}, X^{(0)})\Big]$$
$$= \mathrm{Ker}\Big[K_G(X^{(r-1)}) \xrightarrow{d} K_G^1(X^{(r)}, X^{(r-1)}) \longrightarrow K_G^1(X^{(r)}, X^{(0)})\Big]. \qquad (1)$$

In Lemma 6.2 below, we will show that there is a $K_G(X^{(r-1)})$-module structure on the relative group $K_G^1(X^{(r)}, X^{(r-1)})$ which makes the boundary map $d : K_G(X^{(r-1)}) \to K_G^1(X^{(r)}, X^{(r-1)})$ into a derivation. Then $d(\widetilde{\xi}^k) = k\cdot\widetilde{\xi}^{k-1}\cdot d(\widetilde{\xi})$, so $\widetilde{\xi}^k$ lies in the kernel in (1), and hence $\xi^k = \mathrm{res}_{X^{(0)}}(\widetilde{\xi}^k)$ lies in the image of $K_G(X^{(r)})$. □

It remains to prove:

Lemma 6.2. *Let X be any proper G-complex. Then, for any $r \geq 2$, one can put a $K_G(X^{(r-1)})$-module structure on $K_G^1(X^{(r)}, X^{(r-1)})$ in such a way that for any $\alpha, \beta \in K_G(X^{(r-1)})$,*

$$d(\alpha\beta) = \alpha\cdot d\beta + \beta\cdot d\alpha \in K_G^1(X^{(r)}, X^{(r-1)}).$$

Proof. We can assume $X = X^{(r)}$. Write $Y = X^{(r-1)}$, for short. Fix a map $\Delta : X \to Z \overset{\text{def}}{=} X \times Y \cup Y \times X$ which is homotopic to the diagonal, and such that $\Delta|_Y$ is equal to the diagonal map. Since Z contains the $r+1$-skeleton of $X \times X$, Δ is unique up to homotopy (rel Y). In particular, if $T : Z \to Z$ is the map which switches coordinates, then $T\circ\Delta \simeq \Delta$ (rel Y).

Now, for $\alpha \in K_G(Y)$ and $x \in K_G^1(X, Y)$, let $\alpha\cdot x \in K_G^1(X, Y)$ be the image of $\alpha \times x$ under the following composite

$$\alpha\times x \in K_G^1(Y\times X, Y\times Y) \cong K_G^1(Z, X\times Y) \xrightarrow{\text{incl}^*} K_G^1(Z, Y\times Y) \xrightarrow{\Delta^*} K_G^1(X, Y).$$

Here, the external product $\alpha \times x$ is induced by the pairing $K_G \wedge K_G \to K_{G\times G} \to K_G$ of (2.2); or equivalently is defined to be the internal product of $\mathrm{proj}_1^*(\alpha) \in K_G(Y \times X)$ and $\mathrm{proj}_2^*(x) \in K_G^1(Y \times X, Y \times Y)$. We can thus consider $K_G(X, Y)$ as a $K_G(Y)$-module. In particular, the relation $(\alpha\beta)\cdot x = \alpha\cdot(\beta\cdot x)$ follows since the two composites $(\Delta \times \mathrm{Id}_X)\circ\Delta$ and $(\mathrm{Id}_X \times\Delta)\circ\Delta$ are homotopic as maps from X to

$$(X\times Y\times Y) \cup (Y\times X\times Y) \cup (Y\times Y\times X).$$

Now consider the following commutative diagram:

$$\begin{array}{ccccc}
K_G(Y \times Y) & \xrightarrow{d} & K_G^1(Z, Y \times Y) & \xrightarrow{\cong} & K_G^1((X,Y) \times Y) \oplus K_G^1(Y \times (X,Y)) \\
\downarrow{\scriptstyle\Delta^*} & & \downarrow{\scriptstyle\Delta^*} & & \uparrow{\scriptstyle\cong} \\
K_G(Y) & \xrightarrow{d} & K_G^1(X,Y) & \longleftarrow & K_G^1(Z, Y \times X) \oplus K_G^1(Z, X \times Y)
\end{array}$$

where the isomorphisms hold by excision. For any $\alpha, \beta \in K_G(Y)$, the external product $\alpha \times \beta \in K_G(Y \times Y)$ is sent, by the maps in the top row, to the pair $(d\alpha \times \beta, \alpha \times d\beta)$. This follows from the linearity of the differential (which holds in any multiplicative cohomology theory). And since $T\circ\Delta \simeq \Delta$, as noted above, we have

$$d(\alpha\beta) = \Delta^*(d(\alpha\times\beta)) = \beta\cdot d\alpha + \alpha\cdot d\beta. \quad \square$$

As an immediate consequence of Proposition 6.1, we now get:

Corollary 6.3. *Assume that G is discrete. Fix any family $\mathcal{F}$ of finite subgroups of G of bounded order, and let*

$$\mathbf{V} = \big(V_H\big) \in \varprojlim_{H\in\mathcal{F}} R(H) \qquad \textit{or} \qquad \mathbf{V}' = \big(V'_H\big) \in \varprojlim_{H\in\mathcal{F}} RO(H)$$

be any system of compatible (virtual) representations. Then for any finite dimensional proper G-complex X all of whose isotropy subgroups lie in $\mathcal{F}$, there is an integer $k>0$, and elements $\alpha,\beta\in K_G(X)$ (or $\alpha',\beta'\in KO_G(X)$), such that $\alpha|_x = k\cdot V_{G_x}$ and $\beta|_x = (V_{G_x})^k$ (or $\alpha'|_x = k\cdot V'_{G_x}$ and $\beta'|_x = (V'_{G_x})^k$) for all $x\in X$.

Proof. Let ξ be the image of $\mathbf{V}$ under the ring homomorphism

$$\varprojlim_{H\in\mathcal{F}} R(H) \longrightarrow H^0_G(X;R(-))$$

(or similarly in the orthogonal case); and apply Proposition 6.1. $\square$

Corollary 6.3 can be thought of as a generalization of [8, Theorem 2.7]. It was that result which was the key to proving the completion theorem in [8], and Corollary 6.3 plays a similar role in proving the more general completion theorems here.

In what follows, a *family* of subgroups of a discrete group G will always mean a set of subgroups closed under conjugation and closed under taking subgroups.

Lemma 6.4. *Let X be a proper n-dimensional G-complex. Set*

$$I = \operatorname{Ker}[K^*_G(X) \xrightarrow{\text{res}} K^*_G(X^{(0)})].$$

Then $I^{n+1}=0$.

Proof. Fix any elements $x\in I^n$ and $y\in I$. By induction, we can assume that x vanishes in $K^*_G(X^{n-1})$, and hence that it lifts to an element $x'\in K^*_G(X,X^{(n-1)})$. Recall that $K^*_G(X,X^{(n-1)})$ is a $K^*_G(X)$-module, and the map $K^*_G(X,X^{(n-1)})\to K^*_G(X)$ is $K^*_G(X)$-linear. But $I\cdot K^*_G(X,X^{(n-1)})=0$, since I vanishes on orbits; so $yx'=0$, and hence $yx=0$ in $K^*_G(X)$. $\square$

As in earlier sections, in order to handle the complex and real cases simultaneously, we set $F=\mathbb{C}$ or $\mathbb{R}$, and write $KF^*_G(-)$ and $RF(-)$ for the equivariant K-theory and representation rings over F.

Fix any finite proper G-complex X, and let $f: X\to L$ be any map to a finite dimensional proper G-complex L whose isotropy subgroups have bounded order. Let $\mathcal{F}$ be any family of finite subgroups of G. Regard $KF^*_G(X)$ as a module over the ring $KF_G(L)$. Set

$$I = I_{\mathcal{F},L} = \operatorname{Ker}\Big[KF_G(L) \xrightarrow{\text{res}} \prod_{H\in\mathcal{F}} KF_H(L^{(0)})\Big].$$

For any $n \geq 0$, the composite

$$I^n \cdot KF_G^*(X) \subseteq KF_G^*(X) \xrightarrow{\text{proj}^*} KF_G^*(E_{\mathcal{F}}(G)\times X) \xrightarrow{\text{res}} KF_G^*((E_{\mathcal{F}}(G)\times X)^{(n-1)})$$

is zero, since the image is contained in $IKF_G^*((E_{\mathcal{F}}(G)\times_G X)^{(n-1)})^n = 0$ which vanishes by Lemma 6.4. This thus defines a homomorphism of pro-groups

$$\lambda_{\mathcal{F}}^{X,f} : \{KF_G^*(X)/I^n \cdot K_G^*(X)\}_{n\geq 1} \longrightarrow \{KF_G^*((E_{\mathcal{F}}(G) \times X)^{(n-1)})\}_{n\geq 1}.$$

Theorem 6.5. *Fix $F = \mathbb{C}$ or $\mathbb{R}$. Let G be a discrete group, and let $\mathcal{F}$ be a family of subgroups of G closed under conjugation and under subgroups. Fix a finite proper G-complex X, a finite dimensional proper G-complex Z whose isotropy subgroups have bounded order, and a G-map $f : X \to Z$. Regard $KF_G^*(X)$ as a module over $KF_G(Z)$, and set*

$$I = I_{\mathcal{F},Z}^F = \mathrm{Ker}\Big[KF_G(Z) \xrightarrow{\text{res}} \prod_{H\in\mathcal{F}} KF_H(Z^{(0)})\Big].$$

Then

$$\lambda_{\mathcal{F}}^{X,f} : \{KF_G^*(X)/I^n \cdot KF_G^*(X)\}_{n\geq 1} \longrightarrow \{KF_G^*((E_{\mathcal{F}}(G) \times X)^{(n-1)})\}_{n\geq 1}$$

is an isomorphism of pro-groups. Also, the inverse system

$$\{KF_G^*((E_{\mathcal{F}}(G) \times X)^{(n)})\}_{n\geq 1}$$

satisfies the Mittag-Leffler condition. In particular,

$$\varprojlim{}^1 KF_G^*((E_{\mathcal{F}}(G) \times X)^{(n)}) = 0;$$

and $\lambda_{\mathcal{F}}^{X,f}$ induces an isomorphism

$$KF_G^*(X)\hat{}_I \xrightarrow{\cong} KF_G^*(E_{\mathcal{F}}(G) \times X) \cong \varprojlim KF_G^*((E_{\mathcal{F}}(G) \times X)^{(n)}).$$

Proof. Assume that $\lambda_{\mathcal{F}}^{X,f}$ is an isomorphism. Then the system

$$\{KF_G^*((E_{\mathcal{F}}(G)\times X)^{(n)})\}_{n\geq 1}$$

satisfies the Mittag-Leffler condition because $\{KF_G^*(X)/I^n\}$ does. In particular, $\varprojlim{}^1 KF_G^*((E_{\mathcal{F}}(G)\times X)^{(n)}) = 0$, and so (cf. [3, Proposition 4.1])

$$KF_G^*(E_{\mathcal{F}}(G)\times X) \cong \varprojlim KF_G^*((E_{\mathcal{F}}(G)\times X)^{(n)}).$$

It remains to show that $\lambda_{\mathcal{F}}^{X,f}$ is an isomorphism.

Step 1 Assume first that $X = G/H$, for some finite subgroup $H \subseteq G$. Let $\mathcal{F}|H$ be the family of subgroups of H contained in $\mathcal{F}$, and consider the following commutative diagram:

$$\begin{array}{ccccc} KF_G(Z) & \xrightarrow{f^*} & KF_G^*(G/H) & \xrightarrow{\mathrm{pr}_1} & KF_G^*(E_{\mathcal{F}}(G){\times}G/H) \\ & & \mathrm{ev}_{eH}\downarrow\cong & & q\downarrow\cong \\ RF(H) & \xrightarrow[\cong]{} & KF_H^*(\mathrm{pt}) & \xrightarrow{\mathrm{pr}_2} & KF_H^*(E_{\mathcal{F}|H}(H)). \end{array}$$

Here, pr_2 induces an isomorphism of pro-groups

$$\{KF_H^*(*)/I_{\mathcal{F}}(H)^n{\cdot}KF_H^*(*)\}_{n\geq 1} \longrightarrow \{KF^*((BH)^{(n-1)})\}_{n\geq 1}$$

by the theorem of Jackowski [7, Theorem 5.1], where

$$I_{\mathcal{F}}(H) = \mathrm{Ker}\Big[RF(H) \longrightarrow \prod_{L\in\mathcal{F}|H} RF(L)\Big] \supseteq I' \stackrel{\mathrm{def}}{=} \mathrm{ev}_{f(eH)}(I).$$

(The theorem in [7] is stated only for complex K-theory, but as noted afterwards, the proof applies equally well to the real case.) We want to show that pr_1 induces an isomorphism of pro-groups

$$\{KF_G^*(G/H)/I^n{\cdot}KF_G^*(G/H)\}_{n\geq 1} \longrightarrow \{KF_G^*((E_{\mathcal{F}}(G)\times G/H)^{(n-1)})\}_{n\geq 1}.$$

So we must show that for some k, $I_{\mathcal{F}}(H)^k \subseteq I'$.

This means showing that the ideal $I_{\mathcal{F}}(H)/I'$ is nilpotent; or equivalently (since $R(H)$ is noetherian) that it is contained in all prime ideals of $R(H)/I'$ (cf. [2, Proposition 1.8]). In other words, we must show that every prime ideal of $R(H)$ which contains I' also contains $I_{\mathcal{F}}(H)$. Fix any prime ideal $\mathfrak{P} \subseteq R(H)$ which does *not* contain $I_{\mathcal{F}}(H)$. Set $\zeta = \exp(2\pi i/|H|)$ and $A = \mathbb{Z}[\zeta]$. By a theorem of Atiyah [1, Lemma 6.2], there is a prime ideal $\mathfrak{p} \subseteq A$ and an element $s \in H$ such that

$$\mathfrak{P} = \{v \in R(G) \mid \chi_v(s) \in \mathfrak{p}\}.$$

(This is stated in [1] only in the complex case, but the same arguement applies to prime ideals in the real representation ring.) Also, s is not an element of any $L \in \mathcal{F}$, since $\mathfrak{P} \not\supseteq I_{\mathcal{F}}(H)$. Set $p = \mathrm{char}(A/\mathfrak{p})$ (possibly $p = 0$).

For any $g \in G$ of finite order, we let g_r represent its p-regular component: the unique $g_r \in \langle g\rangle$ such that $p\nmid|g_r|$ and $|(g_r)^{-1}g|$ is a power of p ($g_r = g$ if $p = 0$). By [1, Lemma 6.3], we can replace s by s_r without changing the ideal $\mathfrak{P}$; and can thus assume that $p\nmid|s|$.

Let m' be the least common multiple of the orders of isotropy subgroups in Z, and let m be the largest divisor of m' prime to p ($m = m'$ if $p = 0$). Define $\varphi : \mathrm{tors}(G) \to \mathbb{Z}$ by setting $\varphi(g) = 0$ if $g_r \in L$ for some $L \in \mathcal{F}$, and $\varphi(g) = m$ otherwise. By Corollary 4.3, $\varphi|_L$ is a rational character of L for each $L \in \mathrm{Isotr}(Z)$. So by Corollary 6.3, there is $k > 0$ and an element $\xi \in K_G(Z)$ whose restriction to any orbit has character the restriction of φ^k. In other words, $\xi \in I = I^F_{\mathcal{F},Z}$, and so $\varphi^k|_H$ is the character of an element $v \in I'$. But then $\chi_v(s) = \varphi(s)^k \notin \mathfrak{p}$, so $v \notin \mathfrak{P}$, and thus $\mathfrak{P} \not\supseteq I'$.

Step 2 By Step 1, the theorem holds when $\dim(X) = 0$. So we now assume that $\dim(X) = m > 0$. Assume $X = Y \cup_\varphi (G/H \times D^m)$, for some attaching map $\varphi : G/H \times S^{m-1} \to Y$. We can assume inductively that the theorem holds for Y, $G/H \times S^{m-1}$, and $G/H \times D^m \simeq G/H$.

All terms in the Mayer-Vietoris sequence

$$\longrightarrow KF_G^*(X) \longrightarrow KF_G^*(Y) \oplus KF_G^*(G/H \times D^m) \longrightarrow KF_G^*(G/H \times S^{m-1}) \longrightarrow$$

are $KF_G(X)$-modules, all homomorphisms are $KF_G(X)$-linear and the $KF_G(Z)$-module structure on each term is induced from the $KF_G(X)$-module structure. So if we let $I' \subseteq KF_G(X)$ be the ideal generated by the image of I; then dividing out by $(I')^n$ is the same as dividing out by I^n for all terms. In addition, $KF_G(X)$ is noetherian (in fact, a finitely generated abelian group), and so this Mayer-Vietoris sequence induces an exact sequence of pro-groups

$$\begin{aligned} \longrightarrow \left\{KF_G^*(X)/I^n\right\}_{n\geq 1} \longrightarrow \left\{KF_G^*(Y)/I^n \oplus KF_G^*(G/H \times D^m)/I^n\right\}_{n\geq 1} \\ \longrightarrow \left\{KF_G^*(G/H \times S^{m-1})/I^n\right\}_{n\geq 1} \longrightarrow \end{aligned}$$

by [8, Lemma 4.1]. There is a similar Mayer-Vietoris exact sequence of the pro-groups

$$\left\{KF_G^*((E_{\mathcal{F}}(G) \times -)^{(n-1)})\right\}_{n\geq 1};$$

and the theorem now follows from the 5-lemma for pro-groups together with the induction assumptions. □

References

[1] **M. Atiyah**, *Characters and cohomology of finite groups*, Publ. Math. I.H.E.S. 9 (1961), 23–64.

[2] **M. Atiyah & I. Macdonald**, *Introduction to commutative algebra*, Addison-Wesley (1969).

[3] **M. Atiyah & G. Segal**, *Equivariant K-theory and completion*, J. Diff. Geometry 3 (1969), 1–18.

[4] **P. Baum and A. Connes**: *Chern character for discrete groups*, in: Matsumoto, Miyutami, and Morita (eds.), *A fête of topology; dedicated to Tamura*, 163–232, Academic Press (1988).

[5] **G. Bredon**, *Equivariant cohomology theories*, Lecture notes in mathematics 34, Springer-Verlag (1967).

[6] **J. Davis & W. Lück**, *Spaces over a category and assembly maps in isomorphism conjectures in K-and L-Theory*, MPI-preprint (1996), to appear in K-theory.

[7] **S. Jackowski**, *Families of subgroups and completion*, J. Pure Appl. Algebra 37 (1985), 167–179.

[8] **W. Lück & B. Oliver**: *The completion theorem in K-theory for proper actions of a discrete group*, preprint (1997).

[9] **S. Mac Lane**, *Categories for the working mathematician*, Springer-Verlag (1971).

[10] **G. Segal**, *Categories and cohomology theories*, Topology 13 (1974), 293–312.

[11] **J.-P. Serre**, *Linear representations of finite groups*, Springer-Verlag (1977).

[12] **J. Słomińska**, *On the equivariant Chern homomorphism*, Bull. Acad. Pol. Sci. 24 (1976), 909–913.

WOLFGANG LÜCK
Institut für Mathematik und Informatik
Westfälische Wilhelms-Universtität
Einsteinstr. 62
48149 Münster, Germany
E-mail address: lueck@math.uni-muenster.de
http://wwwmath.uni-muenster.de/math/u/lueck

BOB OLIVER
LAGA
Université Paris Nord
Av. J.-B. Clément
93430 Villetaneuse, France
E-mail address: bob@math.univ-paris13.fr
http://zeus.math.univ-paris13.fr/~bob

Progress in Mathematics, Vol. 196

The Thomified Eilenberg–Moore spectral sequence

Mark Mahowald, Douglas C. Ravenel and Paul Shick

1. Introduction

In this paper we will construct a generalization of the Eilenberg–Moore spectral sequence, which in some interesting cases turns out to be a form of the Adams spectral sequence. We recall the construction of both of these in general terms. Suppose we have a diagram of spectra of the form

$$\begin{array}{ccccccc} X_0 & \longleftarrow & X_1 & \longleftarrow & X_2 & \longleftarrow & \cdots \\ \downarrow{\scriptstyle g_0} & & \downarrow{\scriptstyle g_1} & & \downarrow{\scriptstyle g_2} & & \\ K_0 & & K_1 & & K_2 & & \end{array} \tag{1.1}$$

where X_{s+1} is the fiber of g_s. We get an exact couple of homotopy groups and a spectral sequence with

$$E_1^{s,t} = \pi_{t-s}(K_s) \quad \text{and} \quad d_r\colon E_r^{s,t} \to E_r^{s+r,t+r-1}.$$

This spectral sequence converges to $\pi_*(X)$ (where $X = X_0$) if the homotopy inverse limit $\lim_{\leftarrow} X_s$ is contractible and certain $\lim^1$ groups vanish. When X is connective, it is a first quadrant spectral sequence. For more background, see [Rav86].

In the case of the classical Adams spectral sequence, we have some additional conditions on (1.1), namely:

- Each spectrum K_s is a generalized mod p Eilenberg–Mac Lane spectrum, and
- each map g_s induces a monomorphism in mod p homology.

These conditions enable us to identify the E_2-term as an Ext group over the Steenrod algebra, and to prove convergence when X is connective and p-adically complete.

For the Eilenberg–Moore spectral sequence, let

$$X \xrightarrow{i} E \xrightarrow{h} B \tag{1.2}$$

The second author acknowledges support from NSF grant DMS-9802516 and the Centre de Recerca Matemàtica in Barcelona.

be a fiber sequence with simply connected base space B. Then one uses this (in a manner to be described below) to produce a diagram of the form (1.1) where X_0 is the suspension spectrum of X. This will yield a spectral sequence converging to the stable homotopy of X, but in practice it is not very useful. However if we smash everything in sight with the mod p Eilenberg–Mac Lane spectrum H/p, we get the Eilenberg–Moore spectral sequence converging to $H_*(X)$, where E_2 is a certain Cotor group over $H_*(B)$.

In this paper we will explain a way to twist this construction using a p-local spherical fibration over the total space E. The entire construction can be Thomified to yield a spectral sequence converging to the homotopy of the Thom spectrum for the induced bundle over X. In §2 we recall a geometric construction of the Eilenberg–Moore spectral sequence, and in §3 we explain how it can be Thomified. In §4 we identify the E_2-term under certain circumstances as an Ext group over the Massey–Peterson algebra of the base space of the fibration in question, and in §5 we show that in some other cases we get a BP-theoretic analog of this result. In §6, we show that a special case of the $\mathbf{Z}/(p)$-equivariant Adams spectral sequence of Greenlees can be constructed using the Thomified Eilenberg–Moore spectral sequence.

The authors wish to thank Bill Dwyer, John Greenlees and Brooke Shipley for helpful conversations and correspondence.

2. A geometric construction of the Eilenberg–Moore spectral sequence

We begin by recalling the stable cosimplicial construction associated with the Eilenberg–Moore spectral sequence, due to Larry Smith [Smi69] and Rector [Rec70]. Given the fibration (1.2), for $s \geq 0$ let

$$G_s = E \times \overbrace{B \times \cdots \times B}^{s \text{ factors}}.$$

Define maps $h_t \colon G_{s-1} \to G_s$ for $0 \leq t \leq s$ by

$$h_t(e, b_1, b_2, \ldots, b_{s-1}) = \begin{cases} (e, b_1, b_2, \ldots, b_{s-1}, *) & \text{if } t = 0 \\ (e, b_1, b_2, \ldots, b_t, b_t, b_{t+1}, \ldots, b_{s-1}) & \text{if } 1 \leq t \leq s-1 \\ (e, h(e), b_1, b_2, \ldots, b_{s-1}) & \text{if } t = s. \end{cases}$$

Let $E_0 = E$, $X_0 = X$, $X_1 = E/\mathrm{Im}\, i$, and for $s \geq 1$ we define spectra

$$\begin{aligned} \Sigma^s E_s &= G_s/\mathrm{Im}\, h_0 \cup \cdots \cup \mathrm{Im}\, h_{s-1} \\ \Sigma^{s+1} X_{s+1} &= G_s/\mathrm{Im}\, h_0 \cup \cdots \cup \mathrm{Im}\, h_s \end{aligned}$$

i.e., the spectra X_s and E_s are desuspensions of suspension spectra of the indicated spaces. Then for $s \geq 0$, h_s induces a map $X_s \to E_s$ giving a cofiber sequence

$$X_s \xrightarrow{\quad h_s \quad} E_s \xrightarrow{\quad \partial_s \quad} \Sigma X_{s+1}, \tag{2.1}$$

where ∂_s is projection from the topological quotient of G_s by one subspace to the quotient by a bigger subspace.

For $s \geq 0$ there is a homology isomorphism

$$H_*(E_s) = \Sigma^{-s} H_*(E) \otimes \bar{H}_*(B^{(s)}),$$

where $\bar{H}$ denotes reduced homology. Since B is simply connected, the connectivity of E_s increases without bound with s. Note also that

$$H_*(X_s) = \Sigma^{-s} H_*(X) \otimes \bar{H}_*(B^{(s)}),$$

for $s > 0$, so the homotopy inverse limit of the X_s is contractible. The homology exact couple associated with the cofiber sequences (2.1) leads to the Eilenberg–Moore spectral sequence for the fibration (1.2). The Eilenberg–Moore spectral sequence also converges for non-simply-connected B. Dwyer has proved ([D74]) that the Eilenberg–Moore spectral sequence for the fibration

$$X \to E \to B$$

converges strongly to $H_*(X)$ if and only if $\pi_1(B)$ acts nilpotently on $H_i(E)$ for all $i \geq 0$.

3. The Thomified Eilenberg–Moore spectral sequence

Now suppose that in addition to the fibration (1.2) we also have a p-local stable spherical fibration ξ over E which is oriented with respect to mod p homology. Projection onto the first coordinate gives compatible maps of the G_s to E, and hence a stable spherical fibration over each of them. This means that we can Thomify the entire construction. To each of the quotients X_s and E_s we associate a *reduced Thom spectrum*, which is defined as follows. Given a space A with a spherical fibration and a subspace $B \subset A$, the reduced Thom space for A/B is the space $D_A/(S_A \cup D_B)$ where D_X and S_X denote disk and sphere bundles over the space X. Thus we can associate reduced Thom spectra to the topological quotients E_s and X_{s+1} of G_s.

Let Y, K, Y_s and K_s be the Thomifications of X, E, X_s and E_s. Then the cofiber sequence of (2.1) Thomifies to

$$Y_s \longrightarrow K_s \longrightarrow \Sigma Y_{s+1}. \tag{3.1}$$

and we have

$$H_*(K_s) = \Sigma^{-s} H_*(K) \otimes \bar{H}_*(B^{(s)}).$$

The exact couple of homotopy groups for (3.1) leads to a spectral sequence converging to $\pi_*(Y)$. There is an associated diagram

$$\begin{array}{ccccccc} Y & = & Y_0 & \longleftarrow & Y_1 & \longleftarrow & Y_2 & \longleftarrow & \cdots \\ & & \downarrow{g_0} & & \downarrow{g_1} & & \downarrow{g_2} & & \\ & & K_0 & & K_1 & & K_2 & & \end{array} \tag{3.2}$$

where Y_{s+1} is the fiber of g_s. This is similar to the Adams diagram of (1.1), but $H_*(g_s)$ need not be a monomorphism in general. We will call this the *Thomified Eilenberg–Moore spectral sequence.* We will use the indexing conventions of Adams rather than Eilenberg–Moore, namely

$$E_1^{s,t} = \pi_{t-s}(K_s) \qquad \text{with} \qquad E_r^{s,t} \xrightarrow{d_r} E_r^{s+r,t+r-1}.$$

This puts our spectral sequence in the first rather than the second quadrant.

We will see below (Theorem 4.4(ii) and Corollary 4.5) that under suitable hypotheses (including that the map i of (1.2) induces a monomorphism in homology), the Thomified Eilenberg–Moore spectral sequence coincides with the usual Adams spectral sequence for $\pi_*(Y)$.

The following lemma will be useful.

Lemma 3.3. *For each prime p there is a p-local spherical fibration over $\Omega^2 S^3$ whose Thom spectrum is the mod p Eilenberg–Mac Lane spectrum H/p.*

Proof. For $p = 2$ we can use an ordinary vector bundle. We extend the nontrivial map $S^1 \to BO$ to $\Omega^2 S^3$ using the double loop space structure on BO. It was shown in [Mah79] that the resulting Thom spectrum is $H/2$.

The following argument for odd primes is due to Mike Hopkins. Let $BF(n)_{(p)}$ denote the classifying space for the monoid of homotopy equivalences of the p-local n-sphere. Its fundamental group is $\mathbf{Z}_{(p)}^\times$. A p-local n-dimensional spherical fibration of a space X, i.e., a fibration with fiber $S^n_{(p)}$, is classified by a map $X \to BF(n)_{(p)}$. Its Thom space is the cofiber of the projection map to X. Such fibrations and Thom spectra can be stabilized in the usual way. We denote the direct limit of the $BF(n)_{(p)}$ by $BF_{(p)}$.

Now consider a p-local spherical fibration over S^1 corresponding to an element $u \in \mathbf{Z}_{(p)}^\times$. It Thomifies to the Moore spectrum $S^0 \cup_{1-u} e^1$. If we set $u = 1-p$ (which is a p-local unit) we get the mod p Moore spectrum $V(0)$.

As in the case $p = 2$, we can extend this map $S^1 \to BF_{(p)}$ to $\Omega^2 S^3$ using the double loop space structure on $BF_{(p)}$, and similar arguments to those of [Mah79] identify the resulting Thom spectrum as H/p. □

4. Identifying the E_2-term

Observe that $H_*(K)$ is simultaneously a comodule over A_* and (via the Thom isomorphism and the map h_*) over $H_*(B)$, which is itself a comodule over A_*. Following Massey–Peterson [MP67], we combine these two structures by defining the *Massey–Peterson coalgebra* (they called the dual object the semitensor product)

$$R_* = H_*(B) \otimes A_* \tag{4.1}$$

in which the coproduct is the composite

$$\begin{array}{c} H_*(B) \otimes A_* \\ \Big\downarrow{\scriptstyle \Delta_B \otimes \Delta_A} \\ H_*(B) \otimes H_*(B) \otimes A_* \otimes A_* \\ \Big\downarrow{\scriptstyle H_*(B)\otimes\psi_B\otimes A_*\otimes A_*} \\ H_*(B) \otimes A_* \otimes H_*(B) \otimes A_* \otimes A_* \\ \Big\downarrow{\scriptstyle H_*(B)\otimes A_*\otimes T\otimes A_*} \\ H_*(B) \otimes A_* \otimes A_* \otimes H_*(B) \otimes A_* \\ \Big\downarrow{\scriptstyle H_*(B)\otimes m_A\otimes H_*(B)\otimes A_*} \\ (H_*(B) \otimes A_*) \otimes (H_*(B) \otimes A_*), \end{array} \tag{4.2}$$

where Δ_A and Δ_B are the coproducts on A_* and $H_*(B)$, T is the switching map, $\psi_B\colon H_*(B) \to A_* \otimes H_*(B)$ is the comodule structure map, and m_A is the multiplication in A_*.

Massey–Peterson gave this definition in cohomological terms. They denoted the semitensor algebra R by $H^*(B) \odot A$, which is additively isomorphic to $H^*(B) \otimes A$ with multiplication given by

$$(x_1 \otimes a_1)(x_2 \otimes a_2) = x_1 a_1'(x_2) \otimes a_1'' a_2,$$

where $x_i \in H^*(B)$, $a_i \in A$, and $a_1' \otimes a_1''$ denotes the coproduct expansion of a_1 given by the Cartan formula. Our definition is the homological reformulation of theirs.

Note that given a map $f\colon V \to B$ and a subspace $U \subset V$, $\bar{H}^*(V/U) = H^*(V, U)$ is an R-module since it is an $H^*(V)$-module via relative cup products, even if the map f does not extend to the quotient V/U. In our case we have maps $G_s \to B$ for all $s \geq 0$ given by

$$(e, b_1, \ldots, b_s) \mapsto h_e.$$

These are compatible with all of the maps h_t, so $H_*(Y_s)$ and $H_*(K_s)$ are R_*-comodules, and the maps between them respect this structure.

We will see in the next theorem that under suitable hypotheses, the E_2-term of the Thomified Eilenberg–Moore spectral sequence is $\mathrm{Ext}_{R_*}(\mathbf{Z}/(p), H_*(K))$. When B is an H-space we have a Hopf algebra extension (see [Rav86, A1.1.15] for a definition)

$$A_* \longrightarrow R_* \longrightarrow H_*(B).$$

This gives us a Cartan–Eilenberg spectral sequence ([CE56, page 349] or [Rav86, A1.3.14]) converging to this Ext group with

$$E_2 = \mathrm{Ext}_{A_*}(\mathbf{Z}/(p), \mathrm{Ext}_{H_*(B)}(\mathbf{Z}/(p), H_*(K))). \tag{4.3}$$

Note that the inner Ext group above is the same as $\mathrm{Ext}_{H_*(B)}(\mathbf{Z}/(p), H_*(E))$, the E_2-term of the classical Eilenberg–Moore spectral sequence converging to $H_*(X)$. If the latter collapses from E_2, then the Ext group of (4.3) can be thought of as

$$\mathrm{Ext}_{A_*}(\mathbf{Z}/(p), H_*(Y)),$$

where $H_*(Y)$ is equipped with the Eilenberg–Moore bigrading. This is the usual Adams E_2-term for Y when $H_*(Y)$ is concentrated in Eilenberg–Moore degree 0, but the Ext group of (4.3) is graded differently in general.

Theorem 4.4. (i) *Suppose that B is simply connected. Then the Thomified Eilenberg–Moore spectral sequence associated with the homotopy of* (3.2) *converges to $\pi_*(Y)$. If, in addition, $H^*(K)$ is a free A-module, then*

$$E_2 = \mathrm{Ext}_{R_*}(\mathbf{Z}/(p), H_*(K)),$$

where R_* is the Massey–Peterson coalgebra of (4.1).

(ii) *If, in addition, the map $i\colon X \to E$ induces a monomorphism in mod p homology, then the Thomified Eilenberg–Moore spectral sequence coincides with the classical Adams spectral sequence for Y.*

The hypotheses on $H_*(K)$ may be unnecessary, but they are adequate for our purposes. The result may not be new, but we know of no published proof. Before proving the theorem we give a corollary that indicates that the hypotheses are not as restrictive as they may appear.

Corollary 4.5. *Given a fibration*

$$X \longrightarrow E \longrightarrow B$$

with X p-adically complete, a p-local spherical fibration over E, and B simply connected, there is a spectral sequence converging to $\pi_*(Y)$ (where Y is the Thomification of X) with

$$E_2 = \mathrm{Ext}_{H_*(B)\otimes A_*}(\mathbf{Z}/(p), H_*(K)),$$

where K as usual is the Thomification of E.

Proof. We can apply 4.4 to the product of the given fibration with

$$\mathrm{pt} \to \Omega^2 S^3 \to \Omega^2 S^3,$$

where $\Omega^2 S^3$ is equipped with the p-local spherical fibration of Lemma 3.3. Then the Thomified total space is $K \wedge H/p$, so its cohomology is a free A-module. Thus the E_2-term is

$$\mathrm{Ext}_{H_*(B\wedge H/p)\otimes A_*}(\mathbf{Z}/(p), H_*(K \wedge H/p)) = \mathrm{Ext}_{H_*(B)\otimes A_*}(\mathbf{Z}/(p), H_*(K)).$$

□

Proof of Theorem 4.4 (i) The freeness of $H_*(K)$ over A_* does not make (3.2) an Adams resolution because $H_*(g_s)$ need not be a monomorphism and the cofiber sequence

$$\Sigma^s Y_s \xrightarrow{g_s} \Sigma^s K_s \longrightarrow \Sigma^{s+1} Y_{s+1}$$

need not induce a short exact sequence in homology.

We will finesse this problem by producing a commutative diagram

$$\begin{array}{ccccc} \Sigma^s Y_s & \xrightarrow{g_s} & \Sigma^s K_s & \longrightarrow & \Sigma^{s+1} Y_{s+1} \\ \downarrow{\scriptstyle g_s} & & \downarrow & & \downarrow{\scriptstyle h_{s+1}} \\ \Sigma^s K_s & \longrightarrow & \Sigma^s W_s & \longrightarrow & \Sigma^{s+1} K_{s+1} \end{array} \qquad \text{for } s \geq 0. \tag{4.6}$$

in which the cofiber sequence in the bottom row does induce a short exact sequence in homology with

$$H_*(W_s) = H_*(K_s) \otimes H_*(B). \tag{4.7}$$

By the change-of-rings isomorphism of Milnor–Moore [MM65], this implies that

$$\mathrm{Ext}_{R_*}(\mathbf{Z}/(p), H_*(W_s)) = \mathrm{Ext}_{A_*}(\mathbf{Z}/(p), H_*(K_s)). \tag{4.8}$$

Splicing the short exact sequences in homology from the bottom row of (4.6) gives a long exact sequence

$$0 \longrightarrow H_*(K) \longrightarrow H_*(W_0) \longrightarrow H_*(\Sigma W_1) \longrightarrow \cdots,$$

which gives an algebraic spectral sequence (see [Rav86, A1.3.2]) converging to $\mathrm{Ext}_{R_*}(\mathbf{Z}/(p), H_*(K))$ with

$$E_1 = \mathrm{Ext}_{R_*}(\mathbf{Z}/(p), H_*(W_s)),$$

suitably indexed.

The freeness hypothesis on $H_*(K)$ implies (via (4.7)) that $H_*(W_s)$ is free over R_*. From this fact it follows that the algebraic spectral sequence collapses from E_2, i.e., $\mathrm{Ext}_{R_*}(\mathbf{Z}/(p), H_*(K))$ is the cohomology of the cochain complex

$$\mathrm{Ext}^0_{R_*}(\mathbf{Z}/(p), H_*(W_0)) \longrightarrow \mathrm{Ext}^0_{R_*}(\mathbf{Z}/(p), H_*(\Sigma W_1)) \longrightarrow \cdots.$$

By (4.8) this is the same as

$$\mathrm{Ext}^0_{A_*}(\mathbf{Z}/(p), H_*(K_0)) \longrightarrow \mathrm{Ext}^0_{A_*}(\mathbf{Z}/(p), H_*(\Sigma K_1)) \longrightarrow \cdots$$

and our freeness hypothesis along with (4.6) allows us to identify this cochain complex with the E_1-term of the Thomified Eilenberg–Moore spectral sequence.

Thus the Thomified Eilenberg–Moore spectral sequence has the desired E_2-term if we can produce the diagram (4.6) satisfying (4.7). We shall do this now by geometric construction.

We define the following subspaces of G_s for $s \geq 1$:

$$\begin{aligned} A_s &= \operatorname{Im} h_0 \cup \operatorname{Im} h_2 \cup \cdots \cup \operatorname{Im} h_{s-1}, \\ B_s &= A_s \cup \operatorname{Im} h_1 \\ \text{and} \quad C_s &= B_s \cup \operatorname{Im} h_s. \end{aligned}$$

Then it follows that h_s sends C_{s-1} to B_s and B_{s-1} to A_s, $B_s/A_s = G_{s-1}/B_{s-1}$ and $C_s/B_s = G_{s-1}/C_{s-1}$. Thus for $s \geq 0$ we get the following pointwise commutative diagram in which each row is a cofiber sequence:

$$\begin{array}{ccccc} X & \xrightarrow{i} & E & \xrightarrow{\partial_0} & E/i(X) \\ \downarrow{\scriptstyle i} & & \downarrow{\scriptstyle h_1} & & \downarrow{\scriptstyle h_1} \\ E & \xrightarrow{h_1} & G_1/A_1 & \xrightarrow{\partial_1} & G_1/B_1 \end{array} \qquad \text{for } s = 0$$

and

$$\begin{array}{ccccc} G_{s-1}/C_{s-1} & \xrightarrow{h_s} & G_s/B_s & \xrightarrow{\partial_s} & G_s/C_s \\ \downarrow{\scriptstyle h_s} & & \downarrow{\scriptstyle h_{s+1}} & & \downarrow{\scriptstyle h_{s+1}} \\ G_s/B_s & \xrightarrow{h_{s+1}} & G_{s+1}/A_{s+1} & \xrightarrow{\partial_{s+1}} & G_{s+1}/B_{s+1} \end{array} \qquad \text{for } s \geq 1.$$

We define $\Sigma^{s-1}W_{s-1}$ to be the Thomification of G_s/A_s, and we have previously defined $\Sigma^s K_s$ and $\Sigma^{s+1}X_{s+1}$ to be the Thomifications of G_s/B_s and G_s/C_s, so Thomification converts the diagrams above to (4.6).

Let $p_s\colon G_{s+1} \to G_s \times B$ be the homeomorphism given by

$$p_s(e, b_1, \ldots, b_{s+1}) = ((e, b_2, \ldots, b_{s+1}), b_1).$$

Then we have

$$\begin{aligned} p_s h_0 &= (h_0 \times B)p_{s-1} \\ \text{and} \quad p_s h_t &= (h_{t-1} \times B)p_{s-1} \qquad \text{for } 2 \leq t \leq s. \end{aligned}$$

It follows that

$$G_{s+1}/A_{s+1} = (G_s \times B)/(B_s \times B) = (G_s/B_s) \times B$$

and (4.7) follows.

(ii) If the $H_*(i)$ is monomorphic and $H^*(K_s)$ is a free A-module, then the diagram (3.2) is an Adams resolution for Y. Thus, the identity map on the resolution provides a comparison map from the Thomified Eilenberg–Moore spectral sequence to the Adams spectral sequence. We can identify the inner Ext group of (4.3) with $H_*(Y)$ concentrated in degree 0, the Cartan–Eilenberg spectral sequence collapses and our E_2-term is the usual

$$\mathrm{Ext}_{A_*}(\mathbf{Z}/(p), H_*(Y)).$$

So the comparison map induces an isomorphism on the E_2-term of the spectral sequences, completing the proof of the theorem. □

5. An Adams–Novikov analog

We now describe a case of the Thomified Eilenberg–Moore spectral sequence leading to variants of the Adams–Novikov spectral sequence. Suppose that in the fibration of (1.2), the spherical fibration over E is a complex vector bundle and that $MU_*(K)$ is free as a comodule over $MU_*(MU)$. If in addition $MU_*(i)$ is a monomorphism, then we get the usual Adams–Novikov spectral sequence converging to $\pi_*(Y)$.

We want an analog of 4.4 in the p-local case identifying the E_2-term for more general i. For this we need a BP-theoretic analog of the Massey–Peterson algebra R_* of (4.1), additively isomorphic to

$$\Gamma(B) = BP_*(B) \otimes_{BP_*} \Gamma, \tag{5.1}$$

where $\Gamma = BP_*(BP)$. In order to define a coproduct on this as in (4.2), we need a coalgebra structure on $BP_*(B)$. This does not exist in general, but it does when $H_*(B)$ is torsion free and $BP_*(B)$ is therefore a free BP_*-module. If B is also an H-space, then $BP_*(B)$ is a Hopf algebra over BP_* and $(BP_*, \Gamma(B))$ is a Hopf algbebroid (defined in [Rav86, A1.1.1])

$$(BP_*, \Gamma) \longrightarrow (BP_*, \Gamma(B)) \longrightarrow (BP_*, BP_*(B))$$

is a Hopf algebroid extension as defined in [Rav86, A1.1.15]. This means there is a Cartan–Eilenberg spectral sequence (see [CE56, page 349] or [Rav86, A1.3.14]) converging to $\mathrm{Ext}_{\Gamma(B)}(BP_*, BP_*(K))$ with

$$E_2 = \mathrm{Ext}_{\Gamma}(BP_*, \mathrm{Ext}_{BP_*(B)}(BP_*, BP_*(K))). \tag{5.2}$$

Then we get the following analog of Theorem 4.4, which is proved in the same way.

Theorem 5.3. (i) *Suppose that $BP_*(K)$ is free as a $BP_*(BP)$-comodule and B is simply connected with torsion free homology. Then the Thomified Eilenberg–Moore spectral sequence associated with the homotopy of* (3.2) *converges to $\pi_*(Y)$ with*

$$E_2 = \mathrm{Ext}_{\Gamma(B)}(BP_*, BP_*(K)),$$

where $\Gamma(B)$ is the Massey–Peterson coalgebra of (5.1).

(ii) *If in addition the map* $i\colon X \to E$ *induces a monomorphism in* BP*-homology, then the Thomified Eilenberg–Moore spectral sequence coincides with the Adams–Novikov spectral sequence for* Y.

There is an analog of Corollary 4.5 in which we retain the hypothesis on B while dropping the one on K.

Corollary 5.4. *Given a fibration*

$$X \longrightarrow E \longrightarrow B$$

with X *p-local, a complex vector bundle over* E*, and* B *simply connected with torsion free homology, there is a spectral sequence converging to* $\pi_*(Y)$ *(where* Y *is the Thomification of* X*) with*

$$E_2 = \mathrm{Ext}_{\Gamma(B)}(BP_*, BP_*(K)),$$

where K *as usual is the Thomification of* E.

This can be proved by applying 5.3 to the product of the given fibration with

$$\mathrm{pt} \longrightarrow BU \longrightarrow BU$$

with the universal complex vector bundle over BU.

6. A construction of the equivariant Adams spectral sequence

In this section we provide an alternative construction of a special case of the equivariant Adams spectral sequence, due to Greenlees ([G88] and [G90]). We first recall Greenlees' approach.

Let G be a finite p-group. (Later, we will restrict our attention to the case where G is elementary abelian.) We work in the equivariant stable homotopy category of [LMS86], with all spaces pointed and all homology groups reduced. In this setting, G-free means that the action of G is free away from the base point. Greenlees' version of the equivariant Adams spectral sequence is based on mod p Borel cohomology, defined for a based G-spectrum X as

$$b_G^*(X) = H^*(EG_+ \wedge_G X; \mathbf{Z}/(p)),$$

where, as above, the $\mathbf{Z}/(p)$ coefficient groups will hereafter be suppressed. This is an $RO(G)$-graded cohomology theory, defined as follows for α any virtual real representation of G:

$$b_G^\alpha(X) = H^{|\alpha|}(EG_+ \wedge_G X).$$

Since G is a p-group, all representations are orientable, and the suspension isomorphisms in b_G^* are given by the Thom maps, so the theory is really $\mathbf{Z}$-graded in this case. This cohomology theory b_G is representable in the equivariant stable category. Adams and Greenlees identify the algebra $b_G^*(b_G)$ of natural cohomology operations as

$$b_G^*(b_G) \cong H^*(BG_+)\tilde{\otimes} A,$$

where $\tilde{\otimes}$ denotes the Massey–Peterson semitensor product. Greenlees actually defines the spectral sequence in terms of a variant of Borel cohomology, namely f- or coBorel-cohomology, represented by

$$c_G = b_G \wedge EG_+.$$

Greenlees shows in [G88] that $c_G^*(c_G) \cong b_G^*(b_G)$.

Greenlees' main result is the following cohomology version of the spectral sequence.

Theorem 6.1. ([G88]) *For G a finite p-group, X and Y any G-spectra, with Y p-complete, bounded below, G-free and homologically locally finite, there is a convergent Adams spectral sequence*

$$E_2^{s,t} = \mathrm{Ext}^{s,t}_{c_G^*(c_G)}(c_G^*Y, c_G^*X) \Longrightarrow [X,Y]_*^G,$$

natural in both variables.

One can define a similar spectral sequence based on $b_G^*(-)$, but this requires the additional hypothesis that X is G-free to guarantee proper convergence. A homology version of the spectral sequence can be written using the homology theory represented by the G-spectrum b_G ([G90]), which does calculate $[X,Y]_*^G$ when X or Y is not G-free, provided we take G to be elementary abelian. The hypotheses on Y can just be checked nonequivariantly, if Y is G-free, by looking at the non-equivariant spectrum $EG_+ \wedge_G Y$.

Greenlees' construction involves building a resolution of b_G^*Y by free $b_G^*(b_G)$-modules,

$$0 \longleftarrow b^*Y \overset{\epsilon}{\longleftarrow} P_0 \overset{\delta_0}{\longleftarrow} P_1 \overset{\delta_1}{\longleftarrow} P_1 \longleftarrow \cdots$$

and realizing this resolution geometrically. Apply the functor $[X,-]^G$ to this geometric resolution, obtaining a spectral sequence with

$$E_1 = [\Sigma^{t-s}X, Q_s]^G \Longrightarrow [\Sigma^{t-s}X, Y/\operatorname{holim}_s Y_s]^G,$$

where Q_s is a locally finite wedge of copies of the spectrum representing b_G made free (i.e., a wedge of copies of $c_G = b_G \wedge EG_+$), with $P_s = b_G^*\Sigma^s Q_s$. One identifies the E_2-term in the usual manner, and proves convergence by comparing c_G^*- (or b_G^*-) connectivity with H^*-connectivity to show that $\operatorname{holim}_s Y_s \simeq *$.

We now show how to identify this equivariant Adams spectral sequence as a case of the Thomified Eilenberg–Moore spectral sequence, with certain restrictive hypotheses. From here onward we take G to be $\mathbf{Z}/(p)$, and we will work with the spectrum X G-fixed (so that we will use the $c_G^*(c_G)$-based spectral sequence, rather than its $b_G^*(b_G)$-based analog). Let Z be a p-complete free G-spectrum with a spherical G-fibration

$$F \to E(\xi) \overset{p}{\longrightarrow} Z.$$

Consider the Borel fibration

$$Z \to EG_+ \wedge_G Z \to BG.$$

The spherical G-fibration over Z induces a G-fibration

$$EG_+ \wedge_G F \to EG_+ \wedge_G E(\xi) \to EG_+ \wedge_G Z,$$

so we have the desired fibration over the total space of the Borel fibration.

We smash the Borel fibration with pt $\to \Omega^2 S^3 \to \Omega^2 S^3$ (with the trivial G-action) and apply the Thomified Eilenberg–Moore spectral sequence construction to the resulting fibration. The resulting resolution is $EG_+ \wedge_G H\mathbf{Z}/(p)$-free. Now for a G-fixed spectrum W (like $H\mathbf{Z}/(p)$ here), the Borel construction is very simple: $EG_+ \wedge_G W \simeq BG_+ \wedge W$, so the Thomified Eilenberg–Moore spectral sequence resolution is free over $BG_+ \wedge H\mathbf{Z}/(p)$. Let $T(Z)$ denote the Thom spectrum of the bundle over Z. Then the resulting Thomified Eilenberg–Moore spectral sequence has

$$\begin{aligned} E_2 &= \operatorname{Ext}_{H_*(BG_+)\tilde{\otimes}A_*}(H_*BG_+, H_*(T(EG_+ \wedge_G Z))) \\ &= \operatorname{Ext}_{H_*(BG_+)\tilde{\otimes}A_*}(H_*BG_+, H_*(EG_+ \wedge_G T(Z))) \\ &= \operatorname{Ext}_{b_G^*(b_G)}(b_G^*(T(Z)), b_G^*(S^0)), \end{aligned}$$

by $\mathbf{F}_p$-duality, so that the Thomified Eilenberg–Moore spectral sequence E_2-term agrees with the equivariant Adams spectral sequence E_2-term.

The Thomified Eilenberg–Moore spectral sequence E_1-term here is given by applying the (nonequivariant) functor $\pi_*(-)$ to the Thomified Eilenberg–Moore spectral sequence diagram

$$\begin{array}{ccccccc} Y = & Y_0 & \longleftarrow & Y_1 & \longleftarrow & Y_2 & \longleftarrow \cdots \\ & \downarrow g_0 & & \downarrow g_1 & & \downarrow g_2 & \\ & K_0 & & K_1 & & K_2 & \end{array}$$

where Y is the Thom spectrum of $EG_+ \wedge_G Z \wedge \Omega^2 S^3$. The equivariant $b_G^*(b_G)$-Adams spectral sequence E_1-term arises from applying $[S^0, -]_*^G$ to the geometric resolution

$$\begin{array}{ccccccc} W = & W_0 & \longleftarrow & W_1 & \longleftarrow & W_2 & \longleftarrow \cdots \\ & \downarrow g_0 & & \downarrow g_1 & & \downarrow g_2 & \\ & Q_0 & & Q_1 & & Q_2 & \end{array}$$

where W is the Thom spectrum of Z and Q_s is a wedge of copies of the coBorel spectrum c_G. But the Adams isomorphism ([Ad84], 5.3) shows that

$$[S^0, Q_s]^G = [S^0, K_s]^1,$$

so that the isomorphism of E_2-terms above is induced by one on the E_1 level. This proves the following.

Theorem 6.2. *Let Z be a p-complete based free $\mathbf{Z}/(p)$-spectrum, with a spherical $\mathbf{Z}/(p)$-fibration over Z. The Thomified Eilenberg–Moore spectral sequence for the smash product of the fibrations* $\mathrm{pt} \to \Omega^2 S^3 \to \Omega^2 S^3$ *and*

$$Z \to E\mathbf{Z}/(p)_+ \wedge_{\mathbf{Z}/(p)} Z \to B\mathbf{Z}/(p)$$

agrees with the $b_G^*(b_G)$-based equivariant Adams spectral sequence converging to $\pi_*(T(Z))^{\mathbf{Z}/(p)}$ from E_2 onward.

Unfortunately, the Thomified Eilenberg–Moore spectral sequence is known to converge only in the case where the base space in the fibration is simply connected, from Theorem 4.4, which is not the case for the Borel fibration. Note that we would hope that the case of the Thomified Eilenberg–Moore spectral sequence above would converge to $[EG_+, T(Z)]_*^G$, rather than $[S^0, T(Z)]_*^G$, the target of the $\mathbf{Z}/p$-equivariant Adams spectral sequence. Thus, despite the lack of simple-connectivity for the base space, this special case of the Thomified Eilenberg–Moore spectral sequence does converge if $[S^0, T(Z)]_*^G$ is isomorphic to $[EG_+, T(Z)]_*^G = [S^0, F(EG_+, T(Z))]_*^G$ via the comparison map $T(Z) \to F(EG_+, T(Z))$, which is indeed an equivalence when $T(Z)$ is finite, by the (confirmed) Segal Conjecture ([Car84]). Thus, when Z is finite and G-free, the Thomified Eilenberg–Moore spectral sequence converges to

$$\pi_*(T(EG_+ \wedge_G Z)) \cong \pi_*(EG_+ \wedge_G T(Z)) \cong \pi_*(T(Z)/G) \cong \pi_*(T(Z)^G),$$

as we wish, where we think of $\pi_*(T(Z))^G$ as $[S^0, T(Z)]^G$ with the sphere G-fixed. If Z is not finite, then the Thomified Eilenberg–Moore spectral sequence need not converge. For example, if $T(Z) = EG_+ \wedge H\mathbf{Z}/p$, then $\pi_*(T(Z)) = H_*(G)$, which is bounded below, while $F(EG_+, EG_+ \wedge HG) = F(EG_+, HG)$, which has homotopy $H^*(G)$, which is unbounded. This example was pointed out by the referee.

References

[Ad84] J. F. Adams, Prerequisites (on equivariant stable homotopy theory) for Carlsson's lecture, *Algebraic Topology, Aarhus 1982,* ed. by I. Madsen and B. Oliver, Lecture Notes in Math. **1051**:483–532, Springer-Verlag, 1984.

[CE56] H. Cartan and S. Eilenberg, *Homological Algebra*, Princeton University Press, Princeton, 1956.

[Car84] G. Carlsson, Equivariant stable homotopy theory and Segal's Burnside ring conjecture, *Ann. of Math.* **120**:189–224, 1984.

[D74] W. G. Dwyer, Strong convergence of the Eilenberg–Moore spectral sequence, *Topology* **13**:255–265, 1974.

[G88] J. P. C. Greenlees, Stable maps into free G-spaces, *Trans. Amer. Math. Soc.* **310**:199–215, 1988.

[G90] J. P. C. Greenlees, The power of mod p Borel homology, *Homotopy Theory and Related Topics,* ed. by M. Mimura, Lecture Notes in Math. **1418**:140–151, Springer-Verlag, 1990.

[LMS86] L. G. Lewis, J. P. May, and M. Steinberger (with contributions by J. E. McClure), *Equivariant Stable Homotopy Theory,* Lecture Notes in Math. **1213**, Springer-Verlag, 1986.

[Mah79] M. E. Mahowald, Ring spectra which are Thom complexes, *Duke Math. J.* **46**:549–559, 1979.

[MM65] J. W. Milnor and J. C. Moore, On the structure of Hopf algebras, *Ann. of Math.* **81**(2):211–264, 1965.

[MP67] W. S. Massey and F. P. Peterson, *The* mod *2 cohomology structure of certain fibre spaces, Mem. Amer. Math. Soc.* **74**, American Mathematical Society, Providence, Rhode Island, 1967.

[Rav86] D. C. Ravenel, *Complex Cobordism and Stable Homotopy Groups of Spheres,* Academic Press, New York, 1986.

[Rec70] D. L. Rector, Steenrod operations in the Eilenberg–Moore spectral sequence, *Comment. Math. Helv.* **45**:540–552, 1970.

[Smi69] L. Smith, On the construction of the Eilenberg–Moore spectral sequence, *Bull. Amer. Math. Soc.* **75**:873–878, 1969.

Northwestern University,
Evanston, IL 60208, USA
E-mail address: mark@math.nwu.edu

University of Rochester,
Rochester, NY 14627, USA
E-mail address: drav@math.rochester.edu

John Carroll University,
University Hts, OH 44118, USA
E-mail address: shick@jcvaxa.jcu.edu

Progress in Mathematics, Vol. 196

On the classifying space for proper actions

Guido Mislin

Abstract. We discuss conditions under which the universal proper G-CW-complex $\underline{E}G$ can be chosen to be finite dimensional. The methods we use stem from a general construction introduced in [9], involving spaces parameterized by a partially ordered set. In particular we present a construction, which turns a G-CW-complex X in a canonical way into a proper G-CW-complex $\mathrm{Pr}(X)$ of the same homotopy type, with control on the dimension of the new space.

1. Introduction

Let G be an arbitrary discrete group. There exists up to G-homotopy a unique G-CW-complex $\underline{E}G$ such that the fixed point space $\underline{E}G^H$ is contractible for every finite $H < G$, and empty for infinite H. A G-CW-complex is called *proper* if all point stabilizers are finite (equivalently, if all its G-cells are of the form $G/H \times \sigma$ with H finite in G). The space $\underline{E}G$ is an example of a proper G-CW-complex; it is sometimes referred to as the *classifying space for proper actions*, because it enjoys the following universal property:

- For any proper G-CW-complex X there is a unique G-homotopy class of G-maps $X \to \underline{E}G$.

The following is an explicit description of a standard model for $\underline{E}G$. Let $F(G)$ denote the G-poset of finite non-empty subsets of G, the partial order being the obvious one and the G-action given by left translation. It follows that the geometric realization $|F(G)|$ of $F(G)$ is a G-CW-complex of type $\underline{E}G$. Of course $|F(G)|$ is infinite dimensional for G an infinite group. As we are only interested in $\underline{E}G$ up to G-homotopy, the following definition is useful.

Definition 1.1. We write $\dim_G \underline{E}G$ for the smallest dimension in $\mathbb{N} \cup \{\infty\}$ of a model for $\underline{E}G$.

Clearly $\dim_G \underline{E}G = 0$ if and only if G is finite. In case G is torsion-free, $\underline{E}G = \tilde{K}(G,1)$, the universal cover of a $K(G,1)$. It follows that for a torsion-free group, $\dim_G \underline{E}G < \infty$ is equivalent to $\operatorname{cd} G < \infty$. Since for a general G the space $\underline{E}G$ may be considered as an $\underline{E}H$ for $H < G$, the condition $\dim_G \underline{E}G < \infty$ implies that all torsion-free subgroups of G have finite cohomological dimension, universally bounded by $\dim_G \underline{E}G$, because the cellular chain complex of $\underline{E}G$ yields a free $\mathbb{Z}[H]$-resolution of $\mathbb{Z}$ for each torsion-free subgroup H of G.

A well-known theorem states that a group of finite vcd (virtual cohomological dimension) admits a finite dimensional $\underline{E}G$. The precise relationship between $\operatorname{vcd} G$ and $\dim_G \underline{E}G$ is unknown. The following is a standard conjecture (cf. K. S. Brown [4]).

Conjecture 1.2. *If* $2 < \operatorname{vcd} G < \infty$ *then* $\operatorname{vcd} G = \dim_G \underline{E}G$.

The conjecture holds in the torsion-free case, as is well-known. We excluded the case of $\operatorname{vcd} G = 2$ because there is an example of a group G with $\operatorname{vcd} G = 2$ and $\dim_G \underline{E}G = 3$ (cf. [2]). It is obvious that for a group of finite vcd one has $\operatorname{vcd} G \leq \dim_G \underline{E}G$, since for a torsion-free subgroup $H < G$ of finite index, $\operatorname{cd} H = \operatorname{vcd} G$, and $\dim_G \underline{E}G$ is an upper bound for $\operatorname{cd} H$ as we remarked earlier.

A case for which the minimal dimension of $\underline{E}G$ is well understood is when $\dim_G \underline{E}G = 1$, that is an infinite group acting properly on a tree. Indeed the following theorem holds (cf. [7]).

Theorem 1.3. *For an arbitrary group G the following two conditions are equivalent:*

- $\dim_G \underline{E}G = 1$;
- $\operatorname{cd}_{\mathbb{Q}} G = 1$.

It is always true that $\operatorname{cd}_{\mathbb{Q}} G \leq \dim_G \underline{E}G$ because, upon tensoring with $\mathbb{Q}$, the cellular chain complex of $\underline{E}G$ yields a $\mathbb{Q}[G]$-projective resolution of $\mathbb{Q}$ of length $\dim_G \underline{E}G$. However, the inequality can in general not be replaced by an equality, because according to Bestvina and Mess there exist a torsion-free negatively curved group G with $\operatorname{cd}_{\mathbb{Q}} G < \operatorname{cd} G$ (cf. [1]).

A basic unsolved problem, which served as the main motivation for this note, is the following one.

Problem 1.4. *Let* $K \to G \to Q$ *be a short exact sequence of groups. If* $\dim_Q \underline{E}Q$ *and* $\dim_K \underline{E}K$ *are both finite, is* $\dim_G \underline{E}G$ *finite too?*

In the sequel we will give a partial positive answer to this question. Some of the results presented here have also been obtained by Lück [10] with other techniques.

We thank Jonathan Cornick for his useful comments and the referee for his suggestions.

2. Spaces parameterized by a poset

We briefly recall a construction introduced in [9]. Let Λ be a G-poset (i.e., a partially ordered set on which the group G acts in an order preserving way). Let X be a G-CW-complex and $f\colon X \to \Lambda$ a (continuous) G-map, with Λ considered as a discrete topological space. Write $X(\lambda) \subset X$ for the preimage of $\lambda \in \Lambda$, and for

$$\underline{\lambda} = \lambda_0 < \lambda_1 < \ldots < \lambda_n$$

a chain of length n in Λ, put

$$X(\underline{\lambda}) = \prod_{i=0}^{n} X(\lambda_i).$$

The *geometric realization* $|f|$ of f is a G-CW-complex of the form

$$|f| = \left(\coprod_{n\in\mathbb{N}} \left(\coprod_{\underline{\lambda}\in\Lambda^{(n)}} \sigma^n \times X(\underline{\lambda}) \right) \right) / \sim$$

with $\Lambda^{(n)}$ the set of chains of length n in Λ and σ^n the standard n-simplex (for details see [9]). The assignment $f \mapsto |f|$ defines a functor from the category of G-CW-complexes over Λ to the category of G-CW-complexes. It maps the terminal object $\mathrm{Id}_\Lambda \colon \Lambda \to \Lambda$ to the usual geometric realization, $|\mathrm{Id}_\Lambda| = |\Lambda|$, of the poset Λ. Also, the morphism $f \to \mathrm{Id}_\Lambda$ gives rise to a canonical G-map $|f| \to |\Lambda|$, which is a homotopy equivalence in case all spaces $X(\lambda)$ are contractible. In [9] the poset $\Lambda(G)$ of non-trivial finite subgroups of G was used to construct a model for $\underline{E}G$. This poset is closely related to the *singular part* $(\underline{E}G)^{\mathrm{sing}}$ of $\underline{E}G$, consisting of all points with non-trivial isotropy. This can be seen in the framework of our *spaces parameterized by posets* as follows. If one takes X to be the disjoint union of G-CW-complexes of the form $G \times_N \underline{E}G^H$ with N the normalizer of $H < G$, the union being taken over a set of orbit representatives of points H in $\Lambda(G)$, then there is an obvious G-map

$$f \colon X = \coprod G \times_N \underline{E}G^H \longrightarrow \Lambda(G)$$

for which each point-preimage is contractible. Thus the natural map

$$|f| \to |\Lambda(G)|$$

is a G-map and a homotopy equivalence. We claim that $|f|$ is G-homotopy equivalent to $(\underline{E}G)^{\mathrm{sing}}$ so that the following holds.

Lemma 2.1. *For an arbitrary group G the space $(\underline{E}G)^{\mathrm{sing}}$ is homotopy equivalent to $|\Lambda(G)|$ and G-homotopy equivalent to $\underline{E}G \times |\Lambda(G)|$, with diagonal G-action.*

Proof. We use the well-known fact that a G-map between G-CW-complexes is a G-homotopy equivalence if it induces for every $H < G$ an ordinary homotopy equivalence between H-fixed point spaces. Consider the map $\alpha \colon |f| \to |\Lambda(G)|$ constructed above. It is a homotopy equivalence on the fixed point spaces for any finite subgroup $H < G$ (see [9, Section 9]). Because the G-action on $|f|$ is proper, there is a canonical classifying G-map $\beta \colon |f| \to \underline{E}G$, and it follows that

$$\{\alpha, \beta\} \colon |f| \to |\Lambda(G)| \times \underline{E}G$$

is a G-homotopy equivalence. Moreover the G-map $|f| \to \underline{E}G$ maps into $(\underline{E}G)^{\mathrm{sing}}$ by a G-homotopy equivalence

$$|f| \xrightarrow{\simeq} (\underline{E}G)^{\mathrm{sing}},$$

because for each $H < G$ the map $|f|^H \to \underline{E}G^H$ is a homotopy equivalence (the spaces in question are both either contractible or empty). $\square$

Remark 2.2. A different proof of Lemma 2.1, in the special case when there is a bound on the orders of the finite subgroups of G, can be found in [5, Lemma 2.4].

Note also that Lemma 2.1 has the following obvious consequences, which are useful in studying $\underline{E}G$:

- $H_i(|\Lambda(G)|;\mathbb{Z}) = 0$ for $i > \dim_G \underline{E}G$;
- $H^*_G(|\Lambda(G)|; M) \cong H^*_G((\underline{E}G)^{\mathrm{sing}}; M)$ for any G-module M. (The *Borel cohomology* $H^*_G(X; M)$ of a G-space X with coefficients in the G-module M is the cohomology of the cochain complex $\mathrm{Hom}_G(C_*(\tilde{K}(G,1) \times X), M)$.)

In [9] the poset $\Lambda(G)$ of non-trivial finite subgroups of G was used as the starting point for the construction of $\underline{E}G$. One of the problems encountered in using $\Lambda(G)$ was caused by the fact that $|\Lambda(G)|$ is in general not contractible. In the next section we will show how to construct $\underline{E}G$ out of suitable *contractible* posets.

3. Turning G-actions into proper G-actions

A G-CW-complex is called *simplicial* if it is a simplicial complex with G acting simplicially, such that if $g \in G$ maps a simplex to itself, it fixes it pointwise. The standard model for $\underline{E}G$ described in Section 1 is an example of a simplicial G-CW-complex.

It is an elementary fact that a general G-CW-complex is G-homotopy equivalent to a simplicial one of the same dimension. For technical reasons, it is often more convenient to work with *simplicial* G-CW-complexes rather than with general G-CW-complexes. The following example shall illustrate our point. Let X be a G-CW-complex of dimension d and suppose G is a subgroup of finite index in a larger group L. The L-space $\mathrm{map}_G(L, X)$ has a standard CW-structure with cellular L-action (see the discussion of *Serre's Theorem* in Brown's book [3]). Although this space $\mathrm{map}_G(L, X)$ is homeomorphic in the compactly generated topology to a product of $[L : G]$ copies of X, it is *not* an L-CW-complex. However, it is L-homotopy equivalent to an L-CW-complex of dimension $d \cdot [L : G]$. To prove this, we replace X by a simplicial G-CW-complex Y of the same dimension and same G-homotopy type, and observe that it is easy to define a simplicial structure on $\mathrm{map}_G(L, Y)$ such that it is a simplicial L-CW-complex, of the L-homotopy type of $\mathrm{map}_G(L, X)$. One also checks that if Y is an $\underline{E}G$, then $\mathrm{map}_G(L, Y)$ is an $\underline{E}L$.

Definition 3.1. Let X be a simplicial G-CW-complex. We write $\mathrm{Po}(X)$ for the associated G-poset, whose elements are the simplices of X and whose partial order is given by the inclusion relation between simplices.

The geometric realization $|\mathrm{Po}(X)|$ of $\mathrm{Po}(X)$ is a simplicial G-CW-complex G-homeomorphic to X, with simplicial structure corresponding to the barycentric subdivision of X. We will use the poset $\mathrm{Po}(X)$ to turn X into a proper G-CW-complex, by replacing each G-orbit of a simplex $\sigma \subset X$ by the proper G-space $G \times_{G(\sigma)} \underline{E}G(\sigma)$, with $G(\sigma)$ the stabilizer of σ. The precise construction is as follows.

Definition 3.2. Let X be a simplicial G-CW-complex and write $G(\sigma)$ for the stabilizer of a cell σ. Then $\mathrm{Pr}(X)$ denotes the proper G-CW-complex obtained as the geometric realization $|f|$ of the following map $f\colon Y \to \mathrm{Po}(X)$. Choose a model $\underline{E}G(\sigma)$ for each $\sigma \in \Sigma_0$, where Σ_0 denotes a set of representatives of the G-orbits of simplices in X, and put Y to be the disjoint union

$$\coprod_{\sigma \in \Sigma_0} G \times_{G(\sigma)} \underline{E}G(\sigma).$$

The map $f\colon Y \to \mathrm{Po}(X)$ is now given by $f(g \times x) = g \cdot \sigma$, where x lies in $\underline{E}G(\sigma)$.

Corollary 3.3. *Let X be a simplicial G-CW-complex. Then*

- *the natural G-map $\mathrm{Pr}(X) \to X$ induces a homotopy equivalence of fixed point spaces $\mathrm{Pr}(X)^H \to X^H$ for any finite subgroup $H < G$;*
- $\mathrm{Pr}(X) = \underline{E}G$ *if X^H is contractible for each finite subgroup $H < G$.*

Proof. The first assertion follows from [9, Lemma 8.7], using the fact that $|\mathrm{Po}(X)^H|$ is homeomorphic to X^H. The second one follows then from the first, since $\mathrm{Pr}(X)$ is a proper G-CW-complex. □

As a result, we obtain the following general theorem.

Theorem 3.4. *Let X be a finite dimensional G-CW-complex such that for each finite $H < G$ the fixed point space X^H is contractible. Suppose there is a universal bound $b \in \mathbb{N}$ on $\dim_{G(x)} \underline{E}G(x)$, where x runs over the vertices of X. Then*

$$\dim_G \underline{E}G < \infty.$$

Proof. We may assume that X is a simplicial G-CW-complex. The space $\mathrm{Pr}(X)$ is then a G-CW-complex of the G-homotopy type of $\underline{E}G$ and of dimension bounded by $(b+1)(\dim X + 1) - 1$. □

4. Applications

We now turn to Problem 1.4 concerning group extensions. Recall that $\mathbf{H}_1\mathcal{F}$ stands for the class of groups which admit a finite dimensional contractible G-CW-complex with finite cell stabilizers. For example if G admits a finite dimensional $\underline{E}G$, it certainly belongs to $\mathbf{H}_1\mathcal{F}$; the converse is an open question! The class of groups $\mathbf{H}\mathcal{F}$ of *hierarchically decomposable* groups is defined to be the smallest class containing $\mathbf{H}_1\mathcal{F}$ such that a group G belongs to $\mathbf{H}\mathcal{F}$ if G admits a finite dimensional contractible G-CW-complex with cell stabilizers in $\mathbf{H}\mathcal{F}$ (for a general account of hierarchically decomposable groups the reader is referred to [8]).

Theorem 4.1. *Let $K \to G \to Q$ be a group extension with $\pi\colon G \to Q$ the projection. Then the following holds:*

- *if $\underline{E}Q$ is finite dimensional and if there is a universal bound on the dimension of $\underline{E}\pi^{-1}\pi(H)$ where H ranges over all finite subgroups of G, then G admits a finite dimensional $\underline{E}G$;*
- *if $Q \in \mathbf{H}_1\mathcal{F}$ and if for each finite subgroup $H < G$ there is a finite dimensional contractible $\pi^{-1}\pi(H)$-CW-complex of dimension bounded by a number independent of H, then $G \in \mathbf{H}_1\mathcal{F}$.*

Proof. Take $X = \underline{E}Q$ and consider it as a G-space via $\pi\colon G \to Q$. Then for each finite $H < G$ the space $X^H = X^{\pi(H)}$ is contractible and the result follows from 3.4. The second case is proved similarly, using for X a finite dimensional contractible Q-CW-complex instead of $\underline{E}Q$. □

Corollary 4.2. *Let $K \to G \to Q$ be a short exact sequence of groups and assume that $Q \in \mathbf{H}_1\mathcal{F}$ with Q admitting a bound on the order of its torsion subgroups. Then the following holds:*

- *if $K \in \mathbf{H}_1\mathcal{F}$, then $G \in \mathbf{H}_1\mathcal{F}$;*
- *if $\dim_K \underline{E}K < \infty$ then $\dim_G \underline{E}G < \infty$.*

Proof. Take a finite subgroup $H < G$ and write $\pi\colon G \to Q$ for the projection. Put $L = \pi^{-1}\pi(H)$ and consider the short exact sequence

$$K \to L \to \pi(H).$$

If $K \in \mathbf{H}_1\mathcal{F}$ then so is L, with associated contractible L-CW-complex of the form $\mathrm{map}_K(L, Y)$, where Y is some finite dimensional contractible K-CW-complex. It follows that $\mathrm{map}_K(L, Y)$ is of universally bounded dimension, because there is a bound on the index $[L : K]$ which is independent of H. Thus $G \in \mathbf{H}_1\mathcal{F}$ by 4.1. The proof of the second case is analogous; one uses that $Q \in \mathbf{H}_1\mathcal{F}$ together with the bound on the order of the torsion subgroups of Q implies that $\dim_Q \underline{E}Q < \infty$ (Corollary B of [9]). □

Using a recent result of W. Dicks and P. Kropholler [6] one obtains examples of large torsion groups in $\mathbf{H}_1\mathcal{F}$.

Lemma 4.3. (Dicks–Kropholler) *Let G be a locally finite group of cardinality less than $\aleph_\omega$. Then G admits a finite dimensional $\underline{E}G$.*

Combining 4.2 with 4.3 leads us to

Theorem 4.4. *Let $K \to G \to Q$ be a short exact sequence of groups. Assume that there is a bound on the order of the torsion subgroups of Q and assume that K is a locally finite group of cardinality $< \aleph_\omega$. Then the following holds:*

- *if $Q \in \mathbf{H}_1\mathcal{F}$, then $G \in \mathbf{H}_1\mathcal{F}$;*
- *if $\dim_Q \underline{E}Q < \infty$ then $\dim_G \underline{E}G < \infty$.*

As an example, the theorem can be applied to soluble groups as follows.

Corollary 4.5. *Let G be a soluble group of cardinality $< \aleph_\omega$. Then the following conditions on G are equivalent:*

(1) *G has finite torsion-free rank (Hirsch number);*
(2) $\dim_G \underline{E}G < \infty$;
(3) $\mathrm{cd}_{\mathbb{Q}} G < \infty$.

Proof. It follows from the well-known structure for soluble groups of finite torsion-free rank that there is a short exact sequence $K \to G \to Q$ with K locally finite and Q of finite vcd. Thus $\dim_Q \underline{E}Q < \infty$ and the previous theorem shows that (1)$\Rightarrow$(2). Clearly (2)$\Rightarrow$(3), since one always has $\mathrm{cd}_{\mathbb{Q}} G \leq \dim_G \underline{E}G$. Finally, by a theorem of Stammbach [11] the torsion-free rank of a soluble group equals its (weak) homological dimension over $\mathbb{Q}$ and is therefore bounded by $\mathrm{cd}_{\mathbb{Q}} G$, whence (3)$\Rightarrow$(1). □

References

[1] M. Bestvina and G. Mess, *The boundary of negatively curved groups*, J. Amer. Math. Soc. **4**, No. 3 (1991), 469–481.

[2] N. Brady, I. J. Leary and B. E. A. Nucinkis, *On algebraic and geometric dimensions for groups with torsion*, preprint.

[3] K. S. Brown, *Cohomology of Groups*, Graduate Texts in Math. vol. **87**, Springer, 1982.

[4] K. S. Brown, *Groups of virtually finite dimension*, in: *Homological Group Theory*, edited by C. T. C. Wall, London Math. Soc. Lecture Note Ser. vol. **36**, Cambridge University Press, 1979, 27–70.

[5] F. Connolly and T. Kozniewski, *Finiteness properties of the classifying spaces of proper Γ-actions*, J. Pure Appl. Algebra **41** (1986), 17–36.

[6] W. Dicks and P. H. Kropholler, in preparation.

[7] M. J. Dunwoody, *Accessibility and groups of cohomological dimension one*, Proc. London Math. Soc. **38** (1997), 193–215.

[8] P. H. Kropholler, *Hierarchical decompositions, generalized Tate cohomology, and groups of type FP_∞*, in: *(A. Duncan, N. Gilbert, and J. Howie, eds.) Proceedings of the Edinburgh Conference on Geometric Group Theory, 1993*, London Math. Soc. Lecture Note Ser. vol. **204**, Cambridge University Press, 1995, 190–216.

[9] P. H. Kropholler and G. Mislin, *Groups acting on finite dimensional spaces with finite stabilizers*, Comment. Math. Helv. **73** (1998), 122–136.

[10] W. Lück, *The type of the classifying space for a family of subgroups*, J. Pure Appl. Algebra **149** (2000), 177–203.

[11] U. Stammbach, *On the weak (homological) dimension of the group algebra of solvable groups*, J. London Math. Soc. (2) **2** (1970), 567–570.

Guido Mislin
Department of Mathematics, ETH
CH-8092 Zürich, Switzerland
E-mail address: `mislin@math.ethz.ch`

Progress in Mathematics, Vol. 196

Toric morphisms between p-compact groups

Jesper M. Møller

Abstract. It is well-known that any morphism between two p-compact groups will lift, non-uniquely, to an admissible morphism between the maximal tori. We identify here a class of p-compact group morphisms, the p-toric morphisms, which can be perceived as generalized rational isomorphisms, enjoying the stronger property of lifting uniquely to a morphism between the maximal torus normalizers. We investigate the class of p-toric morphisms and apply our observations to determine the mapping spaces $\mathrm{map}(\mathrm{BSU}(3), \mathrm{BF}_4)$, $\mathrm{map}(\mathrm{BG}_2, \mathrm{BF}_4)$, and $\mathrm{map}(\mathrm{BSU}(3), \mathrm{BG}_2)$ where the classifying spaces have been completed at the prime $p = 3$.

1. Introduction

The classification up to homotopy of maps between classifying spaces of compact Lie groups is a traditional project of algebraic topology [18, 26]. One line of development started with the investigations 25 years ago by Hubbuck [15, 16] and Adams-Mahmud [1]. They noted the close relationship between maps between classifying spaces and admissible homomorphisms between maximal tori. The regular admissible homomorphisms, in particular, turned out to have especially nice properties. It is the purpose of this paper to study regular admissible morphisms, here called toric admissible morphisms, in light of the more recent theory by Dwyer-Wilkerson [9] of p-compact groups. As case studies, we classify homotopy homomorphisms $\mathrm{SU}(3) \to \mathrm{F}_4$, $\mathrm{G}_2 \to \mathrm{F}_4$, and $\mathrm{SU}(3) \to \mathrm{G}_2$ at the prime $p = 3$.

In order to describe the content in more detail, let X_1 and X_2 be p-compact groups, for the sake of this introduction assumed to be connected, with maximal tori $T(X_1) \to X_1$ and $T(X_2) \to X_2$, respectively. For any morphism $f\colon X_1 \to X_2$ there is a lift $T(f)\colon T(X_1) \to T(X_2)$, unique up the action of the Weyl group of X_2, such that the diagram

$$\begin{array}{ccc} T(X_1) & \xrightarrow{T(f)} & T(X_2) \\ {\scriptstyle i_1}\downarrow & & \downarrow{\scriptstyle i_2} \\ X_1 & \xrightarrow[f]{} & X_2 \end{array}$$

commutes up to conjugacy. As a consequence of uniqueness, the morphism $T(f)$ is *admissible* in the sense that for any element w_1 of the Weyl group of X_1 there exists and element w_2 of the Weyl group of X_2 such that $T(f)w_1 = w_2T(f)$. In general, w_2 is not uniquely determined by w_1, but if it is, we say that f is p-toric (2.1). (As we shall see (2.4), f is p-toric, if and only if the centralizer $C_{X_2}(fi_1T(X_1))$ of the maximal torus of X_1 in X_2 is a maximal torus of X_2. This explains the name.) In that case, the correspondence $w_1 \to w_2$ is a homomorphism of Weyl groups and, by Theorem 3.5, there is a *unique* lift $N(f)\colon N(X_1) \to N(X_2)$ to a map between the maximal torus normalizers such that the diagram

$$\begin{array}{ccc} N(X_1) & \xrightarrow{N(f)} & N(X_2) \\ \downarrow & & \downarrow \\ X_1 & \xrightarrow[f]{} & X_2 \end{array}$$

commutes up to conjugacy, i.e. a p-toric morphism lifts uniquely to a morphism between the maximal torus normalizers.

In many concrete cases the generic morphism is p-toric. As a first example, we consider the case where the domain $X_1 = \mathrm{SU}(3)$, the codomain $X_2 = \mathrm{F}_4$, and the prime $p = 3$. The compact Lie group F_4 contains a unique copy of $\mathrm{SU}(3,3) = \mathrm{SU}(3) \times_{Z(\mathrm{SU}(3))} \mathrm{SU}(3)$ as a subgroup of maximal rank (4.10). Any morphism $\mathrm{SU}(3) \to \mathrm{SU}(3,3)$ is of the form

$$\psi^{(u,v)}\colon \mathrm{SU}(3) \xrightarrow{\Delta} \mathrm{SU}(3) \times \mathrm{SU}(3) \xrightarrow{\psi^u \times \psi^v} \mathrm{SU}(3) \times \mathrm{SU}(3) \to \mathrm{SU}(3,3)$$

where u and v are 3-adic units or zero (2.17). Composing with the inclusion $e\colon \mathrm{SU}(3,3) \to \mathrm{F}_4$ we obtain the morphism $e\psi^{(u,v)}\colon \mathrm{SU}(3) \to \mathrm{F}_4$. Observe that we have $e\psi^{(u,v)} = e\psi^{(-u,-v)}$ since the inclusion e is invariant under the action of the Weyl group $W_{\mathrm{F}_4}(\mathrm{SU}(3,3))$ [11, 4.3] [24, 8.4] which is of order two generated by the self-map $\psi^{-1} \times_{Z(\mathrm{SU}(3))} \psi^{-1}$ of $\mathrm{SU}(3,3)$ (4.15). These maps $e\psi^{(u,v)}$, $u, v \in \mathbf{Z}_3^* \cup \{0\}$, with the relation $e\psi^{(u,v)} = e\psi^{(-u,-v)}$, turn out to describe $\mathrm{Rep}(\mathrm{SU}(3), \mathrm{F}_4) = [B\mathrm{SU}(3)^\wedge_3, (B\mathrm{F}_4)^\wedge_3]$ completely.

Theorem 1.1. *The map*

$$e \circ -\colon W_{\mathrm{F}_4}(\mathrm{SU}(3,3)) \backslash \mathrm{Rep}(\mathrm{SU}(3), \mathrm{SU}(3,3)) \to \mathrm{Rep}(\mathrm{SU}(3), \mathrm{F}_4)$$

is a bijection when $p = 3$.

See (4.16, 5.7, 6.7) for information about the centralizers [9, 3.5] of these maps. The proof of Theorem 1.1 is divided into three cases: Monomorphisms $\mathrm{SU}(3) \to \mathrm{F}_4$ (4.13), p-toric monomorphisms $\mathrm{PU}(3) \to \mathrm{F}_4$ (5.4), and, the technically most demanding case, non-p-toric monomorphisms $\mathrm{PU}(3) \to \mathrm{F}_4$ (6.1).

As a second example, we consider the case where $X_1 = \mathrm{G}_2$ and $X_2 = \mathrm{F}_4$ and $p = 3$ and reprove a result from Jackowski-McClure-Oliver [19]. To state the theorem, we recall that the compact Lie group G_2 contains a unique copy of $\mathrm{SU}(3)$

as a subgroup of maximal rank (8.5). Thus we may restrict morphisms defined on G_2 to this subgroup $SU(3) \subset G_2$.

Theorem 1.2. [19, 3.4] *The restriction map*

$$\operatorname{Rep}(G_2, F_4) \to \operatorname{Rep}(SU(3), F_4)$$

is a bijection when $p = 3$.

See (7.2) for information about the centralizers of the homotopy morphisms from G_2 to F_4 at the prime $p = 3$.

When working with this paper, I made use of a MAGMA program written by K. Andersen for computing admissible homomorphisms. I also wish to thank C. McGibbon for a clarifying remark.

2. Toric morphisms

In this section I introduce the concept of a p-toric morphism, relate it to other, more familiar, types of morphisms between p-compact groups, and provide examples of morphisms that are p-toric and others that are not.

Let X_1 and X_2 be p-compact groups (or extended p-compact tori [10, 3.12]) with maximal tori $T_1 = T(X_1) \to X_1$, $T_2 = T(X_2) \to X_2$ and Weyl groups $W_1 = W(X_1)$ and $W_2 = W(X_2)$ [9], respectively. Write $\operatorname{Rep}(X_1, X_2)$ for the set $[BX_1, BX_2]$ of conjugacy classes of loop space morphisms [9, §3].

Definition 2.1. 1. *A loop space morphism* $T_1 \to X_2$ *is* p-toric *(or* regular [1, 2.22], [19, 1.3]*) if its centralizer* $C_{X_2}(T_1)$ *is a* p*-compact toral group* [9, 6.3].

2. *A loop space morphism* $X_1 \to X_2$ *is* p*-toric if its composition with* $T_1 \to X_1$ *is* p*-toric.*

Note that the centralizer $C_{X_2}(T_1)$ in (2.1.1) is known to be a p-compact group [9, §6] [10, 2.5].

We shall now consider some alternative criteria for a morphism to be p-toric. For any loop space morphism $f\colon X_1 \to X_2$ between p-compact groups or extended p-compact tori there exists [9, 8.11] [10, 2.14] a loop space morphism $T(f)\colon T_1 \to T_2$ between the maximal tori such that

$$\begin{array}{ccc} T_1 & \xrightarrow{T(f)} & T_2 \\ {\scriptstyle i_1}\downarrow & & \downarrow{\scriptstyle i_2} \\ X_1 & \xrightarrow[f]{} & X_2 \end{array} \tag{2.2}$$

commutes. Moreover, the conjugacy class of $T(f)$ in $\operatorname{Rep}(T_1, T_2)$ is unique up to the action by the Weyl group W_2 of the target [22, 3.5]. The Adams-Mahmud map

$$\operatorname{Rep}(X_1, X_2) \to W_2 \backslash \operatorname{Rep}(T_1, T_2),$$

taking $f \in \operatorname{Rep}(X_1, X_2)$ to the W_2-orbit of $T(f) \in \operatorname{Rep}(T_1, T_2)$, is instrumental in the proofs of 1.1 and 1.2. Note that, by uniqueness of $T(f)$, the image of the Adams-Mahmud map is contained in $W_2 \backslash \operatorname{Adm}(T_1, T_2)$ where

$$\operatorname{Adm}(T_1, T_2) = \{\varphi \in \operatorname{Rep}(T_1, T_2) \mid \varphi W_1 \subseteq W_2 \varphi\} \tag{2.3}$$

is the set of *admissible* homomorphisms. For each element w_1 of the Weyl group W_1 of the domain there are in general several solutions for $w_2 \in W_2$ in the equation $T(f)w_1 = w_2 T(f)$. As we shall shortly see (2.4), the p-toric morphisms are characterized (for connected X_2) as the ones for which w_2 is uniquely determined by w_1.

Let

$$W_2^{T(f)} = \{w_2 \in W_2 \mid w_2 \cdot T(f) = T(f)\}$$

denote the stabilizer subgroup at $T(f)$ for the action of W_2 on $\operatorname{Rep}(T_1, T_2)$. The conjugacy class of this subgroup does not depend of the choice of $T(f)$ but only on f.

In case X_1 and X_2 are extended p-compact tori, there is a short exact sequence of loop spaces

$$T_2 \to C_{X_2}(T(f)T_1) \to W_2^{T(f)}$$

from which we see that

$$f \colon X_1 \to X_2 \text{ is } p\text{-toric} \Leftrightarrow W_2^{T(f)} = \pi_0(C_{X_2}(T_1)) \text{ is a finite } p\text{-group.}$$

In case X_1 and X_2 are p-compact groups, $C_{X_2}(T_1) \to X_2$ is a monomorphism of maximal rank [10, §4], so

$$f \colon X_1 \to X_2 \text{ is } p\text{-toric} \Leftrightarrow C_{X_2}(T_1)_0 \to X_2 \text{ is a maximal torus for } X_2$$

where subscript 0 indicates identity component. If X_2 is assumed to be *connected*, a stronger statement is possible.

Proposition 2.4. *Assume that X_2 is a connected p-compact group. The following are equivalent*

1. *f is p-toric.*
2. *$C_{X_2}(T_1) \to X_2$ is a maximal torus for X_2.*
3. *$W_2^{T(f)}$ is trivial.*

for any p-compact group morphism $f \colon X_1 \to X_2$.

Proof. For general reasons, the centralizer $C_{X_2}(T_1)$ is a connected [21, 3.11] [10, 7.8] p-compact group [10, 2.5] and the evaluation morphism $C_{X_2}(T_1) \to X_2$ a monomorphism of maximal rank [10, 4.3]. Also, any p-compact group with trivial Weyl group is [9, 9.7] [21, 3.7, 3.8] a p-compact torus. These general facts, in combination with [9, 8.11] [21, 3.6], easily imply the proposition. □

Consequently, for any p-toric morphism $f\colon X_1 \to X_2$ with connected target, there is for each element w of the Weyl group of the domain a *unique* element $\chi(f)(w)$ of the Weyl group of the target so that $T(f)w = \chi(f)(w)T(f) \in \mathrm{Rep}(T_1, T_2)$, and $\chi(f)\colon W_1 \to W_2$ is a group homomorphism.

In general, for a possible non-connected target X_2, we consider an enlarged version of diagram (2.2) in the form of the diagram

$$\begin{array}{ccc} T_1 & \xrightarrow{T(f)} & T_2 \\ & & \downarrow{\scriptstyle i_2'} \\ \downarrow{\scriptstyle i_1} & & N_2 \\ & & \downarrow{\scriptstyle j_2} \\ X_1 & \xrightarrow[f]{} & X_2 \end{array} \tag{2.5}$$

where $j_2\colon N_2 \to X_2$ is the normalizer [9, 9.8] of the maximal torus. Using that $C_{N_2}(T_1) \to C_{X_2}(T_1)$ is a maximal torus normalizer [22, 3.4.3], we get

$$f \text{ is } p\text{-toric} \Leftrightarrow T_1 \xrightarrow{i_1} X_1 \xrightarrow{f} X_2 \text{ is } p\text{-toric} \tag{2.6}$$
$$\Leftrightarrow C_{N_2}(T_1) \to C_{X_2}(T_1) \text{ is an isomorphism} \tag{2.7}$$
$$\Rightarrow T_1 \xrightarrow{T(f)} T_2 \xrightarrow{i_2'} N_2 \text{ is } p\text{-toric} \tag{2.8}$$
$$\Leftrightarrow W_2^{T(f)} \text{ is a finite } p\text{-group.} \tag{2.9}$$

When $p > 2$, also the converse of the third implication holds because, for odd p, a p-compact group is a p-compact toral group if and only if its Weyl group is a finite p-group [23, 7.9].

In some cases, see e.g. [22, 5.1] or (3.5) below, it is possible to lift f to a loop space morphism $N(f)$ between the maximal torus normalizers such that

$$\begin{array}{ccc} N_1 & \xrightarrow{N(f)} & N_2 \\ \downarrow{\scriptstyle j_1} & & \downarrow{\scriptstyle j_2} \\ X_1 & \xrightarrow[f]{} & X_2 \end{array} \tag{2.10}$$

commutes up to conjugacy. In this situation

$$f \text{ is } p\text{-toric} \Rightarrow N(f) \text{ is } p\text{-toric} \tag{2.11}$$

and for $p > 2$ also the converse holds. (Use (2.7, 2.8) to see this.)

In the following examples and elsewhere

- $\mathrm{TRep}(X_1, X_2) \subset \mathrm{Rep}(X_1, X_2)$ denotes the set of conjugacy classes of p-toric morphisms
- $\mathrm{Mono}(X_1, X_2) \subset \mathrm{Rep}(X_1, X_2)$ denotes the set of conjugacy classes of monomorphisms

- $\mathrm{TMono}(X_1, X_2) = \mathrm{Mono}(X_1, X_2) \cap \mathrm{TRep}(X_1, X_2)$
- $\varepsilon_{\mathbf{Q}}(X_1, X_2) \subset \mathrm{Rep}(X_1, X_2)$ is the set of rational isomorphisms [22, 2.1]
- $\varepsilon_{\mathbf{Q}}(X_1) = \varepsilon_{\mathbf{Q}}(X_1, X_1)$ is the monoid of rational automorphisms of X_1
- $\mathrm{Out}(X_1)$ is the group of conjugacy classes of automorphisms of X_1 (the invertible elements of the monoid $\mathrm{Rep}(X_1, X_1)$).

Above, a loop space morphism between extended p-compact tori is a monomorphism if its discrete approximation [10, 3.12] is a group monomorphism.

Example 2.12. *If X_1 and X_2 have the same rank* [9, 5.11],

$$\mathrm{Mono}(X_1, X_2) \subset \mathrm{TRep}(X_1, X_2) \supset \varepsilon_{\mathbf{Q}}(X_1, X_2)$$

because any monomorphism [9, 3.2] *(rational isomorphism* [22, 2.1]*) restricts to an isomorphism (epimorphism) between maximal tori* [21, 3.6] [22, 3.6].

If X_1 and X_2 are locally isomorphic, simple p-compact groups [22, 2.7, 5.4]

$$\mathrm{TRep}(X_1, X_2) = \mathrm{Rep}(X_1, X_2) - \{0\} = \varepsilon_{\mathbf{Q}}(X_1, X_2)$$

because f is p-toric or a rational isomorphism if and only if $T(f)$ is non-trivial if and only if f is non-trivial [22, 6.7].

Example 2.13. *For any p-compact group X and any integer $m > 0$,*

$$\mathrm{TRep}(X, X^m) = (\mathrm{TRep}(X, X))^m.$$

If X is simple,

$$\mathrm{TRep}(X, X^m) = (\mathrm{Rep}(X, X) - \{0\})^m = \varepsilon_{\mathbf{Q}}(X)^m \overset{p||W|}{=} \mathrm{Out}(X)^m,$$

where the last identity holds under the assumption that p divides the order of the Weyl group [22, 5.5, 5.6].

Proposition 2.14. *Assume that X_1 is connected and that $z\colon Z_1 \to X_1$ is a central monomorphism* [9, 3.5]. *Then there are bijections*

- $\mathrm{Rep}(X_1/Z_1, X_2) \to \{f \in \mathrm{Rep}(X_1, X_2) \mid f \circ z \text{ is trivial}\}$
- $\mathrm{TRep}(X_1/Z_1, X_2) \to \{f \in \mathrm{TRep}(X_1, X_2) \mid f \circ z \text{ is trivial}\}$

induced by the epimorphism $X_1 \to X_1/Z_1$ [9, 3.2, 8.3]. *In fact, the mapping space* $\mathrm{map}(B(X_1/Z_1), BX_2)$ *is homotopy equivalent to a union of connected components of* $\mathrm{map}(BX_1, BX_2)$.

Proof. The epimorphism of X_1 to X_1/Z_1 induces a homotopy equivalence between $\mathrm{map}(B(X_1/Z_1), BX_2)$ and a collection of components of $\mathrm{map}(BX_1, BX_2)$ [22, 2.10]. This shows the injection of sets of representations, and, when applied with X_1 replaced by T_1, it shows that a morphism $X_1 \to X_2$ is p-toric if and only if its composition with the epimorphism $X_1 \to X_1/Z_1$ is p-toric. □

Proposition 2.15. *Assume that X_1 is simply connected, X_2 is connected, and that $z\colon Z_2 \to X_2$ is a central monomorphism. Then there are bijections*

- $\mathrm{Rep}(X_1, X_2) \to \mathrm{Rep}(X_1, X_2/Z_2)$
- $\mathrm{TRep}(X_1, X_2) \to \mathrm{TRep}(X_1, X_2/Z_2)$

induced by the epimorphism $X_2 \to X_2/Z_2$.

Proof. Obstruction theory (remember that BX_1 is 3-connected [6]) shows that $\mathrm{Rep}(X_1, X_2) = \mathrm{Rep}(X_1, X_2/Z_2)$ and the existence of a short exact sequence of p-compact groups [9, 3.2]

$$K \to C_{X_2}(X_1) \to C_{X_2/Z_2}(X_1)$$

where BK is one component of the homotopy fixed point set $BZ_2^{hX_1}$; in particular K is a p-compact toral group. It follows that $C_{X_2}(X_1)$ is a p-compact toral group if and only if $C_{X_2/Z_2}(X_1)$ is. □

Example 2.16. *For any simply connected, simple p-compact group X and any central monomorphism $Z \to X^m$,*

$$\mathrm{TRep}(X, X^m/Z) = \mathrm{TRep}(X, X^m) = \varepsilon_{\mathbf{Q}}(X)^m \overset{p||W|}{=} \mathrm{Out}(X)^m$$

where the last identity holds if p divides the order of the Weyl group [22, 5.5, 5.6].

Example 2.17. *Let p be an odd prime and let $\mathrm{SU}(p,p)$ denote the quotient of $\mathrm{SU}(p) \times \mathrm{SU}(p)$ be the central subgroup generated by $(\zeta E, \zeta^{-1}E)$ where $\zeta \neq 1$ is a pth root of unity. Then* (2.15)

$$\mathrm{Rep}(\mathrm{SU}(p), \mathrm{SU}(p,p)) = \mathrm{Rep}(\mathrm{SU}(p), \mathrm{SU}(p)) \times \mathrm{Rep}(\mathrm{SU}(p), \mathrm{SU}(p))$$
$$\mathrm{TRep}(\mathrm{SU}(p), \mathrm{SU}(p,p)) = \mathrm{Out}(\mathrm{SU}(p)) \times \mathrm{Out}(\mathrm{SU}(p))$$

where [20, 2.5, 3.5] [24, 4.8] $\mathrm{Rep}(\mathrm{SU}(p), \mathrm{SU}(p)) - \{0\} = \mathrm{Out}(\mathrm{SU}(p)) = \mathbf{Z}_p^*$, *the group of units in the ring of p-adic integers. Relative to this identification*

$$\mathrm{Mono}(\mathrm{SU}(p), \mathrm{SU}(p,p)) = \{(u,v) \in (\mathbf{Z}_p^* \cup \{0\})^2 \mid u + v \in \mathbf{Z}_p^*\} \tag{2.18}$$

for [24, 5.2] *the morphism $\psi^{(u,v)}$ defined as the composition*

$$\mathrm{SU}(p) \xrightarrow{\Delta} \mathrm{SU}(p) \times \mathrm{SU}(p) \xrightarrow{\psi^u \times \psi^v} \mathrm{SU}(p) \times \mathrm{SU}(p) \to \mathrm{SU}(p,p)$$

is a monomorphism if and only if $u + v \in \mathbf{Z}_p^*$. The monoid $\mathrm{Rep}(\mathrm{SU}(p,p), \mathrm{SU}(p,p))$ *is* (2.14, 2.15) *isomorphic to a submonoid of* $\mathrm{Rep}(\mathrm{SU}(p) \times \mathrm{SU}(p), \mathrm{SU}(p) \times \mathrm{SU}(p))$ *and, in particular,*

$$\mathrm{Out}(\mathrm{SU}(p,p)) = \{(u,v) \in \mathbf{Z}_p^* \times \mathbf{Z}_p^* \mid u \equiv v \bmod p\} \rtimes \langle \tau \rangle$$

where τ is the automorphism that swaps the two $\mathrm{SU}(p)$*-factors.*

The set of monomorphisms (2.18) *consists of two orbits, represented by $\psi^{(1,1)}$ and $\psi^{(1,0)}$, under the action of the automorphism group* $\mathrm{Out}(\mathrm{SU}(p,p))$. *It follows that the centralizers of the monomorphisms $\psi^{(u,v)}$ are*

$$C_{\mathrm{SU}(p,p)}(\psi^{(u,v)}\mathrm{SU}(p)) \cong \begin{cases} Z(\mathrm{SU}(p)) & \text{if } u \neq 0 \text{ and } v \neq 0 \\ \mathrm{SU}(p) & \text{if } u = 0 \text{ or } v = 0 \end{cases} \tag{2.19}$$

i.e. that $\psi^{(u,v)}$ is centric [7] *precisely when it is p-toric. (To prove that $\psi^{(1,1)}$ is centric one uses the fact that $Z(\mathrm{SU}(p)) \xrightarrow{\Delta} Z(\mathrm{SU}(p) \times \mathrm{SU}(p)) \to Z(\mathrm{SU}(p,p))$ is an isomorphism of centers.) In the non-toric case, observe that the projection* *morphism* $\mathrm{SU}(p) \times \mathrm{SU}(p) \to \mathrm{SU}(p,p)$ *restricts to* $\psi^{(1,0)}$ *on the first factor and to* $\psi^{(0,1)}$ *on the second factor. This gives a factorization*

$$\mathrm{SU}(p) \to C_{\mathrm{SU}(p,p)}(\psi^{(1,0)}\mathrm{SU}(p)) \to \mathrm{SU}(p,p)$$

of $\psi^{(0,1)}$ *through the centralizer of* $\psi^{(1,0)}$ *where the first map is an isomorphism. We conclude that if* $f\colon \mathrm{SU}(p) \to \mathrm{SU}(p,p)$ *is a non-toric monomorphism, so is the evaluation monomorphism* $\mathrm{SU}(p) = C_{\mathrm{SU}(p,p)}(f\mathrm{SU}(p)) \to \mathrm{SU}(p,p)$. *The Weyl group,* $W_{\mathrm{SU}(p,p)}(\psi^{(u,v)}\mathrm{SU}(p))$, *of any monomorphism* $\psi^{(u,v)}$ *is trivial* [24, 8.5].

Finally, we note that by (2.14),

$$\mathrm{Rep}(\mathrm{PU}(p), \mathrm{SU}(p,p)) = \{(u,v) \in (\mathbf{Z}_p^* \cup \{0\})^2 \mid u+v \in p\mathbf{Z}_p\}$$
$$\mathrm{TRep}(\mathrm{PU}(p), \mathrm{SU}(p,p)) = \{(u,v) \in (\mathbf{Z}_p^*)^2 \mid u+v \in p\mathbf{Z}_p\}$$

so that

$$\mathrm{Rep}(\mathrm{PU}(p), \mathrm{SU}(p,p)) = \{0\} \cup \mathrm{Mono}(\mathrm{PU}(p), \mathrm{SU}(p,p))$$

and

$$\mathrm{Mono}(\mathrm{PU}(p), \mathrm{SU}(p,p)) = \mathrm{TRep}(\mathrm{PU}(p), \mathrm{SU}(p,p)).$$

Lemma 2.20. *Let* $f\colon X \to Y_1$ *be any morphism and* $g\colon Y_1 \to Y_2$ *a monomorphism between* p*-compact groups. Then*

$$g \circ f\colon X \to Y_2 \text{ is } p\text{-toric} \Rightarrow f\colon X \to Y_1 \text{ is } p\text{-toric.}$$

Proof. Let T be a maximal torus of X_1. Since composition with Bg, $C_{Y_1}(fiT) \to C_{Y_2}(gfiT)$, is a monomorphism, $C_{Y_2}(gfiT)$ is a p-compact toral group if $C_{Y_1}(fiT)$ is a p-compact toral group [21, 3.5.(1)]. □

The converse of (2.20) is not true in general; take for instance Y_1 to be the maximal torus of Y_2.

3. Lifting p-toric morphisms

In this section I show that all p-toric morphisms between two p-compact groups lift uniquely to p-toric morphisms between the maximal torus normalizers.

Recall that X_1 and X_2 are p-compact groups or extended p-compact tori and that $j_1\colon N_1 \to X_1$ and $j_2\colon N_2 \to X_2$ are normalizers of the respective maximal tori, $i_1\colon T_1 \to X_1$ and $i_2\colon T_2 \to X_2$.

By the very definition of a p-toric morphism, the maps j_1 and j_2 induce maps

$$\mathrm{TRep}(X_1, X_2) \to \mathrm{TRep}(N_1, X_2) \leftarrow \mathrm{TRep}(N_1, N_2) \tag{3.1}$$

of sets of p-toric representations. Our first objective is to prove that the arrow to the right is a bijection. This will enable us to define a map from $\mathrm{TRep}(X_1, X_2)$ to $\mathrm{TRep}(N_1, N_2)$. Note the favorable input provided by the information [22, 3.2] that

$$\mathrm{TRep}(T_1, X_2) \leftarrow \mathrm{TRep}(T_1, N_2) \tag{3.2}$$

is a bijection and

$$C_{X_2}(T_1) \leftarrow C_{N_2}(T_1) \tag{3.3}$$

an isomorphism for any p-toric morphism $T_1 \to N_2$.

For any set $S \subset \mathrm{Rep}(X_1, X_2)$, write $\mathrm{map}(BX_1, BX_2)_S$ for the space of all maps $BX_1 \to BX_2$ homotopic to a member of S.

Lemma 3.4. *The map, induced by j_2,*

$$\mathrm{map}(BN_1, BX_2)_{\mathrm{TRep}(N_1,X_2)} \leftarrow \mathrm{map}(BN_1, BN_2)_{\mathrm{TRep}(N_1,N_2)}$$

is a homotopy equivalence.

Proof. The map of the lemma is the map on homotopy fixed point spaces

$$\mathrm{map}(BN_1, BY_2)_{\mathrm{TRep}(N_1,Y_2)} = \left(\mathrm{map}(BT_1, BY_2)_{\mathrm{TRep}(T_1,Y_2)}\right)^{hW_1}, \quad Y_2 = N_2, X_2,$$

induced by the map

$$\mathrm{map}(BT_1, BX_2)_{\mathrm{TRep}(T_1,X_2)} \leftarrow \mathrm{map}(BT_1, BN_2)_{\mathrm{TRep}(T_1,N_2)}$$

which is known to be a homotopy equivalence (3.2, 3.3). □

This lemma immediately leads to the main result of this section.

Theorem 3.5. *(Cf. [1, 2.22]) Let X_1 and X_2 be p-compact groups and $f\colon X_1 \to X_2$ a p-toric morphism. Then there exists a morphism $N(f)\colon N_1 \to N_2$ between extended p-compact tori such that*

$$\begin{array}{ccc} N_1 & \xrightarrow{N(f)} & N_2 \\ {\scriptstyle j_1}\downarrow & & \downarrow{\scriptstyle j_2} \\ X_1 & \xrightarrow[f]{} & X_2 \end{array}$$

commutes up to conjugacy. Moreover,

- *$N(f)$ is unique up to conjugacy*
- *$N(f)$ is p-toric*
- *$C_{X_2}(fj_1N_1) \leftarrow C_{N_2}(N(f)N_1)$ is an isomorphism of loop spaces*

Proof. The map

$$N\colon \mathrm{TRep}(X_1, X_2) \to \mathrm{TRep}(N_1, N_2) \tag{3.6}$$

is defined as the composition of the map $\mathrm{TRep}(X_1, X_2) \to \mathrm{TRep}(X_1, N_2)$ with the inverse of the bijection $\mathrm{TRep}(N_1, X_2) \leftarrow \mathrm{TRep}(N_1, N_2)$ from (3.1). That $N(f)$ is p-toric is (2.11) and the isomorphism of centralizers is (3.4). □

Example 3.7. *If X is simple and $N \to X$ the normalizer of the maximal torus, the map $\mathrm{TRep}(X, X^m) \to \mathrm{TRep}(N, N^m)$ is injective if $\varepsilon_{\mathbf{Q}}(X) \to \mathrm{Rep}(N, N)$ is injective (2.13); e.g. if $X = \mathrm{PU}(p)$, $X = G_2$ and $p = 3$, or $X = \mathrm{DI}_2$ and $p = 3$* [24].

The above theorem is intended as a tool to facilitate the computation of $\mathrm{TRep}(X_1, X_2)$ in concrete cases. We now address injectivity of (3.6).

Remark 3.8. *According to the homology decomposition theorem of Jackowski-McClure [17] and Dwyer-Wilkerson [8], the exists an $\mathbf{F}_p$-equivalence*

$$\operatorname{hocolim}_{\mathbf{A}^{\mathrm{op}}} BC_{X_1}(\nu) \to BX_1$$

where the homotopy colimit is taken over some full subcategory $\mathbf{A}$ of the Quillen category $\mathbf{A}(X_1)$. Let us assume that

- *Any object $\nu : V \to X_1$ of $\mathbf{A}$ admits a factorization $\mu \colon V \to T_1$ through the maximal torus and*
- *$N\colon \mathrm{TRep}(C_{X_1}(\nu), X_2) \to \mathrm{TRep}(C_{N_1}(\mu), N_2)$ is injective for all objects $\nu : V \to X_1$ of $\mathbf{A}$*

and let now f and f' be two p-toric morphisms with $N(f) = \varphi = N(f')$ for some $\varphi \in \mathrm{TRep}(C_{N_1}(\mu), N_2)$. Then the two possible compositions

$$C_{X_1}(\nu) \xrightarrow{e(\nu)} X_1 \overset{f}{\underset{f'}{\rightrightarrows}} X_2$$

are again p-toric morphisms for $C_{X_2}(fe(\nu)C_{T_1}(\mu)) = C_{X_2}(fi_1T_1)$ is a p-compact torus and similarly for the other morphism f'. Since also,

$$N(f \circ e(\nu)) = \varphi \circ e(\mu) = N(f' \circ e(\nu))$$

we have $f \circ e(\nu) \simeq f' \circ e(\nu)$ for all objects ν of $\mathbf{A}$ by hypothesis. (Here, $e(\nu) : C_X(\nu) \to X$ stands for the evaluation monomorphism.) The obstructions to constructing a homotopy between Bf and Bf' lie in

$$\lim\nolimits^i_{\mathbf{A}} \pi_i(\mathrm{map}(BC_{X_1}(\nu), BX_2)_{B(f\circ e(\nu))}), \quad i \geq 1$$

which is an abelian group for $i > 1$ but just a set if $i = 1$ and the fundamental groups are non-abelian.

It is possible that (3.8) can be generalized to a more general situation using the preferred lifts of [25].

While (3.8) applies to the case where X_1 is center-free, the following lemma can be helpful if X_1 has a non-trivial center [10] [21].

Consider the following situation

$$\begin{array}{ccccc} Z & & & & \\ {\scriptstyle z}\downarrow & \overset{z_1}{\searrow} & & \overset{z_2}{\searrow} & \\ X & & Y_1 & \xrightarrow[g]{} & Y_2 \end{array}$$

of p-compact groups and loop space morphisms. Let $\mathrm{Rep}(X, Y_1)_{z \to z_1} = \{f \in \mathrm{Rep}(X, Y_1) \mid f \circ z = z_1\}$ denote the set of conjugacy classes of morphisms under Z and $\mathrm{map}(BX, BY)_{z \to z_1}$ the corresponding mapping space.

Lemma 3.9. *(Cf. [9, 8.4].) Assume that* $z\colon Z \to X$ *is a central monomorphism into the connected* p*-compact group* X *and that composition with* Bg *is an isomorphism* $\underline{g}\colon C_{Y_1}(z_1Z) \to C_{Y_2}(z_2Z)$ *of centralizers. Then composition with* Bg*,*

$$Bg \circ -\colon \operatorname{map}(BX, BY_1)_{z\to z_1} \to \operatorname{map}(BX, BY_2)_{z\to z_2}$$

is a homotopy equivalence.

Proof. The fibration [9, 8.3] [21, 4.1] $BZ \to BX \to B(X/Z)$ allows us to view $BX = BZ_{h(X/Z)}$ as a homotopy orbit space [9, 9.10] and

$$\operatorname{map}(BX, BY_i) = \operatorname{map}(BZ_{h(X/Z)}, BY_1) = \operatorname{map}(BZ, BY_i)^{h(X/Z)}, \quad i = 1, 2,$$

as homotopy fixed point spaces. Composition with $Bg\colon BY_1 \to BY_2$,

$$\begin{aligned}\operatorname{map}(BX, BY_1)_{z\to z_1} = \operatorname{map}(BZ, BY_1)_{Bz_1}^{h(X/Z)} &\to \operatorname{map}(BZ, BY_2)_{Bz_2}^{h(X/Z)} \\ &= \operatorname{map}(BX, BY_2)_{z\to z_2}\end{aligned}$$

is a homotopy equivalence because [9, 10.2] it is induced by the map

$$\operatorname{map}(BZ, BY_1)_{Bz_1} = BC_{Y_1}(z_1) \to BC_{Y_2}(z_2) = \operatorname{map}(BZ, BY_2)_{Bz_2}$$

which by assumption is a homotopy equivalence. □

Here is a typical application of (3.9). In the diagram

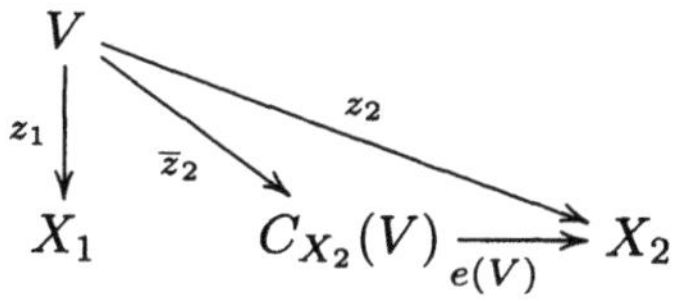

V is an elementary abelian p-group, z_1 a central monomorphism, z_2 a monomorphism , and $\overline{z}_2$ the canonical factorization of z_2 through its centralizer [9, 8.2]. Since the evaluation monomorphism $e(V)\colon C_{X_2}(V) \to X_2$ clearly [9, 8.2] satisfies the hypothesis of (3.9) we see that

$$\operatorname{map}(BX_1, BC_{X_2}(V))_{z_1\to\overline{z}_2} \to \operatorname{map}(BX_1, BX_2)_{z_1\to z_2} \tag{3.10}$$

is a homotopy equivalence.

Definition 3.11. *Let* R *be a subset of* $\operatorname{Rep}(X_1, X_2)$*. We say that* R *is* T*-determined if the implication*

$$f|T(X_1) = g|T(X_1) \Rightarrow f = g$$

holds for all $f \in R$ *and all* $g \in \operatorname{Rep}(X_1, X_2)$*.*

Example 3.12. *If the order of* $W(X_1)$ *is prime to* p*, then*

$$\operatorname{Rep}(X_1, X_2) = W(X_2)\backslash \operatorname{Adm}(T(X_1), T(X_2)) \tag{3.13}$$

where $\operatorname{Adm}(T(X_1), T(X_2))$ *consists of the admissible homomorphisms (2.3). Thus* $\operatorname{Rep}(X_1, X_2)$ *is* T*-determined in this case. The bijection (3.13) follows by exploiting the* $H^*\mathbf{F}_p$*-equivalence* $BN(X_1) \to BX_1$ *[23, 3.12].*

Remark 3.14. *Let $S_1 \to G_1 \to \pi_0(G_1)$ and $S_2 \to G_2 \to \pi_0(G_2)$ be two extensions of finite groups, $\pi_0(G_1)$ and $\pi_0(G_2)$, by p-compact tori, S_1 and S_2. Let* $\mathrm{Hom}(G_1, G_2) = [BG_1, *; BG_2]$ *denote the set of* based *and*

$$\mathrm{Rep}(G_1, G_2) = [BG_1, BG_2] = \pi_0(G_2)\backslash \mathrm{Hom}(G_1, G_2)$$

the set of free *homotopy classes of maps of BG_1 into BG_2.*

The two functors π_1 and π_2 define a map

$$\mathrm{Hom}(G_1, G_2) \to \mathrm{Hom}_{(\pi_0(G_1),\pi_0(G_2))}(S_1, S_2) \tag{3.15}$$

into the set $\mathrm{Hom}_{(\pi_0(G_1),\pi_0(G_2))}(S_1, S_2)$ of pairs $(\chi, \phi) \in \mathrm{Hom}(\pi_0(G_1), \pi_0(G_2)) \times \mathrm{Hom}(S_1, S_2)$ such that ϕ is χ-equivariant. The fibre over (χ, ϕ) is either empty or in bijection with the set

$$\pi_0(\mathrm{map}(BS_1, BS_2)^{\pi_0(G_1)}_{B\phi}) = H^2(\pi_0(G_1); \pi_2(BS_2)) = H^1_\chi(\pi_0(G_1); \check{S}_2) \tag{3.16}$$

where $\pi_0(G_1)$ acts on $\check{S}_2$, the discrete approximation to S_2, through χ.

*If we put $w_2 \cdot (\chi, \phi) = (w_2 \chi w_2^{-1}, w_2\phi)$ for all $w_2 \in \pi_0(G_2)$ and all $(\chi, \phi) \in \mathrm{Hom}_{(\pi_0(G_1),\pi_0(G_2))}(S_1, S_2)$ then (*3.15*) becomes $\pi_0(G_2)$-equivariant, so it descends to a map*

$$\mathrm{Rep}(G_1, G_2) \to \pi_0(G_2)\backslash \mathrm{Hom}_{(\pi_0(G_1),\pi_0(G_2))}(S_1, S_2) \tag{3.17}$$

of $\pi_0(G_2)$-orbit sets. The fibre over the orbit $\pi_0(G_2)(\chi, \phi)$ is either empty or in bijection with the orbit set

$$\pi_0(G_2)^{(\chi,\phi)} \backslash H^1_\chi(\pi_0(G_1), \check{S}_2)$$

*for the action of the stabilizer group $\pi_0(G_2)^{(\chi,\phi)}$, consisting of all $w_2 \in \pi_0(G_2)$ such that $w_2\chi = \chi w_2$ and $w_2\phi = \phi$, on the fibre (*3.16*).*

Proposition 3.18. *Let (χ, ϕ) be an element of $Hom_{(\pi_0(G_1),\pi_0(G_2))}(S_1, S_2)$ and suppose that the stabilizer subgroup $\pi_0(G_2)^{(\chi,\phi)}$ acts transitively on the cohomology group $H^1_\chi(\pi_0(G_1), \check{S}_2)$. Then at most one element of* $\mathrm{Rep}(G_1, G_2)$ *is mapped to the orbit $\pi_0(G_2)(\chi, \phi)$ under the map (*3.17*).*

For later reference, I record here a non-realizability result.

Lemma 3.19. *(Cf.* [19, 1.8]*) Let $f\colon X_1 \to X_2$ be a p-compact group morphism where p is odd and X_1 is connected. Assume that*

- *$\pi_1(T(f))$ is injective, and*
- *p divides the order of the Weyl group W_1.*

Then p does not divide $\pi_1(T(f))$ in $\mathrm{Hom}(\pi_1(T_1), \pi_1(T_2))$.

Proof. By fixed point theory [10, 2.10, 2.14], f lifts to a morphism $N_p(f)$: $\mathrm{Syl}_p(N_1) \to \mathrm{Syl}_p(N_2)$ of the p-normalizers. The assumption that $\pi_1(T(f))$ be injective implies, since W_1 is faithfully represented in $\pi_1(T_1)$ [9, 9.7], that $\pi_0(N_p(f))$ embeds the Sylow p-subgroup of W_1 into W_2.

Choose a monomorphism $\mu : \mathbf{Z}/p \to \mathrm{Syl}_p(N_1)$ such that also $\pi_0(\mu) : \mathbf{Z}/p \to \mathrm{Syl}_p(W_1)$ is injective. This is possible since the epimorphism $\mathrm{Syl}_p(N_1) \to \mathrm{Syl}_p(W_1)$

admits a section when p is odd [2]. Note that the composition $N_p(f)\mu$ is a monomorphism since it induces a monomorphism on component groups. Consider now the commutative diagram

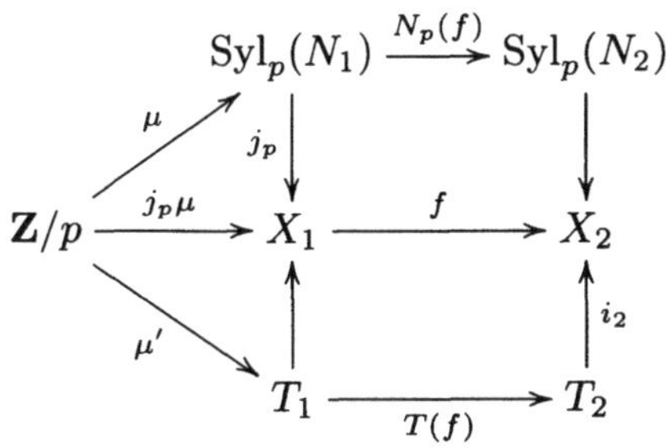

where μ' is a lift of $j_p\mu$ [9, 4.7, 5.6]. Since $N_p(f)\mu$ is monomorphic, so is $i_2T(f)\mu'$ by commutativity of the diagram. However, this map would be trivial were $\pi_1(T(f))$ divisible by p. □

The rest of the paper consists of an analysis of the special case where $X_1 =$ SU(3) or G_2, $X_2 = F_4$, and the prime $p = 3$.

4. Embeddings of SU(3) in F_4

In this section we apply the concepts of the previous sections to investigate monomorphisms from SU(3) to F_4 at the prime $p = 3$. First, a few facts about the Quillen category $\mathbf{A}(\mathrm{F}_4)$ of F_4. (See [28] for more details.)

Lemma 4.1. [14, 7.4][28, 8.2.2] *Let E^1 be an elementary abelian group of order 3^1. The set* $\mathrm{Mono}(E^1, \mathrm{F}_4)$ *of conjugacy classes of monomorphisms of E^1 into* F_4 *has three elements e_1^1, e_2^1, e_3^1. The centralizers of these three elements are connected 3-compact groups with Weyl groups of order* 36, 48, *and* 48, *respectively. The centralizer $C_{\mathrm{F}_4}(e_1^1)$ of e_1^1 is isomorphic to* SU(3, 3)*. The automorphism group* $\mathrm{Aut}(E^1)$ *acts trivially on* $\mathrm{Mono}(E^1, \mathrm{F}_4)$.

Lemma 4.2. [14, 7.4][28, 8.2.4], [27, 7.5] *Let E^2 be an elementary abelian group of order 3^2. The set* $\mathrm{Mono}(E^2, \mathrm{F}_4)/\mathrm{Aut}(E^2)$ *of isomorphism classes of conjugacy classes of monomorphisms of E^2 into* F_4 *has* 5 *elements, $e_1^2, e_2^2, e_3^2, e_4^2, e_5^2$, with Quillen automorphism groups of order* 8, 4, 12, 12, 48, *and with centralizer Weyl groups of order* 4, 6, 6, 8, 3, *respectively. The centralizer, $C_{\mathrm{F}_4}(e_5^2)$, of e_5^2 is a 3-compact toral group of maximal rank with component group $\pi_0(C_{\mathrm{F}_4}(e_5^2))$ of order 3. There are no maps in the Quillen category from e_2^1 or e_3^1 to e_5^2.*

*Proofs of (*4.1*) and (*4.2*).* With computer assistance it is easy to determine, using [24, 2.6] and [22, 3.2], that $\mathrm{Mono}(E^1, \mathrm{F}_4)$ is a trivial $\mathrm{Aut}(E^1)$-set containing three elements whose centralizers are connected 3-compact groups with Weyl groups of order 36, 48, 48, respectively. See [19, 3.3] for the precise structure of $C_{\mathrm{F}_4}(a)$. Since each centralizer of E^1 is connected, any monomorphism $E^2 \to \mathrm{F}_4$ will factor through the maximal torus. □

The Quillen automorphism group referred to in (4.2) consists of all automorphism of E^2 that leaves $e_i^2 \in \mathrm{Mono}(E^2, \mathrm{F}_4)$ invariant.

We now show that for any monomorphism of SU(3) or SU(3,3) to F_4 the triangles

$$\begin{array}{ccc} & E^1 & \\ {\scriptstyle z}\swarrow & & \searrow{\scriptstyle e_1^1} \\ \mathrm{SU}(3) & \rightarrowtail & \mathrm{F}_4 \end{array} \qquad \begin{array}{ccc} & E^1 & \\ {\scriptstyle z}\swarrow & & \searrow{\scriptstyle e_1^1} \\ \mathrm{SU}(3,3) & \rightarrowtail & \mathrm{F}_4 \end{array} \tag{4.3}$$

where $z\colon E^1 \to \mathrm{SU}(3)$ and $z\colon E^1 \to \mathrm{SU}(3,3)$ are centers, will commute up to conjugacy. This observation is the key to the classification of monomorphisms of $\mathrm{SU}(3) \rightarrowtail \mathrm{F}_4$.

Lemma 4.4. 1. $\mathrm{Mono}(\mathrm{SU}(3), \mathrm{F}_4)_{z\to e_1^1} = \mathrm{Mono}(\mathrm{SU}(3), \mathrm{F}_4)$.
2. $\mathrm{Mono}(\mathrm{SU}(3,3), \mathrm{F}_4)_{z\to e_1^1} = \mathrm{Mono}(\mathrm{SU}(3,3), \mathrm{F}_4)$.

The proof of this lemma uses *admissible homomorphisms* (2.3) which we now discuss.

Let $\mathbf{Z}_3$ denote the ring of 3-adic integers. The Weyl group $W_1 = W(\mathrm{SU}(3))$ of SU(3) is [24, 3.8, 3.13] $\langle \sigma, \tau \rangle \subseteq \mathrm{Aut}(\Sigma_0(\mathbf{Z}_3^3))$ where $\Sigma_0(\mathbf{Z}_3^3)$ is the free $\mathbf{Z}_3$-module with basis $(1,-1,0), (0,1,-1) \in \mathbf{Z}_3^3$ and σ and τ have matrices

$$\sigma = \begin{pmatrix} 0 & -1 \\ 1 & -1 \end{pmatrix}, \quad \tau = \begin{pmatrix} 0 & -1 \\ -1 & 0 \end{pmatrix}$$

with respect to this basis. The Weyl group $W(\mathrm{F}_4) = W(\mathrm{F}_4) < \mathrm{GL}(4, \mathbf{Z}_3)$ of F_4 is [3] [24, 3.13] the group (of order $1152 = 384 \cdot 3$)

$$W(\mathrm{F}_4) = W(B_4)E \cup W(B_4)H_1 \cup W(B_4)H_2 \tag{4.5}$$

where $W(B_4)$ is the reflection group (of order $384 = 2^4{\cdot}4!$) of all signed permutation matrices, and H_1 and H_2 are the matrices

$$H_1 = \frac{1}{2}\begin{pmatrix} -1 & 1 & 1 & 1 \\ 1 & -1 & 1 & 1 \\ 1 & 1 & -1 & 1 \\ 1 & 1 & 1 & -1 \end{pmatrix}, \quad H_2 = \frac{1}{2}\begin{pmatrix} 1 & 1 & 1 & 1 \\ 1 & 1 & -1 & -1 \\ 1 & -1 & 1 & -1 \\ 1 & -1 & -1 & 1 \end{pmatrix}$$

satisfying $H_1^2 = E = H_2^2, H_2H_1 = -H_2, H_1H_2 = \mathrm{diag}(-1,1,1,1)H_1$.

We say that a linear map $A\colon \Sigma_0(\mathbf{Z}_3^3) \to \mathbf{Z}_3^4$ is *admissible* if $AW(\mathrm{SU}(3)) \subseteq W(\mathrm{F}_4)A$. The linear map $A(u,v)\colon \Sigma_0(\mathbf{Z}_3^3) \to \mathbf{Z}_3^4$, $u, v \in \mathbf{Z}_3$, for instance, with matrix

$$\boxed{A(u,v) = \begin{pmatrix} -u & v \\ u & v-u \\ 0 & v+u \\ -2v & v \end{pmatrix} = u\begin{pmatrix} -1 & 0 \\ 1 & -1 \\ 0 & 1 \\ 0 & 0 \end{pmatrix} + v\begin{pmatrix} 0 & 1 \\ 0 & 1 \\ 0 & 1 \\ -2 & 1 \end{pmatrix}} \tag{4.6}$$

with respect to the chosen basis for $\Sigma_0(\mathbf{Z}_3^3)$ and the canonical basis for $\mathbf{Z}_3^4$, is admissible. Indeed, $A(u,v)$ is χ-equivariant where $\chi\colon W(\mathrm{SU}(3)) \to W(\mathrm{F}_4)$ is the group homomorphism given by

$$\chi(\sigma) = \frac{1}{2}\begin{pmatrix} -1 & -1 & 1 & -1 \\ 1 & -1 & -1 & -1 \\ -1 & 1 & -1 & -1 \\ 1 & 1 & 1 & -1 \end{pmatrix}, \quad \chi(\tau) = \frac{1}{2}\begin{pmatrix} -1 & -1 & 1 & 1 \\ -1 & 1 & -1 & 1 \\ 1 & -1 & -1 & 1 \\ 1 & 1 & 1 & 1 \end{pmatrix} \tag{4.7}$$

The next lemma classifies the admissible homomorphisms. Note that $A(u,v)$ and $-A(u,v)$ lie in the same orbit under the action of $W(\mathrm{F}_4)$ as $-E \in W(\mathrm{F}_4)$.

Lemma 4.8. 1. *Let $A\colon \Sigma_0(\mathbf{Z}_3^3) \to \mathbf{Z}_3^4$ be a linear map. Then A is admissible with respect to $W(\mathrm{SU}(3))$ and $W(\mathrm{F}_4)$ if and only if $A \in W(\mathrm{F}_4)A(u,v)$ for some 3-adic integers $u, v \in \mathbf{Z}_3$.*

2. *$A(u,v)$ is split injective if and only if $u+v$ is a 3-adic unit.*
3. *The map*

$$\begin{aligned} \langle(-1,-1)\rangle \backslash (\mathbf{Z}_3)^2 &\to W(\mathrm{F}_4)\backslash \mathrm{Hom}_{\mathbf{Z}_3}(\Sigma_0(\mathbf{Z}_3^3), \mathbf{Z}_3^4) \\ \pm(u,v) &\to W(\mathrm{F}_4)A(u,v) \end{aligned}$$

is injective.

Proof. 1. Using a computer, it is possible to show that up to inner automorphisms, any admissible homomorphism $\Sigma_0(\mathbf{Z}_3^3) \to \mathbf{Z}_3^4$ must be χ-equivariant. Given this, one simply solves the system of linear equations $Aw = \chi(w)A$ for A where w runs through a generating set for $W(\mathrm{SU}(3))$.
2. The matrix $A(u,v)$ is equivalent to the matrix

$$\begin{pmatrix} u-2v & 0 \\ 0 & 2v-u \\ 3u & 0 \\ -u & v \end{pmatrix}$$

which is split injective if and only if $u-2v$ or, equivalently, $(u-2v)+3v = u+v$ is a 3-adic unit.
3. The claim is that for any w in $W(\mathrm{F}_4)$ the set of solutions to the homogeneous system of linear equations

$$wA(u_1,v_1) - A(u_2,v_2) = 0$$

in the four unknowns (u_1,v_1,u_2,v_2) is contained in the diagonal $(u_1,v_1) = (u_2,v_2)$ or in the anti-diagonal $(u_1,v_1) = -(u_2,v_2)$. This is easily verified on a computer. □

Our interest in the admissible homomorphisms lies in the fact that the induced homomorphism $\pi_1(T(f))$ is admissible for any lift $T(f)\colon T(\mathrm{SU}(3)) \to T(\mathrm{F}_4)$ to the maximal tori of any morphism $f\colon \mathrm{SU}(3) \to \mathrm{F}_4$. Thus we must have

$$\pi_1(T(f)) \in W(\mathrm{F}_4)A(u,v)$$

for some 3-adic integers u and v. However, as we shall shortly see, not all the homomorphisms $A(u,v)$ are induced in this way from morphisms $\mathrm{SU}(3) \to \mathrm{F}_4$.

The proof of (4.4) follows immediately from (4.8.1).

Proof of Lemma 4.4. 1. Let $f\colon \mathrm{SU}(3) \to \mathrm{F}_4$ be any monomorphism. Then $\pi_1(T(f))$ is admissible, so we may assume that $\pi_1(T(f)) = A(u,v)$ for some 3-adic integers $u, v \in \mathbf{Z}_3$. The restriction $fz\colon E^1 \to \mathrm{F}_4$ of f to the center $z\colon E^1 \to \mathrm{SU}(3)$ of $\mathrm{SU}(3)$ is given by

$$A(u,v)\begin{pmatrix} -1 \\ 1 \end{pmatrix} = \begin{pmatrix} u+v \\ u+v \\ u+v \\ 0 \end{pmatrix} \tag{4.9}$$

where we have reduced modulo 3. Since fz is a monomorphism, $u+v \not\equiv 0 \bmod 3$ and then the stabilizer in $W(\mathrm{F}_4)$ of $(u+v, u+v, u+v, 0) \in (\mathbf{Z}/3)^4$ has order 36. Thus $fz \simeq e_1^1 \in \mathrm{Mono}(E^1, \mathrm{F}_4)$.

2. Let $f\colon \mathrm{SU}(3,3) \to \mathrm{F}_4$ be any monomorphism and choose some monomorphism $g\colon \mathrm{SU}(3) \to \mathrm{SU}(3,3)$ such that $gz = z$, e.g. $g = \psi^{(1,0)}$. Then $fz = fgz = e_1^1$. □

Let $e\colon \mathrm{SU}(3,3) = C_{\mathrm{F}_4}(e_1^1) \to \mathrm{F}_4$ denote the inclusion of the centralizer of e_1^1 into F_4; this map is described in detail in [19, 3.3].

Corollary 4.10. *The maps*

$$\mathrm{Mono}(\mathrm{SU}(3), \mathrm{SU}(3,3))_{z\to z} \xrightarrow{e\circ -} \mathrm{Mono}(\mathrm{SU}(3), \mathrm{F}_4)$$

$$\mathrm{Out}(\mathrm{SU}(3,3))_{z\to z} \xrightarrow{e\circ -} \mathrm{Mono}(\mathrm{SU}(3,3), \mathrm{F}_4)$$

are bijections.

Proof. By (3.9) and (4.4),

$$\mathrm{Mono}(\mathrm{SU}(3), \mathrm{SU}(3,3)_{z\to z} = \mathrm{Mono}(\mathrm{SU}(3), \mathrm{F}_4)_{z\to e_1^1} = \mathrm{Mono}(\mathrm{SU}(3), \mathrm{F}_4)$$

and similarly for morphisms from $\mathrm{SU}(3,3)$. □

Lemma 4.11. *Let $\psi^{(u,v)}\colon \mathrm{SU}(3) \to \mathrm{SU}(3,3)$ be the morphism (2.17) indexed by $u, v \in \mathbf{Z}_3^* \cup \{0\}$. Then $W(\mathrm{F}_4)\pi_1(T(e\psi^{(u,v)})) = W(\mathrm{F}_4)A(u,v)$.*

Proof. The monomorphism $e\colon \mathrm{SU}(3,3) \to \mathrm{F}_4$ is [19, 3.3] realizable on the level of compact Lie groups as an inclusion $\mathrm{SU}(3,3) \hookrightarrow \mathrm{F}_4$ such that the restriction $\Sigma_0(\mathbf{Z}^3) \times \Sigma_0(\mathbf{Z}^3) \to \Sigma_2(\mathbf{Z}^4)$ to the integral lattices of the composite morphism $\mathrm{SU}(3) \times \mathrm{SU}(3) \twoheadrightarrow \mathrm{SU}(3,3) \hookrightarrow \mathrm{F}_4$ takes $(x_1, x_2, x_3; y_1, y_2, y_3)$ to $(x_1 + y_3, x_2 + y_3, x_3 + y_3, y_1 - y_2)$. Thus

$$\begin{pmatrix} 1 & 0 & 0 & -1 \\ -1 & 1 & 0 & -1 \\ 0 & -1 & 0 & -1 \\ 0 & 0 & 2 & -1 \end{pmatrix} \begin{pmatrix} u & 0 \\ 0 & u \\ v & 0 \\ 0 & v \end{pmatrix} = \begin{pmatrix} u & -v \\ -u & u-v \\ 0 & -u-v \\ 2v & -v \end{pmatrix} = -A(u,v)$$

represents $\pi_1(T(e\psi^{(u,v)}))$. □

Lemma 4.12. *Let u and v be 3-adic integers and $A(u,v)$ the corresponding admissible homomorphism.*

1. *There exists a morphism $f\colon \mathrm{SU}(3) \to \mathrm{F}_4$ such that*
$$W(\mathrm{F}_4)\pi_1(T(f)) = W(\mathrm{F}_4)A(u,v)$$
if and only if both u and v are in the set $\mathbf{Z}_3^* \cup \{0\}$.
2. *There exists a monomorphism $f\colon \mathrm{SU}(3) \to \mathrm{F}_4$ such that $W(\mathrm{F}_4)\pi_1(T(f)) = W(\mathrm{F}_4)A(u,v)$ if and only if $u,v \in \mathbf{Z}_3^* \cup \{0\}$ and $u+v \in \mathbf{Z}_3^*$.*

Proof. We have already seen (4.11) that $A(u,v)$ is realizable for all $u,v \in \mathbf{Z}_3^* \cup \{0\}$.

Suppose, conversely, that $\pi_1(T(f)) = A(u,v)$ for some 3-adic integers, u and v, and some morphism $f\colon \mathrm{SU}(3) \to \mathrm{F}_4$. If f is a monomorphism, then $f = e\psi^{(u,v)}$ for some $u,v \in \mathbf{Z}_3^* \cup \{0\}$ with $u+v \in \mathbf{Z}_3^*$ by (4.10). If f is not a monomorphism, $A(u,v)$ is not split injective [24, 5.2] [21, 3.6.1], so $u+v$ is not a 3-adic unit (4.8.2). □

Theorem 4.13. 1. $\mathrm{Mono}(\mathrm{SU}(3), \mathrm{F}_4)$ *is T-determined.*

2. *The map*
$$\langle(-1,-1)\rangle \backslash \{(u,v) \in (\mathbf{Z}_3^* \cup \{0\})^2 | u+v \in \mathbf{Z}_3^*\} \to \mathrm{Mono}(\mathrm{SU}(3), \mathrm{F}_4)$$
$$\pm(u,v) \to e\psi^{(u,v)}$$
is a bijection.

Proof. 1. The restriction map $\mathrm{Mono}(\mathrm{SU}(3), \mathrm{F}_4) \to \mathrm{Mono}(T(\mathrm{SU}(3)), \mathrm{F}_4)$ can be identified to the map
$$\{(u,v) \in (\mathbf{Z}_3^* \cup \{0\})^2 \mid u+v \equiv 1 \bmod 3\} \to W(\mathrm{F}_4)\backslash \mathrm{Hom}(\Sigma_0(\mathbf{Z}_3^3), \mathbf{Z}_3^4)$$
$$(u,v) \to W(\mathrm{F}_4)A(u,v)$$
which is injective by (4.8.3).
2. This is immediate from (2.18) and (4.10). □

Here is an alternative formulation of (4.10): Consider the commutative diagrams

$$\begin{array}{ccc}
\mathrm{Mono}(\mathrm{SU}(3), \mathrm{SU}(3,3))_{z\to z} & \longrightarrow & \langle \psi^{-1} \times \psi^{-1} \rangle \backslash \mathrm{Mono}(\mathrm{SU}(3), \mathrm{SU}(3,3)) \\
& \underset{eo-}{\overset{\cong}{\searrow}} & \downarrow{\scriptstyle eo-} \\
& & \mathrm{Mono}(\mathrm{SU}(3), \mathrm{F}_4)
\end{array}$$

$$\begin{array}{ccc}
\mathrm{Out}(\mathrm{SU}(3,3))_{z\to z} & \longrightarrow & \langle \psi^{-1} \times \psi^{-1} \rangle \backslash \mathrm{Out}(\mathrm{SU}(3,3) \\
& \underset{eo-}{\overset{\cong}{\searrow}} & \downarrow{\scriptstyle eo-} \\
& & \mathrm{Mono}(\mathrm{SU}(3,3), \mathrm{F}_4)
\end{array}$$

where the slanted arrows are bijections. The vertical arrows exist because $e(\psi^{-1} \times \psi^{-1}) = e$ by [19, 3.3]. Noting (2.17) that

$$\begin{aligned} \mathrm{Mono}(\mathrm{SU}(3), \mathrm{SU}(3,3))_{z \to z} &= \{(u,v) \in (\mathbf{Z}_3^* \cup \{0\})^2 \mid u + v \equiv 1 \bmod 3\} \\ \mathrm{Out}(\mathrm{SU}(3,3))_{z \to z} &= \{(u,v) \in (\mathbf{Z}_3^*)^2 \mid u \equiv 1 \equiv v \bmod 3\} \rtimes \langle \tau \rangle \end{aligned}$$

we see that the vertical arrow in each of the diagrams is a bijection, too, and hence that the vertical arrow of the upper (lower) diagram is a bijection of right Out(SU(3))- (Out(SU(3, 3))-) sets. Thus the action

$$\mathrm{Mono}(\mathrm{SU}(3,3), \mathrm{F}_4) \times \mathrm{Out}(\mathrm{SU}(3,3)) \to \mathrm{Mono}(\mathrm{SU}(3,3), \mathrm{F}_4) \tag{4.14}$$

is transitive and the stabilizer subgroup at the centric monomorphism e, i.e. the Weyl group [11, 4.3] [24, 8.4]

$$W_{\mathrm{F}_4}(e\mathrm{SU}(3,3)) = \langle \psi^{-1} \times \psi^{-1} \rangle \tag{4.15}$$

is cyclic of order two.

The next lemma lists the centralizers of all monomorphisms $\mathrm{SU}(3) \rightarrowtail \mathrm{F}_4$. We let ψ^{-1} denote the automorphism $\psi^{-1} \times_{Z(\mathrm{SU}(3))} \psi^{-1}$ of $T(\mathrm{SU}(3) \times_{Z(\mathrm{SU}(3))} \mathrm{SU}(3)$ [22, 4.3].

Lemma 4.16. *Let $(u,v) \in (\mathbf{Z}_3^* \cup \{0\})^2$ and $u + v \in \mathbf{Z}_3^*$. If $uv \neq 0$, then*

$$\begin{aligned} C_{\mathrm{F}_4}(e\psi^{(u,v)}\mathrm{SU}(3)) &= Z(\mathrm{SU}(3)) \\ C_{\mathrm{F}_4}(e\psi^{(u,v)}T(\mathrm{SU}(3))) &= T(\mathrm{F}_4) \end{aligned}$$

If $uv = 0$, then

$$\begin{aligned} C_{\mathrm{F}_4}(e\psi^{(u,v)}\mathrm{SU}(3)) &= Z(\mathrm{SU}(3)) \times_{Z(\mathrm{SU}(3))} \mathrm{SU}(3) \\ C_{\mathrm{F}_4}(e\psi^{(u,v)}T(\mathrm{SU}(3))) &= T(\mathrm{SU}(3)) \times_{Z(\mathrm{SU}(3))} \mathrm{SU}(3) \end{aligned}$$

In all cases, $C_{\mathrm{F}_4}(\psi^{-1}) = \psi^{-1}$.

Proof. It only remains to determine the map $C_{\mathrm{F}_4}(\psi^{-1})$ induced by ψ^{-1} since the centralizers themselves are given by (2.19, 3.9). Let us, for example, consider the case where $(u,v) = (0,1)$. Consider the morphism μ: $(\mathrm{SU}(3) \times T(\mathrm{SU}(3))) \times T(\mathrm{SU}(3)) \to \mathrm{SU}(3) \times T(\mathrm{SU}(3)) \to \mathrm{SU}(3,3)$ constructed from the multiplication on the maximal torus and the projection map. Since

$$e\mu((1 \times 1) \times \psi^{-1}) = e(\psi^{-1} \times \psi^{-1})\mu((1 \times 1) \times \psi^{-1}) = e\mu((\psi^{-1} \times \psi^{-1}) \times 1)$$

it follows from (4.17) that $C_{\mathrm{F}_4}(\psi^{-1}) = \psi^{-1}$ on $C_{\mathrm{F}_4}(e\psi^{(0,1)}T(\mathrm{SU}(3)))$. The other cases are similar. □

Lemma 4.17. *If the diagram of p-compact groups*

$$\begin{array}{ccccc} X_1 \times X_2 & \xleftarrow{1 \times f_2} & X_1 \times X_2' & \xrightarrow{f_1 \times 1} & X_1' \times X_2' \\ \Big\downarrow{\scriptstyle \mu} & & & & \Big\downarrow{\scriptstyle \mu'} \\ Y & & \xrightarrow[h]{\qquad\qquad} & & Y' \end{array}$$

commutes up to conjugacy, so does the induced diagram

$$\begin{array}{ccc} X_1 & \xrightarrow{\operatorname{ad}(\mu)} & C_Y(X_2) \\ {\scriptstyle f_1}\downarrow & & \downarrow{\scriptstyle C_h(f_2)} \\ X_1' & \xrightarrow{\operatorname{ad}(\mu')} & C_{Y'}(X_2') \end{array}$$

where the horizontal arrows are adjoints of μ and μ'.

Corollary 4.18. *Let N be a (topological) group with subgroups $g_1\colon G_1 \to N$ and $g_2\colon G_2 \to N$. Suppose that $n \in N$ is an element such that conjugation with n, $c(n)(m) = nmn^{-1}$, $m \in N$, takes G_1 into G_2. Then conjugation with n^{-1} takes the centralizer $C_N(G_2)$ into $C_N(G_1)$ and the diagram*

$$\begin{array}{ccc} BC_N(G_1) & \longrightarrow & \operatorname{map}(BG_1, BN)_{Bg_1} \\ {\scriptstyle Bc(n^{-1})}\uparrow & & \uparrow{\scriptstyle \overline{Bc(n)}} \\ BC_N(G_2) & \longrightarrow & \operatorname{map}(BG_2, BN)_{Bg_2} \end{array}$$

commutes up to homotopy.

Proof. We have $\mu(c(n) \times 1) = c(n)\mu(1 \times c(n^{-1}))$ where μ is group multiplication and where the induced map $Bc(n)\colon BN \to BN$ is homotopic to the identity. □

5. Toric representations of PU(3) in F$_4$

In this section I classify the p-toric morphisms from PU(3) to F$_4$ viewed as 3-compact groups. The first step is the determination of the admissible homomorphisms.

Let X be a connected p-compact group with maximal torus $i\colon T \to X$. We want to describe the integral lattice of the central quotients of X. Suppose that Z is a subgroup of the discrete approximation $\check{T} = (\pi_1(T) \otimes \mathbf{Q})/\pi_1(T)$ such that the composition $Z \to \check{T} \to X$ is a central monomorphism. Then we may form the p-compact group X/Z [9, 8.3] with induced maximal torus $i/Z\colon T/Z \to X/Z$ [21, 4.6] that fits into the commutative diagram

$$\begin{array}{ccccccccc} 0 & \longrightarrow & \kappa^{-1}(0) & \longrightarrow & \kappa^{-1}(Z) & \longrightarrow & Z & \longrightarrow & 0 \\ & & \| & & \downarrow & & \downarrow & & \\ 0 & \longrightarrow & \pi_1(T) & \longrightarrow & \pi_1(T) \otimes \mathbf{Q} & \xrightarrow{\kappa} & \check{T} & \longrightarrow & 0 \\ & & \downarrow & & \downarrow{\scriptstyle\cong} & & \downarrow & & \\ 0 & \longrightarrow & \pi_1(T/Z) & \longrightarrow & \pi_1(T/Z) \otimes \mathbf{Q} & \longrightarrow & \check{T}/Z & \longrightarrow & 0 \end{array}$$

with exact rows. From this we get an isomorphism

$$\begin{array}{ccccccccc} 0 & \longrightarrow & \pi_1(T) & \longrightarrow & \pi_1(T/Z) & \longrightarrow & Z & \longrightarrow & 0 \\ & & \Big\| & & \Big\downarrow & & \Big\downarrow & & \\ 0 & \longrightarrow & \kappa^{-1}(0) & \longrightarrow & \kappa^{-1}(Z) & \longrightarrow & Z & \longrightarrow & 0 \end{array}$$

of extensions of $W_T(X) = W_{T/Z}(X/Z)$-modules.

In particular, let $\Sigma_0(\mathbf{Z}_3^3) \subseteq \Sigma_0(\mathbf{Q}_3^3)$ be the free $\mathbf{Z}_3$-submodule with basis $e_1 = (1,-1,0)$ and $e_2 = (0,1,-1)$; this is the integral lattice for SU(3). Put $f = \frac{1}{3}(e_1 - e_2)$ and let $P\Sigma_0(\mathbf{Z}_3^3)$ be the free $\mathbf{Z}_3$-submodule of $\mathbf{Q}_3^3$ with basis $\{e_1, f\}$. Then there is an exact sequence

$$0 \to \Sigma_0(\mathbf{Z}_3^3) \xrightarrow{\iota} P\Sigma_0(\mathbf{Z}_3^3) \to \mathbf{Z}/3 \to 0$$

of $\mathbf{Z}_3[\Sigma_3]$-modules and $P\Sigma_0(\mathbf{Z}_3^3)$ corresponds to the maximal torus for PU(3).

Note that there is an extension, $B(u,v)$, of $A(u,v)$,

$$\begin{array}{ccc} \Sigma_0(\mathbf{Z}_3^3) & \xrightarrow{\ \iota\ } & P\Sigma_0(\mathbf{Z}_3^3) \\ {\scriptstyle A(u,v)}\Big\downarrow & \swarrow{\scriptstyle B(u,v)} & \\ L_4 & & \end{array}$$

if and only if $u+v$ is divisible by 3 and in that case the extension is unique and given by

$$B(u,v) = A(u,v)\begin{pmatrix} 1 & 1 \\ 0 & -3 \end{pmatrix}^{-1} = \begin{pmatrix} -u & -\frac{1}{3}(u+v) \\ u & \frac{1}{3}(2u-v) \\ 0 & -\frac{1}{3}(u+v) \\ -2v & -v \end{pmatrix}$$

where u and v are 3-adic integers and $u+v \in 3\mathbf{Z}_3$. Moreover, the inclusion ι is $W(\mathrm{SU}(3)) = W(\mathrm{PU}(3))$-equivariant and $B(u,v)$ is χ-equivariant where χ is the group homomorphism from (4.7).

Lemma 5.1. 1. *A $\mathbf{Z}_3$-linear map $B\colon P\Sigma_0(\mathbf{Z}_3^3) \to \mathbf{Z}_3^4$ is admissible with respect to $W(\mathrm{PU}(3))$ and $W(\mathrm{F}_4)$ is and only if $B \in W(\mathrm{F}_4)B(u,v)$ where u and v are 3-adic integers whose sum is divisible by 3.*
2. *$B(u,v)$ is split-injective when u and v are 3-adic units.*
3. *The map*

$$\begin{aligned} \langle(-1,-1)\rangle \backslash \{(u,v) \in (\mathbf{Z}_3)^2 | u+v \in 3\mathbf{Z}_3\} &\to W(\mathrm{F}_4)\backslash \operatorname{Hom}_{\mathbf{Z}_3}(P\Sigma_0(\mathbf{Z}_3^3), \mathbf{Z}_3^4) \\ \pm(u,v) &\to W(\mathrm{F}_4)B(u,v) \end{aligned}$$

is injective.

Proof. 1. B is admissible if and only if $B \circ \iota$ is, i.e. if and only if B is an extension of $A(u,v)$ (4.8.1) for some 3-adic integers, u and v.

2. If u and v are units then

$$\begin{pmatrix} -u^{-1} & 0 & u^{-1} & 0 \\ 2u^{-1} & 0 & 2u^{-1} & -v^{-1} \end{pmatrix}$$

is a left inverse of $B(u,v)$.

3. If $B(u_1,v_1) \in W(\mathrm{F}_4)B(u_2,v_2)$ then also $A(u_1,v_1) \in W(\mathrm{F}_4)A(u_2,v_2)$ and then (4.8.3) (u_1,v_1) and (u_2,v_2) are equal up to sign. □

When $u, v \in \mathbf{Z}_3^* \cup \{0\}$ with sum $u+v \in 3\mathbf{Z}_3$ there is a unique conjugacy class, $\overline{\psi}^{(u,v)}$, that makes the diagram

$$\begin{array}{ccccc} \mathrm{SU}(3) & \xrightarrow{\psi^{(u,v)}} & \mathrm{SU}(3,3) & \xrightarrow{e} & \mathrm{F}_4 \\ \downarrow & \nearrow_{\overline{\psi}^{(u,v)}} & & & \\ \mathrm{PU}(3) & & & & \end{array}$$

commutes up to conjugation. By construction,

$$W(\mathrm{F}_4)\pi_1\big(T(e \circ \overline{\psi}^{(u,v)})\big) = W(\mathrm{F}_4)B(u,v)$$

in $W(\mathrm{F}_4)\backslash \mathrm{Hom}_{\mathbf{Z}_3}(P\Sigma_0(\mathbf{Z}_3^3), \mathbf{Z}_3^4)$.

Lemma 5.2. *Let u and v be 3-adic integers with sum $u+v \in 3\mathbf{Z}_3$ and let $B(u,v)\colon P\Sigma_0(\mathbf{Z}_3^3) \to \mathbf{Z}_3^4$ be the corresponding admissible homomorphism.*

1. *There exists a morphism $f\colon \mathrm{PU}(3) \to \mathrm{F}_4$ such that $W(\mathrm{F}_4)\pi_1(T(f)) = W(\mathrm{F}_4)B(u,v)$ if and only if $u = 0 = v$ or $u,v \in \mathbf{Z}_3^*$.*
2. *There exists a monomorphism $f\colon \mathrm{PU}(3) \to \mathrm{F}_4$ such that $W(\mathrm{F}_4)\pi_1(T(f)) = W(\mathrm{F}_4)B(u,v)$ if and only if $u,v \in \mathbf{Z}_3^*$.*

Proof. We have already seen that $W(\mathrm{F}_4)B(u,v)$ is realizable by a morphism $f\colon \mathrm{PU}(3) \to \mathrm{F}_4$ if $u = 0 = v$ or $u,v \in \mathbf{Z}_3^*$; if both u and v are non-zero then f is a monomorphism by (5.1.2). Conversely, if $W(\mathrm{F}_4)B(u,v)$ is realizable, so is $W(\mathrm{F}_4)A(u,v)$ and then (4.12) $u,v \in \mathbf{Z}_3^* \cup \{0\}$, $u+v \notin \mathbf{Z}_3^*$. □

Alternatively, (5.2) says that any non-trivial morphism $\mathrm{PU}(3) \to \mathrm{F}_4$ is a monomorphism.

Proposition 5.3. *(Cf.* [1, 2.27.(ii)]*) Suppose that u and v are 3-adic units with $u+v \in 3\mathbf{Z}_3$. Then*

$$T(\mathrm{PU}(3)) \xrightarrow{B(u,v)} T(\mathrm{F}_4) \xrightarrow{i_2} \mathrm{F}_4$$

is toric if and only if $(u,v) \notin \mathbf{Z}_3^*(2,1) \cup \mathbf{Z}_3^*(1,-1)$.

Proof. Explicit (computer aided) computations of $W(\mathrm{F}_4)^{B(u,v)} = W(\mathrm{F}_4)^{A(u,v)}$. □

The two generic non-3-toric morphisms

$$B(2,1)=\begin{pmatrix} -2 & -1 \\ 2 & 1 \\ 0 & -1 \\ -2 & -1 \end{pmatrix} \quad \text{and} \quad B(1,-1)=\begin{pmatrix} -1 & 0 \\ 1 & 1 \\ 0 & 0 \\ 2 & 1 \end{pmatrix}$$

are related by the equation $\varepsilon B(2,1) = 2B(1,-1)$ where

$$\varepsilon = \begin{pmatrix} 0 & 0 & -1 & 1 \\ 0 & 0 & -1 & -1 \\ -1 & -1 & 0 & 0 \\ -1 & 1 & 0 & 0 \end{pmatrix}$$

is the admissible automorphism of $\mathbf{Z}_3^4$ corresponding to the exotic automorphism of F_4. (In general, $W(\mathrm{F}_4)(\varepsilon A(u,v)) = W(\mathrm{F}_4)(A(2v,-u))$, cf. [1, 2.11].)

Theorem 5.4. 1. $\mathrm{TRep}(\mathrm{PU}(3),\mathrm{F}_4)$ *is T-determined.*
2. *The map*

$$\langle(-1,-1)\rangle \backslash (\{(u,v)\in(\mathbf{Z}_3^*)^2 \mid u+v\in 3\mathbf{Z}_3\}) \setminus (\mathbf{Z}_3^*(2,1)\cup\mathbf{Z}_3^*(1,-1))) \\ \to \mathrm{TRep}(\mathrm{PU}(3),\mathrm{F}_4)$$

taking $\pm(u,v)$ *to* $e\circ\overline{\psi}^{(u,v)}$, *is a bijection.*

Consider the set $\mathrm{Rep}(N(\mathrm{PU}(3)),N(\mathrm{F}_4))$ of conjugacy classes of maps from the maximal torus normalizer $N(PU(3))$ of $\mathrm{PU}(3)$ to the maximal torus normalizer $N(\mathrm{F}_4)$ of F_4. As we have seen (3.17), there is a map

$$\mathrm{Rep}(N(\mathrm{PU}(3)),N(\mathrm{F}_4)) \to W(\mathrm{F}_4)\backslash \mathrm{Hom}_{(W(\mathrm{PU}(3)),W(\mathrm{F}_4))}(T(\mathrm{PU}(3)),T(\mathrm{F}_4))$$

induced by the functors π_1 and π_2. It is easy to calculate directly that the cohomology group $H^2(\langle\chi(\sigma)\rangle;\pi_1(T(\mathrm{F}_4)))$ is trivial. Then also

$$H^2_\chi(W(\mathrm{PU}(3));\pi_1(T(\mathrm{F}_4)))=0 \tag{5.5}$$

for $\langle\sigma\rangle$ is a Sylow 3-subgroup of the Weyl group of $\mathrm{PU}(3)$ and we get

Lemma 5.6. *There is at most one element of* $\mathrm{Rep}(N(\mathrm{PU}(3)),N(\mathrm{F}_4))$ *corresponding to the orbit* $W(\mathrm{F}_4)(\chi,B(u,v))$, $(u,v)\in(\mathbf{Z}_3^*)^2$, $u+v\in 3\mathbf{Z}_3$.

Proof of Theorem 5.4. Let $f_1,f_2\in T\,\mathrm{Rep}(\mathrm{PU}(3),\mathrm{F}_4)$ be two toric representations and suppose that their restrictions to the maximal torus of $\mathrm{PU}(3)$ agree. Under the map

$$\mathrm{TRep}(\mathrm{PU}(3),\mathrm{F}_4) \to \mathrm{TRep}(N(\mathrm{PU}(3)),N(\mathrm{F}_4)) \\ \to W(\mathrm{F}_4)\backslash \mathrm{Hom}_{(W(\mathrm{PU}(3)),W(\mathrm{F}_4))}(T(\mathrm{PU}(3)),T(\mathrm{F}_4))$$

f_1 and f_2 go to the same element of the target and it follows (5.6) that the lifts (3.5) $N(f_1)$ and $N(f_2)$ are conjugate, i.e. that f_1 and f_2 have conjugate restrictions to the maximal torus normalizer $N(\mathrm{PU}(3))$. In fact, $N(f_1)=B(u,v)\rtimes\chi=N(f_2)$ for some $(u,v)\in(\mathbf{Z}_3^*)^2\setminus(\mathbf{Z}_3^*(2,1)\cup\mathbf{Z}_3^*(1,-1))$.

We may approximate BPU(3) by a homotopy colimit over a category $\mathbf{I} = \mathbf{I}(\mathrm{SL}(2,\mathbf{F}_3), S_3)$ (a full subcategory of the Quillen category that may be described as formed from the inclusion of a Sylow 3-subgroup S_3 into the special linear group $\mathrm{SL}(2,\mathbf{F}_3)$) with just two objects, $\lambda\colon E^1 \to \mathrm{PU}(3)$ and $\nu\colon E^2 \to \mathrm{PU}(3)$, where E^1 and E^2 are elementary abelian groups of order 3 and 3^2, respectively [17, 6.8, 7,7]; see [24, §4] for the notation used here. Since f_1 and f_2 agree on the centralizers, $C_{\mathrm{PU}(3)}(\lambda E^1) = N_3(\mathrm{PU}(3))$ and $C_{\mathrm{PU}(3)}(\nu E^2) = E^2$, it only remains to compute the relevant Wojtkowiak obstruction groups [29]. For this we need information about the centralizer $C_{\mathrm{F}_4}(f_i E^2)$ and $C_{\mathrm{F}_4}(f_i N_3(\mathrm{PU}(3)))$.

We must have $f_1|E^2 = e_5^2 = f_2|E^2$ for only $e_5^2 \in \mathrm{Mono}(E^2, \mathrm{F}_4)$ can contain in its automorphism group the automorphism group $\mathrm{SL}(2,\mathbf{F}_3)$ of (E^2,ν). Thus $C_{\mathrm{F}_4}(f_i E^2)$ is a p-compact toral group of maximal rank with E^1 as its component group (4.2).

The centralizer $C_{\mathrm{F}_4}(f_i N_3(\mathrm{PU}(3)))$ is (3.4) the p-compact toral group

$$C_{\check{T}(\mathrm{F}_4)\rtimes W(\mathrm{F}_4)}(\check{T}(\mathrm{PU}(3)) \rtimes \langle\sigma\rangle) = \check{T}(\mathrm{F}_4)^{\langle\chi(\sigma)\rangle} = t(\mathrm{F}_4)^{\langle\chi(\sigma)\rangle} = E^2$$

where $t(\mathrm{F}_4) \subset \check{T}(\mathrm{F}_4)$ denotes the maximal elementary abelian subgroup of the discrete approximation $\check{T}(\mathrm{F}_4)$ to $T(\mathrm{F}_4)$ and

The obstructions to a homotopy between the two maps $Bf_1, Bf_2 : \mathrm{BPU}(3) \to \mathrm{BF}_4$ lie in the abelian groups $\lim^1_{\mathbf{I}} \underline{\pi}_1$ and $\lim^2_{\mathbf{I}} \underline{\pi}_2$ where $\underline{\pi}_1$ and $\underline{\pi}_2$ are the abelian $\mathbf{I}$-groups

$$\mathbf{Z}/2 \circlearrowright E^2 \xrightarrow{\mathrm{SL}(2,\mathbf{F}_3)/S_3} E^1 \circlearrowleft \mathrm{SL}(2,\mathbf{F}_3)$$

$$\mathbf{Z}/2 \circlearrowright 0 \xrightarrow{\mathrm{SL}(2,\mathbf{F}_3)/S_3} \mathbf{Z}_3^4 \circlearrowleft \mathrm{SL}(2,\mathbf{F}_3)$$

given by the homotopy groups of the above centralizers. The group $\mathrm{SL}(2,\mathbf{F}_3)$ has no normal subgroups of index two, so it necessarily acts trivially on E^1. It now follows from [24, 10.7.5] that both obstruction groups are trivial and we conclude that f_1 and f_2 are conjugate. This shows that $\mathrm{TRep}(\mathrm{PU}(3), \mathrm{F}_4)$ is T-determined.

Let now $f\colon \mathrm{PU}(3) \to \mathrm{F}_4$ be any toric monomorphism. Then there is (5.1.3, 5.3) a unique, up to sign, pair of units $(u,v) \in (\mathbf{Z}_3^*)^2$, $u+v \in 3\mathbf{Z}_3$, $(u,v) \notin \mathbf{Z}_3^*(2,1) \cup \mathbf{Z}_3^*(1,-1)$, such that $W(\mathrm{F}_4)\pi_1(T(f)) = W(\mathrm{F}_4)B(u,v)$ and then $f = \overline{\psi}^{(u,v)}$ since the p-toric monomorphisms are T-determined. ☐

Lemma 5.7. *Let* $(u,v) \in (\mathbf{Z}_3^*)^2$, $u+v \in 3\mathbf{Z}_3$, $(u,v) \notin \mathbf{Z}_3^*(2,1) \cup \mathbf{Z}_3^*(1,-1)$. *Then*

$$C_{\mathrm{F}_4}(e\psi^{(u,v)}\mathrm{SU}(3)) = \check{T}(\mathrm{F}_4)^{\chi(W(\mathrm{SU}(3)))}$$
$$C_{\mathrm{F}_4}(e\psi^{(u,v)}T(\mathrm{SU}(3))) = T(\mathrm{F}_4)$$

and $C_{\mathrm{F}_4}(\psi^{-1}) = \psi^{-1}$ *in both cases.*

Proof. Since $e\psi^{(u,v)}$ is toric, the centralizer in F_4 of $e\psi^{(u,v)}T(\mathrm{SU}(3))$ equals the maximal torus of F_4. Proceed as in (4.16) to show that $C_{\mathrm{F}_4}(\psi^{-1}) = \psi^{-1}$.

The centralizer $BC_{\mathrm{F}_4}(e\overline{\psi}^{(u,v)}\mathrm{PU}(3))$ is the homotopy colimit of the $\mathbf{I}$-space

$$\mathbf{Z}/2 \circlearrowright B(0) \xrightarrow{\mathrm{SL}(2,\mathbf{F}_3)/S_3} B(1) \circlearrowleft \mathrm{SL}(2,\mathbf{F}_3)$$

where $B(0) = B\check{T}(\mathrm{F}_4)^{\langle\chi(\sigma)\rangle}$ and $B(0) = BC_{\mathrm{F}_4}(e_5^2)$. We need to be more specific about the group actions that occur here.

The 3-normalizer $N_3(\mathrm{PU}(3)) = C_{N(\mathrm{PU}(3))}(\check{T}(\mathrm{PU}(3))^{\langle\sigma\rangle})$ is the centralizer in $N(\mathrm{PU}(3))$ of $\check{T}(\mathrm{PU}(3))^{\langle\sigma\rangle} = E^1$. Since the conjugation by $(0,\tau)$ restricts to the non-trivial automorphism of $\check{T}(\mathrm{PU}(3))^{\langle\sigma\rangle}$ we see that the induced action on $N_3(\mathrm{PU}(3)) = T(\mathrm{PU}(3)) \rtimes \langle\sigma\rangle$ is given by conjugation with $(0,\tau) \in \check{N}(\mathrm{PU}(3)) = \check{T}(\mathrm{PU}(3)) \rtimes W(\mathrm{PU}(3))$.

Since $\check{B}(u,v) \rtimes \chi\colon \check{N}_3(\mathrm{PU}(3)) \to \check{N}(\mathrm{F}_4)$ is χ-equivariant with the Weyl groups acting by conjugation, we see (4.17) that $\mathbf{Z}/2$-acts on

$$\check{T}(\mathrm{F}_4)^{\langle\chi(\sigma)\rangle} = C_{\check{N}(\mathrm{F}_4)}(\check{N}_3(\mathrm{PU}(3)))$$

as conjugation with $(0,\chi(\tau))$. With this information it is now easy to see, using [24, 10.7.5], that

$$\lim\nolimits^0_{\mathbf{I}}\pi_1 = (\check{T}(\mathrm{F}_4)^{\langle\chi(\sigma)\rangle})^{\langle\chi(\tau)\rangle} = \check{T}(\mathrm{F}_4)^{\chi(W(\mathrm{SU}(3)))}$$

is the only non-trivial contribution from the $\mathbf{I}$-groups $\underline{\pi}_1$ and $\underline{\pi}_2$ to the Bousfield-Kan spectral sequence. This means that the morphisms

$$C_{\mathrm{F}_4}(e\overline{\psi}^{(u,v)}\mathrm{PU}(3)) \to C_{\mathrm{F}_4}(N(e\overline{\psi}^{(u,v)})(N(\mathrm{PU}(3))))$$
$$\leftarrow C_{N(\mathrm{F}_4)}(N(e\overline{\psi}^{(u,v)})(N(\mathrm{PU}(3))))$$

are isomorphisms. Consider the corresponding group homomorphism

$$\mu : \check{T}(\mathrm{F}_4)^{\chi(W(\mathrm{SU}(3)))} \times \check{N}(\mathrm{SU}(3)) rightarrow \check{N}(\mathrm{F}_4)$$

which is the inclusion on the first factor and equals $\check{N}(e\psi^{(u,v)})$ on the second factor. Since $\psi^{-1} \rtimes 1$ is inner on $\check{N}(\mathrm{F}_4)$, we have $\mu(1 \times (\psi^{-1} \rtimes 1)) = (\psi^{-1} \rtimes 1)\mu(1 \times (\psi^{-1} \rtimes 1)) = \mu(\psi^{-1} \times (1 \rtimes 1))$ up to inner automorphism. This shows (4.17) that $C_{\mathrm{F}_4}(\psi^{-1}) = \psi^{-1}$ is the non-trivial automorphism of $C_{\mathrm{F}_4}(e\psi^{(u,v)}\mathrm{SU}(3)) = E^1$. □

6. Non-toric morphisms of PU(3) to F_4

The non-toric morphisms of PU(3) to F_4 require special treatment. It is the object of this section to show that also the non-toric morphisms are T-determined, i.e. to complete the proof of the following theorem.

Theorem 6.1. 1. $\mathrm{Mono}(\mathrm{PU}(3),\mathrm{F}_4)$ *is T-determined.*

2. *The map*

$$\langle(-1,-1)\rangle \backslash \{(u,v) \in (\mathbf{Z}_3^*)^2 | u+v \in 3\mathbf{Z}_3\} \to \mathrm{Mono}(\mathrm{PU}(3),\mathrm{F}_4)$$
$$\pm(u,v) \to e\overline{\psi}^{(u,v)}$$

is a bijection.

Since the toric morphisms were dealt with in (5.4) only the non-toric ones need be considered in order to finish the proof of (6.1).

The first lemma, which is of a general nature, assures the existence of a kind of preferred lifts in certain situations.

Let G be a p-compact toral group sitting in short exact sequence $S \xrightarrow{i_1} G \to \pi_0(G)$ where S is a p-compact torus and $\pi_0(G)$ *cyclic* p-group. Let $j\colon N \to X$ be the maximal torus normalizer of a p-compact group, X, and let $i_2\colon T \to N$ be the inclusion of the identity component. Suppose that we are given a morphisms, B and f, such that the diagram

$$\begin{array}{ccc} S & \xrightarrow{B} & T \\ {\scriptstyle i_1}\big\downarrow & & \big\downarrow{\scriptstyle ji_2} \\ G & \xrightarrow[f]{} & X \end{array}$$

commutes up to conjugacy and B is admissible in the sense that for any $\xi \in \pi_0(G)$ there exists some w in the Weyl group for X such that $B\xi = wB$.

Lemma 6.2. *Assuming that the component group $\pi_0(G)$ is cyclic there is a unique representation $\phi \in \operatorname{Rep}(G, N)$ such that the diagram*

$$\begin{array}{ccccc} & & & & T \\ & & & {\scriptstyle B}\nearrow & \big\downarrow{\scriptstyle i_2} \\ & & & & N \\ & & & {\scriptstyle \phi}\nearrow & \big\downarrow{\scriptstyle j} \\ S & \xrightarrow[i_1]{} & G & \xrightarrow[f]{} & X \end{array}$$

commutes up to conjugacy and such that the morphism

$$C_j\colon C_N(\phi G) \to C_X(fG),$$

induced by j, is a maximal torus normalizer for the centralizer $C_X(fG)$ of G in X.

Proof. The $\pi_0(G)$-map induced by j

$$\begin{array}{ccc} BC_N(i_2BS)_{h\pi_0(G)} & \longrightarrow & BC_X(ji_2BS)_{h\pi_0(G)} \\ & \searrow \quad \swarrow & \\ & B\pi_0(G) & \end{array}$$

between the $\pi_0(G)$-spaces $BC_N(i_2BS) = \mathrm{map}(BS, BN)_{i_2B}$ and $BC_X(ji_2BS) = \mathrm{map}(BS, BX)_{ji_2B}$ is a maximal torus normalizer. There is an induced map

$$\mathrm{map}(BG, BN)_{i_1 \to i_2B} = BC_N(i_2BS)^{h\pi_0(G)} \to BC_X(ji_2BS)^{h\pi_0(G)} = \mathrm{map}(BG, BX)_{i_1 \to ji_2B} \quad (6.3)$$

of homotopy fixed point spaces.

According to [25, 4.6], the section $Bf \in BC_X(ji_2BS)^{h\pi_0(G)}$ admits, since $\pi_0(G)$ is assumed to be cyclic, a unique lift $B\phi \in BC_N(i_2BS)^{h\pi_0(G)}$ such that the restriction of (6.3) to the corresponding components,

$$BC_N(\phi G) = \mathrm{map}(BG, BN)_{B\phi} \to \mathrm{map}(BG, BX)_{Bf} = BC_X(fG)$$

is a maximal torus normalizer for the p-compact group $C_X(fG)$. □

After these general and preparatory remarks, we now return to the discussion of non-toric morphisms from PU(3) to F_4.

Let $f\colon \mathrm{PU}(3) \to \mathrm{F}_4$ be a morphism of 3-compact groups such that

$$f|T(\mathrm{PU}(3)) = W(\mathrm{F}_4)B(2,1) \in [BT(\mathrm{PU}(3)), \mathrm{BF}_4].$$

By (6.2), there is a unique $\phi(2,1) \in \mathrm{Rep}(N_3(\mathrm{PU}(3)), N(\mathrm{F}_4))$, extending $B(2,1)$, such that $C_{N(\mathrm{F}_4)}(\phi(2,1)N_3(\mathrm{PU}(3)))$ is a maximal torus normalizer for the centralizer $C_{\mathrm{F}_4}(fN_3(\mathrm{PU}(3)))$. We shall now determine this map $\phi(2,1)$.

Let $\check{N}_3 = \check{T}_1 \rtimes \langle\sigma\rangle$ and $\check{N}_2 = \check{T}_2 \rtimes W_2$ be the discrete approximations to the the 3-normalizer $N_3(\mathrm{PU}(3))$ and the maximal torus normalizer $N(\mathrm{F}_4)$, respectively. Also, let $\check{B}(2,1)\colon \check{T}_1 \to \check{T}_2$ be a discrete approximation to $B(2,1)$. The stabilizer subgroup $W(\mathrm{F}_4)^{\check{B}(2,1)}$ at $\check{B}(2,1)$ for the action of $W(\mathrm{F}_4)$ on $\mathrm{Hom}(\check{T}_1, \check{T}_2)$ is isomorphic to the permutation group Σ_3 and generated by the two Weyl group elements

$$w_1 = \begin{pmatrix} 0 & 0 & 0 & 1 \\ -1 & 0 & 0 & 0 \\ 0 & 0 & 1 & 0 \\ 0 & -1 & 0 & 0 \end{pmatrix} \quad \text{and} \quad w_2 = \begin{pmatrix} 0 & -1 & 0 & 0 \\ -1 & 0 & 0 & 0 \\ 0 & 0 & 1 & 0 \\ 0 & 0 & 0 & 1 \end{pmatrix}$$

of order 3 and 2, respectively.

Lemma 6.4. *The discrete approximation $\check{\phi}(2,1)\colon \check{T}_1 \rtimes \langle\sigma\rangle \to \check{T}_2 \rtimes W(\mathrm{F}_4)$ to $\phi(2,1)$ is conjugate to $\check{B}(2,1) \rtimes \chi$.*

Proof. For general reasons, the discrete approximation $\check{\phi}(2,1)$ to $\phi(2,1)$ has the form $\check{\phi}(2,1)(t,1) = (\check{B}(2,1)(t),1)$ and $\check{\phi}(2,1)(0,\sigma) = (a, \lambda(\sigma))$ where $\lambda : \langle\sigma\rangle \to W(\mathrm{F}_4)$ is a group homomorphism, $\check{B}(2,1)$ is λ-equivariant, and a is a 1-cocycle in $Z^1(\langle\lambda(\sigma)\rangle ; \check{T}(\mathrm{F}_4))$.

Since the homomorphism $\check{B}(2,1)$ is χ-equivariant we know that $\lambda(\sigma)$ is an element of order 3 in the coset $\chi(\sigma)W_2^{\check{B}(2,1)}$. This leaves the three possibilities $\chi(\sigma), \chi(\sigma)w_1$, and $\chi(\sigma)w_1^2$ for $\lambda(\sigma)$. Since w_2 conjugates $\chi(\sigma)$ into $\chi(\sigma)w_1^2$ we can ignore the third possibility. We now rule out the second possibility.

Assume for the moment that $\lambda(\sigma) = \chi(\sigma)w_1$. Explicit computation shows that $H^0(\langle\chi(\sigma)w_1\rangle ; \check{T}(\mathrm{F}_4))$ is a 3-discrete torus of rank 2 and that the group $H^0(\langle\chi(\sigma)w_1\rangle ; \check{T}(\mathrm{F}_4))$ is cyclic of order 3 generated by the cohomology class of the 1-cocycle

$$a = \begin{pmatrix} 0 \\ 0 \\ 1 \\ 0 \end{pmatrix} \in t(\mathrm{F}_4) \subset \check{T}(\mathrm{F}_4)$$

which is fixed by $W(\mathrm{F}_4)^{\check{B}(2,1)}$. It follows that the centralizer

$$\begin{aligned} &C_{\check{T}(\mathrm{F}_4)\rtimes W(\mathrm{F}_4)}(\check{\phi}(2,1)\check{N}_3) \\ &\quad = C_{\check{T}(\mathrm{F}_4)\rtimes W(\mathrm{F}_4)}(\check{B}(2,1)\check{T}(\mathrm{PU}(3))) \cap C_{\check{T}(\mathrm{F}_4)\rtimes W(\mathrm{F}_4)}(a, \chi(\sigma)w_1) \\ &\quad = (\check{T}(\mathrm{F}_4) \rtimes W(\mathrm{F}_4)^{\check{B}(2,1)}) \cap C_{\check{T}(\mathrm{F}_4)\rtimes W(\mathrm{F}_4)}(a, \chi(\sigma)w_1) \\ &\quad = C_{\check{T}(\mathrm{F}_4)\rtimes W(\mathrm{F}_4)^{\check{B}(2,1)}}(a, \chi(\sigma)w_1) \\ &\quad = \check{T}(\mathrm{F}_4)^{\langle\chi(\sigma)w_1\rangle} \rtimes W(\mathrm{F}_4)^{\check{B}(2,1)} \end{aligned}$$

is the (discrete) maximal torus normalizer for SU(3) and hence (6.2) that the centralizer $C_{\mathrm{F}_4}(fN_3(\mathrm{PU}(3)))$ is isomorphic to the N-determined 3-compact group SU(3) [24, 1.2]. Thus $\phi(2,1)\colon N_3(\mathrm{PU}(3)) \to \mathrm{F}_4$ extends to a morphism

$$N_3(\mathrm{PU}(3)) \times \mathrm{SU}(3) \to \mathrm{F}_4$$

which is a non-toric monomorphism on the second factor and we get a factorization

$$N_3(\mathrm{PU}(3)) \to C_{\mathrm{F}_4}(\mathrm{SU}(3)) = \mathrm{SU}(3) \to \mathrm{F}_4$$

of $\phi(2,1)$ through another non-toric monomorphism of SU(3) to F_4. The restriction of this map to the maximal tori

$$T(\mathrm{PU}(3)) \to T(\mathrm{SU}(3)) \to T(\mathrm{F}_4)$$

provides a factorization, up to left action by $W(\mathrm{F}_4)$, of $B(2,1)$ as the composition of an isomorphism followed by $A(u,0)$ or $A(0,u)$, $u \in \mathbf{Z}_3^*$, and hence we have that the set

$$W(\mathrm{F}_4) \cdot A(2,1) \cdot \mathrm{GL}(\Sigma_0(\mathbf{Q}_3^3)) \subset \mathrm{Hom}_{\mathbf{Q}_3}(\Sigma_0(\mathbf{Q}_3^3), \mathbf{Q}_3^4)$$

contains $A(1,0)$ or $A(0,1)$. It is easy to verify, using a computer, that this is not the case, so we have arrived at a contradiction.

Thus $\lambda(\sigma) = \chi(\sigma)w_1$ can not occur and we are left with $\lambda(\sigma) = \chi(\sigma)$ as the only possibility. As $H^1(\langle\chi(\sigma)\rangle ; \check{T}(\mathrm{F}_4)) = 0$ (5.5), $\check{\phi}(2,1) = \check{B}(2,1) \rtimes \chi$ is, up to conjugation, the only extension of the pair $(\check{B}(2,1), \chi)$ to a homomorphism $\check{T}_1 \rtimes \langle\sigma\rangle \to \check{T}(\mathrm{F}_4) \rtimes W(\mathrm{F}_4)$. □

A similar statement holds for the non-toric morphism $B(1,-1)$ which differs from $B(2,1)$ by an automorphism of F_4.

Proof of Theorem 6.1. It suffices to show that $f_1 \simeq f_2$ whenever $f_1, f_2 : \mathrm{PU}(3) \to \mathrm{F}_4$ are monomorphisms such that $f_1|T(\mathrm{PU}(3)) = W(\mathrm{F}_4)B(2,1) = f_2|T(\mathrm{PU}(3))$. We already know (6.4) that the two morphisms become conjugate when restricted to $N_3(\mathrm{PU}(3))$. Therefore, the situation is now exactly as in the proof of Theorem 5.4: In order to compute the relevant Wojtkowiak obstruction groups [29] we need information about the centralizer $C_{\mathrm{F}_4}(f_iE^2)$ and $C_{\mathrm{F}_4}(f_iN_3(\mathrm{PU}(3)))$.

Again, we must have $f_1|E^2 = e_5^2 = f_2|E^2$ and $C_{\mathrm{F}_4}(f_iE^2)$ is a p-compact toral group of maximal rank with $\mathbf{Z}/3$ as its component group (4.2).

Also, we know (6.2, 6.4) that the centralizer in $\check{T}(\mathrm{F}_4) \rtimes W(\mathrm{F}_4)$ of $\check{\phi}(2,1)$ is the (discrete) maximal torus normalizer for $C_{\mathrm{F}_4}(f_iN_3(\mathrm{PU}(3)))$. Since

$$\begin{aligned}
C_{\check{T}(\mathrm{F}_4)\rtimes W(\mathrm{F}_4)}(\check{\phi}(2,1)\check{N}_3) &= C_{\check{T}(\mathrm{F}_4)\rtimes W(\mathrm{F}_4)}(\check{B}(2,1)\check{T}_1) \cap C_{\check{T}(\mathrm{F}_4)\rtimes W(\mathrm{F}_4)}(\chi(\sigma)) \\
&= (\check{T}(\mathrm{F}_4) \rtimes W(\mathrm{F}_4)^{\check{B}(2,1)}) \cap (\check{T}(\mathrm{F}_4)^{\chi(\sigma)} \rtimes C_{W(\mathrm{F}_4)}(\chi(\sigma))) \\
&= \check{T}(\mathrm{F}_4)^{\chi(\sigma)} \rtimes C_{W(\mathrm{F}_4)^{\check{B}(2,1)}}(\chi(\sigma)) \\
&= t(\mathrm{F}_4)^{\chi(\sigma)} \rtimes \langle w_1 \rangle
\end{aligned}$$

is a finite group (of order 27 and with center of order 3) it follows that also $C_{\mathrm{F}_4}(f_iN_3(\mathrm{PU}(3)))$ is this finite, but non-abelian, 3-group.

The obstructions to a homotopy between the two maps $Bf_1, Bf_2 : B\mathrm{PU}(3) \to B\mathrm{F}_4$ lie in the set $\lim^1_{\mathbf{I}}\underline{\pi}_1$ and in the abelian group $\lim^2_{\mathbf{I}}\underline{\pi}_2$ where $\underline{\pi}_1$ and $\underline{\pi}_2$ are the $\mathbf{I}$-groups

$$\mathbf{Z}/2 \circlearrowright \pi \xrightarrow{\mathrm{SL}(2,\mathbf{F}_3)/S_3} E^1 \circlearrowleft \mathrm{SL}(2,\mathbf{F}_3)$$

$$\mathbf{Z}/2 \circlearrowright 0 \xrightarrow{\mathrm{SL}(2,\mathbf{F}_3)/S_3} \mathbf{Z}_3^4 \circlearrowleft \mathrm{SL}(2,\mathbf{F}_3)$$

given by the homotopy groups of the above centralizers, e.g. $\pi = t(\mathrm{F}_4)^{\chi(\sigma)} \rtimes \langle w_1 \rangle$. The group $\lim^2_{\mathbf{I}}\underline{\pi}_2$ is trivial for general reasons [24, 10.7.5]. That also $\lim^1_{\mathbf{I}}\underline{\pi}_1 = *$ follows from (6.5) below since both the central $\mathbf{I}$-subgroup

$$\mathbf{Z}/2 \circlearrowright 0 \longrightarrow \mathbf{Z}/3 \circlearrowleft \mathrm{SL}(2,\mathbf{F}_3)$$

as well as the quotient $\mathbf{I}$-group

$$\mathbf{Z}/2 \circlearrowright \pi \longrightarrow 0 \circlearrowleft \mathrm{SL}(2,\mathbf{F}_3)$$

where $\mathrm{SL}(2,\mathbf{F}_3)$ necessarily acts trivially, have vanishing $\lim^1$ by [24, 10.7] and (6.6). □

The following observations were used to compute the non-abelian $\lim^1$.

Let $\mathbf{I}$ be a small category. Define an $\mathbf{I}$-group to be a functor from the category $\mathbf{I}$ to the category of groups. Let $A \to E \to G$ be a central extension of $\mathbf{I}$-groups meaning that A, E, and G are $\mathbf{I}$-groups, the arrows are natural transformations, and that the evaluation at each object of $\mathbf{I}$ yields a central extension of groups.

Lemma 6.5. *Any central extension of* $\mathbf{I}$*-groups* $A \to E \to G$ *induces an exact sequence*

$$* \to \lim^0_{\mathbf{I}} A \to \lim^0_{\mathbf{I}} E \to \lim^0_{\mathbf{I}} G \to \lim^1_{\mathbf{I}} A \to \lim^1_{\mathbf{I}} E \to \lim^1_{\mathbf{I}} G \to \lim^2_{\mathbf{I}} A$$

of sets. Moreover, the fibres of the map $\lim^1_{\mathbf{I}} E \to \lim^1_{\mathbf{I}} G$ *are precisely the orbits for an induced action of the abelian group* $\lim^1_{\mathbf{I}} A$ *on the set* $\lim^1_{\mathbf{I}} E$.

Corollary 6.6. *Let* $\mathbf{I}$ *be a finite group acting on a finite group* π. *If the* π *is a* p*-group and* p *does not divide the order of* $\mathbf{I}$, *then* $\lim^1_{\mathbf{I}} \pi = *$.

Proof. This follows, using the preceding lemma, by induction over the order of π since any non-trivial p-group has a non-trivial center. □

Proof of Theorem 1.1. Modulo the action of the Weyl group $W_{\mathrm{F}_4}(\mathrm{SU}(3,3))$ of order two (4.15), the sets

$$\mathrm{Rep}(\mathrm{SU}(3), \mathrm{SU}(3,3)) = \{0\} \cup \mathrm{Mono}(\mathrm{SU}(3), \mathrm{SU}(3,3)) \cup \mathrm{Mono}(\mathrm{PU}(3), \mathrm{SU}(3,3))$$

and

$$\mathrm{Rep}(\mathrm{SU}(3), \mathrm{F}_4) = \{0\} \cup \mathrm{Mono}(\mathrm{SU}(3), \mathrm{F}_4) \cup \mathrm{Mono}(\mathrm{PU}(3), \mathrm{F}_4)$$

are (4.13, 6.1) in correspondence. □

Lemma 6.7. *Let* $(u,v) \in \mathbf{Z}_3^*(2,1) \cup \mathbf{Z}_3^*(1,-1)$. *Then*

$$C_{\mathrm{F}_4}(e\psi^{(u,v)}\mathrm{SU}(3)) = \check{T}(\mathrm{F}_4)^{\chi(W(\mathrm{SU}(3)))}$$

$$C_{\mathrm{F}_4}(e\psi^{(u,v)}T(\mathrm{SU}(3))) = T(\mathrm{SU}(3)) \times_{Z(\mathrm{SU}(3))} \mathrm{SU}(3)$$

and $C_{\mathrm{F}_4}(\psi^{-1}) = \psi^{-1}$ *in both cases.*

Proof. We shall apply the Bousfield-Kan spectral sequence [4] to the mapping space $\mathrm{map}(\mathrm{BPU}(3), \mathrm{BF}_4)_{e\overline{\psi}^{(u,v)}}$ where BPU(3) is viewed as the homotopy colimit of the $\mathbf{I}$-space

$$\mathbf{Z}/2 \circlearrowright B(0) \xrightarrow{\mathrm{SL}(2,\mathbf{F}_3)/S_3} B(1) \circlearrowleft \mathrm{SL}(2,\mathbf{F}_3) \tag{6.8}$$

where $B(0) = BC_{\mathrm{F}_4}(e\overline{\psi}^{(u,v)} N_3(\mathrm{PU}(3)))$ and $B(1) = BC_{\mathrm{F}_4}(e_5^2)$. It represents no loss of generality to assume that $(u,v) = (2,1)$.

As we saw in the proof of (5.7), $\mathbf{Z}/2$-acts on $\check{N}_3(\mathrm{PU}(3)) = \check{T}(\mathrm{PU}(3)) \rtimes \langle\sigma\rangle$ as conjugation with $(0,\tau) \in \check{N}(\mathrm{PU}(3)) = \check{T}(\mathrm{PU}(3)) \rtimes W(\mathrm{PU}(3))$. But this is again the restriction to

$$\check{\varphi}(2,1)(\check{N}_3(\mathrm{PU}(3))) \subset \check{T}(\mathrm{F}_4) \rtimes W(\mathrm{F}_4)$$

of conjugation by $(0, \chi(\tau))$. Thus (4.18) the $\mathbf{Z}/2$-action on

$$C_{\mathrm{F}_4}(e\overline{\psi}^{(2,1)} N_3(\mathrm{PU}(3))) = C_{\check{T}(\mathrm{F}_4)\rtimes W(\mathrm{F}_4)}(\check{\varphi}(2,1)\check{N}_3) = \check{T}(\mathrm{F}_4)^{\langle\sigma\rangle} \rtimes \langle w_1\rangle$$

is through conjugation with $(0, \chi(\tau))$.

Note also that the multiplication map

$$\mu\colon C_{\check{T}(\mathrm{F}_4)\rtimes W(\mathrm{F}_4)}(\check{\varphi}(2,1)\check{N}_3)\times \check{N}_3 \to \check{T}(\mathrm{F}_4)\rtimes W(\mathrm{F}_4)$$

satisfies

$$\mu(\psi^{-1}\times 1)=\psi^{-1}\mu(\psi^{-1}\times 1)=\mu(1\times N_3(\psi^{-1}))$$

up to inner automorphism. This means that the induced action on the centralizer $C_{\mathrm{F}_4}(e\overline{\psi}^{(2,1)}N_3(\mathrm{PU}(3)))$ is $C_{\mathrm{F}_4}(N_3(\psi^{-1}))=\psi^{-1}\rtimes 1$.

Recall from [5] that there is an essentially unique monomorphism $\iota\colon \mathrm{DI}_2\to \mathrm{F}_4$ inducing a monomorphism $t(\iota)\colon t(\mathrm{DI}_2)\to t(\mathrm{F}_4)$ and a group monomorphism $\chi:\mathrm{GL}(2,\mathbf{F}_3)=W(\mathrm{DI}_2)\to W(\mathrm{F}_4)$ extending (4.7). Now, $t(\iota)$ is isomorphic to e_5^2 and from the commutative diagram

$$\begin{array}{ccc} t(\mathrm{DI}_2) & \xrightarrow{\;w^{-1}\;} & t(\mathrm{DI}_2)\\ {\scriptstyle t(\iota)}\downarrow & & \downarrow{\scriptstyle t(\iota)}\\ \check{T}(\mathrm{F}_4)\rtimes W(\mathrm{F}_4) & \xrightarrow[(0,\chi(w))]{} & \check{T}(\mathrm{F}_4)\rtimes W(\mathrm{F}_4) \end{array}$$

we see (4.18) that $w\in \mathrm{GL}(2,\mathbf{F}_3)$ acts on $C_{\check{N}(\mathrm{F}_4)}(t(\mathrm{DI}_2))=\check{T}(\mathrm{F}_4)\rtimes W(\mathrm{F}_4)^{t(\mathrm{DI}_2)}$ as conjugation with the element $(0,\chi(w))$ of the semi-direct product. The restriction to $\mathrm{SL}(2,\mathbf{F}_3)$ of this action gives the action on $C_{N(\mathrm{F}_4)}(t(\mathrm{DI}_2))=C_{\mathrm{F}_4}(t(\mathrm{DI}_2))$ in (6.8).

The conclusion of this is that

$$\lim\nolimits_{\mathbf{I}}^0\underline{\pi}_1=(\check{T}(\mathrm{F}_4)^{\langle\chi(\sigma)\rangle}\rtimes\langle w_1\rangle)^{\langle\chi(\tau)\rangle}=\check{T}(\mathrm{F}_4)^{\chi(W(\mathrm{SU}(3)))}$$

is the only non-trivial contribution from the groups $\lim_{\mathbf{I}}^{-i}\underline{\pi}_j$, $i+j\geq 0$, of the Bousfield-Kan spectral sequence. Consequently, $C_{\mathrm{F}_4}(e\psi^{(2,1)}\mathrm{SU}(3))$ is isomorphic to this group of order 3. The action of $C_{\mathrm{F}_4}(\psi^{-1})$, which is the restriction of the action of $C_{\mathrm{F}_4}(N_3(\psi^{-1}))$, is given by ψ^{-1}.

The centralizer

$$C_{\check{T}(\mathrm{F}_4)\rtimes W(\mathrm{F}_4)}(e\psi^{(2,1)}\check{T}(\mathrm{SU}(3)))=\check{T}(\mathrm{F}_4)\rtimes W(\mathrm{F}_4)^{A(2,1)}$$

is the (discrete) maximal torus normalizer for

$$C_{\mathrm{F}_4}(e\psi^{(2,1)}\check{T}(\mathrm{SU}(3)))$$

and the centralizer

$$C_{\check{T}(\mathrm{F}_4)\rtimes W(\mathrm{F}_4)}(e\psi^{(0,1)}\check{T}(\mathrm{SU}(3)))=\check{T}(\mathrm{F}_4)\rtimes W(\mathrm{F}_4)^{A(0,1)}$$

is the (discrete) maximal torus normalizer for

$$C_{\mathrm{F}_4}(e\psi^{(0,1)}\check{T}(\mathrm{SU}(3)))=\mathrm{SU}(3)\times_{Z(\mathrm{SU}(3))}T(\mathrm{SU}(3))$$

[22, 3.4.3]. Since the two stabilizer subgroups $W(\mathrm{F}_4)^{A(2,1)}$ and $W(\mathrm{F}_4)^{A(0,1)}$ are conjugate in $W(\mathrm{F}_4)$, the two maximal torus normalizers are isomorphic and hence the two centralizers are isomorphic, too, by N-determinism [23] [24].

The group homomorphism $\mu\colon (\check{T}(\mathrm{F}_4)\rtimes W(\mathrm{F}_4)^{A(2,1)} \to \check{T}(\mathrm{F}_4)\rtimes W(\mathrm{F}_4)$ which is the inclusion on the first factor and equals $A(2,1)$ on the second factor satisfies

$$\mu((1\rtimes 1)\times \psi^{-1}) = (\psi^{-1}\rtimes 1)\mu((1\rtimes 1)\times\psi^{-1}) = \mu((\psi^{-1}\rtimes 1)\times 1$$

up to inner automorphisms. This shows (4.17) that $C_{\mathrm{F}_4}(\psi^{-1}) = \psi^{-1}$. □

7. Morphisms from G_2 to F_4 at the prime $p=3$

Using the Jackowski-McClure decomposition of $B\,\mathrm{G}_2$ and the Bousfield-Kan spectral sequence we classify morphisms $\mathrm{G}_2\to\mathrm{F}_4$ viewed as 3-compact groups and compute their centralizers.

The Weyl group of G_2, $W(\mathrm{G}_2) < \mathrm{GL}(\Sigma_0(\mathbf{Z}_3^3))$ is the product of the Weyl group $W(\mathrm{SU}(3)) = \langle\sigma,\tau\rangle$ of $\mathrm{SU}(3)$ and the central group $\langle -1\rangle$ of order 2. The group morphism χ from (4.7) extends to a group homomorphism $\chi\colon W(\mathrm{G}_2)\to W(\mathrm{F}_4)$ simply by putting $\chi(-1) = -1$. Let $\mathbf{I} = \mathbf{I}(W(\mathrm{G}_2), W(\mathrm{SU}(3)))$ denote the category

$$\langle -1\rangle \circlearrowright 0 \xrightarrow{W(\mathrm{G}_2)/W(\mathrm{SU}(3))} 1 \circlearrowleft W(\mathrm{G}_2)$$

of the central inclusion of $W(\mathrm{SU}(3))$ into $W(\mathrm{G}_2)$. Then $B\,\mathrm{G}_2$ is [24, §7] $H^*\mathbf{F}_3$-equivalent to the homotopy colimit of an $\mathbf{I}^{\mathrm{op}}$-space

$$\langle\psi^{-1}\rangle \circlearrowright B(0) \xleftarrow{W(\mathrm{SU}(3))^{\mathrm{op}}\backslash W(\mathrm{G}_2)^{\mathrm{op}}} B(1) \circlearrowleft W(\mathrm{G}_2)^{\mathrm{op}} \qquad (7.1)$$

where $B(0) = B\mathrm{SU}(3)$ and $B(1) = BT(\mathrm{SU}(3))$.

Theorem 7.2. *The restriction map*

$$\mathrm{Rep}(\mathrm{G}_2,\mathrm{F}_4)\to\mathrm{Rep}(\mathrm{SU}(3),\mathrm{F}_4)$$

is bijective. The centralizer $C_{\mathrm{F}_4}(e\psi^{(u,v)}\mathrm{G}_2)$, $u,v\in\mathbf{Z}_3^*\cup\{0\}$, is isomorphic to $\mathrm{SU}(2)$ *if $uv=0$ and trivial otherwise.*

Proof. We must show that any morphism $\mathrm{SU}(3)\to\mathrm{F}_4$ extends uniquely to G_2. Since this is true for the trivial morphism by [22, 6.7], we only need here to consider non-trivial morphisms.

Let $(u,v)\in(\mathbf{Z}_3^*\cup\{0\})^2$, $(u,v)\neq(0,0)$. Since $e\psi^{(u,v)}\colon \mathrm{SU}(3)\to\mathrm{F}_4$ is invariant under ψ^{-1}, this map $e\psi^{(u,v)}$ and its restriction to the maximal torus form a homotopy coherent set of maps out of the $\mathbf{I}^{\mathrm{op}}$-space (7.1). Thus it suffices to show that $\lim_{\mathbf{I}}^{-i}\underline{\pi}_j(u,v) = 0$ for $i+j\geq -1$ where $\underline{\pi}_j(u,v)$ is the $\mathbf{I}$-group

$$\mathbf{Z}/2 \circlearrowright \pi_j(0) \xrightarrow{W(\mathrm{G}_2)/W(\mathrm{SU}(3))} \pi_j(1) \circlearrowleft W(\mathrm{G}_2)$$

where the group $\pi_j(0) = \pi_j(u,v)(0) = \pi_j(BC_{\mathrm{F}_4}(e\psi^{(u,v)}\mathrm{SU}(3)))$ and the group $\pi_j(1) = \pi_j(u,v)(1) = \pi_j(BC_{\mathrm{F}_4}(e\psi^{(u,v)}T(\mathrm{SU}(3))))$. Since the abelian $\mathbf{I}$-groups $\underline{\pi}_j(u,v)$ are in fact $\mathbf{Z}_3[\mathbf{I}]$-modules and $W(\mathrm{SU}(3))$ is normal in $W(\mathrm{G}_2)$, it follows from [24, 10.7.5] that $\lim_{\mathbf{I}}^0\underline{\pi}_j(u,v) = \pi_j(u,v)(0)^{\mathbf{Z}/2} = \pi_j(BC_{\mathrm{F}_4}(e\psi^{(u,v)}\mathrm{SU}(3)))^{\mathbf{Z}/2}$ is the subgroup that is invariant under the action of ψ^{-1} and that the higher limits are automatically trivial. By (4.16, 5.7, 6.7), $\pi_j(u,v)(0)^{\mathbf{Z}/2}$ is trivial except when either $u=0$ or $v=0$ when it equals the invariants $\pi_j(\mathrm{BSU}(3))^{\langle B\psi^{-1}\rangle}$.

We now examine the case $(u,v)=(0,1)$ more closely. According to Dynkin [12, 13] the Lie group F_4 contains a copy of (a central quotient of) $SU(2)\times \mathrm{G}_2$. The restriction to G_2 of this inclusion $\mathrm{SU}(2)\times\mathrm{G}_2\to\mathrm{F}_4$ equals, up to an automorphism of F_4, the map $e\psi^{(0,1)}$ for otherwise the restriction to the other factor, the inclusion of SU(2) into F_4, would factor through the trivial 3-compact group. The homotopy class of the restriction

$$\mathrm{BSU}(2)\times\mathrm{BSU}(3)\to\mathrm{BSU}(2)\times\mathrm{BG}_2\to\mathrm{BF}_4$$

to $\mathrm{SU}(2)\times\mathrm{SU}(3)$ is determined by its adjoint in

$$\begin{aligned}\pi_0(\mathrm{map}(\mathrm{BSU}(2),\mathrm{map}(\mathrm{BSU}(3),\mathrm{BF}_4)_{B(e\psi^{(0,1)})}) &=\\ \pi_0(\mathrm{map}(\mathrm{BSU}(2),\mathrm{BSU}(3))) &= \mathrm{Rep}(\mathrm{SU}(2),\mathrm{SU}(3))\end{aligned}$$

so. Since SU(3) contains (7.3) an essentially unique copy of SU(2), we conclude that the diagram of 3-compact groups

$$\begin{array}{ccc}\mathrm{SU}(2)\times\mathrm{SU}(3) & \xrightarrow{S\iota(2,3)\times 1} & \mathrm{SU}(3)\times\mathrm{SU}(3)\\ \downarrow & & \downarrow\\ \mathrm{SU}(2)\times\mathrm{G}_2 & \longrightarrow & \mathrm{F}_4\end{array}$$

commutes up to conjugacy. After taking adjoint maps we end up with

$$\begin{array}{ccc}\mathrm{BSU}(2) & \longrightarrow & \mathrm{map}(\mathrm{BG}_2,\mathrm{BF}_4)_{B(e\psi^{(0,1)})}\\ {\scriptstyle BS\iota(2,3)}\downarrow & & \downarrow\\ \mathrm{BSU}(3) & \xrightarrow{\simeq} & \mathrm{map}(\mathrm{BSU}(3),\mathrm{BF}_4)_{B(e\psi^{(0,1)})}\end{array}$$

which commutes up to homotopy and where the lower horizontal arrow represents (4.16) a homotopy equivalence homotopy equivariant under the action $\langle B\psi^{-1}\rangle$. By the above computations with the Bousfield-Kan spectral sequence,

$$\pi_*(\mathrm{map}(\mathrm{BG}_2,\mathrm{BF}_4),B(e\psi^{(0,1)}))=\pi_*(\mathrm{map}(\mathrm{BSU}(3),\mathrm{BF}_4),B(e\psi^{(0,1)}))^{\langle B\psi^{-1}\rangle},$$

and linked with (7.4) this shows that the upper horizontal map is a homotopy equivalence as well. □

The morphism $e\psi^{(u,v)}\colon \mathrm{G}_2\to\mathrm{F}_4$ where $u,v\in\mathbf{Z}_3^*$ with sum $u+v\in 3\mathbf{Z}_3$, is an example a non-trivial non-monomorphism defined on a center-free 3-compact group.

The following two results were needed for the proof of Theorem 7.2.

Lemma 7.3. *Let $S\iota(2,3)\colon \mathrm{SU}(2)\to\mathrm{SU}(3)$ be the canonical inclusion. The map*

$$\begin{aligned}\mathrm{Rep}(\mathrm{SU}(2),\mathrm{SU}(2))&\to\mathrm{Rep}(\mathrm{SU}(2),\mathrm{SU}(3))\\ \psi^u&\to S\iota(2,3)\psi^u\end{aligned}$$

is a bijection that identifies $\mathrm{Out}(\mathrm{SU}(2)) = \mathbf{Z}_3^*/\langle -1\rangle$ *and* $\mathrm{Mono}(\mathrm{SU}(2), \mathrm{SU}(3))$.

Proof. This follows from (3.13) that allows to identify both $\mathrm{Rep}(\mathrm{SU}(2), \mathrm{SU}(2))$ and $\mathrm{Rep}(\mathrm{SU}(2), \mathrm{SU}(3))$ to $\mathbf{Z}_3/\langle -1\rangle$. □

Since $\psi^{-1} S\iota(2,3) = S\iota(2,3)\psi^{-1} = S\iota(2,3)$, the image of $\pi_*(\mathrm{BSU}(2))$ in $\pi_*(\mathrm{BSU}(3))$ is invariant under the action of the group $\langle B\psi^{-1}\rangle$.

Lemma 7.4. *There is an isomorphism, induced by* $S\iota(2,3)$,

$$\pi_*(\mathrm{BSU}(2)) \to \pi_*(\mathrm{BSU}(3))^{\langle B\psi^{-1}\rangle}$$

between the homotopy of $\mathrm{BSU}(2)$ *and the* $\langle B\psi^{-1}\rangle$*-invariant subgroup of the homotopy of* $\mathrm{BSU}(3)$.

Proof. There is a short exact sequence of homotopy groups

$$0 \to \pi_*(\mathrm{SU}(2)) \to \pi_*(\mathrm{SU}(3)) \to \pi_*(S^5) \to 0$$

of $\mathbf{F}_3$-complete spaces induced by the fibration of $\mathrm{SU}(3)$ onto S^5 with fibre $\mathrm{SU}(2)$. This fibration splits since $\pi_4(\mathrm{SU}(2)) \otimes \mathbf{Z}_3 = 0$. The homomorphism ψ^{-1}, complex conjugation of matrices, restricts to the identity on the fibre and induces the degree -1-map on the base. Using that the 3-completion of S^5 is an H-space we see that the degree -1 self-map induces multiplication by -1 on the homotopy groups $\pi_*(S^5) \otimes \mathbf{Z}_3$ and the claim follows. □

8. Morphisms from $\mathrm{SU}(3)$ to G_2 at the prime $p = 3$

The classification of morphisms $\mathrm{SU}(3) \to \mathrm{G}_2$ of 3-compact groups proceeds very much like the classification of morphisms $\mathrm{SU}(3) \to \mathrm{F}_4$.

Lemma 8.1. *The set* $\mathrm{Mono}(E^1, \mathrm{G}_2)$ *contains two elements,* e_1^1, e_2^1, *with centralizer Weyl groups of order* 2, 6, *and Quillen automorphism groups of order* 2, 2, *respectively. The centralizer* $C_{\mathrm{G}_2}(\mathrm{e}_2^1)$ *is isomorphic to* $\mathrm{SU}(3)$.

The set $\mathrm{Mono}(E^2, \mathrm{G}_2)/\mathrm{Aut}(E^2)$ *contains a unique element,* $\mathrm{e}_2^2 = t(\mathrm{G}_2)$, *with Quillen automorphism group* $W(\mathrm{G}_2)$ *of order* 12.

Let $\chi_1 \colon W(\mathrm{SU}(3)) \to W(\mathrm{G}_2)$ be the inclusion and $\chi_2 \colon W(\mathrm{SU}(3)) \to W(\mathrm{G}_2)$ the injection given by $\chi_2(\sigma) = \sigma$ and $\chi_2(\tau) = -\tau$. Then the identity map $A_1 \colon \Sigma_0(\mathbf{Z}_3^3) \to \Sigma_0(\mathbf{Z}_3^3)$ is χ_1-equivariant and the $\mathbf{Z}_3$-linear map $A_2 : \Sigma_0(\mathbf{Z}_3^3) \to \Sigma_0(\mathbf{Z}_3^3)$ with matrix

$$A_2 = \begin{pmatrix} 1 & -2 \\ 2 & -1 \end{pmatrix}$$

is χ_2-equivariant.

Lemma 8.2. *A* $\mathbf{Z}_3$*-linear map* $\Sigma_0(\mathbf{Z}_3^3) \to \Sigma_0(\mathbf{Z}_3^3)$ *is admissible with respect to* $W(\mathrm{SU}(3))$ *and* $W(\mathrm{G}_2)$ *is and only it belongs to* $W(\mathrm{G}_2)(uA_1)$ *or* $W(\mathrm{G}_2)(uA_2)$ *for some scalar* $u \in \mathbf{Z}_3$.

Proof. Computerized calculations show that any admissible homomorphism must, up to inner automorphisms, be either χ_1- or χ_2-equivariant. Next, one solves the two systems of linear equations $Aw = \chi_i(w)A$, $w \in W(\mathrm{SU}(3))$, $i = 1, 2$. □

Proposition 8.3. *Any non-trivial morphism* $f\colon \mathrm{SU}(3) \to \mathrm{G}_2$ *is a monomorphism.*

Proof. Let $f\colon \mathrm{SU}(3) \to \mathrm{G}_2$ be any non-trivial morphism and $T(f) : T(\mathrm{SU}(3)) \to T(\mathrm{G}_2)$ a lift of f to the maximal tori. Then $W(\mathrm{G}_2)\pi_1(T(f))$ equals $W(\mathrm{G}_2)(uA_1)$ or $W(\mathrm{G}_2)(uA_2)$ for some 3-adic integer, u. In fact, since the order of $W(\mathrm{SU}(3))$ is divisible by 3, u must be a unit (3.19). In the first case, $W(\mathrm{G}_2)\pi_1(T(f)) = W(\mathrm{G}_2)(uA_1)$, f is a monomorphism. And if $W(\mathrm{G}_2)\pi_1(T(f)) = W(\mathrm{G}_2)(uA_2)$, the kernel of $\check{T}(f)$ equals the center of SU(3) and f factors through a monomorphism $\overline{f}\colon \mathrm{PU}(3) \to \mathrm{G}_2$. However, such a monomorphism can not exist since the Quillen category of PU(3) contains an object $E^2 \rightarrowtail \mathrm{PU}(3)$ with Quillen automorphism group $\mathrm{SL}(2, \mathbf{F}_3)$ of order 24 exceeding the order of the Quillen automorphism group of $e_2^2 \in \mathrm{Mono}(E^2, \mathrm{G}_2)$. □

Consider now the diagram

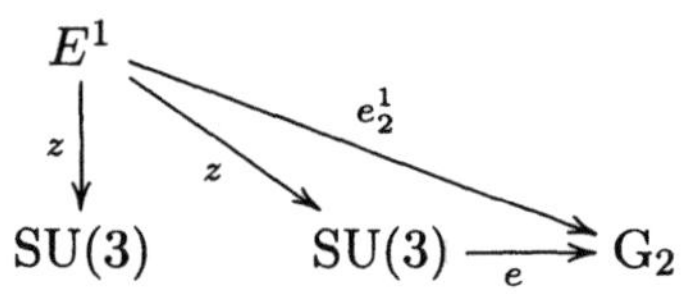

where the SU(3) to the right stands for $C_{\mathrm{G}_2}(e_2^1)$ and z stands for center. Here, $e\psi^{-1} = e$ since $C_{\mathrm{G}_2}(\psi^{-1}) = \psi^{-1}$.

Lemma 8.4. *For any monomorphism* $f\colon \mathrm{SU}(3) \to \mathrm{G}_2$, $fz = e_2^1$.

Proof. Since $\pi_1(T(f)) = uA_1$, $u \in \mathbf{Z}_3^*$, the reduction mod 3, $t(f) : t(\mathrm{SU}(3)) \to t(\mathrm{G}_2)$, takes the center, $(1, -1)$, of SU(3) to the element $u(1, -1) \in t(\mathrm{G}_2)$ whose stabilizer subgroup is $W(\mathrm{SU}(3))$. □

It follows (3.9) that

$$\mathrm{Mono}(\mathrm{SU}(3), \mathrm{SU}(3))_{z \to z} = \mathrm{Mono}(\mathrm{SU}(3), \mathrm{G}_2)_{z \to e_2^1} = \mathrm{Mono}(\mathrm{SU}(3), \mathrm{G}_2)$$

or, alternatively, that the map

$$\begin{aligned} \langle -1 \rangle \backslash \mathbf{Z}_3^* &\to \mathrm{Mono}(\mathrm{SU}(3), \mathrm{G}_2) \\ \pm u &\to e\psi^u \end{aligned}$$

is a bijection. Also, any monomorphism $f\colon \mathrm{SU}(3) \to \mathrm{G}_2$ is centric [7] in the sense that the map given by composition with Bf,

$$\mathrm{map}(B\mathrm{SU}(3), B\mathrm{SU}(3))_{B1} \to \mathrm{map}(B\mathrm{SU}(3), B\,\mathrm{G}_2)_{Bf}$$

is a homotopy equivalence. Clearly, f is toric as well (2.12).

Theorem 8.5. 1. $\mathrm{Rep}(\mathrm{SU}(3), \mathrm{G}_2) = \{0\} \cup \mathrm{Mono}(\mathrm{SU}(3), \mathrm{G}_2)$ *is* T*-determined.*

2. *The action*

$$\mathrm{Mono}(\mathrm{SU}(3), \mathrm{G}_2) \times \mathrm{Out}(\mathrm{SU}(3)) \to \mathrm{Mono}(\mathrm{SU}(3), \mathrm{G}_2)$$

is transitive and the stabilizer at $f \in \mathrm{Mono}(\mathrm{SU}(3), \mathrm{G}_2)$ *is equal to* $\langle \psi^{-1} \rangle = W_{\mathrm{G}_2}(f\mathrm{SU}(3))$.

Proof. This is clear from the explicit description of the set $\mathrm{Rep}(\mathrm{SU}(3), \mathrm{G}_2)$. For instance, the restriction map

$$\mathrm{Mono}(\mathrm{SU}(3), \mathrm{G}_2) \to \mathrm{Mono}(T(\mathrm{SU}(3)), \mathrm{G}_2)$$

can be identified to the map

$$\{u \in \mathbf{Z}_3^* \mid u \equiv 1 \bmod 3\} \to W(\mathrm{G}_2)(uA_1)$$

which clearly is injective. □

References

[1] J.F. Adams and Z. Mahmud, *Maps between classifying spaces*, Invent. Math. **35** (1976), 1–41.

[2] Kasper K. S. Andersen, *The normalizer splitting conjecture for p-compact groups*, Fund. Math. **161** (1999), no. 1-2, 1–16, Algebraic topology (Kazimierz Dolny, 1997).

[3] N. Bourbaki, *Éléments de mathématique. Fasc. XXXIV. Groupes et algèbres de Lie. Chapitre IV: Groupes de Coxeter et systèmes de Tits. Chapitre V: Groupes engendrés par des réflexions. Chapitre VI: systèmes de racines*, Hermann, Paris, 1968, Actualités Scientifiques et Industrielles, No. 1337.

[4] A.K. Bousfield and D.M. Kan, *Homotopy limits, completions and localizations*, 2nd ed., Lecture Notes in Mathematics, vol. 304, Springer-Verlag, Berlin–Heidelberg–New York–London–Paris–Tokyo, 1987.

[5] C. Broto and J.M. Møller, *Embeddings of DI_2 in F_4*, Hopf Topology Archive, 1999.

[6] W. Browder, *Torsion in H-spaces*, Ann. of Math. (2) **74** (1961), 24–51.

[7] W.G. Dwyer and D.M. Kan, *Centric maps and realizations of diagrams in the homotopy category*, Proc. Amer. Math. Soc. **114** (1992), 575–584.

[8] W.G. Dwyer and C.W. Wilkerson, *A cohomology decomposition theorem*, Topology **31** (1992), 433–443.

[9] ———, *Homotopy fixed point methods for Lie groups and finite loop spaces*, Ann. of Math. (2) **139** (1994), 395–442.

[10] ———, *The center of a p-compact group*, The Čech Centennial. Contemporary Mathematics, vol. 181 (Providence, Rhode Island) (M. Cenkl and H. Miller, eds.), American Mathematical Society, 1995, pp. 119–157.

[11] ———, *Product splittings for p-compact groups*, Fund. Math. **147** (1995), 279–300.

[12] E. B. Dynkin, *Semisimple subalgebras of semisimple Lie algebras*, Mat. Sbornik N.S. **30(72)** (1952), 349–462 (3 plates).

[13] ———, *Semisimple subalgebras of semisimple Lie algebras*, Amer. Math. Soc. Translations, Series 2, **6** (1957), 111–244.

[14] Jr. Griess, Robert L., *Elementary abelian p-subgroups of algebraic groups*, Geom. Dedicata **39** (1991), no. 3, 253–305.

[15] J. R. Hubbuck, *Homotopy homomorphisms of Lie groups*, (1974), 33–41. London Math. Soc. Lecture Note Ser., No. 11.

[16] ———, *Mapping degrees for classifying spaces. I*, Quart. J. Math. Oxford Ser. (2) **25** (1974), 113–133.

[17] S. Jackowski and J.E. McClure, *Homotopy decompostion of classifying spaces via elementary abelian subgroups*, Topology **31** (1992), 113–132.

[18] Stefan Jackowski, James McClure, and Bob Oliver, *Homotopy theory of classifying spaces of compact Lie groups*, Algebraic topology and its applications, Math. Sci. Res. Inst. Publ., vol. 27, Springer, New York, 1994, pp. 81–123.

[19] ———, *Maps between classifying spaces revisited*, The Čech centennial (Boston, MA, 1993), Contemp. Math., vol. 181, Amer. Math. Soc., Providence, RI, 1995, pp. 263–298.

[20] ———, *Self-homotopy equivalences of classifying spaces of compact connected Lie groups*, Fund. Math. **147** (1995), no. 2, 99–126.

[21] J. M. Møller and D. Notbohm, *Centers and finite coverings of finite loop spaces*, J. Reine Angew. Math. **456** (1994), 99–133.

[22] Jesper M. Møller, *Rational isomorphisms of p-compact groups*, Topology **35** (1996), no. 1, 201–225.

[23] ———, *Deterministic p-compact groups*, Stable and unstable homotopy (Toronto, ON, 1996), Amer. Math. Soc., Providence, RI, 1998, pp. 255–278.

[24] ———, *N-determined p-compact groups*, Preprint, May 1999.

[25] ———, *Normalizers of maximal tori*, Math. Z. **231** (1999), 51–74.

[26] Dietrich Notbohm, *Classifying spaces of compact Lie groups and finite loop spaces*, Handbook of algebraic topology, North-Holland, Amsterdam, 1995, pp. 1049–1094.

[27] D. L. Rector, *Noetherian cohomology rings and finite loop spaces with torsion*, J. Pure Appl. Algebra **32** (1984), no. 2, 191–217.

[28] Antonio Viruel, *Homotopy uniqueness of classifying spaces of compact Lie groups*, Ph.D. thesis, Universitat Autònoma de Barcelona, 1996.

[29] Zdzisław Wojtkowiak, *On maps from* holim F *to* Z, Algebraic topology, Barcelona, 1986, Lecture Notes in Math., vol. 1298, Springer, Berlin, 1987, pp. 227–236.

Matematisk Institut
Universitetsparken 5
DK–2100 København
E-mail address: `moller@math.ku.dk`

Progress in Mathematics, Vol. 196

On the vanishing of certain K-theory Nil-groups

Hans Jørgen Munkholm[1] and Stratos Prassidis[1,2,3]

Abstract. Let Γ_i, $i = 0, 1$, be two groups containing C_p, the cyclic group of prime order p, as a subgroup of index 2. Let $\Gamma = \Gamma_0 *_{C_p} \Gamma_1$. We show that the Nil-groups appearing in Waldhausen's splitting theorem for computing $K_j(\mathbb{Z}\Gamma)$ ($j \leq 1$) vanish. Thus, in low degrees, the K-theory of $\mathbb{Z}\Gamma$ can be computed by a Mayer–Vietoris type exact sequence involving the K-theory of the integral group rings of the groups Γ_0, Γ_1 and C_p.

1. Introduction

We prove the vanishing of Waldhausen's Nil-groups, in degrees less than or equal to 0, associated to certain amalgamated free products of groups ([12], [13]).

In more detail, let C_p denote the cyclic group of prime order p, and let Γ_0, Γ_1 be two groups, each containing C_p as a subgroup of index 2. Our main result concerns Waldhausen's Nil-groups associated to the amalgamated free product of groups $\Gamma = \Gamma_0 *_{C_p} \Gamma_1$. We write $B_i = \mathbb{Z}[\Gamma_i - C_p]$, $i = 0, 1$, for the $\mathbb{Z}C_p$-sub-bimodule generated by $\Gamma_i - C_p$.

Main Theorem. *With the above notation*

$$\widetilde{Nil}_j(\mathbb{Z}C_p; B_0, B_1) = 0, \quad j \leq 0.$$

Remark. For $j \leq -1$, this is a special case of results obtained in [10]. The extension to the case $j = 0$ was prompted by a question put to the second author (by Jim Davis) in connection with the results appearing in [3].

Using the Main Theorem and Waldhausen's splitting theorem, we can get information about the (lower) K-theory of Γ.

Corollary. *There are exact sequences*

$$K_1(\mathbb{Z}C_p) \to K_1(\mathbb{Z}\Gamma_0) \oplus K_1(\mathbb{Z}\Gamma_1) \to K_1(\mathbb{Z}\Gamma) \to K_0(\mathbb{Z}C_p) \to \cdots,$$

and

$$Wh(C_p) \to Wh(\Gamma_0) \oplus Wh(\Gamma_1) \to Wh(\Gamma) \to \widetilde{K}_0(\mathbb{Z}C_p) \to \cdots.$$

---

[1] Supported in part by the SNF (Denmark) under grant number 9502188.
[2] Supported in part by a Vanderbilt University Summer Research Fellowship.
[3] Supported in part by National Science Foundation Grant DMS-9504479.

Remark. For each prime p, this covers precisely three different groups Γ. In fact, each Γ_i is cyclic of order $2p$ or dihedral of order $2p$.

The proof involves an extension of the methods developed in [10]. There, the Nil-groups in question were shown to be related to the Nil-groups of certain additive categories given in [8]. And this fact was used to establish naturality properties and certain Mayer–Vietoris properties.

For the present proof, we recall the classical Rim square associated to C_p, i.e., the cartesian square of rings

$$\begin{array}{ccc} \mathbb{Z}C_p & \longrightarrow & \mathbb{Z}[\zeta_p] \\ \downarrow & & \downarrow \\ \mathbb{Z} & \longrightarrow & \mathbb{F}_p \end{array}$$

where ζ_p is a primitive p^{th} root of unity and $\mathbb{F}_p$ is the finite field of p elements. The methods of [10] can be extended to provide a long exact sequence of Nil-groups coming from this square. The three smaller rings in the diagram are Noetherian and have finite cohomological dimension (called regular in [1]). Hence, by Waldhausen's vanishing result, the Nil-groups associated to those rings vanish. Using the exact sequence, we can then derive vanishing results for the Nil-groups associated to the triple $(\mathbb{Z}C_p; B_0, B_1)$.

The second author would like to thank IMADA at Odense Universitet for its hospitality during his Spring 1997 sabbatical.

2. Preliminaries

We assume that all rings considered have a unit which is preserved by all ring homomorphisms, and that finitely generated free modules have well-defined rank. For any ring R, $\mathcal{M}_R$ denotes the category of right R-modules, $\mathcal{P}_R$ the subcategory of finitely generated projective right R-modules, and $\mathcal{F}_R$ the subcategory of finitely generated right free R-modules. For $\mathcal{A} = \mathcal{M}$, $\mathcal{P}$, or $\mathcal{F}$, $\mathcal{A}^n_R$ denotes the category $\mathcal{A}_R \times \mathcal{A}_R \times \cdots \times \mathcal{A}_R$ (n times).

We will use the notation established in [10], and write $\mathbf{R} = (R; B_0, B_1)$ for a triple where R is a ring and B_i, $i = 0, 1$, are two R-bimodules. Moreover, $\mathbb{F}_{\mathcal{A}}(\mathbf{R})$ denotes the twisted polynomial extension category defined in [8] and [10], for $\mathcal{A} = \mathcal{P}, \mathcal{F}$. To recall its definition, from [8], we first note that the triple $\mathbf{R}$ gives rise to a functor $\alpha_R \colon \mathcal{M}^2_R \to \mathcal{M}^2_R$ defined by

$$\alpha_R(M_0, M_1) = (M_1 \otimes_R B_0, M_0 \otimes_R B_1), \quad \alpha_R(f_0, f_1) = (f_1 \otimes 1, f_0 \otimes 1).$$

Now, the objects of $\mathbb{F}_{\mathcal{A}}(\mathbf{R})$ are simply those of $\mathcal{A}^2_R$, and

$$\mathbb{F}_{\mathcal{A}}(\mathbf{R})(u, v) = \bigoplus_{i=0}^{\infty} \mathcal{M}^2_R(u, \alpha^i_R(v)) = \left\{ \sum_{i=0}^{\infty} p_i t^i \ : \ p_i \in \mathcal{M}^2_R(u, \alpha^i_R(v)) \right\}$$

where we write $p_i \colon u \to \alpha^i_R(v)$ for the ith component of the morphism. Thus the morphism sets are graded abelian groups, and the powers of the formal variable t

are there simply to keep track of degrees. In order to give a different description of these morphism sets, we set $B_i = B_0$ for all even $i \geq 0$, $B_i = B_1$ for all odd $i > 0$, and put

$$B_i^{(j)} = B_i \otimes_R B_{i+1} \otimes_R \cdots \otimes_R B_{i+j-1}$$

for all $i, j \geq 0$. In particular, $B_i^{(0)} = R$, $B_i^{(1)} = B_i$. Similarly, if (Q_0, Q_1) is an object in $\mathbb{F}_{\mathcal{A}}(\mathbf{R})$, we put $Q_i = Q_0$ for all even $i \geq 0$, and $Q_i = Q_1$ for all odd $i > 0$. With this notation

$$\alpha_R^i(Q_0, Q_1) = (Q_i \otimes_R B_{i+1}^{(i)}, Q_{i+1} \otimes_R B_i^{(i)}).$$

Thus, if $u = (P_0, P_1)$ and $v = (Q_0, Q_1)$ are objects in $\mathbb{F}_{\mathcal{A}}(\mathbf{R})$, then

$$\begin{aligned}\mathbb{F}_{\mathcal{A}}(\mathbf{R})(u,v) &= \bigoplus\nolimits_{i \geq 0} [\mathcal{M}_R(P_0, Q_i \otimes_R B_{i+1}^{(i)}) \oplus \mathcal{M}_R(P_1, Q_{i+1} \otimes_R B_i^{(i)})] \\ &= \left\{ \sum\nolimits_{i=0}^{\infty} (p_{(0,i)} \oplus p_{(1,i)}) t^i \, : \, p_{(k,i)} \in \mathcal{M}_R(P_k, Q_{i+k} \otimes_R B_{i+k+1}^{(i)}),\ k = 0, 1 \right\},\end{aligned}$$

and there is a forgetful functor ("evaluation at $t = 0$")

$$\eta_{\mathcal{A}} \colon \mathbb{F}_{\mathcal{A}}(\mathbf{R}) \to \mathcal{A}_R^2, \ \mathcal{A} = \mathcal{P}, \mathcal{F}.$$

In [10], it was shown that $\mathbb{F}_{\mathcal{A}}$ is a functor on a category $\mathcal{T}$ of triples $\mathbf{R} = (R; B_0, B_1)$ as above with suitable, rather obvious, morphisms. In particular, if $h \colon R \to S$ is a ring homomorphism, there is a functor

$$h_* \colon \mathbb{F}_{\mathcal{A}}(\mathbf{R}) \to \mathbb{F}_{\mathcal{A}}(\mathbf{S})$$

where $\mathbf{S} = \left(S; \overline{\overline{B_0}}, \overline{\overline{B_1}}\right)$ with $\overline{\overline{B_i}} = S \otimes_R B_i \otimes_R S$ $(i = 0, 1)$ given by two-sided reduction of scalars along h.

Triples of the form $\mathbf{R}$ arise naturally from certain co-cartesian diagrams ([12], [13]). To wit, let

$$\begin{array}{ccc} R & \xrightarrow{\alpha_0} & A_0 \\ {\scriptstyle\alpha_1}\downarrow & & \downarrow{\scriptstyle\beta_0} \\ A_1 & \xrightarrow{\beta_1} & \Lambda \end{array}$$

be a co-cartesian diagram of rings and assume further that the maps α_i, $i = 0, 1$, are pure inclusions, i.e., they are inclusions and they induce R-bimodule splittings

$$A_i = R \oplus B_i, \quad i = 0, 1$$

where we have identified R with its image under α_i. There result a triple

$$\mathbf{R} = (R; B_0, B_1) \in \mathcal{T};$$

a splitting of Λ as an R-bimodule

$$\Lambda = R \oplus B_0 \oplus B_1 \oplus (B_0 \otimes_R B_1) \oplus (B_1 \otimes_R B_0) \oplus (B_0 \otimes_R B_1 \otimes_R B_0) \oplus \cdots;$$

and an induced filtration of Λ as a ring

$F_0\Lambda = R,$

$F_1\Lambda = R\oplus B_0\oplus B_1,$

$F_2\Lambda = R\oplus B_0\oplus B_1 \oplus (B_0\otimes_R B_1) \oplus (B_1\otimes_R B_0),$

$F_3\Lambda = R\oplus B_0\oplus B_1 \oplus (B_0\otimes_R B_1) \oplus (B_1\otimes_R B_0) \oplus (B_0\otimes_R B_1\otimes_R B_0) \oplus (B_1\otimes_R B_0\otimes_R B_1),$

......

Moreover, by [13], there is an exact sequence (for $j \in \mathbb{N}$)

$$\cdots \to K_j(A_0)\oplus K_j(A_1) \to K_j(\Lambda) \to K_{j-1}(R)\oplus \widetilde{Nil}^W_{j-1}(R; B_0, B_1) \to \cdots . \quad (*)$$

In other words, the Nil-groups measure the failure of exactness of a K-theory Mayer–Vietoris sequence associated to a co-cartesian diagram of rings with the extra purity assumption.

For any amalgamated free product of groups, $\Gamma = \Gamma_0 *_G \Gamma_1$, the integral group ring $\mathbb{Z}\Gamma$ fits into such a co-cartesian diagram of rings

$$\begin{array}{ccc} R = \mathbb{Z}G & \longrightarrow & A_0 = \mathbb{Z}\Gamma_0 \\ \downarrow & & \downarrow \\ A_1 = \mathbb{Z}\Gamma_1 & \longrightarrow & \Lambda = \mathbb{Z}\Gamma \end{array}$$

with $B_i = \mathbb{Z}[\Gamma_i - G]$, $i = 0, 1$.

In this case, each B_i is free both as a left and a right R-module, but we shall start more generally by considering a triple $\mathbf{R}$ which is associated to a co-cartesian diagram of rings for which the bimodules B_i are only assumed to be flat as left R-modules. We set

$$NK_j(\mathbf{R}) = \operatorname{Ker}((\eta_{\mathcal{F}})_j \colon K_j(\mathbb{F}_{\mathcal{F}}(\mathbf{R})) \to K_j(\mathcal{F}^2_R))$$

(for $j \leq 0$, the K_j-group of an additive category is understood as the K_j-group of its idempotent completion). Then, for $j \leq 1$, there is a natural isomorphism

$$NK_j(\mathbf{R}) \to \widetilde{Nil}^W_{j-1}(R; B_0, B_1)$$

([8], Theorem 2.11, for $j = 1$; [10], Proposition 13, for the lower K-groups) identifying the kernel of the “augmentation” induced map for $\mathbb{F}_{\mathcal{F}}(\mathbf{R})$ with Waldhausen's Nil-groups of one degree less. Since $\mathbb{F}_{\mathcal{F}}(\mathbf{R})$ is cofinal in $\mathbb{F}_{\mathcal{P}}(\mathbf{R})$, and $\mathcal{F}^2_R$ is cofinal in $\mathcal{P}^2_R$, one also has the identification

$$NK_j(\mathbf{R}) = \ker((\eta_{\mathcal{P}})_j).$$

The main purpose of the comparison between the kernel of $(\eta_{\mathcal{P}})_j$ and Waldhausen's Nil-groups is that we can use vanishing results for the former to derive similar results for the latter. Thus, the next result follows immediately from [12], [13].

Lemma 2.1. *Let R be a regular Noetherian ring and $\mathbf{R} = (R; B_0, B_1)$ be a triple associated to a co-cartesian diagram of rings such that B_i is flat as a left R-module for $i = 0, 1$. Then*

$$NK_j(R; B_0, B_1) = 0, \quad j \leq 1.$$

Proof. In fact, by Theorem 4, p. 138, of [13], $\widetilde{Nil}^{W}_{j-1}(R; B_0, B_1)$ is zero for $j \leq 1$. □

Remark. The assumption of the Lemma can be weakened to coherent regular rings but we will not use the stronger version in this paper.

The main result of the present section is Proposition 2.4, which extends the vanishing result of Lemma 2.1 to $j \geq 2$ in case $B_0 \cong B_1 \cong R$. The case $j = 2$ is the one we actually need (in the proof of Theorem 3.15).

We start by establishing the appropriate terminology. Let $\mathbf{R} = (R; B_0, B_1)$ be a triple in $\mathcal{T}$. Then $\rho = (R, R)$ is a basic object in $\mathbb{F}_{\mathcal{P}}(\mathbf{R})$, in the sense of Bass ([2], p. 197), i.e., each object u of $\mathbb{F}_{\mathcal{P}}(\mathbf{R})$ is isomorphic to a direct summand of $\rho^n = (R^n, R^n)$ for some integer n. We write $R_\rho = \mathrm{End}_{\mathbb{F}_{\mathcal{P}}(\mathbf{R})}(\rho)$ for the endomorphism ring of ρ. There is a split inclusion of rings $\iota\colon R\times R \to R_\rho$ by considering pairs of elements of R as endomorphisms of degree zero of ρ. The splitting σ is given by the forgetful map to the zero degree component of any endomorphism. A morphism of degree i, $\phi = (\phi_0, \phi_1)t^i\colon \rho \to \alpha^i_R(\rho)$, can be identified with the element $(\phi_0(1), \phi_1(1)) \in B^{(i)}_{i+1} \oplus B^{(i)}_i$. Multiplication in R_ρ, i.e., composition of endomorphisms, is then given by concatenation with the added convention that $B_iB_i = 0$, $i = 0, 1$. Considering the degree mod 2 of components one obtains a natural splitting of R_ρ as an $R\times R$-bimodule

$$R_\rho = R_{even} \oplus R_{odd}.$$

The component R_{even} is a subring of R_ρ, and R_{odd} is an R_{even}-bimodule.

The ring R_ρ is also $\mathbb{N}$-graded. The abelian group of degree i is $R_{\rho,i} = B^{(i)}_{i+1} \oplus B^{(i)}_i$, which also has a natural diagonal $R\times R$-bimodule structure. In case the triple $\mathbf{R}$ is associated to a co-cartesian diagram of rings, then R_ρ is the associated grading of the filtration of $\Lambda\times R$. Another grading of the ring Λ is given in [11].

Lemma 2.2. *With the above notation, there is an isomorphism $F_j\colon K_j(R_\rho) \to K_j(\mathbb{F}_{\mathcal{P}}(\mathbf{R}))$ making the diagram*

$$\begin{array}{ccc} K_j(R_\rho) & \xrightarrow{F_j} & K_j(\mathbb{F}_{\mathcal{P}}(\mathbf{R})) \\ {\scriptstyle K_j(\sigma)}\downarrow & & \downarrow{\scriptstyle (\eta_{\mathcal{P}})_j} \\ K_j(R\times R) & \longrightarrow & K_j(\mathcal{P}^2_R) \end{array}$$

commute for $j \geq 1$. The horizontal map at the bottom is the natural isomorphism.

Proof. Since ρ is a basic object in $\mathbb{F}_{\mathcal{P}}(\mathbf{R})$, the functor

$$F\colon \mathcal{F}_{R_\rho} \to \mathbb{F}_{\mathcal{P}}(\mathbf{R}), \quad F(R_\rho^n) = \rho^n$$

is a full, faithful and cofinal functor. Thus, it induces an isomorphism on K_j-groups for $j \geq 1$ ([7], Theorem 1.1; also [5], Proposition 1.1; [6], p. 225).

The bottom arrow is an isomorphism, in degrees $j \geq 1$, because $R \times R$ can be thought of as the endomorphism ring of the basic object (R, R) in the category $\mathcal{F}_R^2$ which is cofinal in $\mathcal{P}_R^2$.

Commutativity of the diagram is clear. □

We now restrict our attention to the case where $B_i \cong R$, $i = 0, 1$, as R-bimodules. In this case, the ring R_ρ has a description as a matrix ring. In fact, let S be the subring of $M_2(R[x])$ given by

$$S = \begin{pmatrix} R[x^2] & xR[x^2] \\ xR[x^2] & R[x^2] \end{pmatrix}$$

and let $\epsilon\colon S \to R{\times}R$ be the natural augmentation map

$$\begin{pmatrix} a(x^2) & xb(x^2) \\ xc(x^2) & d(x^2) \end{pmatrix} \longmapsto (a(0), d(0)).$$

Proposition 2.3. *Let $B_0 \cong B_1 \cong R$ as R-bimodules. Then there is a ring isomorphism*

$$\kappa\colon R_\rho \to S$$

which commutes with the augmentation maps, i.e., $\epsilon \circ \kappa = \sigma$.

Proof. Because of the assumption on B_i, the degree i component $R_{\rho,i}$, is isomorphic to $R{\times}R$ as an $R{\times}R$-bimodule with the degree i endomorphism $(id_R, id_R)t^i$ corresponding to the element $(1,1)$. We define $R{\times}R$-bimodule maps

$$\kappa|R_{\rho,2i}\colon R_{\rho,2i} \to \begin{pmatrix} Rx^{2i} & 0 \\ 0 & Rx^{2i} \end{pmatrix}, \qquad (1,1) \mapsto \begin{pmatrix} x^{2i} & 0 \\ 0 & x^{2i} \end{pmatrix};$$

$$\kappa|R_{\rho,2i+1}\colon R_{\rho,2i+1} \to \begin{pmatrix} 0 & Rx^{2i+1} \\ Rx^{2i+1} & 0 \end{pmatrix}, \qquad (1,1) \mapsto \begin{pmatrix} 0 & x^{2i+1} \\ x^{2i+1} & 0 \end{pmatrix}.$$

The resulting map κ is the required ring isomorphism. By construction, it commutes with the augmentation homomorphisms. □

The above result reduces the problem of computing the NK-groups to a problem in the K-theory of certain matrix rings.

Proposition 2.4. *Let R be a regular Noetherian ring and assume that $B_0 \cong B_1 \cong R$ as R-bimodules. Then for all $j \in \mathbb{Z}$, $NK_j(\mathbf{R}) = 0$.*

Proof. For $j \leq 0$ the result follows from [10]. Let $j \geq 1$. By Lemma 2.2, it is enough to prove the vanishing of the kernel of the map induced on the K-groups by the augmentation σ. If R is regular Noetherian, then $R[x]$ is regular Noetherian by Hilbert's Basis and Syzygy Theorems. Then $M_2(R[x])$ is Noetherian (because it is finitely generated as an $R[x]$-module) and it has finite cohomological dimension. Also, $M_2(R[x])$ is a free S-module with basis

$$\left\{\begin{pmatrix}1 & 0\\ 0 & 1\end{pmatrix}, \begin{pmatrix}0 & 1\\ 1 & 0\end{pmatrix}\right\}.$$

Then S is a regular Noetherian ring ([9], p. 96, Proposition 2.30) and the same is true for R_ρ. Since R_ρ is a graded ring with zero grading $R\times R$, the augmentation induced map

$$\sigma_j\colon K_j(R_\rho) \to K_j(R\times R)$$

is an isomorphism ([9], Theorem 2.37, p. 98). Therefore $NK_j(\mathbf{R})$, being the kernel of σ_j, vanishes. □

Corollary 2.5. *Let* $\mathbf{F} = (\mathbb{F}_p;\mathbb{F}_p,\mathbb{F}_p)$ *where* $\mathbb{F}_p$ *is the field of* p *elements with* p *prime. Then*

$$NK_j(\mathbf{F}) = 0, \quad j \in \mathbb{N}.$$

In particular, $(\eta_{\mathcal{P}})_j\colon K_j(\mathbb{F}_{\mathcal{P}}(\mathbf{R})) \to K_j(\mathcal{P}^2_{\mathbb{R}}))$ *is an isomorphism.*

3. Mayer–Vietoris Sequences

Let $h\colon R \to S$ be a ring homomorphism. Then h induces a functor

$$h^*\colon \mathcal{M}_S \to \mathcal{M}_R$$

which maps an S-module M to the R-module with underlying abelian group M and R-structure induced by h. We are interested in the image of the functor h^*.

Definition 3.1. Let $h\colon R \to S$ be a ring epimorphism. A right R-module M is called h-extended if there is a right S-module structure on M with $mr = mh(r)$ for all $m \in M$, $r \in R$. In other words, M is h-extended if M is in the image of h^*.

The main technical property of such extended modules is expressed in the following Lemma.

Lemma 3.2. *Let* $h\colon R \to S$ *be a ring epimorphism and* M *an* h*-extended right* R*-module. Then there is a natural right* S*-module isomorphism*

$$k\colon M\otimes_R S \to M.$$

Proof. It is easy to check that there is a well defined homomorphism given by $k(m\otimes s) = ms$, and that $\ell(m) = m\otimes 1_S$ defines an inverse. □

For an R-bimodule B we write $\overline{B} = S\otimes_R B$ for the S-R-bimodule obtained by left-sided reduction of scalars. Also, recall that $\overline{\overline{B}} = S\otimes_R B\otimes_R S$. We immediately get the following result.

Corollary 3.3. *Let B_i, $i = 0, 1$, be R-bimodules. If the S-R-bimodules $\overline{B_0}$ and $\overline{B_1}$ are h-extended as right R-modules, then*

$$\overline{\overline{B_0}}\otimes_S\overline{\overline{B_1}} = (S\otimes_R B_0\otimes_R S)\otimes_S(S\otimes_R B_1\otimes_R S) \cong S\otimes_R B_0\otimes_R B_1 = \overline{B_0\otimes_R B_1}$$

as S-bimodules. In particular,

$$h^*\left(\overline{\overline{B_i^{(j)}}}\right) \cong \overline{B_i^{(j)}}$$

as S-bimodules.

We will study the extension properties of a pull-back diagram of rings. We start with a cartesian diagram of rings

$$\begin{array}{ccc} R & \xrightarrow{h_2} & R_2 \\ {\scriptstyle h_1}\downarrow & & \downarrow{\scriptstyle f_2} \\ R_1 & \xrightarrow{f_1} & R_0 \end{array}$$

where we assume that h_1 and h_2 are epimorphisms. The diagram induces a pull-back diagram of categories ([1], Ch. IX, Theorem 5.1)

$$\begin{array}{ccc} \mathcal{P}_R & \longrightarrow & \mathcal{P}_{R_2} \\ \downarrow & & \downarrow \\ \mathcal{P}_{R_1} & \longrightarrow & \mathcal{P}_{R_0}. \end{array}$$

Notice that the diagram induces an exact sequence of R-bimodules

$$0 \to R \xrightarrow{(\,h_1\ h_2\,)} R_1\oplus R_2 \xrightarrow{\binom{f_1}{-f_2}} R_0 \to 0 \qquad (E)$$

where the action of R on R_j, $j = 0, 1, 2$, is induced by the maps in the cartesian square.

First we recall a routine algebraic lemma which uses the following notation. Let $h\colon R \to S$ be a ring homomorphism, Q and P right R-modules, and B an R-bimodule. Then there is a right R-module homomorphism

$$Q\otimes_R B \to Q\otimes_R S\otimes_R B, \quad q\otimes b \mapsto q\otimes 1_s\otimes b$$

and an induced abelian group homomorphism

$$\overline{h}\colon Hom_R(P, Q\otimes_R B) \to Hom_R(P, Q\otimes_R S\otimes_R B).$$

Lemma 3.4. *If P and Q are projective, and B is left flat, then the sequence*

$$0 \to Hom_R(P, Q\otimes_R B) \xrightarrow{(\,\overline{h_1}\ \overline{h_2}\,)} Hom_R(P, Q\otimes_R R_1\otimes_R B)\oplus Hom_R(P, Q\otimes_R R_2\otimes_R B)$$
$$\xrightarrow{\binom{\overline{f_1}}{-\overline{f_2}}} Hom_R(P, Q\otimes_R R_0\otimes_R B) \to 0$$

is exact.

Proof. The assumptions on Q and B show that the induced sequence

$$0 \to Q\otimes_R B \to Q\otimes_R R_1\otimes_R B\oplus Q\otimes_R R_2\otimes_R B \to Q\otimes_R R_0\otimes_R B \to 0$$

is exact. The result follows because P is projective. □

Corollary 3.5. *Let P and Q be projective right R-modules and B an R-bimodule which is flat as a left R-module. Assume further that $R_j\otimes_R B$ is h_j-extended as a right R-module ($j = 1, 2$), and that $R_0\otimes_R B$ is f_1h_1-extended ($= f_2h_2$-extended) as a right R-module. Then the exact sequence (E) induces an exact sequence of abelian groups*

$$0 \to Hom_R(P, Q\otimes_R B) \xrightarrow{(\overline{h_1}\ \overline{h_2})}$$

$$Hom_{R_1}(P\otimes_R R_1, Q\otimes_R R_1\otimes_R B)\oplus Hom_{R_2}(P\otimes_R R_2, Q\otimes_R R_2\otimes_R B) \xrightarrow{\binom{\overline{f_1}}{-\overline{f_2}}}$$

$$Hom_{R_0}(P\otimes_R R_0, Q\otimes_R R_0\otimes_R B) \longrightarrow 0.$$

Proof. This follows from Lemma 3.4 using the adjointness isomorphisms

$$Hom_R(P, Q\otimes_R R_i\otimes_R B) \cong Hom_{R_i}(P\otimes_R R_i, Q\otimes_R R_i\otimes_R B)$$

($i = 0, 1, 2$). □

We now consider a triple $\mathbf{R} = (R; B_0, B_1)$ such that $R_j\otimes_R B_i$ is h_j-extended as a right R-module ($j = 1, 2$ and $i = 0, 1$) and $R_0\otimes_R B_i$ is f_1h_1-extended as a right R-module ($i = 0, 1$). It follows that $R_j\otimes_R B_i$ is f_j-extended as a right R_j-module ($i = 0, 1$, $j = 1, 2$). We further assume that the modules B_i are flat as left R-modules ($i = 0, 1$). Then we get corresponding objects in $\mathcal{T}$,

$$\mathbf{R}_j = (R_j; R_j\otimes_R B_0, R_j\otimes_R B_1), \quad j = 0, 1, 2;$$

a pull-back of additive categories (defining $\mathbb{P}$)

$$\begin{array}{ccc} \mathbb{P} & \xrightarrow{h_2'} & \mathbb{F}_{\mathcal{P}}(\mathbf{R}_2) \\ {\scriptstyle h_1'}\downarrow & & \downarrow{\scriptstyle f_2'} \\ \mathbb{F}_{\mathcal{P}}(\mathbf{R}_1) & \xrightarrow{f_1'} & \mathbb{F}_{\mathcal{P}}(\mathbf{R}_0) \end{array}$$

where f_j' is induced by f_j ($j = 1, 2$); and a functor

$$\phi\colon \mathbb{F}_{\mathcal{P}}(\mathbf{R}) \to \mathbb{P}$$

induced by the universal properties of the pull-back.

In [10], it has been shown that if a ring homomorphism is surjective, then the map induced in the twisted polynomial extension categories is E-surjective in the sense of [1] (Definition 2.4, p. 356). Thus by [2], A.13, p. 151, the above square induces a commutative diagram of exact sequences in K-theory

$$\begin{array}{ccccccc} K_2(\mathbb{F}_{\mathcal{P}}(\mathbf{R}_1))\oplus K_2(\mathbb{F}_{\mathcal{P}}(\mathbf{R}_2)) & \to & K_2(\mathbb{F}_{\mathcal{P}}(\mathbf{R}_0)) & \to & K_1(\mathbb{P}) & \to & K_1(\mathbb{F}_{\mathcal{P}}(\mathbf{R}_1))\oplus K_1(\mathbb{F}_{\mathcal{P}}(\mathbf{R}_2)) \\ \downarrow & & \downarrow & & \downarrow{\scriptstyle\kappa} & & \downarrow \\ K_2(\mathcal{P}^2_{R_1})\oplus K_2(\mathcal{P}^2_{R_2}) & \to & K_2(\mathcal{P}^2_{R_0}) & \to & K_1(\mathcal{P}^2_R) & \to & K_1(\mathcal{P}^2_{R_1})\oplus K_1(\mathcal{P}^2_{R_2}). \end{array}$$

The vertical maps are induced by the obvious functors between two pull-back diagrams.

Lemma 3.6. *The functor ϕ is full and faithful.*

Proof. Let $u = (P_0, P_1)$, $v = (Q_0, Q_1)$ be two objects in $\mathbb{F}_{\mathcal{P}}(\mathbf{R})$. We must show that the map induced by ϕ

$$\phi' \colon \mathbb{F}_{\mathcal{P}}(\mathbf{R})(u, v) \to \mathbb{P}(\phi(u), \phi(v))$$

is a group isomorphism, and start by setting the notation. For $k = 0, 1$, and $i \geq 0$,

$$\overline{h^{(i)}_{j,k}} \colon \mathcal{P}_R(P_k, Q_{i+k}\otimes_R B^{(i)}_{i+k+1}) \to \mathcal{P}_R(P_k\otimes_R R_j, Q_{i+k}\otimes_R R_j\otimes_R B^{(i)}_{i+k+1})$$

and

$$\overline{f^{(i)}_{j,k}} \colon \mathcal{P}_R(P_k\otimes_R R_j, Q_{i+k}\otimes_R R_j\otimes_R B^{(i)}_{i+k+1}) \to \mathcal{P}_R(P_k\otimes_R R_0, Q_{i+k}\otimes_R R_0\otimes_R B^{(i)}_{i+k+1})$$

are the maps induced by h_j, f_j, $j = 1, 2$.

<u>ϕ' is a monomorphism.</u> An element β in the morphism set $\mathbb{F}_{\mathcal{P}}(\mathbf{R})(u, v)$ can be written

$$\beta = \sum\nolimits_{i\geq 0} (p_{(0,i)}\oplus p_{(1,i)})t^i$$

and its image under ϕ' has the form

$$\phi'(\beta) = \sum\nolimits_{i\geq 0} ((\overline{h^{(i)}_{1,0}}(p_{(0,i)}) \oplus \overline{h^{(i)}_{2,0}}(p_{(0,i)})) \oplus ((\overline{h^{(i)}_{1,1}}(p_{(1,i)}) \oplus \overline{h^{(i)}_{2,1}}(p_{(1,i)}))t^i.$$

If $\phi'(\beta) = 0$, then for all $k = 0, 1$ and $i \geq 0$,

$$\overline{h^{(i)}_{1,k}}(p_{(k,i)}) \oplus \overline{h^{(i)}_{2,k}}(p_{(k,i)}) = 0.$$

Using Corollary 3.5, we see that each $p_{(k,i)} = 0$. Thus $\beta = 0$.

<u>ϕ' is an epimorphism.</u> Let $\gamma \in \mathbb{P}(\phi(u), \phi(v))$. Then $\gamma = \gamma_1\oplus\gamma_2$ where

$$\gamma_j \in \mathbb{F}_{\mathcal{P}}(\mathbf{R}_j)((h_j)_*(u), (h_j)_*(v)), \quad j = 1, 2.$$

Since γ is a morphism in the pull-back category

$$(f_1)_*(\gamma_1) = (f_2)_*(\gamma_2) \text{ in } \mathbb{F}_{\mathcal{P}}(\mathbf{R}_0)((f_1h_1)_*(u), (f_1h_1)_*(v)). \qquad (1)$$

As before, each γ_j can be written as a direct sum of homomorphisms

$$\gamma_j = \sum\nolimits_{i\geq 0} (p_{(0,i)j}\oplus p_{(1,i)j})t^i$$

and condition (1) implies that, for $k = 0, 1$, $i \geq 0$,

$$\overline{f_{1,k}^{(i)}}(p_{(k,i)1}) = \overline{f_{2,k}^{(i)}}(p_{(k,i)2}), \quad \text{i.e.,} \quad (p_{(k,i)1}, p_{(k,i)2}) \in \operatorname{Ker}\left(\overline{f_{1,k}^{(i)}} - \overline{f_{2,k}^{(i)}}\right).$$

By Corollary 3.5, there is $p_{(k,i)} \in Hom_R(P_k, Q_{i+k} \otimes_R B_{i+k+1}^{(i)})$ such that $\overline{h_{j,k}^{(i)}}(p_{(k,i)}) = p_{(k,i)j}$. Set

$$\beta = \sum\nolimits_{i \geq 0} (p_{(0,i)} \oplus p_{(1,i)}) t^i.$$

Then $\phi'(\beta) = \gamma$. □

We recall the definition of an elementary morphism in an additive category. Let $\mathbf{A}$ be an additive category, u an object of $\mathbf{A}$. An automorphism a of u is called elementary if there is a decomposition $u = u_0 \oplus u_1$ such that a takes the form

$$a = \begin{pmatrix} 1 & b \\ 0 & 1 \end{pmatrix}$$

for some $b\colon u_1 \to u_0$. Also, $K_1(\mathbf{A})$ can be defined as the group generated by all pairs (u, a), where u is an object of $\mathbf{A}$ and a an automorphism of u, divided by the subgroup generated by pairs (v, e) with e elementary.

We will prove the analogue of E-surjectivity for functors induced by ring epimorphism on the twisted polynomial extension category of finitely generated projective modules ([1], p. 449). Let $\mathbf{R} = (R; B_0, B_1)$ be a triple. We start with an observation on the morphism sets of objects in $\mathbb{F}_{\mathcal{F}}(\mathbf{R})$. Let $u = (F_0, F_1)$ and $v = (G_0, G_1)$ be two objects in $\mathbb{F}_{\mathcal{F}}(\mathbf{R})$ of ranks (m_0, m_1) and (n_0, n_1). As before, we write $G_i = G_0$ for all even $i \geq 0$, $G_i = G_1$ for all odd $i > 0$ and we write n_i for the rank of G_i. For any R-bimodule B, we write $M_{m \times n}(B)$ for the abelian group of $m \times n$ matrices with entries in B.

Lemma 3.7. *With the above notation, a choice of bases of the free modules involved induces an isomorphism of abelian groups*

$$\mathbb{F}_{\mathcal{F}}(\mathbf{R})(u, v) \cong \bigoplus\nolimits_{i \geq 0} \left[M_{m_0 \times n_i}(B_{i+1}^{(i)}) \oplus M_{m_1 \times n_{i+1}}(B_i^{(i)}) \right].$$

Proof. This is standard matrix calculation. □

Let $h\colon R \to S$ be a ring epimorphism and $\mathbf{R} = (R; B_0, B_1)$. Let $\mathbf{S} = \left(S; \overline{\overline{B_0}}, \overline{\overline{B_1}}\right)$. The map h induces a functor

$$h_*\colon \mathbb{F}_{\mathcal{F}}(\mathbf{R}) \to \mathbb{F}_{\mathcal{F}}(\mathbf{S}).$$

Let $\overline{u_j}$, $j = 1, 2$, be objects in $\mathbb{F}_{\mathcal{F}}(\mathbf{S})$. Then there are objects u_j in $\mathbb{F}_{\mathcal{F}}(\mathbf{R})$ such that $h_*(u_j) \cong \overline{u_j}$, $j = 1, 2$. A choice of isomorphisms induces an abelian group homomorphism

$$h_*\colon \mathbb{F}_{\mathcal{F}}(\mathbf{R})(u_1, u_2) \to \mathbb{F}_{\mathcal{F}}(\mathbf{S})(\overline{u_1}, \overline{u_2}).$$

The next result is an easy corollary of Lemma 3.7.

Corollary 3.8. *With the above notation, h_* is an epimorphism.*

Proof. Let B be any R-bimodule. Since h is a ring epimorphism, the map

$$h'\colon B \to \overline{\overline{B}}, \qquad b \mapsto 1_S \otimes b \otimes 1_S$$

is an abelian group epimorphism. Thus for any $m, n > 0$, the induced map on the matrix group

$$h'_{m\times n}\colon M_{m\times n}(B) \to M_{m\times n}\left(\overline{\overline{B}}\right)$$

is an epimorphism. The result follows from the identifications proved in Lemma 3.7. □

The next Lemma is on the E-surjectivity of the functor h_*.

Lemma 3.9. *Let $\overline{u}$ be an object of $\mathbb{F}_{\mathcal{P}}(\mathbf{S})$ and g an elementary automorphism of $\overline{u}$. Then there is an object $\overline{v}$ in $\mathbb{F}_{\mathcal{P}}(\mathbf{S})$, an object w in $\mathbb{F}_{\mathcal{P}}(\mathbf{R})$ and an elementary automorphism f of w such that $h_*(w) \cong \overline{u}\oplus\overline{v}$ and under the isomorphism $h_*(f)$ is conjugate to $g\oplus 1_{\overline{v}}$.*

Proof. Since g is elementary, there is a splitting $\overline{u} = \overline{u_1}\oplus\overline{u_2}$ such that

$$g = \begin{pmatrix} 1 & \gamma \\ 0 & 1 \end{pmatrix}$$

where $\gamma \in \mathbb{F}_{\mathcal{P}}(\mathbf{S})(\overline{u_1}, \overline{u_2})$. Choose objects $\overline{v_j}$, $j = 1, 2$, in $\mathbb{F}_{\mathcal{P}}(\mathbf{S})$ such that $\overline{u_j}\oplus\overline{v_j}$ is in $\mathbb{F}_{\mathcal{F}}(\mathbf{S})$. Set $\overline{v} = \overline{v_1}\oplus\overline{v_2}$. Then $\overline{u}\oplus\overline{v}$ is an object in $\mathbb{F}_{\mathcal{F}}(\mathbf{S})$ and under the isomorphism

$$\overline{u}\oplus\overline{v} \cong (\overline{u_1}\oplus\overline{v_1}) \oplus (\overline{u_2}\oplus\overline{v_2})$$

$g\oplus 1_{\overline{v}}$ corresponds to

$$g\oplus 1_{\overline{v}} = \begin{pmatrix} 1 & \gamma' \\ 0 & 1 \end{pmatrix}$$

where

$$\gamma'\colon \overline{u_1}\oplus\overline{v_1} \xrightarrow{\left(\begin{smallmatrix} \gamma & 0 \\ 0 & 0 \end{smallmatrix}\right)} \overline{u_2}\oplus\overline{v_2}.$$

Since $\overline{u_j}\oplus\overline{v_j}$, $j = 1, 2$, are objects in $\mathbb{F}_{\mathcal{F}}(\mathbf{S})$, the result follows from Corollary 3.8. □

Using the above basic results, we will show that the functor ϕ is cofinal.

Lemma 3.10. *The functor $\phi\colon \mathbb{F}_{\mathcal{P}}(\mathbf{R}) \to \mathbb{P}$ is cofinal.*

Proof. Let $u = (u_1, u_2, g)$ be an object in $\mathbb{P}$. We can add an object $v = (v_1, v_2, g')$, with g' an isomorphism of degree zero, to u such that $u_j\oplus v_j$ is an object of $\mathbb{F}_{\mathcal{F}}(\mathbf{R}_j)$, $j = 1, 2$. Thus we can assume that u has the property that u_j is an object in $\mathbb{F}_{\mathcal{F}}(\mathbf{R}_j)$, $j = 1, 2$. Then g is an isomorphism between $f'_1(u_1)$ and $f'_2(u_2)$ in $\mathbb{F}_{\mathcal{F}}(\mathbf{R}_0)$. By choosing bases for the free modules involved, we can assume that g is an automorphism. First we will show that (u_1, u_2, g) is in the image of ϕ (up to equivalence) if g is an elementary automorphism in $\mathbb{F}_{\mathcal{P}}(\mathbf{R}_0)$. In this case, after more stabilization, $g = f'_1(g_1)$ for some automorphism g_1 of u_1 (Lemma 3.9) because f_1 is onto. Then the pair $(g_1, 1)$ induces an isomorphism between $(u_1, u_2, 1)$ and (u_1, u_2, g). But $(u_1, u_2, 1)$ is in the image of ϕ. The general case follows because the morphism in the object $(u_1 \oplus u_1, u_2 \oplus u_2, g \oplus g^{-1})$ can be written as a composition of elementary matrices. □

The following theorem is the main technical result of this paper. From it, the truly main result, Theorem 3.15, follows essentially by manipulation of definitions.

Theorem 3.11. *The functor ϕ induces an isomorphism*

$$\phi_j \colon K_j(\mathbb{F}_{\mathcal{P}}(\mathbf{R})) \to K_j(\mathbb{P}), \quad j \geq 1.$$

Proof. The functor ϕ is full and faithful and cofinal. Thus $\mathbb{F}_{\mathcal{P}}(\mathbf{R})$ can be identified with a full cofinal subcategory of $\mathbb{P}$. The result follows from [7], Theorem 1.1. □

Corollary 3.12. *Let $NK_1(\mathbf{R}_j) = 0$ for $j = 1, 2$ and $NK_2(\mathbf{R}_0) = 0$. Then $NK_1(\mathbf{R}) = 0$.*

Proof. The NK_1-group associated to $\mathbf{R}$ is given as the kernel of the composition

$$K_1(\mathbb{F}_{\mathcal{P}}(\mathbf{R})) \xrightarrow{\phi_1} K_1(\mathbb{P}) \xrightarrow{\kappa} K_1(\mathcal{P}^2_R).$$

If we use ϕ_1 from Theorem 3.11 to identify $K_1(\mathbb{F}_{\mathcal{P}}(\mathbf{R}))$ with $K_1(\mathbb{P})$, then $NK_1(\mathbf{R})$ is identified as the kernel of κ in the diagram preceding Lemma 3.6. The vanishing assumptions guarantee that the immediate neighbors of κ are monomorphims (actually isomorphisms). Also, the leftmost vertical map is a split epimorphism (in fact, an obvious splitting exists at the level of categories). Thus, by the five lemma, κ is a monomorphism. □

We finally specialize to the case of interest. Let $\Gamma = \Gamma_0 *_G \Gamma_1$ where G is a finite normal subgroup of Γ_i, $i = 0, 1$. Let $B_i = \mathbb{Z}[\Gamma_i - G]$, $i = 0, 1$, be the two $\mathbb{Z}G$-bimodules which appear in the definition of Waldhausen's Nil-groups in this case. Let N be the norm element in $\mathbb{Z}G$, i.e., N is the sum of all the group elements, and $\langle N \rangle$ the ideal generated by N. Let $n = |G|$. Notice that $\mathbb{Z}$ is isomorphic to the quotient of $\mathbb{Z}G$ by the ideal generated by the elements of the form $g - 1$, $g \in G$ and $\mathbb{Z}/n\mathbb{Z}$ is the quotient of $\mathbb{Z}G$ by the ideal generated by the elements N and $g - 1$, $g \in G$. Then we have a cartesian square

$$\begin{array}{ccc} \mathbb{Z}G & \xrightarrow{p_1} & \mathbb{Z}G/\langle N \rangle \\ \big\downarrow{\scriptstyle p_2} & & \big\downarrow{\scriptstyle q_1} \\ \mathbb{Z} & \xrightarrow{q_2} & \mathbb{Z}/n\mathbb{Z}. \end{array}$$

Lemma 3.13. *With the above notation, the right module $\mathbb{Z}G/\langle N \rangle \otimes_{\mathbb{Z}G} B_i$ (respectively $\mathbb{Z} \otimes_{\mathbb{Z}G} B_i$) is p_1 (respectively p_2) extendable, $i = 0, 1$. That implies that B_i is $q_1 p_1$-extendable.*

Proof. Notice that if $\gamma \in \Gamma_i$ then $\gamma N = N \gamma$ because G is normal in Γ_i. That implies that the ideal generated by N acts trivially, from the right, on $\mathbb{Z}G/\langle N \rangle \otimes_{\mathbb{Z}G} B_i$. For the other ring, notice that all the elements of $\mathbb{Z}G$ of the form $g - g'$, $g, g' \in G$ act trivially on the right on $\mathbb{Z} \otimes_{\mathbb{Z}G} B_i$. □

In the special case that $n = p$ is a prime then $\mathbb{Z}G/\langle N\rangle \cong \mathbb{Z}[\zeta_p]$, $\mathbb{Z}$ with a primitive p-th root of unity attached. This ring is regular Noetherian. The rings $\mathbb{Z}$ and $\mathbb{Z}/p\mathbb{Z}$ are also regular Noetherian rings and therefore the corresponding NK_j - groups vanish for $j \leq 1$, cf. Lemma 2.1.

Let C_p have index 2 in a group G. We will describe the $\mathbb{F}_p$-bimodule structure of $\mathbb{F}_p \otimes_{\mathbb{Z}C_p} \mathbb{Z}[G - C_p]$. The action of $\mathbb{Z}C_p$ on $\mathbb{F}_p$ is given via the epimorphism

$$\mathbb{Z}C_p \to \mathbb{Z}C_p/\langle N, g-1, g \in C_p\rangle \cong \mathbb{F}_p.$$

Lemma 3.14. *With the above notation, there is an $\mathbb{F}_p$-bimodule isomorphism*

$$\alpha\colon \mathbb{F}_p \otimes_{\mathbb{Z}C_p} \mathbb{Z}[G - C_p] \to \mathbb{F}_p.$$

Proof. The $\mathbb{Z}C_p$-bimodule $\mathbb{Z}[G - C_p]$ has a decomposition

$$\mathbb{Z}[G - C_p] = \bigoplus_{h \in G - C_p} \mathbb{Z}h$$

as an abelian group. The action of C_p is given by permuting the summands according to the action of C_p on $G - C_p$. Let h, h' be two elements in $G - C_p$. Then there is $g \in C_p$ such that $gh = h'$. Therefore, in $\mathbb{F}_p \otimes_{\mathbb{Z}C_p} \mathbb{Z}[G - C_p]$,

$$x \otimes h' = x \otimes (gh) = x \otimes h, \quad \text{for all } x \in \mathbb{F}_p.$$

Define α on $\mathbb{Z}h$ by setting $\alpha(x \otimes h) = x$ and extending linearly. Then α is the required isomorphism. □

The next theorem is a combination of the above observations and Theorem 3.11.

Theorem 3.15. *In the above notation, if $n = p$ is a prime, then*

$$\widetilde{Nil}_0^W(\mathbb{Z}C_p; B_0, B_1) = NK_1(\mathbb{Z}C_p; B_0, B_1) = 0.$$

Proof. In this case the pull-back diagram above becomes

$$\begin{array}{ccc} \mathbb{Z}C_p & \longrightarrow & \mathbb{Z}[\zeta_p] \\ \downarrow & & \downarrow \\ \mathbb{Z} & \longrightarrow & \mathbb{F}_p. \end{array}$$

The rings $\mathbb{Z}$, $\mathbb{Z}[\zeta_p]$, and $\mathbb{F}_p$ are regular. By Lemma 2.1 the NK_1-groups of the induced triples vanish for the three rings above. The bimodules B_i, $i = 0, 1$, are extendable over $\mathbb{F}_p$ (Lemma 3.13). Thus Lemma 3.14 applies and Corollary 2.5 implies that

$$NK_2(\mathbb{F}_p; \mathbb{F}_p \otimes_{\mathbb{Z}C_p} B_0, \mathbb{F}_p \otimes_{\mathbb{Z}C_p} B_1) = 0.$$

Then the result follows from Corollary 3.12. □

Combining with the results in [10] and Waldhausen's exact sequence (*) of Section 2, we have

Corollary 3.16. *With the above assumptions, there are exact sequences*

$$K_1(\mathbb{Z}C_p) \to K_1(\mathbb{Z}\Gamma_0)\oplus K_1(\mathbb{Z}\Gamma_1) \to K_1(\mathbb{Z}\Gamma) \to K_0(\mathbb{Z}C_p) \to \cdots,$$

and

$$Wh(C_p) \to Wh(\Gamma_0)\oplus Wh(\Gamma_1) \to Wh(\Gamma) \to \widetilde{K}_0(\mathbb{Z}C_p) \to \cdots.$$

As a particular application we have the following result which was used in the calculations in [3].

Corollary 3.17. $Wh(S_3 *_{\mathbb{Z}/3\mathbb{Z}} S_3) = 0$ *where* S_3 *is the symmetric group on 3 letters.*

Proof. From Theorem 3.15, we know that Waldhausen's Nil group vanish. We also know that the lower K-theory of S_3 and $\mathbb{Z}/3\mathbb{Z}$ vanish. The result follows from Corollary 3.16. □

References

[1] H. Bass, *Algebraic K-theory*, Benjamin, 1968.

[2] H. Bass, *Unitary algebraic K-theory*, Proceedings of the Conference in Algebraic K-theory, Battelle 1972, Lecture Notes in Math. vol. 343, 57–265, Springer, 1973.

[3] E. Berkove, F. T. Farrell, D. J. Pineda, and K. Pearson, *The Farrell–Jones isomorphism conjecture for finite co-volume hyperbolic actions and the algebraic K-theory of Bianchi groups*, Trans. Amer. Math. Soc., to appear.

[4] S. M. Gersten, *K-theory of free rings*, Comm. Algebra **1**, 39–64 (1974).

[5] S. M. Gersten, *The localization theorem for projective modules*, Comm. Algebra **2**, 307–350 (1974).

[6] D. R. Grayson, *Higher algebraic K-theory II (after D. Quillen)*, Algebraic K-theory, Evanston 1976, Lecture Notes in Math. vol. 551, 217–240, Springer, 1976.

[7] D. R. Grayson, *Localization for flat modules in algebraic K-theory*, J. Algebra **61**, 463–496 (1979).

[8] F. X. Connolly and T. Koźniewski, *Nil-groups in K-theory and surgery theory*, Forum Math. **7**, 45–76 (1995).

[9] H. Inassaridze, *Algebraic K-theory*, Math. Appl. vol. 311, Kluwer, 1995.

[10] S. Prassidis, *On the vanishing of lower K-groups*, IMADA, Odense Universitet preprint, 1997.

[11] J. Stallings, *Whitehead torsion of free products*, Ann. of Math. **82**, 354–363 (1965).

[12] F. Waldhausen, *Whitehead groups of generalized free products*, Proceedings of the Conference in Algebraic K-theory, Battelle 1972, Lecture Notes in Math. vol. 342, Springer, 1973.

[13] F. Waldhausen, *Algebraic K-theory of generalized free products*, Ann. of Math. **108**, 135–256 (1978).

IMADA, SDU – Odense Universitet, Campusvej 55, DK 5230 Odense, Denmark

Coordenação de Pós Graduação em Matemática, Rua Mário Santos Braga, s/n, 7º andar, Valonguinho, Niterói, RJ 24020–005, Brazil

Progress in Mathematics, Vol. 196

Lusternik-Schnirelmann cocategory: A Whitehead dual approach

Aniceto Murillo and Antonio Viruel

Abstract. In this paper a new approach to the Lusternik–Schnirelmann cocategory of a space is presented. This approach is based on a dual of the Whitehead definition of category. Using this new definition we are able to prove all the classical properties satisfied by the original Ganea's concept of cocategory.

1. Introduction

Since the very first moment the notion of category was introduced by Lusternik and Schnirelmann in 1934 [14], there have been several successful attempts to describe this homotopy invariant in a more functorial, and therefore manageable, form. The story can be simplified like this: For reasonable classes of spaces X (for instance, having the homotopy type of CW-complexes) the following notions of the Lusternik–Schnirelmann category of X, cat X, are all the same:

– The original one, i.e., the least integer n (or infinite) for which there exists an open covering of $n+1$ subspaces contractible in X.

–The Whitehead approach [19]: A space X has category less or equal than n if the diagonal map $\Delta\colon X \longrightarrow \Pi_{n+1}X$ can be deformed into the fat wedge of order n, T^nX.

–The inductive category [7]: cat $X \leq n+1$ if there exists a cofibration $A \longrightarrow Y \longrightarrow Z$ such that cat $Y \leq n$ and Z dominates X.

–The Ganea's characterization of category: cat $X \leq n$ if the n-th Ganea's fibration admits a homotopy section [8].

The understanding of these different approaches to the notion of category is one of the reasons why this subject has had such an enormous development in the past years.

Unlike this, the situation in the Eckmann-Hilton dual of this concept is not at all as satisfactory. On one hand this duality applies (when it does!) to the based homotopy category so it is difficult to present a dual of the original definition of category. On the other, it is not clear what the dual of a categorical covering would be. However, M. Hopkins [11] defined several notions of cocategory which

Authors are partially supported by DGES grant PB97-1095.

are a sort of dual of the initial one. Unfortunately, they do not seem to be too manageable to apply to daily situations.

For inductive category and Ganea's approach there are precise definitions of their dual notions, namely:

Definition 1.1. [7] *We say that the inductive cocategory of X is zero,* indcocat $X = 0$, *if X is contractible. Then,* indcocat $X \leq n+1$ *if there exists a fibration $F \longrightarrow E \longrightarrow B$ such that* indcocat $E \leq n$ *and F dominates X.*

Definition 1.2. [8] *The n-th Ganea cofibration of X, $X \xrightarrow{q_n} G_nX \longrightarrow C_nX$, is defined inductively as follows: q_0 is the cofibration $X \xrightarrow{q_0} CX \longrightarrow \Sigma X$. Next consider F the fibre of $G_{n-1}X \longrightarrow C_{n-1}X$ and factor q_{n-1} through F to get a map $X \longrightarrow F$. The associated cofibration to this map $X \xrightarrow{q_n} G_nX \longrightarrow C_nX$ is by definition the n-th Ganea cofibration of X. This can be better viewed in the following diagram:*

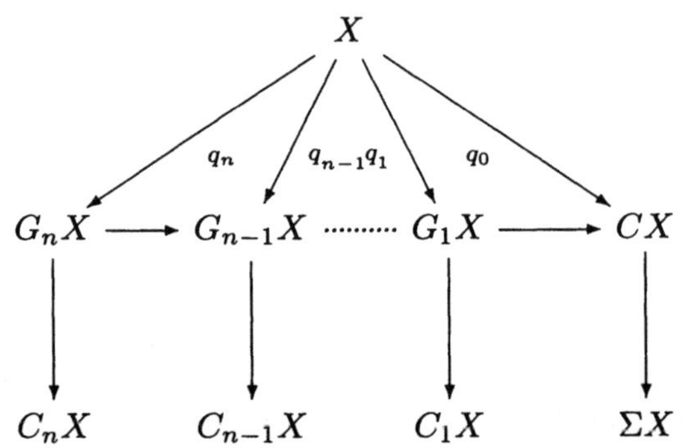

Then, indcocat X *is the least integer n for which q_n has a homotopy retraction.*

Indeed, these two definitions are equivalent for reasonable spaces [8]. However, with respect to the Whitehead approach, the only reference we are aware of is the work of M. Hovey [12] where he presents a notion of cocategory based in a dual concept of the fat wedge of a space, defined as the component of the base point of an inverse homotopy limit. The author proves some of the properties satisfied by the inductive cocategory.

The purpose of this paper is to present a Whitehead approach to the notion of cocategory which is easy enough to handle so that it enables us to deduce the dual of all the classical homotopy properties satisfied by the category. We should remark that the list of these properties are known to be true for the inductive cocategory but up to now we do not know whether it coincides with our notion of cocategory (we call it cocat in the sequel). In fact the most we can say is (see theorem 3.12) cocat $X \leq$ indcocat X. With respect to Hovey's notion we also point out (see remark 3.16) that our invariant is smaller than his and both coincide in the rational category. We now present how the paper is organized and summarize very briefly its content.

In the second section we collect all the basic results we shall make use of, particularly we focus our attention on the study of the cojoin of two maps.

Our definition of cocategory is based in a map $i_n\colon \vee_{n+1} X \longrightarrow W_n X$ which mimics the dual of the inclusion $T^n X \hookrightarrow \Pi_{n+1} X$ of the fat wedge into the product (this is the analogous of the map p_n of [12]). This definition together with basic properties and a few examples form the third section. Also in §3, the definition of wcocat, the weak cocategory of a space, can be found.

In §4 we present, as a key point of this paper, a deep study of the map i_n and extract important information.

As a consequence of this one immediately has:

$$\text{nil}\ \Omega X \leq \text{wcocat}\ X \leq \text{cocat}\ X.$$

In section §5 we prove that the last inequality may be strict. It is important to remark that this, together with the fact that cocat $X \leq$ indcocat X, makes the weak cocategory into a sharper upper bound for nil ΩX or the Whitehead product-length of X. We end up investigating how localization affects our notion of cocategory for simply connected spaces.

We would like to thank to Prof. Daniel Tanré for helpful conversations.

2. Basic results

Throughout the paper all considered spaces are based, and have the homotopy type of CW-complexes. Also, since we work in the homotopy category, we shall often not distinguish, unless necessary or explicitly stated, between homotopy classes and the maps which represent them. Hence, equality will often mean "homotopic to" or "have the same homotopy type as".

Homotopy theory is plenty of surprising facts. One of those is that honest limits do not exist in the homotopy category as these universal objects do not respect homotopy type. Homotopy limits were defined in order to solve partially this problem while still having nice universal properties [2]. We shall mostly deal with very simple homotopy limits, namely, homotopy pullbacks and pushouts. Briefly, we recall here the definition and some general facts about them. The homotopy pullback

$$\begin{array}{ccc} P & \longrightarrow & Y \\ \big\downarrow & \lrcorner & \big\downarrow{\scriptstyle g} \\ X & \xrightarrow{\ f\ } & Z \end{array}$$

of any diagram $X \xrightarrow{f} Z \xleftarrow{g} Y$ is defined as the honest pull back

$$\begin{array}{ccccc} & & P & \longrightarrow & Y \\ & & \big\downarrow & p.b. & \big\downarrow{\scriptstyle g} \\ X & \xrightarrow{\ \phi_X\ } & \widehat{X} & \xrightarrow{\ \hat{f}\ } & Z \end{array}$$

in which the bottom row is just the decomposition of f into a homotopy equivalence ϕ_X and a fibration $\hat{f}$. Homotopy pushout can be defined in a similar fashion. These definitions do not depend on the homotopy type of the consider maps. Moreover, the same object is obtained by decomposing g instead of f, or both.

With such a definition we cannot expect the universal property of pullbacks (resp. pushouts) to hold. Nevertheless there is a weak version of it;

Theorem 2.1. [15] *A homotopy commutative diagram*

$$\begin{array}{ccc} P & \xrightarrow{f_1} & Y \\ {\scriptstyle g_1}\downarrow & & \downarrow{\scriptstyle g} \\ X & \xrightarrow{f} & Z \end{array} \qquad H : gf_1 \simeq fg_1$$

is a homotopy pullback iff for any homotopy commutative diagram

$$\begin{array}{ccc} D & \xrightarrow{f_2} & Y \\ {\scriptstyle g_2}\downarrow & & \downarrow{\scriptstyle g} \\ X & \xrightarrow{f} & Z \end{array} \qquad G : gf_2 \simeq fg_2$$

the following hold:

(1) *There exists $D \xrightarrow{w} P$, called a whisker map, such that $K : f_2 \simeq f_1 w$, $L : g_1 w \simeq g_2$ and $gK + Hw + fL \simeq G$. That is, the following diagram is homotopy commutative*

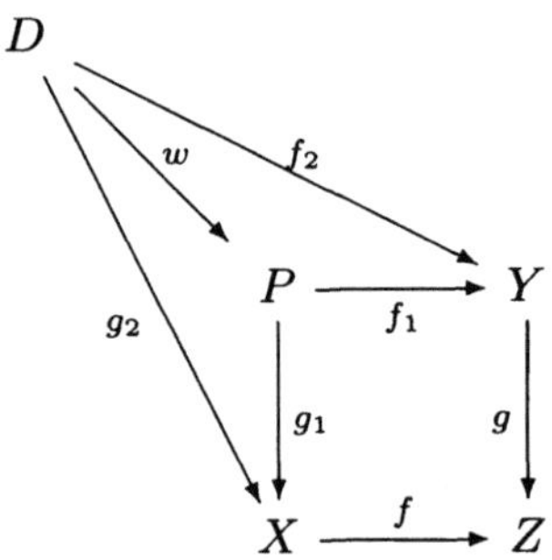

(2) *If there exists another map $D \xrightarrow{w'} P$ with the same properties as that of w in (1), then there exists $M : w \simeq w'$ such that $K + f_1 M \simeq K'$ and $g_1 M + L' \simeq L$.*

We call this the weak universal property of homotopy pullbacks (WUPHPB).

A similar weak universal property of homotopy pushout (WUPHPO) can be stated.

The first and classical examples of homotopy pullbacks and pushouts are homotopy fibrations and cofibrations respectively. The reader could find very profitable the reading of the papers of J.P. Doeraene [4] and M. Mather [15].

Recall that the join of two maps can be defined in these terms.

Definition 2.2. *The join of a diagram* $X \xrightarrow{f} Z \xleftarrow{g} Y$, *denoted by* $X *_Z Y$ *if the maps* f *and* g *are known, is defined as the homotopy pushout of* $X \longleftarrow E \longrightarrow Y$, *where* E *is the homotopy pullback of the original diagram. The whisker map provided by the WUPHPO is denoted* $X *_Z Y \xrightarrow{f *_Z g} Z$.

If $Z \simeq *$, and therefore f and g are nullhomotopic maps, the join is simply denoted by $X * Y$, and it is easily calculated;

Lemma 2.3. $X*Y \simeq \Sigma(X\wedge Y)$, *where* $X\wedge Y$ *means the cofibre of* $X\vee Y \longrightarrow X\times Y$.

Now, as the Eckmann-Hilton duality suggests, a new concept can be defined interchanging homotopy pushouts and pullbacks;

Definition 2.4. *The cojoin of a diagram* $X \xleftarrow{f} Z \xrightarrow{g} Y$, *denoted by* $X\natural_Z Y$ *if the maps* f *and* g *are known, is defined as the homotopy pullback of* $X \longrightarrow E \longleftarrow Y$, *where* E *is the homotopy pushout of the original diagram. The whisker map provided by the WUPHPB is denoted* $Z \xrightarrow{f\natural_Z g} X\natural_Z Y$.

Again, if $Z \simeq *$, the cojoin is simply denoted by $X\natural Y$. We can prove a formula dual to that of lemma 2.3. First recall the following result due to O. Cornea,

Proposition 2.5. [3] *Given* $F \longrightarrow E \longrightarrow B$ *and* $F' \longrightarrow E' \longrightarrow B'$ *(homotopy) fibrations, the following is a homotopy pullback*

$$\begin{array}{ccc} F \times F' \times \Omega(B\flat B') & \longrightarrow & E \vee * \\ \downarrow & & \downarrow \\ * \vee E' & \longrightarrow & B \vee B' \end{array}$$

Therefore the formula is

Lemma 2.6. $X\natural Y \simeq \Omega(X\flat Y)$, *where* $X\flat Y$ *is the hotomopy fibre of the standard inclusion* $X \vee Y \longrightarrow X \times Y$.

Proof. The pushout of the diagram $X \longleftarrow * \longrightarrow Y$ is $X\vee Y$. Hence the formula is obtained applying proposition 2.5 to the trivial fibrations $* \longrightarrow X \longrightarrow X$ and $* \longrightarrow Y \longrightarrow Y$. □

Another interesting property we shall use intensively in the sequel is

Lemma 2.7. *Given a homotopy commutative diagram:*

$$\begin{array}{ccccc} X & \xleftarrow{f} & Z & \xrightarrow{g} & Y \\ \downarrow{\scriptstyle\alpha} & & \downarrow{\scriptstyle\beta} & & \downarrow{\scriptstyle\gamma} \\ A & \xleftarrow{h} & C & \xrightarrow{i} & B \end{array}$$

there exists a natural map $\alpha\sharp_\beta\gamma$ such that the following diagram is homotopy commutative

$$\begin{array}{ccc} Z & \xrightarrow{f\sharp_Z g} & X\sharp_Z Y \\ \beta\downarrow & & \downarrow \alpha\sharp_\beta\gamma \\ C & \xrightarrow{h\sharp_C i} & A\sharp_C B \end{array}$$

Proof. Let E_1 and E_2 be the homotopy pushouts of the diagrams $X \xleftarrow{f} Z \xrightarrow{g} Y$ and $A \xleftarrow{h} C \xrightarrow{i} B$ respectively. Then the desired map appears as the (dotted) whisker map $w = \alpha\sharp_\beta\gamma$, in the following commutative diagram:

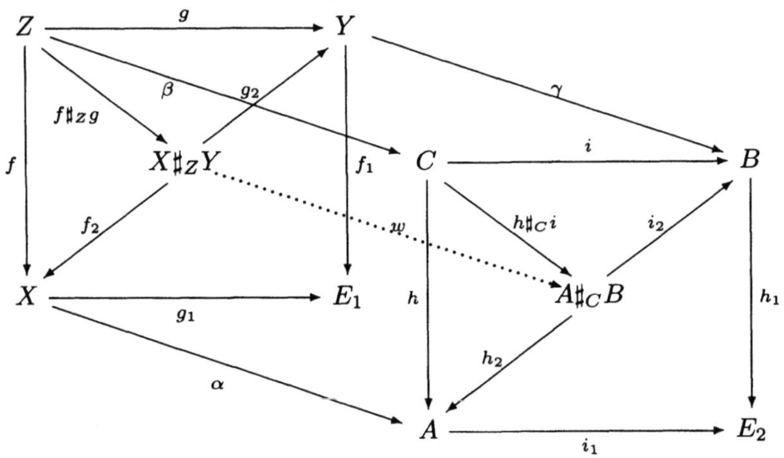

Indeed, in view of the WUPHPB, the commutativity of this follows from the fact that

$$\begin{aligned} i_2(h\sharp_C i)\beta &= i\beta \\ &= \gamma g \\ &= \gamma g_2(f\sharp_Z g) \\ &= i_2 w(f\sharp_Z g) \end{aligned}$$

and in the same way $h_2(h\sharp_C i)\beta = h\beta h_2 w(f\sharp_Z g)$ □

3. Cocategory of a space

We begin with the definition of the map i_n dual to the inclusion $k_n\colon T^nX \hookrightarrow \Pi_{n+1}X$ of the fat wedge into the product.

Definition 3.1. *Define recursively W_nX and $i_n\colon \vee_{n+1} X \to W_nX$ as follows: let W_0X be a point and i_0 is the constant map. Then*

$$W_n = \vee_n X\sharp_{\vee_{n+1}X}(W_{n-1}X \vee X)$$

via the maps $\vee_n X \xleftarrow{(1,*)} \vee_{n+1} X \xrightarrow{i_{n-1}\vee 1} W_{n-1}X \vee X$, *and* $i_n\colon \vee_{n+1} X \to W_n X$ *defined in view of the diagram:*

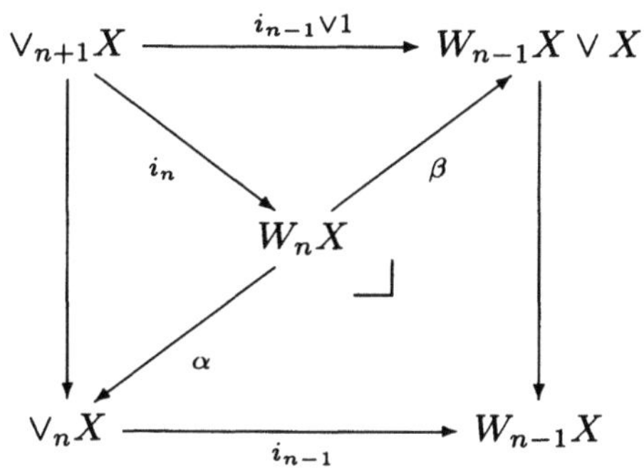

(1)

That is, $i_n = (1,*)\sharp_{\vee_{n+1}X}(i_{n-1} \vee 1)$.

Remark 3.2. (1) $W_1X = X \times X$ *and* $i_1\colon X \vee X \to X \times X$ *is just the inclusion.*

(2) *Observe that* W_nX *is in fact the Eckmann-Hilton dual of* T^nX, *the fat wedge of* X*: Indeed, write* $T^{n-1}X$ *as the homotopy pullback*

$$\begin{CD} T^{n-1}X @>{(1,*)}>> T^{n-1}X \times X \\ @V{k_{n-1}}VV @VV{(k_{n-1},*)}V \\ \Pi_n X @>>{(1,*)}> \Pi_{n+1}X \end{CD}$$

and note that T^nX *is the pushout (homotopy or strict since* k_{n-1} *is a cofibration)*

$$\begin{CD} T^{n-1}X @>{(1,*)}>> T^{n-1}X \times X \\ @V{k_{n-1}}VV @VVV \\ \Pi_n X @>>> T^nX \end{CD}$$

That is to say,

$$T^nX = \Pi_nX *_{\Pi_{n+1}X} (T^{n-1}X \times X)$$

(3) *Note that since* i_n *is a whisker (and therefore not unique!) map, to define it without ambiguity and at the same time exhibit* W_n *as a functor one has to be more precise and fix representatives of the different objects chosen in the process.*

Another functorial property of W_n is given in the following

Proposition 3.3. *Given maps $f_i\colon X \longrightarrow Y$, for $i = 1, \dots, n+1$, there exists a map $\alpha_n\colon W_nX \longrightarrow W_nY$ making commutative the diagram,*

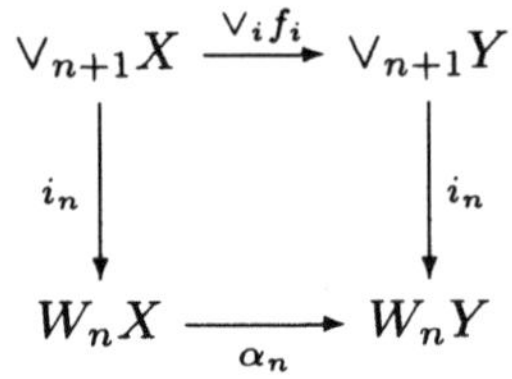

If $f_i = f$ for all i we call $\alpha_n = W_n(f)$.

Proof. The map α_n is defined inductively by lemma 2.7 as follows: $\alpha_1 = f_1 \times f_2$ and $\alpha_n = (\vee_n f_i)\sharp_{\vee_{n+1} f_i}(\alpha_{n-1} \vee f_{n+1})$. □

Our notion of Lusternik-Schnirelmann cocategory is:

Definition 3.4. *Given a space X,* cocat $X \leq n$ *if the folding map $\sigma\colon \vee_{n+1} X \to X$ has a homotopy extension to W_nX, i.e., there exists $\varphi\colon W_nX \to X$ so that the following commutes:*

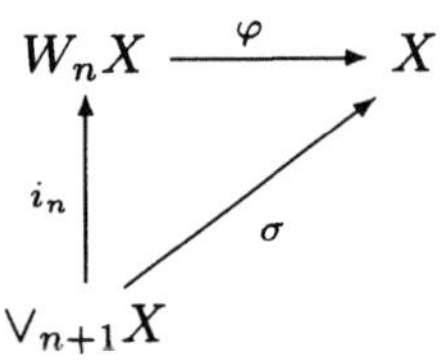

In fact one has

Proposition 3.5. *If* cocat $X \leq n$ *then* cocat $X \leq n+1$.

Proof. Let $\varphi\colon W_nX \to X$ so that $\varphi i_n = \sigma$ and consider the diagram as in (1)

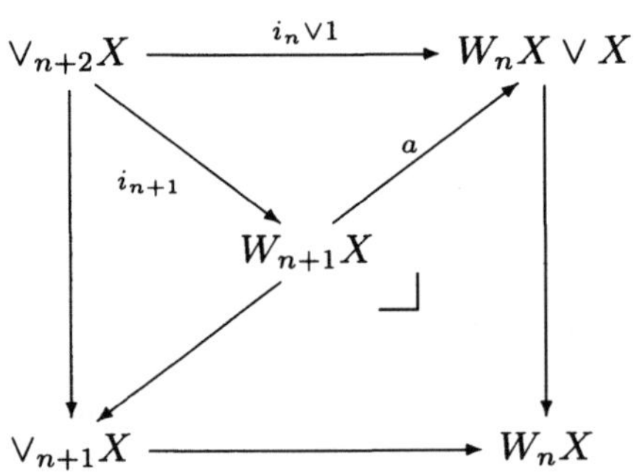

Note that the composition $\vee_{n+2}X \xrightarrow{i_n \vee 1} W_nX \vee X \xrightarrow{(\varphi,1)} X$ is just the folding map and therefore $(\varphi, 1_X)\,a$ is the extension we need. □

Proposition 3.6. *If Y dominates X then* cocat $X \leq$ cocat Y.

Proof. Let $\varphi : W_nY \to Y$ be a map for which $\varphi i_n = \sigma$ and assume $X \xrightarrow{f} Y$, $Y \xrightarrow{g} X$ are such that $gf = 1_X$. Then, in view of the functorial property of i_n above described, the following diagram is commutative:

$$\begin{array}{ccccccc}
W_nX & \xrightarrow{W_nf} & W_nY & \xrightarrow{\varphi} & Y & \xrightarrow{g} & X \\
\uparrow{\scriptstyle i_n} & & \uparrow{\scriptstyle i_n} & \nearrow{\scriptstyle \sigma} & & & \uparrow{\scriptstyle \sigma} \\
\vee_{n+1}X & \xrightarrow[\vee_{n+1}f]{} & \vee_{n+1}Y & & \xrightarrow[\vee_{n+1}g]{} & & \vee_{n+1}X
\end{array}$$

But the bottom row is just $1_{\vee_{n+1}X}$ and therefore $g\,\varphi\,W_nf\,i_n = \sigma$, that is to say, cocat $X \leq n$. □

Corollary 3.7. *The cocategory of a space is a homotopy invariant.*

We obviously have our first

Example 3.8. *Let X be any space, then*

(1) cocat $X = 0$ *iff X is contractible.*
(2) cocat $X = 1$ *iff X is an H-space.*

We can also define the weak version of cocategory:

Definition 3.9. *Let $F_n \xrightarrow{j_n} \vee_{n+1}X \xrightarrow{i_n} W_nX$ be the fibration sequence associated to i_n. Define the weak cocategory of X,* wcocat X, *as the least integer n (or infinite) for which $\sigma\,j_n = *$ with σ the folding map.*

Similar considerations as in 3.6 and 3.7 show that wcocat is a homotopy invariant. Moreover,

Proposition 3.10. wcocat $X \leq$ cocat X.

Proof. Assume cocat $X \leq n$, i.e., there exists $\varphi\colon W_nX \to X$ so that $\varphi\,i_n = \sigma$. Then $\sigma\,j_n = \varphi\,i_n\,j_n = *$. □

The fact that this is indeed a weaker notion shall be proved in §5. We now relate our invariant with the inductive cocategory or the equivalent definition based in the Ganea's cofibrations, both stated in the introduction. For that, first we find a bridge between W_nX and G_nX:

Proposition 3.11. *For any $n \geq 0$ there exists a map $f_n : W_nX \longrightarrow G_nX$ such that the following diagram is commutative:*

$$\begin{array}{ccc}
\vee_{n+1}X & \xrightarrow{i_n} & W_nX \\
\downarrow{\scriptstyle \sigma} & & \downarrow{\scriptstyle f_n} \\
X & \xrightarrow{q_n} & G_nX
\end{array}$$

Proof. By induction on n. For $n = 0$ there is nothing to prove. Assume that we have already constructed the map $f_{n-1} : W_{n-1}X \longrightarrow G_{n-1}X$, then the following diagram is commutative

$$\begin{array}{ccccc} \vee_n X & \xleftarrow{(1,*)} & \vee_{n+1}X & \xrightarrow{i_{n-1}\vee 1_X} & W_{n-1}X \vee X \\ \downarrow & & \downarrow{\scriptstyle \sigma} & & \downarrow{\scriptstyle f_{n-1}\vee q_{n-1}} \\ * & \longleftarrow & X & \xrightarrow{q_{n-1}} & G_{n-1}X. \end{array}$$

Applying lemma 2.7 to the diagram above we get the desired f_n, taking into account that $G_{n+1}(X) = * \sharp_X G_n(X)$ for $n \geq 0$. □

Now, the relation between cocat X and indcocat X is clear;

Theorem 3.12. cocat $X \leq$ indcocat X

Proof. Using 3.11, any retraction $r_n\colon G_n(X) \longrightarrow X$ of the map q_n gives us a map $\varphi = r_n f_n$ such that $\varphi i_n = \sigma_{n+1}$. Thus, cocat $X \leq n$. □

It is important to remark that at the present we are unable to prove or disprove if cocat equals indcocat. The way of proceeding to show that all the notion of category coincide cannot be applied in the dual context. This is due to the fact that the J-axiom of Doeranne [4] (or Mather's cube theorem [15]) does not dualize.

We use 3.12 in the following

Example 3.13. cocat $S^2 = 2$. *In fact* cocat $S^2 > 1$ *since it is not an H-space. We now show that* indcocat $S^2 \leq 2$. *For that simply observe that the Hopf fibration* $S^1 \to S^3 \to S^2$ *induces another fibration sequence* $S^2 \to BS^1 \to BS^3$. *Thus, since* BS^1 *is an H-space we have* cocat $BS^1 = 1$. *Hence, in view of the definition of inductive cocategory,* cocat $S^2 =$ indcocat $S^2 = 2$.

We end up this section with another basic property

Proposition 3.14. cocat $(X \times Y) = \max\{\text{cocat } X, \text{cocat } Y\}$.

Proof. Since X and Y are dominated by $X \times Y$, both cocat X and cocat Y are less or equal than cocat $(X \times Y)$. Let $n = \max\{\text{cocat } X, \text{cocat } Y\}$. Then, there exists maps $\varphi\colon W_n X \to X$ and $\psi\colon W_n Y \to Y$ for which $\varphi i_n = \sigma$, $\psi i_n = \sigma$. Now, using proposition 3.3, we see that i_n behaves nicely with respect to the product:

$$\begin{array}{ccc} W_n(X \times Y) & \xrightarrow{(W_n p_X, W_n p_Y)} & W_n X \times W_n Y \\ \uparrow{\scriptstyle i_n} & & \uparrow{\scriptstyle i_n \times i_n} \\ \vee_{n+1}(X \times Y) & \xrightarrow{(\vee_{n+1} p_X, \vee_{n+1} p_Y)} & \vee_{n+1} X \times \vee_{n+1} Y \end{array}$$

This, together with the aid of this picture,

$$\begin{array}{ccccc} \vee_{n+1}(X \times Y) & \xrightarrow{f} & \vee_{n+1}X \times \vee_{n+1}Y & \xrightarrow{i_n \times i_n} & W_nX \times W_nY \\ & {\scriptstyle\sigma}\searrow & \downarrow{\scriptstyle \sigma\times\sigma} & \swarrow{\scriptstyle \varphi\times\psi} & \\ & & X \times Y & & \end{array}$$

being $f = (\vee_{n+1}p_X, \vee_{n+1}p_Y)$, let us conclude that the following commutes:

$$\begin{array}{ccc} W_n(X \times Y) & \xrightarrow{g} & X \times Y \\ \uparrow{\scriptstyle i_n} & \nearrow{\scriptstyle \sigma} & \\ \vee_{n+1}(X \times Y) & & \end{array}$$

where $g = (\varphi \times \psi)(W_np_X, W_np_Y)$. Hence, $\operatorname{cocat}(X \times Y) \leq n$. □

Remark 3.15. *One might be tempted to think that a relation between* $\operatorname{cocat}(X \vee Y)$ *and* $\operatorname{cocat} X + \operatorname{cocat} Y$ *should exist, at least for simply connected spaces, as the Eckmann-Hilton duality suggests. However, this is not the case. Indeed, the fact* $\operatorname{cat}(X \times Y) \leq \operatorname{cat} X + \operatorname{cat} Y$ *is a consequence of a very special behavior of categorical coverings of the product (see* [6] *or* [18] *for details). We postpone an example for this until 4.9.*

Remark 3.16. *We point out here that our notion of cocategory is smaller that the one defined by M. Hovey* [12] *which we now recall: Let $\mathcal{N}$ be the opposite category of proper subsets of* $\{1, \dots, n\}$ *and let $G_n\colon Top \longrightarrow Top^{\mathcal{N}}$ be the functor defined by $G_n(X) = \vee_{i\in A}X$ and $G_n(X)(A \supset B)\colon \vee_{i\in A} \longrightarrow \vee_{j\in B}$ the projection. Then P^nX is defined as the component of the base point of* $\operatorname{holim} G_n(X)$. *Considering spaces as constant diagrams in $Top^{\mathcal{N}}$, the obvious map $\vee_{i=1}^n X \longrightarrow G_n(X)$ induces a natural map $\tau_n\colon \vee_{i=1}^n X \longrightarrow P^nX$. Then Hovey's notion of cocategory of a space X is the least integer n (or infinite) for which there exists $g\colon P^{n+1}X \longrightarrow X$ such that the following diagram commutes:*

$$\begin{array}{ccc} P^{n+1}X & \xrightarrow{g} & X \\ {\scriptstyle \tau_n}\uparrow & \nearrow{\scriptstyle \sigma} & \\ \vee_{n+1}X & & \end{array}$$

At present we do not know whether $P^{n+1}X$ and W_nX coincide for $n \geq 2$ (they do in the rational category!). However there exists a map $\mu_n\colon W_nX \longrightarrow P^{n+1}X$

such that the diagram

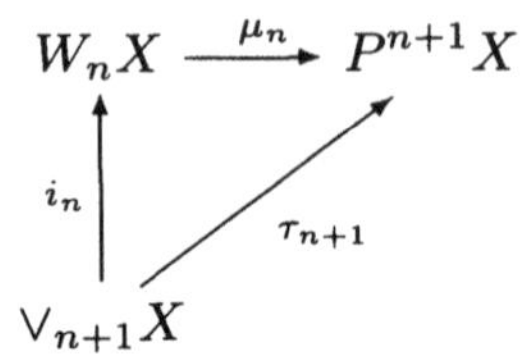

commutes. Indeed μ_n *is induced by maps* $\omega_n : W_nX \longrightarrow \vee_{i\in A} X$ *(for each subset* $A \subset \{1, \dots, n+1\}$*) which are defined inductively by:*

$$\omega_n = \begin{cases} p\beta, & \text{if } n+1 \notin A, \\ (\omega_{n-1} \vee 1_X)\alpha, & \text{if } n+1 \in A, \end{cases}$$

with α *and* β *as in diagram* (1), *where* $p : \vee_{i=1}^{n} X \longrightarrow \vee_{i\in A} X$ *is the projection, and* $\omega_{n-1} \colon W_{n-1}X \longrightarrow \vee_{A-\{n+1\}} X$.

Hence, if $g \colon P^{n+1}X \longrightarrow X$ *makes Hovey's invariant less or equal than* n*, then* $g\mu$ *allows us to conclude than* cocat $X \leq n$.

4. Investigating the map i_n

As we pointed out in the introduction, a careful study of the map i_n shall let us deduce the dual of all the classical homotopy properties of the LS category (properties which are known to be satisfied by the inductive cocategory [7], [8]). We start off by showing how the connectivity degree of i_n increases with n.

First, we recall a result of O. Cornea, which relate cofibrations and fibrations.

Proposition 4.1. [3] *Let* $F \longrightarrow E \longrightarrow B$ *be a fibration and let* $Z \longrightarrow B' \longrightarrow B$ *a cofibration. Let* E' *be the homotopy pullback of the diagram* $B' \longrightarrow B \longleftarrow E$*. Then there is a cofibration sequence* $\Sigma(Z \wedge F) \longrightarrow B'/E' \longrightarrow B/E$.

Then the following technical lemma is an easy consequence (as in §3, call F_n the homotopy fibre of i_n)

Lemma 4.2. *There is a cofibration sequence of the form:* $\Sigma(X \wedge F_n) \longrightarrow A \xrightarrow{f_n} B$ *where* $A = (W_nX \vee X)/W_{n+1}X$ *and* $B = W_nX/\vee_{n+1} X$.

Proof. It follows from 4.1 applied to the fibration $F_n \longrightarrow \vee_{n+1} X \xrightarrow{i_n} W_nX$ and the cofibration $X \longrightarrow W_nX \vee X \longrightarrow W_nX$. □

Now, recall that if Z is n-connected and Y is m-connected $(m, n \geq 0)$, then $Z \wedge Y$ is $(m+n+1)$-connected. Then, we have

Theorem 4.3. *Let* X *be* q*-connected, then* i_n *is* $(n+1)q$*-connected.*

Proof. By induction on n. It is trivial for $n = 0$. Assume that the result has been proved up to $n-1$. Then, the map f_{n-1} constructed in lemma 4.2 induces isomorphism in homology till degree $(n+1)q+2$ as $\Sigma(X \wedge F_{n-1})$ is $((n+1)q+2)$-connected by hypothesis. Also, it fits in the following commutative diagram of cofibrations

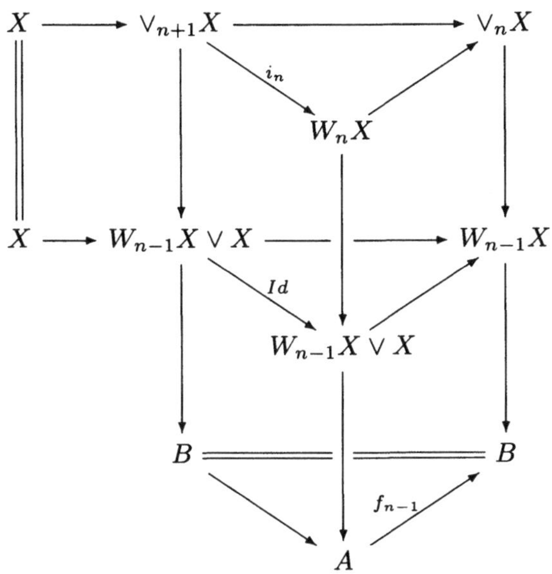

where $A = (W_{n-1}X \vee X)/W_nX$ and $B = W_{n-1}X/\vee_n X$. This diagram gives rise to the following diagram in homology:

$$\begin{array}{ccccccccc}
H_{r+1}(W_{n-1}X \vee X) & \to & H_{r+1}(B) & \to & H_r(\vee_{n+1}X) & \to & H_r(W_{n-1}X \vee X) & \to & H_r(B) \\
\| & & \uparrow{\scriptstyle (f_{n-1})_{r+1}} & & \downarrow{\scriptstyle (i_n)_*} & & \| & & \uparrow{\scriptstyle (f_{n-1})_r} \\
H_{r+1}(W_{n-1}X \vee X) & \to & H_{r+1}(A) & \to & H_r(W_nX) & \to & H_r(W_{n-1}X \vee X) & \to & H_r(A)
\end{array}$$

If $r \leq (n+1)q+1$, then $(f_{n-1})_{r+1}$ and $(f_{n-1})_r$ are isomorphism, and by the Five's Lemma, $(i_n)_r$ is so. □

This proposition allows us to calculate the cocategory of spaces with a few homotopy groups as the following corollary shows,

Corollary 4.4. *Let X be a q-connected space such that $\pi_r X = 0$ if $r > (n+1)q$. Then* cocat $X \leq n$. *In particular, n-th Postnikov stages of simply connected spaces have cocategory bounded above by n.*

Proof. For any space Z, we will denote by $Z^{(r)}$ the r-th stage of the Postnikov system of Z. By theorem 4.3, i_n induces homotopy equivalences between $(\vee_{n+1}X)^{(r)}$ and $(W_nX)^{(r)}$ for any $r \leq (n+1)q$. If we choose $r = (n+1)q$, we have the following

commutative diagram

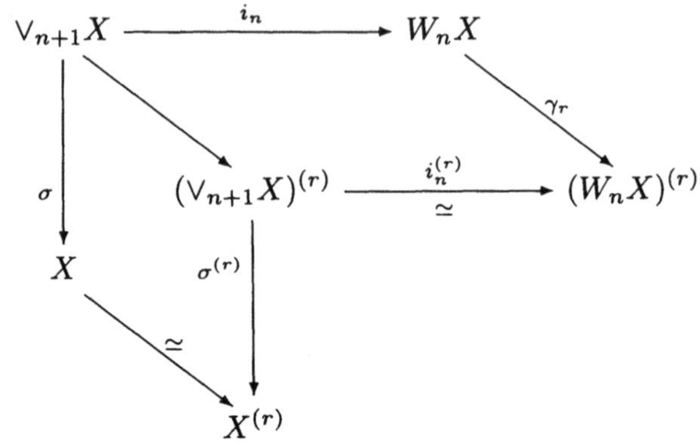

Therefore, the desired extension to the folding map $\varphi : W_nX \longrightarrow X$ appears as the composition $\varphi = \sigma^{(r)}(i_n^{(r)})^{-1}\gamma_r$, which proves the corollary. □

Another important consequence of theorem 4.3 is the following result, known for the inductive cocategory [9]

Theorem 4.5. *Let $X\langle q\rangle$ be the q-connected cover of X, $q \geq 0$. Then* cocat $X\langle q\rangle \leq$ cocat X.

Proof. Assume cocat $X = n$ and let $\varphi \colon W_nX \to X$ be an extension of the folding map. Then consider the following diagram (where the dotted ψ has to be defined)

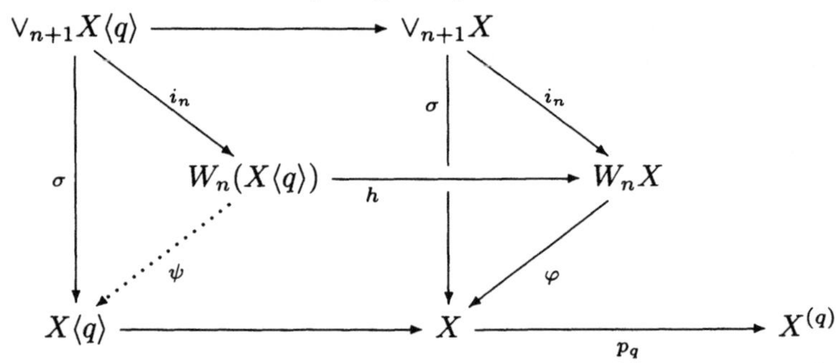

in which the bottom row is a fibration. By theorem 4.3, $\vee_{n+1}X\langle q\rangle \xrightarrow{i_n} W_n(X\langle q\rangle)$ is q-connected so is $W_n(X\langle q\rangle)$. Therefore $p_q\,\varphi\, h = *$ and the dotted arrow ψ making the whole set commutative does exist. Hence, cocat $X\langle q\rangle \leq n$. □

Next, we shall see how the space W_nX mimics the same property as the product with respect to Whitehead products. From now on, given a space X and elements $\alpha_i \in \pi_*(X)$, $i = 1,\dots,n$, we denote by $[\alpha_1,\dots,\alpha_n]$ the iterated Whitehead product $[[[\cdots[\alpha_1,\alpha_2]\alpha_3]\cdots]\alpha_n]$. Then we have,

Theorem 4.6. *Let $\alpha_1,\dots,\alpha_{n+1} \in \pi_*(X)$ and let $j_k \colon X \hookrightarrow \vee_{n+1}X$ the inclusion of the k-th copy of X, $k = 1,\dots,n+1$. Then, in $\pi_*(W_nX)$,*

$$\pi_*(i_n)[j_1\alpha_1,\dots,j_{n+1}\alpha_{n+1}] = 0$$

Proof. We prove it by induction on n. For $n = 1$ is a well known fact. Indeed, for a given space Y and elements $\alpha \in \pi_p(Y)$, $\beta \in \pi_q(Y)$, the Whitehead product $[\alpha, \beta] = 0$ if and only if the map $S^p \vee S^q \xrightarrow{(\alpha,\beta)} Y$ has a homotopy extension to $S^p \times S^q$. Hence, given elements $\alpha_1 \in \pi_p(X)$, $\alpha_2 \in \pi_q(X)$, $\pi_*(i_1)[j_1\alpha_1, j_2\alpha_2] = 0$ since the map $(i_1j_1\alpha_1, i_1j_2\alpha_2)\colon S^p \vee S^q \longrightarrow W_1X = X \times X$ has obviously an extension to the product: $\alpha_1 \times \alpha_2$.

Assume the assertion is true for $n-1$ and let $\alpha_1, \dots, \alpha_{n+1} \in \pi_*(X)$. Again, if $[\alpha_1, \dots, \alpha_n] \in \pi_p(X)$ and $\alpha_{n+1} \in \pi_q(X)$ then

$$\pi_*(i_n)[j_1\alpha_1, \dots, j_{n+1}\alpha_{n+1}] = \pi_*(i_n)\big[[j_1\alpha_1, \dots, j_n\alpha_n], j_{n+1}\alpha_{n+1}\big]$$

will vanish if the map

$$S^p \vee S^q \xrightarrow{[j_1\alpha_1, \dots, j_n\alpha_n] \vee j_{n+1}\alpha_{n+1}} \vee_{n+1}X \xrightarrow{i_n} W_nX$$

has an extension to $S^p \times S^q$. For explicitly defining such an extension consider the diagram (where the dotted arrows are yet to be defined and q is the projection)

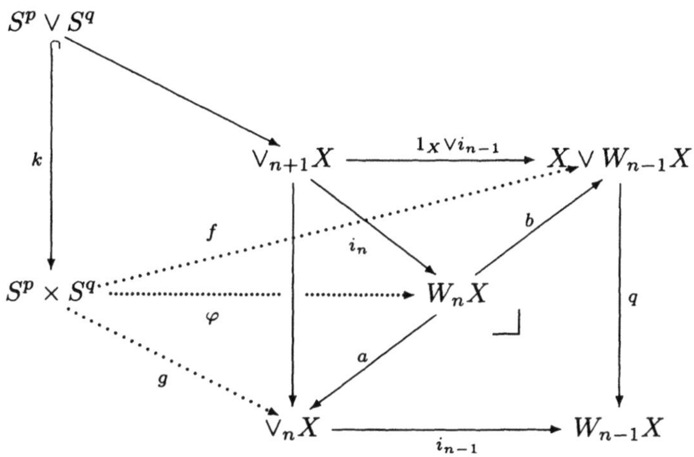

and define f and g as follows:

$$f\colon S^p \times S^q \xrightarrow{p_2} S^q \xrightarrow{\alpha_{n+1}} X \xrightarrow{j} W_{n-1}X \vee X$$

$$g\colon S^p \times S^q \xrightarrow{p_1} S^p \xrightarrow{[j_1\alpha_1, \dots, j_n\alpha_n]} \vee_n X$$

where p_1, p_2 are projections and j is the inclusion. Next, observe that, by induction hypothesis, $qf = * = i_{n-1}[j_1\alpha_1, \dots, j_n\alpha_n]$, and therefore there exists the dotted arrow φ of the diagram above making the whole picture homotopy commutative.

It remains to see that the square

$$\begin{array}{ccc} S^p \vee S^q & \xrightarrow{[j_1\alpha_1,\dots,j_n\alpha_n]\vee j_{n+1}\alpha_{n+1}} & \vee_{n+1}X \\ {\scriptstyle k}\big\downarrow & & \big\downarrow{\scriptstyle i_n} \\ S^p \times S^q & \xrightarrow[\varphi]{} & W_nX \end{array}$$

is homotopy commutative. But, for that, in view of the WUPHPB, it is enough to see that (call $\beta = [j_1\alpha_1, \dots, j_n\alpha_n]$)

$$b\, i_n\, (\beta \vee j_{n+1}\alpha_{n+1}) = b\,\varphi\, k,$$
$$a\, i_n\, (\beta \vee j_{n+1}\alpha_{n+1}) = a\,\varphi\, k.$$

Indeed:

$$\begin{aligned} b\, i_n\, (\beta \vee j_{n+1}\alpha_{n+1})|_{S^p} &= (i_{n-1} \vee 1_X)(\beta \vee j_{n+1}\alpha_{n+1})|_{S^p} \\ &= i_{n-1}\beta = \text{(by hypothesis)} = * \\ &= f\, k|_{S^p} = b\,\varphi\, k|_{S^p}, \\ b\, i_n\, (\beta \vee j_{n+1}\alpha_{n+1})|_{S^q} &= (i_{n-1} \vee 1_X)(\beta \vee j_{n+1}\alpha_{n+1})|_{S^q} = \alpha_{n+1} \\ &= f\, k|_{S^q} = b\,\varphi\, k|_{S^q}. \end{aligned}$$

And, on the other hand,

$$a\, i_n\, (\beta \vee j_{n+1}\alpha_{n+1})|_{S^p} = \beta = a\,\varphi\, k|_{S^p},$$
$$a\, i_n\, (\beta \vee j_{n+1}\alpha_{n+1})|_{S^q} = * = a\,\varphi\, k|_{S^q}.$$

□

As an immediate consequence we obtain the following: Given a space X recall that wlength X (Whitehead product length of X) is the largest integer n for which there is a non trivial Whitehead product of order n. This invariant is a classical bound for the inductive cocategory [Ga2] and from the theorem above we may sharpen this bound:

Corollary 4.7. wlength $X \leq$ wcocat X.

Proof. Assume wcocat $X \leq n$. Observe that, any $(n+1)$-order Whitehead product, in view of its naturality, can be written as:

$$[\alpha_1, \dots, \alpha_{n+1}] = \pi_*(\sigma)[j_1\alpha_1, \dots, j_{n+1}\alpha_{n+1}].$$

However, from proposition above we have $i_n[j_1\alpha_1, \dots, j_{n+1}\alpha_{n+1}] = 0$ and therefore $[j_1\alpha_1, \dots, j_{n+1}\alpha_{n+1}]$ factors through the fiber of i_n (call it $j\colon F_n \hookrightarrow \vee_{n+1}X$). That is to say, for some map $h\colon S^m \longrightarrow F_n$,

$$[j_1\alpha_1, \dots, j_{n+1}\alpha_{n+1}] = \pi_*(\sigma)[j_1\alpha_1, \dots, j_{n+1}\alpha_{n+1}] = \pi_*(\sigma)\, j\, h = *,$$

since $\sigma j = *$. □

Remark 4.8. *As we shall see in the next section, the corollary above is in fact a sharper bound since the inequality* wcocat $X <$ cocat X *may occurs.*

Corollary 4.7 leads to some interesting remarks:

Remark 4.9. (1) *We are now able to exhibit the promised example which shows that there is no relation between* cocat $(X \vee Y)$ *and* cocat $X +$ cocat Y*, even for simply connected spaces: On one hand, in view of the corollary above,* cocat $(S^2 \vee S^2) = \infty$ *since there are non trivial Whitehead products of arbitrarily high order, while* cocat $S^2 = 2$*. On the other hand, consider the space* $X = K(\mathbb{Z}/p, n) \vee K(\mathbb{Z}/q, n)$ *with* $1 < n$ *and* $(p, q) = 1$*. Then the inclusion* $K(\mathbb{Z}/p, n) \vee K(\mathbb{Z}/q, n) \hookrightarrow K(\mathbb{Z}/p, n) \times K(\mathbb{Z}/q, n)$ *induces an isomorphism in cohomology and thus it is an equivalence. Hence, X is an H-space, i.e.,* cocat $X = 1$*. Therefore,* $1 =$ cocat $K(\mathbb{Z}/p, n) \vee K(\mathbb{Z}/q, n) <$ cocat $K(\mathbb{Z}/p, n) +$ cocat $K(\mathbb{Z}/q, n) = 2$.

(2) *Another consequence is that we can see now how the hypothesis of q-connectivity in 4.4 is necessary. Let X be a space with perfect fundamental group. Thus, there are non trivial commutators of any length in $\pi_1(X)$. Therefore, by 4.7 and independently of its higher homotopy groups,* cocat $X = \infty$.

Another special behavior of the map i_n occurs when dealing with universal commutators:

Definition 4.10. *Let X be a co-H-space with comultiplication ν and inverse η. We define the universal commutator $c\colon X \to X \vee X$ as the composite,*

$$X \xrightarrow{\nu \vee \nu} X \vee X \vee X \vee X \xrightarrow{\tau_{2,3}} X \vee X \vee X \vee X \xrightarrow{h} X \vee X \vee X \vee X \xrightarrow{\sigma \vee \sigma} X \vee X$$

in which $\tau_{2,3}$ is the $(2,3)$ transposition and $h = 1_X \vee \eta \vee 1_X \vee \eta$. This is suggested from the fact that, if X is a cogroup, given maps $\alpha, \beta\colon X \to Y$, the commutator (α, β) in the group $[X, Y]$ is precisely $(\alpha, \beta) = \sigma(\alpha \vee \beta)c$. Now define recursively the n-th universal commutator as follows: $c_0 = 1_X$, $c_1 = c$ and $c_n\colon X \to \vee_{n+1} X$, where $c_n = (c_{n-1} \vee 1_X)c$. Again, the commutator $(\alpha_1, \dots, \alpha_{n+1})$ in $[X, Y]$ is $\sigma(\alpha_1 \vee \dots \vee \alpha_{n+1})c_n$.

Then we have

Theorem 4.11. $c_n i_n = *$.

Proof. By induction on n. For $n = 1$ it is a well known fact [1]. Assume the assertion is true for $n-1$ and consider the diagram

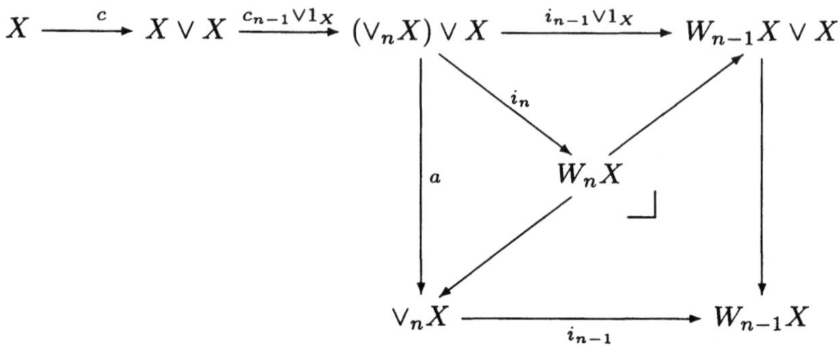

Therefore, to conclude, it is enough to see that

(i) $(i_{n-1} \vee 1_X)c_n = *$ and
(ii) $a\, c_n = *$.

Indeed,

(i) We have that

$$\begin{aligned}(i_{n-1} \vee 1_X)c_n &= (i_{n-1} \vee 1_X)(c_{n-1} \vee 1_X)c \\ &= (i_{n-1}c_{n-1} \vee 1_X)c = (\text{by hypothesis}) = (* \vee 1_X)c \\ &= (p_1 i_1)c = (\text{again by hypothesis}) = *\end{aligned}$$

Here p_1 denotes the projection over the first factor.

(ii) $a\, c_n = (c_{n-1} \vee 1_X)c = \sigma(c_{n-1} \vee *)c$, i.e., the following composite,

$$X \xrightarrow{c} X \vee X \xrightarrow{c_{n-1}\vee *} (\vee_n X) \vee (\vee_n X) \xrightarrow{\sigma} \vee_n X$$

But this is, by definition, the commutator $(c_{n-1}, *)$ in the group $[X, \vee_n X]$ and therefore it is trivial.

□

As a consequence we obtain the following result which is the Eckmann-Hilton dual of the classical Whitehead bound for the category [20], [1]. Recall that for a group G, the nilpotency index of G, nil G, is the least integer n for which all the commutators of order $n+1$ of G vanish.

Theorem 4.12. *Given a cogroup X, nil$[X, Y] \leq$* wcocat *Y.*

Proof. Assume wcocat $Y \leq n$ and let $(f_1, \ldots, f_{n+1})$ be an $(n+1)$-commutator in $[X, Y]$. Consider the diagram

$$\begin{array}{ccccccc}
 & & F_n(X) & \xrightarrow{\beta_n} & F_n(Y) & & \\
 & \overset{\varphi}{\nearrow} & \downarrow j & & \downarrow j & & \\
X & \xrightarrow{c_n} & \vee_{n+1} X & \xrightarrow[f_1 \vee \ldots \vee f_{n+1}]{} & \vee_{n+1} Y & \xrightarrow{\sigma} & Y \\
 & & \downarrow i_n & & \downarrow i_n & & \\
 & & W_n X & \xrightarrow[\alpha_n]{} & W_n Y & &
\end{array}$$

in which α_n is produced by $f_1 \vee \ldots \vee f_{n+1}$ (see proposition 3.3), j is the inclusion of the fibre F_n of i_n and β_n is induced by α_n and $f_1 \vee \ldots \vee f_{n+1}$. In view of theorem above $i_n\, c_n = *$ and therefore c_n factors through $F_n(X)$ (the dotted φ of the diagram). Hence $(f_1, \cdots, f_{n+1}) = \sigma(f_1 \vee \ldots \vee f_{n+1})c_n = \sigma\, j\, \beta\, \varphi = 0$ since $j\, \sigma = *$ □

Remark 4.13. *Recall that given a space X,* nil ΩX *is the biggest integer n for which the universal commutator map of order n $\Omega X \times \ldots \Omega X \to \Omega X$ is essential. A result of* [1] *asserts that* wlength $X \leq$ nil ΩX *while on the other hand it is easy to see that* nil $\Omega X = \sup_Z \operatorname{nil}[Z, \Omega X]$. *But this is the same as* $\sup_Z \operatorname{nil}[\Sigma Z, X]$ *which, in view of theorem 4.12, is bounded above by* wcocat X. *Hence we have the following:*

$$\text{wlength } X \leq \text{nil } \Omega X \leq \text{wcocat } X \leq \text{cocat } X$$

Also, observe that corollary 4.7 could be deduced from this remark.

5. Some examples

Along this section we shall construct spaces for which the inequality

$$\text{wcocat } X \leq \text{cocat } X$$

is strict. All the examples are constructed in a similar way. Let BS^1 be the classifying space of S^1. It is well known that $H^*(BS^1; \mathbb{Z}) = \mathbb{Z}[\iota]$, where ι has degree two. Let γ_n denote the map $BS^1 \xrightarrow{\gamma_n} K(\mathbb{Z}, 2n)$ which detects the cohomological class ι^n, that is, if ι_{2n} is the fundamental class generating $H^{2n}(K(\mathbb{Z}, 2n), \mathbb{Z})$, the map γ_n is characterized by $\gamma_n(\iota_{2n}) = \iota^n$.

Call $X(n)$ the homotopy fiber of the map γ_n, that is, we have a fibration:

$$X(n) \xrightarrow{h_n} BS^1 \xrightarrow{\gamma_n} K(\mathbb{Z}, 2n). \qquad (2)$$

The spaces $X(n)$ are the source of all our examples. The following lemmas show that cocat $X(n) = 2$ for $n > 1$ (note that $X(1)$ is a point).

Lemma 5.1. cocat $X(n) \leq 2$

Proof. As BS^1 is an H-space then indcocat $BS^1 = 1$, and by (2), we get that indcocat $X(n) \leq$ indcocat $BS^1 + 1 = 2$. Using that cocat $X(n) \leq$ indcocat $X(n)$ we get the desired result. □

Lemma 5.2. *The spaces $X(n)$ are not H-spaces for $n > 1$.*

Proof. To prove the result, we show that $H^*(X;\mathbb{Z})$ is not a Hopf-algebra. Assume that $X(n)$ is an H-space, and therefore $H^*(X;\mathbb{Z})$ is a Hopf-algebra. Let $\alpha = h_n(\iota)$ be the generator of $H^2(X;\mathbb{Z})$. As ι is primitive in $H^*(BS^1;\mathbb{Z})$, α is so in $H^*(X;\mathbb{Z})$ and therefore it generates a Hopf-subalgebra of $H^*(X;\mathbb{Z})$. The Hopf-algebra generated by α in $H^*(X;\mathbb{Z})$ is a monogenic Hopf-algebra over a ring of characteristic 0, and its generator is even dimensional, thus it has to be $\mathbb{Z}[\alpha]$. But

$$\alpha^n = h_n(\iota^n) = h_n(\gamma_n(\iota_{2n})) = 0$$

which is impossible. □

An easy consequence is

Corollary 5.3. cocat $X(n) = 2$ *if* $n > 1$.

Now we study $\Omega X(n)$ in order to calculate wcocat $X(n)$.

Lemma 5.4. *If $n > 1$, then $\Omega X(n) \simeq S^1 \times K(\mathbb{Z}, 2(n-1))$.*

Proof. Looping the fibration (2) we get a new fibration

$$\Omega X(n) \xrightarrow{\Omega i_n} S^1 \xrightarrow{\Omega\gamma_n} K(Z, 2n-1).$$

As $n > 1$, $K(Z, 2n-1)$ is simply connected and thus $\Omega\gamma_n$ is trivial. Hence $\Omega X(n) \simeq S^1 \times K(\mathbb{Z}, 2(n-1))$. □

Then we can calculate wcocat $X(n)$ when $n \geq 3$.

Proposition 5.5. wcocat $X(n) = 1$ *for* $n \geq 3$.

Proof. Let F_1 be the homotopy fiber of $i_1 \colon X(n) \vee X(n) \longrightarrow X(n) \times X(n)$. To prove the result we show that $[F_1, X(n)] = 1$ for $n \geq 3$. But this follows from this sequence of bijections:

$$\begin{aligned}
[F_1, X(n)] &= [\Sigma(\Omega X(n) \wedge \Omega X(n)), X(n)] \\
&\cong [\Omega X(n) \wedge \Omega X(n), \Omega X(n)] \\
&\cong [\Omega X(n) \wedge \Omega X(n), K(\mathbb{Z}, 2(n-1))] \oplus [\Omega X(2) \wedge \Omega X(2), S^1] \\
&\cong H^{2n-1}(\Omega X(n) \wedge \Omega X(n); \mathbb{Z}) \oplus H^1(\Omega X(n) \wedge \Omega X(n); \mathbb{Z})
\end{aligned}$$

Finally, as $n \geq 3$ those cohomology groups are trivial. □

Corollary 5.6. $1 = \text{nil}\ \Omega X(n) = \text{wcocat}\ X(n) < \text{cocat}\ X(n) = 2$ *for* $n \geq 3$.

Finally, the case of $X(2)$ is exceptional.

Proposition 5.7. wcocat $X(2) = 2$.

Proof. To show that wcocat $X(2) = 2$, it is enough to prove that the map $F_1 \longrightarrow X(2) \vee X(2) \xrightarrow{\sigma} X(2)$ is not trivial. In order to prove that, first we prove that any map from F_1 to $X(2)$ factors uniquely through the map $K(\mathbb{Z},3) \xrightarrow{\delta} X(2)$ that appears extending the fibration (2). This follows from the sequence of bijections:

$$\begin{aligned}[F_1, X(2)] &\cong [\Omega X(2) \wedge \Omega X(2), \Omega X(2)] \\ &\cong [\Omega X(2) \wedge \Omega X(2), K(\mathbb{Z},2)] \oplus [\Omega X(2) \wedge \Omega X(2), S^1] \\ &\cong H^2(\Omega X(2) \wedge \Omega X(2); \mathbb{Z}) \oplus H^1(\Omega X(2) \wedge \Omega X(2); \mathbb{Z}) \\ &\cong H^2(\Omega X(2) \wedge \Omega X(2); \mathbb{Z})\end{aligned}$$

and

$$\begin{aligned}[F_1, K(\mathbb{Z},3)] &\cong [\Omega X(2) \wedge \Omega X(2), \Omega K(\mathbb{Z},3)] \\ &\cong [\Omega X(2) \wedge \Omega X(2), K(\mathbb{Z},2)] \\ &\cong H^2(\Omega X(2) \wedge \Omega X(2); \mathbb{Z})\end{aligned}$$

which are compatible with the map δ. Thus we have the following commutative diagram

$$\begin{array}{ccccc} \Omega X(2)^2 & \longrightarrow & F_1 & \longrightarrow & X(2) \vee X(2) \\ \downarrow & & \downarrow \theta & & \downarrow \sigma \\ S^1 & \longrightarrow & K(\mathbb{Z},3) & \xrightarrow{\delta} & X(2) \end{array} \tag{3}$$

where the composition $F_1 \longrightarrow X(2) \vee X(2) \xrightarrow{\sigma} X(2)$ is trivial if and only if the map θ is so. To see that θ is not trivial we show that $\theta^*(\iota_3) \neq 0$ where ι_3 is the fundamental class that generates $H^3(K(\mathbb{Z},3);\mathbb{Z})$. To do that we compare the Serre spectral sequences associated to the fibrations in (3).

The low dimensional cohomology of the spaces in (3) is

$$\begin{aligned} H^{\leq 3}(X(2);\mathbb{Z}) &= \mathbb{Z}[\alpha_2], & H^{\leq 3}(K(\mathbb{Z},3);\mathbb{Z}) &= \Lambda_{\mathbb{Z}}[\iota_3], \\ H^{\leq 3}(S^1;\mathbb{Z}) &= \Lambda_{\mathbb{Z}}[\iota_1], & H^{\leq 3}(X(2) \vee X(2);\mathbb{Z}) &= \mathbb{Z}[\beta_2, \beta_2'], \\ H^{\leq 3}(F_1;\mathbb{Z}) &= \Lambda_{\mathbb{Z}}[\epsilon_3], & H^{\leq 3}(\Omega X(2)^2;\mathbb{Z}) &= \Lambda_{\mathbb{Z}}[x_1, x_1'] \end{aligned}$$

where the differentials (in the Serre s.s.) are $d_2(\iota_1) = \alpha_2$, and $d_2(x_1) = \beta_2$ and $d_2(x_1') = \beta_2'$. Therefore, in the corresponding spectral sequences, the classes ι_3 and ϵ_3 are represented by the classes $\iota_1 \otimes \alpha_2$ and $(x_1 + x_1') \otimes (\alpha_2 + \alpha_2')$ respectively. By the edge morphisms we get that $\theta^*(\iota_3) = \epsilon_3$.

Hence the map $F_1 \longrightarrow X(2) \vee X(2) \xrightarrow{\sigma} X(2)$ is not trivial and the result is proved. □

6. Cocategory and localization

In what follows all spaces will be 1-connected and localization shall mean $\mathbb{Z}_P$-localization in which P is a (non necessary non empty) set of prime numbers and $\mathbb{Z}_P$ denotes the P-localized integers. Here we study the behavior of cocat respect to localization and show that in the same fashion that the category of a space decreases when localized, cocategory does. We start with an easy lemma which shows that generalized inverse limits behaves as usual limits with respect to localization:

Lemma 6.1. *Let $A \longrightarrow B \longleftarrow C$ be a diagram of abelian groups, then*

$$(\text{inv-lim}^s(A \longrightarrow B \longleftarrow C)) \otimes \mathbb{Z}_P \cong \text{inv-lim}^s(A \otimes \mathbb{Z}_P \longrightarrow B \otimes \mathbb{Z}_P \longleftarrow C \otimes \mathbb{Z}_P)$$

Proof. Let

$$0 \longrightarrow A \longrightarrow E^0 \longrightarrow E^1 \longrightarrow \cdots$$

be an injective resolution of the abelian group A. Then

$$0 \longrightarrow A \otimes \mathbb{Z}_P \longrightarrow E^0 \otimes \mathbb{Z}_P \longrightarrow E^1 \otimes \mathbb{Z}_P \longrightarrow \cdots$$

is an injective resolution of the abelian group $A \otimes \mathbb{Z}_P$ since $\_ \otimes \mathbb{Z}_P$ is exact and again each $E^i \otimes \mathbb{Z}_P$ is injective. Finally the result follows from this together with the fact that for usual limits we have the isomorphism

$$(\text{inv-lim}(A \longrightarrow B \longleftarrow C)) \otimes \mathbb{Z}_P \cong \text{inv-lim}(A \otimes \mathbb{Z}_P \longrightarrow B \otimes \mathbb{Z}_P \longleftarrow C \otimes \mathbb{Z}_P).$$

□

The next result shows that P-localization commutes with the W_n construction.

Proposition 6.2. *$W_n(X_P)$ is P-local. Moreover, $W_n(X_P) = W_n(X)_P$ and the natural map $W_n(X) \longrightarrow W_n(X_P)$ is the P-localization map.*

Proof. By induction on n. For $n = 0$ we have that $W_0(X) = W_0(X_P) = W_0(X)_P = *$ and there is nothing to prove.

Assume the proposition holds up to $n-1$ and recall that $W_n(X)$ is obtained as the homotopy pullback of the diagram $\vee_n X \xrightarrow{i_{n-1}} W_{n-1}(X) \longleftarrow W_{n-1}(X) \vee X$. Therefore, by the induction hypothesis and lemma 6.1, P-localization induces the following commutative diagram in which the front and back faces are homotopy pullbacks, and f is the whisker map that closes the diagram in view of the

WUPHPB:

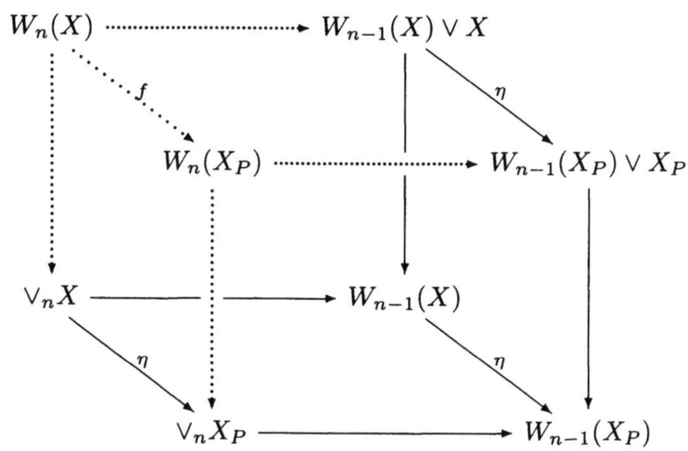

Now $W_n(X_P)$ is P-local since it is an inverse homotopy limit of local spaces. To finish the proof all we show that f induces iso in $\pi_* \otimes \mathbb{Z}_P$. In order to do so we compare the Bousfield-Kan spectral sequences (BKss) associated to both pullbacks (see [2]). The 2nd stage of the BKss associated to the pullback diagram $\underline{D} = \vee_n X \longrightarrow W_{n-1}(X) \longleftarrow W_{n-1}(X) \vee X$ is given by $E_2^{s,t}(\underline{D}) = \text{inv-lim}^s(\pi_t(\underline{D}))$ and it converges to $\pi_* \operatorname{holim} \underline{D}$. By lemma 6.1 and since $\_ \otimes \mathbb{Z}_P$ is an exact the spectral sequence with 2nd stage $E_2^{s,t} \otimes \mathbb{Z}_P = \text{inv-lim}^s(\pi_t(\underline{D}) \otimes \mathbb{Z}_P)$ converges to $\pi_* \operatorname{holim} \underline{D} \otimes \mathbb{Z}_P$.

Now, the P-localization map gives an isomorphism

$$\eta_\sharp : \pi_*(\operatorname{holim} \underline{D}) \otimes \mathbb{Z}_P \longrightarrow \pi_*(\operatorname{holim} \underline{D}_P)$$

which also induces an isomorphism between the 2nd stage of the spectral sequences

$$E_2^{*,*} \otimes \mathbb{Z}_P(\underline{D}) \xrightarrow{\eta_\sharp} E_2^{*,*}(\underline{D}_P).$$

Hence the E_∞ terms are isomorphic, that is,

$$\pi_*(W_n(X)) \otimes \mathbb{Z}_P \xrightarrow{f_\sharp} \pi_*(W_n(X_P)).$$

To finish, since $W_n(X_P)$ is local, the map f factors through $W_n(X)_P$ as the following diagram shows:

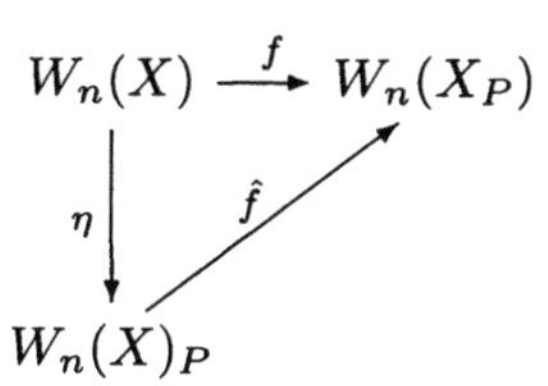

An easy analysis shows that $\hat{f}$ induces isomorphism in homotopy, which finishes the proof. □

Finally, the main theorem of this section is

Theorem 6.3. cocat $X_P \leq$ cocat X.

Proof. Assume $n =$ cocat X and let $\varphi : W_n(X) \longrightarrow X$ so that $\varphi\, i_n = \sigma$. Applying P-localization to this we obtain the following diagram:

$$\begin{array}{ccc} (\vee_{n+1}X)_P & \xrightarrow{i_n} & W_n(X)_P \\ {\scriptstyle \sigma_P}\downarrow & \nearrow{\scriptstyle \varphi_P} & \\ X_P & & \end{array}$$

Now recall that $(\vee_{n+1}X)_P = \vee_{n+1}(X_P)$ is an equivalence and therefore the map $\psi = \varphi_P\, \hat{f}^{-1}$, with $\hat{f}$ as in proposition above, makes the diagram.

$$\begin{array}{ccc} \vee_{n+1}X_P & \xrightarrow{i_n} & W_n(X_P) \\ {\scriptstyle \sigma}\downarrow & \nearrow{\scriptstyle \psi} & \\ X_P & & \end{array}$$

commutative □

References

[1] I. Berstein and T. Ganea, *Homotopical nilpotency*, Illinois J. Math. **5** (1961), 99–130.

[2] A.K. Bousfield and D.M. Kan, *Homotopy limits, completions and localizations*, Lectures Notes in Mathematics **304**, 1972

[3] O. Cornea, *Strong LS-category equals cone-length*, Topology **34** (1995), 377–381.

[4] J.P. Doeraene, *Homotopy pullbacks, homotopy pushouts and joins*, Bull. Soc. Mat. Bel. **5** (1998), 15–37.

[5] Y. Félix and S. Halperin, *Rational LS-category and its applications*, Trans. Amer. Math. Soc. **273** (1982), 1–37.

[6] R. Fox, *On the Lusternik-Schnirelmann category*, Ann. of Math. **42** (1941), 333–370.

[7] T. Ganea, *Lusternik-Schnirelmann category and cocategory*, Proc. London Math. Soc. **10** (1960) 623–639.

[8] T. Ganea, *Fibrations and cocategory*, Comment. Math. Helvetici **35** (1961), 15–24.

[9] T. Ganea, *A generalization of the homology and the homotopy suspension*, Comment. Math. Helvetici **39** (1965), 295–322.

[10] W. Gilbert, *Some examples for weak category and conilpotency*, Illinois J. of Math. **12** (1968), 421–432.

[11] M.J. Hopkins, *Formulations of cocategory and the iterated suspension*, In Astérisque (Homotopie algébrique et algébre locale) **113–114** (1984), 238–246

[12] M. Hovey, *Lusternik-Schnirelmann cocategory*, Illinois J. of Math. **37** (1993), 224–239.

[13] I. James, *On category in the sense of Lusternik-Schnirelmann*, Topology **17** (1978), 331–348 .

[14] L. Lusternik and L. Schnirelmann, *Méthodes topologiques dans les problèmes variationnels*, Herman, Paris 1986.

[15] M. Mather, *Pullbacks in homotopy theory*, Canadian J. of Math. **28** (1976), 225–263.

[16] D. Tanré, *Homotopie Rationnelle: Modéles de Chern, Quillen, Sullivan*, Lectures Notes in Mathematics **1025** 1983.

[17] G.H. Toomer, *Topological localization, category and cocategory*, Canadian J. of Math. **27** (1975), 319–322.

[18] A. Viruel, *La categoría de Lusternik-Schnirelmann*, Tesina de Licenciatura 1992.

[19] J.H.C. Whitehead, *The homology suspension*, Colloque de topologie algébrique á (1956), 89–95 .

[20] J.H.C. Whitehead, *On mapping into group-like spaces*, Comment. Math. Helvetici **28** (1964), 320–328.

[21] S. Abhyankar, *Local rings of high embedding dimension*, Amer. J. Math., **89** (1967), 1073–1077.

Departamento de Algebra, Geometría y Topología,
Universidad de Málaga,
Ap. 59,
29080-Málaga, Spain
E-mail address: `aniceto@agt.cie.uma.es, viruel@agt.cie.uma.es`

Progress in Mathematics, Vol. 196

Rational self-equivalences for H-spaces

Salvina Piccarreta

Abstract. Let X be an H-space and a 1-connected CW-complex. We describe the rational homotopy type of the H-space $\mathrm{aut}(X)$ of the self-equivalences of X and its subspaces $\mathrm{aut}_\sharp(X)$ and $\mathrm{aut}_1(X)$; bounds for their H-nilpotency are given. As a particular case G/U is considered for G and U belonging to certain classes of classical Lie groups.

1. Introduction

Given a space X, we denote by $\mathrm{aut}(X)$ the subspace of $\mathrm{map}(X,X)$ consisting of maps which are homotopy equivalences. With respect to composition this space is an H-group, strictly associative and with a strict unity. We are interested in the following subspaces of $\mathrm{aut}(X)$: $\mathrm{aut}_1(X)$ is the path-connected component of $\mathrm{aut}(X)$ containing the identity of X, and $\mathrm{aut}_\sharp(X)$ is the subspace of maps inducing the identity on homotopy groups. Both subspaces are H-groups.

Clearly, $\pi_0(\mathrm{aut}(X))$ is a group with respect to the operation induced by composition, which is denoted by $\mathcal{E}(X)$ and called *group of homotopy self-equivalences* of X. We consider also the subgroup $\mathcal{E}_\sharp(X) = \pi_0(\mathrm{aut}_\sharp(X))$.

In his work, Sullivan associates to the rationalization $X_\mathbb{Q}$ of X a differential graded algebra $\mathcal{M}_X$; moreover, if X is a finite 1-connected CW-complex, he gives a graded Lie algebra isomorphism:

$$H_*(\mathrm{Der}_1(\mathcal{M}_X)) \cong \pi_*(\mathrm{aut}_1(X)) \otimes \mathbb{Q}$$

where the product structure on $\pi_*(\mathrm{aut}_1(X))$ is given by the Samelson product and $H_*(\mathrm{Der}_1(\mathcal{M}_X))$ is an algebraic object depending only on $\mathcal{M}_X$. Analogous properties hold for the rational homotopy type of $\mathrm{aut}_\sharp(X)$ (see [4]).

2. The derivations Lie algebra

We begin with some standard conventions. If V_i is a vector space over the rationals $\mathbb{Q}$ for each positive integer i, then we call the collection $V = \{V_i\}$ a graded vector space. If $v_1, \dots, v_k, \dots$ is a basis of V we write $V = \langle v_1, \dots, v_k, \dots \rangle$. By $x \in V$ we mean $x \in V_p$ for some p, and write $|x| = p$ for the degree of x.

The author was partially supported by CNR (Consiglio Nazionale delle Ricerche).

Let $\mathcal{M} = (\Lambda(V), d)$ be a free, differential graded commutative algebra over $\mathbb{Q}$. A derivation of degree i on $\mathcal{M}$, $\theta\colon \mathcal{M} \to \mathcal{M}$, is a map of degree i, satisfying the Leibnitz rule:

$$\theta(x \cdot y) = \theta(x) \cdot y + (-1)^{i|x|} x \cdot \theta(y).$$

Let X be a finite 1-connected CW-complex and let $\mathcal{M} = (\Lambda(V), d)$ be its Sullivan minimal model. There is a natural isomorphism of DG Lie algebras [4]:

$$H_*(\mathrm{Der}_\sharp(\mathcal{M}), \partial) \cong \pi_*(\mathrm{aut}_\sharp X) \otimes \mathbb{Q}, \tag{1}$$

where the graded algebra structure on the right hand term is given by the Samelson product and $\mathrm{Der}_\sharp(\mathcal{M})$ is the DG Lie algebra defined as follows:

- $\mathrm{Der}_\sharp(\mathcal{M}) = \sum_{i\geq 0}(\mathrm{Der}_\sharp(\mathcal{M})_i)$, where $\mathrm{Der}_\sharp(\mathcal{M})_i$ is the rational vector space of all derivations of degree $-i$ if $i > 0$ and $\mathrm{Der}_\sharp(\mathcal{M})_0 = \mathrm{DER}_\sharp(\mathcal{M}) = \{\theta \mid \theta$ derivation of degree 0 commuting with d and such that $\theta(v) \in \mathcal{M}^{i>0}/(\mathcal{M}^{i>0} \cdot \mathcal{M}^{i>0})$ for all $v \in V\}$;
- product $\langle\ ,\ \rangle\colon \mathrm{Der}_\sharp(\mathcal{M})_k \times \mathrm{Der}_\sharp(\mathcal{M})_j \longrightarrow \mathrm{Der}_\sharp(\mathcal{M})_{k+j}$ given by:
  1. $\langle\phi, \psi\rangle = [\phi, \psi]$ as in $\mathrm{Der}_1(\mathcal{M})$, when $|\phi| = k > 0$, $|\psi| = j > 0$;
  2. $\langle\phi, \psi\rangle = \Phi(\Phi(\phi, \psi), \Phi(-\phi, -\psi))$, when $|\phi| = 0$, $|\psi| = 0$, where
  $$\Phi(\phi, \psi) = \phi + \psi + \frac{1}{2}[\phi, \psi] + \frac{1}{12}[\phi, [\phi, \psi]] + \cdots$$
  is the Baker–Campbell–Hausdorff formula;
  3. $\langle\phi, \psi\rangle = \sum_{n\geq 1} \frac{(-1)^n}{n!}[\phi, [\phi, \ldots (n \text{ times}) \ldots, [\phi, \psi] \ldots]$,
  when $|\phi| = 0$ and $|\psi| > 0$;
- differential of degree -1,
$$\partial = [-, d]\colon \mathrm{Der}_\sharp(\mathcal{M})_i \longrightarrow \mathrm{Der}_\sharp(\mathcal{M})_{i-1}.$$

This isomorphism is the extension of the natural isomorphism of differential graded Lie algebras [5]:

$$H_*(\mathrm{Der}_1(\mathcal{M}), \partial) \cong \pi_*(\mathrm{aut}_1(X)) \otimes \mathbb{Q}, \tag{2}$$

where the graded Lie algebra structure on the right hand term is given by the Samelson product and $\mathrm{Der}_1(\mathcal{M}) = \sum_{i>0}(\mathrm{Der}_\sharp(\mathcal{M})_i)$, by the natural isomorphism of groups given by the exponential map [5]:

$$e\colon \mathrm{DER}_\sharp(\mathcal{M}) \to \mathcal{E}_\sharp(\mathcal{M}). \tag{3}$$

If X is a 1-connected CW-complex, the H-nilpotency of $\mathrm{aut}_\sharp(X_\mathbb{Q})$ (respectively of $\mathrm{aut}_1(X_\mathbb{Q})$) equals the maximum length of the Samelson products in $\pi_*(\mathrm{aut}_\sharp(X_\mathbb{Q}))$ (respectively in $\pi_*(\mathrm{aut}_1(X_\mathbb{Q}))$) and then [4]:

$$H\text{-nil}(\mathrm{aut}_1(X_\mathbb{Q})) = \mathrm{nil}\, H_*(\mathrm{Der}_1(\mathcal{M})), \tag{4}$$

$$H\text{-nil}(\mathrm{aut}_\sharp(X_\mathbb{Q})) = \mathrm{nil}\, H_*(\mathrm{Der}_\sharp(\mathcal{M})), \tag{5}$$

$$\mathrm{nil}(\mathcal{E}_\sharp(X_\mathbb{Q})) = \mathrm{nil}\, H_0(\mathrm{Der}_\sharp(\mathcal{M})). \tag{6}$$

3. Rational H-spaces

If X is a rational H-space, its Sullivan model $\mathcal{M} = \Lambda(v_1, \ldots, v_k)$ has differential zero, so that $H_*(\mathrm{Der}_\sharp(\mathcal{M})) = \mathrm{Der}_\sharp(\mathcal{M})$ and $H^*(\mathrm{Der}_1(\mathcal{M})) = \mathrm{Der}_1(\mathcal{M})$. Moreover, in this case $\mathrm{Der}_\sharp(\mathcal{M})$ (respectively $\mathrm{Der}_1(\mathcal{M})$) is generated by the elementary derivations θ^j_χ (resp. by the elementary derivations with positive degree), $j = 1, \ldots, k$, where χ is a monomial in $\Lambda(V)$, $|\chi| \leq |v_j|$, χ decomposable if $|\chi| = |v_j|$, defined by:

$$\begin{cases} \theta^j_\chi(v_j) = \chi \\ \theta^j_\chi = 0 \qquad \text{on the other generators.} \end{cases}$$

Thus,

Proposition 3.1. *If X has the rational homotopy type of an H-space, then*

$$\dim \pi_i(\mathrm{aut}_1(X_{\mathbb{Q}})) = \sum_{j-k=i} \dim(\pi_j(X) \otimes H^k(X; \mathbb{Q}))$$

$$\text{and} \quad \rho(\mathcal{E}_\sharp(X_{\mathbb{Q}})) = \sum_k \left(\dim(\pi_k(X) \otimes H^k(X; \mathbb{Q})) - \dim(\pi_k(X) \otimes \mathbb{Q})\right). \qquad (*)$$

(Note that the summands in $(*)$ are nonnegative; in fact,

$$\dim H^k(X; \mathbb{Q}) \geq \dim(\pi_k(X) \otimes \mathbb{Q}),$$

as it can be easily deduced from the Sullivan model of X.)

$\rho(\mathcal{E}_\sharp(X_{\mathbb{Q}}))$ denotes the number of generators of $\mathrm{Der}_\sharp(\mathcal{M})_0$; if X has the homotopy type of a finite CW-complex (and $\mathcal{M}$ has generators in odd degrees) it is equal to the Hirsch rank of $\mathcal{E}_\sharp(X)$, and consequently (cf. Theorem 1 in [1]):

$$\rho(\mathcal{E}_\sharp(X)) = \sum_{l=1}^{n} \mathrm{rank}(\pi_{2l-1}(X) \otimes \mathbb{Q}) \cdot (\beta_{2l-1} - \mathrm{rank}(\pi_{2l-1}(X) \otimes \mathbb{Q})).$$

We calculate explicitly the H-nilpotency of some classes of H-spaces using the equalities (4) and (5).

Remark 3.2. Suppose that S is a set of elementary derivations such that the composition of elements in S is an element of S. If S is a generating set for $H_*(\mathrm{Der}_\sharp(\mathcal{M}))$, then the elements in $H_*(\mathrm{Der}_\sharp(\mathcal{M}))$ are in S or they are sums of elements of S; thus, the linearity of the bracket implies that $\mathrm{nil}(H_*(\mathrm{Der}_\sharp(\mathcal{M}))) \leq n-1$, if the brackets of n elements of S are trivial.

The following proposition is an extension of a result of Arkowitz and Lupton [2], which gives a rule to calculate $\mathrm{nil}(\mathcal{E}_\sharp(\mathcal{M}))$, when $\mathcal{M}$ satisfies some special conditions:

Proposition 3.3. *Let $\mathcal{M} = (\Lambda(v_1, \ldots, v_k), 0)$ with $|v_i| = n_i$. Suppose the degrees of the generators form an arithmetic progression, $n_j = a + (j-1)u$, for an odd integer $a \geq 1$ and $u \geq 2$ even, with a and u relatively prime. Then*

$$\mathrm{nil}(\mathrm{Der}_\sharp(\mathcal{M})) = \mathrm{nil}(\mathrm{Der}_1(\mathcal{M})) = k - 1.$$

Proof. We prove first that $\mathrm{nil}(\mathrm{Der}_1(\mathcal{M})) \leq \mathrm{nil}(\mathrm{Der}_\sharp(\mathcal{M})) \leq k-1$. By Remark 3.2 it suffices to show that if $\theta_\chi^N \neq 0$, $\chi = v_{j_1} \cdot \ldots \cdot v_{j_l}$, is a bracket of s elementary derivations, then $s \leq k-1$. Let h_0 be the number of these derivations with degree zero, h_1 be the number of derivations with positive degree not multiple of u, and $s - h_0 - h_1$ be the number of derivations with positive degree multiple of u. Note that given an integer h, if $|v_m| = |v_{j_1} \cdot \ldots \cdot v_{j_l}| + h$, then $a(l-1) \equiv -h \bmod u$.

It follows that $l \geq uh_0 + h_1 + 1$ and

$$|\chi| \leq |v_N| - u(s - h_0 - h_1) - h_1 \leq a + (k-1)u - us + uh_0 + uh_1 - h_1,$$

and consequently

$$a + (k-1)u - us + uh_0 + uh_1 - h_1 \geq |\chi| \geq |v_1 \cdot \ldots \cdot v_l| \geq \sum_{i=0}^{l-1}(a + iu) = al + \frac{l(l-1)}{2}u,$$

$$2(k-1)u - 2us + 2uh_0 + 2uh_1 - 2h_1 \geq 2uah_0 + 2ah_1 + \frac{u}{2}(u^2h_0^2 + h_1^2 + 2uh_0h_1 + uh_0 + h_1).$$

Clearly s is maximal when $l = 1$, i.e., when $h_i = 0$ for $i = 0, 1$, and in this case $s \leq k-1$. On the other hand, $\mathrm{nil}(\mathrm{Der}_1(\mathcal{M})) \geq \mathrm{nil}(\mathrm{Der}_\sharp(\mathcal{M})) \geq k-1$, because

$$\langle \theta_{v_{k-1}}^k \cdots \theta_{v_1}^2 \rangle = [\theta_{v_{k-1}}^k \cdots \theta_{v_1}^2] = \theta_{v_1}^k$$

is a nontrivial bracket/product of $k-1$ derivations. □

For the Sullivan model of a generic finite dimensional H-space, bounds can be determined by:

Proposition 3.4. *Let* $\mathcal{M} = (\Lambda(v_1, \ldots, v_k), 0)$ *be a minimal algebra satisfying* $|v_i| = 2n_i + 1$, $n_1 \leq \cdots \leq n_k$. *Then*

$$r - 1 \leq \mathrm{nil}(\mathrm{Der}_\sharp(\mathcal{M})) = \mathrm{nil}(\mathrm{Der}_1(\mathcal{M})) \leq n_k - n_1,$$

where r is the number of different n_i's.

Proof. We prove first that $\mathrm{nil}(\mathrm{Der}_1(\mathcal{M})) \leq \mathrm{nil}(\mathrm{Der}_\sharp(\mathcal{M})) \leq n_k - n_1$. It suffices to show that if $\theta_\chi^N \neq 0$, $\chi = v_{j_1} \cdot \ldots \cdot v_{j_l}$, is the bracket of s elementary derivations, then $s \leq n_k - n_1$. Let $s = \sum_{i=0}^2 h_i$, where h_2 is the number of these derivations with even, positive degree, h_1 is the number of derivations with odd degree, and h_0 is the number of derivations with degree 0. Then, noting that $|v_m| - |v_{j_1} \cdot \ldots \cdot v_{j_t}| + 1 \equiv t \bmod 2$, we have

$$l \geq \sum_{i=0}^{2}(2-i)h_i + 1 \geq 2s - \sum_{i=0}^{2} ih_i + 1 \quad \text{and}$$

$$|\chi| \leq |v_N| - \sum_{i=0}^{2} ih_i \leq 2n_k + 1 - \sum_{i=0}^{2} ih_i \leq 2n_k + l - 2s; \quad \text{therefore}$$

$$2n_k + l - 2s \geq |\chi| \geq |v_1 \cdot \ldots \cdot v_l| \geq l|v_1| \geq 2n_1 l + l \quad \Rightarrow \quad 2s \leq 2n_k - 2ln_1.$$

Clearly, s is maximal when $l = 1$, i.e., when $h_i = 0$ for $i = 0, 1$. In this case, $s \leq n_k - n_1$. Moreover, there are r indecomposables, $v_{i_1}, \ldots, v_{i_r}$ such that $n_{i_k} < n_{i_{k+1}}$, so we have the nontrivial bracket/product of $r-1$ derivations:

$$\langle \theta^{i_r}_{v_{i_{r-1}}} \cdots \theta^{i_2}_{v_{i_1}} \rangle = [\theta^{i_r}_{v_{i_{r-1}}} \cdots \theta^{i_2}_{v_{i_1}}] = \theta^{i_r}_{v_{i_1}}.$$

Thus, $\mathrm{nil}(\mathrm{Der}_\sharp(\mathcal{M})) \geq \mathrm{nil}(\mathrm{Der}_1(\mathcal{M})) \geq r - 1$. □

Remark 3.5. Both inequalities can be strict. For $u \geq 4$, Proposition 3.3 gives a minimal algebra such that $r - 1 = \mathrm{nil}(\mathrm{Der}_\sharp(\mathcal{M})) < n_k - n_1$.

Consider now the minimal model

$$(\Lambda(x_1, x_2, x_3, x_4, x_5, x_6, x_7, x_8, x_9, y_{10}, y_{11}, y_{12}, z_{13}), 0)$$

with $|z_i| = 27$, $|y_i| = 9$, $|x_i| = 3$. Since

$$[\theta^{13}_{y_{10}y_{11}y_{12}}, \theta^{12}_{x_7x_8x_9}, \theta^{11}_{x_4x_5x_6}, \theta^{10}_{x_1x_2x_3}] = \pm\theta^{13}_{x_1x_2x_3x_4x_5x_6x_7x_8x_9}$$

is nontrivial, $\mathrm{nil}(\mathrm{Der}_\sharp(\mathcal{M})) \geq 4 > 2 = r - 1$.

Corollary 3.6. *If $X_\mathbb{Q}$ is a finite dimensional H-space, then*

$$r - 1 \leq H\text{-nil}(\mathrm{aut}_\sharp(X_\mathbb{Q})) = H\text{-nil}(\mathrm{aut}_1(X_\mathbb{Q})) \leq \frac{m-n}{2}$$

where $\pi_i(X) \otimes \mathbb{Q} = 0$, except for $n \leq i \leq m$, and r is the number of nontrivial groups $\pi_i(X) \otimes \mathbb{Q}$.

Proposition 3.7. *Let $u \geq 2$ be a fixed even integer and $\mathcal{M} = (\Lambda(v_1, \ldots, v_k), 0)$, with $|v_i| = iu$, $i = 1, \ldots, k$. Then $\mathrm{nil}(\mathrm{Der}_\sharp(\mathcal{M})) = k - 1$.*

Proof. We prove first that $\mathrm{nil}(\mathrm{Der}_1(\mathcal{M})) \leq \mathrm{nil}(\mathrm{Der}_\sharp(\mathcal{M})) \leq k-1$. It is sufficient to show that if $\theta^N_\chi \neq 0$, $\chi = v_{j_1} \cdot \ldots \cdot v_{j_l}$, is the bracket of s elementary derivations, then $s \leq n_k - n_1$. Let h be the number of these derivations with degree 0. Then $l \geq h + 1$ and $|\chi| \leq |v_k| - u(s-h) \leq uk - us + uh$. It follows that

$$uk - us + uh \geq |\chi| \geq |v_1 \cdot \ldots \cdot v_l| \geq ul \geq uh + u \qquad \Rightarrow \qquad s \leq k - 1.$$

On the other hand, $\mathrm{nil}(\mathrm{Der}_\sharp(\mathcal{M})) \geq \mathrm{nil}(\mathrm{Der}_1(\mathcal{M})) \geq k - 1$, as we have the following nontrivial bracket/product of $k-1$ derivations: $\langle \theta^k_{v_{k-1}}, \ldots, \theta^2_{v_1} \rangle = [\theta^k_{v_{k-1}}, \ldots, \theta^2_{v_1}] = \theta^k_{v_1}$. □

Examples

1. The Sullivan model of the group $SU(n)$ is $\mathcal{M} = \Lambda(v_1, \ldots, v_{n-1})$ with $|v_i| = 2i + 1$ and $d = 0$ $(n \geq 2)$. By Proposition 3.3,

$$H\text{-nil}(\mathrm{aut}_\sharp(SU(n)_\mathbb{Q})) = H\text{-nil}(\mathrm{aut}_1(SU(n)_\mathbb{Q})) = n - 2.$$

2. The Sullivan model of the symplectic group $Sp(n)$ is $\mathcal{M} = \Lambda(v_1, \ldots, v_n)$ with $|v_i| = 4i - 1$ and $d = 0$ $(n \geq 1)$. Again, by Proposition 3.3,

$$H\text{-nil}(\mathrm{aut}_\sharp(Sp(n)_\mathbb{Q})) = H\text{-nil}(\mathrm{aut}_1(Sp(n)_\mathbb{Q})) = n - 1.$$

3. The Sullivan model of the classifying spaces $BU(k)$ is $\mathcal{M} = \Lambda(v_1, \dots, v_k)$ with $|v_i| = 2i$ and $d = 0$ $(k \geq 1)$. By Proposition 3.7,

$$H\text{-nil}(\mathrm{aut}_\sharp(BU(k)_\mathbb{Q})) = H\text{-nil}(\mathrm{aut}_1(BU(k)_\mathbb{Q})) = k - 1.$$

4. The Sullivan model of the classifying spaces $BSp(k)$ is $\mathcal{M} = \Lambda(v_1, \dots, v_k)$ with $|v_i| = 4i$ and $d = 0$ $(k \geq 1)$. Again, by Proposition 3.7,

$$H\text{-nil}(\mathrm{aut}_\sharp(BSp(k)_\mathbb{Q})) = H\text{-nil}(\mathrm{aut}_1(BSp(k)_\mathbb{Q})) = k - 1.$$

The following result is analogous to Proposition 4.4 in [2].

Proposition 3.8. *Let $\mathcal{M} = \Lambda(v_1, \dots, v_k, \dots)$ with $|v_1| \leq \dots \leq |v_k| \leq \cdots$ and let $(i_1, \dots, i_s)$ be a sequence of positive integers with $i = \max(i_1, \dots, i_s)$. Assume that for every $j = 1, \dots, s$ there is an elementary derivation $\theta^{i_j}_{\chi_j} \in \mathrm{Der}_\sharp(\mathcal{M})$ such that $|v_{i_j}|$ is odd and χ_j is a monomial in $v_1, \dots, v_k, \dots$. Define a monomial $\alpha \in \mathcal{M}$ as follows: list all the v_r which occur as factors in $\chi_1, \dots, \chi_s$ (allowing repetitions). If some of $v_{i_1}, \dots, \hat{v}_i, \dots v_{i_s}$ do not appear in the list, set $\alpha = 0$. Otherwise, delete one occurrence of each of $v_{i_1}, \dots, \hat{v}_i, \dots v_{i_s}$ from the list and set α equal to the product of the remaining members. Then the s-fold bracket $[\theta^{i_s}_{\chi_s}, \dots, \theta^{i_2}_{\chi_2}, \theta^{i_1}_{\chi_1}]$ is either 0 or $\theta^i_{\pm\alpha}$.*

In [3] rational models (not always minimal) for a large class of homogeneous spaces are given; models for the same spaces can be found in other contexts and with different notations.

Proposition 3.9.

$$H\text{-nil}\,(\mathrm{aut}_\sharp\,((U(n)/U(k))_\mathbb{Q})) = H\text{-nil}\,(\mathrm{aut}_1\,((U(n)/U(k))_\mathbb{Q})) = n - k - 1,$$

$$H\text{-nil}\,(\mathrm{aut}_\sharp\,((U(n)/SO(n))_\mathbb{Q})) = H\text{-nil}\,(\mathrm{aut}_1\,((U(n)/SO(n))_\mathbb{Q})) = m - 1,$$

$$H\text{-nil}\,(\mathrm{aut}_\sharp\,((U(n)/Q(m))_\mathbb{Q})) = H\text{-nil}\,(\mathrm{aut}_1\,((U(n)/Q(m))_\mathbb{Q})) = m - 2,$$

$$\begin{aligned} &H\text{-nil}\,(\mathrm{aut}_\sharp\,((SO(2m+1)/SO(2k+1))_\mathbb{Q})) = \\ &\qquad H\text{-nil}\,(\mathrm{aut}_1\,((SO(2m+1)/SO(2k+1))_\mathbb{Q})) = m - k - 1, \end{aligned}$$

$$H\text{-nil}\,(\mathrm{aut}_\sharp\,((Q(n)/Q(k))_\mathbb{Q})) = H\text{-nil}\,(\mathrm{aut}_1\,((Q(n)/Q(k))_\mathbb{Q})) = n - k - 1,$$

$$\begin{aligned} &H\text{-nil}\,(\mathrm{aut}_\sharp\,((SO(2m)/SO(2k+1))_\mathbb{Q})) = \\ &\qquad H\text{-nil}\,(\mathrm{aut}_1\,((SO(2m)/SO(2k+1))_\mathbb{Q})) = \\ &\qquad \begin{cases} m - k - 1 & \text{if } m < 2k+2 \text{ or } m \text{ odd} \\ m - k - 2 & \text{if } m \geq 2k+2 \text{ and } m \text{ even.} \end{cases} \end{aligned}$$

Proof. The proof is an immediate consequence of Proposition 3.3 in all cases except $SO(2m)/SO(2k+1)$, which can anyway be obtained by straightforward calculations. □

Proposition 3.10. $\mathrm{nil}(\mathcal{E}(X_{\mathbb{Q}})) = r$, *where r is the biggest integer such that*

$$l \geq 1 + ra + \frac{ru(ru+1)}{2}$$

where

$$\begin{aligned} &a = 2k+1,\ u = 2,\ l = n-k \text{ if } X = U(n)/U(k),\\ &a = 1,\ u = 2,\ l = m \text{ if } X = U(n)/SO(n),\\ &a = 1,\ u = 2,\ l = m-1 \text{ if } X = U(n)/Q(m),\\ &a = 4k+3,\ u = 4,\ l = m-k \text{ if } X = SO(2m+1)/SO(2k+1),\\ &a = 4k+3,\ u = 4,\ l = n-k \text{ if } X = Q(n)/Q(k), \end{aligned}$$

and

$$r_1 \leq nil(\mathcal{E}_\sharp(SO(2m+1)/SO(2k+1))) \leq r_2,$$

where r_1 is the maximum positive integer such that

$$3r_1 + 2r_1(4r_1+1) \leq m-k-2$$

and r_2 is the maximum positive integer such that

$$32(r_2)^2 + 10 - 44r_2 + 16kr_2 \leq 4m - 5.$$

References

[1] M. Arkowitz, C. R. Curjel, *Groups of Homotopy classes*, Lecture Notes in Math. **4**, Springer, Heidelberg, Berlin, New York, 1967.

[2] M. Arkowitz, G. Lupton, *On the Nilpotency of Subgroups of Self-Homotopy Equivalences*, Progress in Math. **136**, Birkhäuser, Basel, 1996, 1–22.

[3] W. H. Greub, S. Halperin, J. R. Vanstone, *Connections, Curvature and Cohomology*, vol. III, Academic Press, New York, 1975.

[4] P. Salvatore, *Rational Homotopy Nilpotency of Self-Equivalences*, Topology Appl. **77** (1997), 37–50.

[5] D. Sullivan, *Infinitesimal Computation in Topology*, Inst. Hautes Études Sci. Publ. Math. **47** (1977), 269–331.

[6] T. Yamanoshita, *On the spaces of self-homotopy equivalences of certain CW-complexes*, J. Math. Soc. Japan **37** (1985), 455–470.

Dipartimento di Matematica,
Università di Milano,
via Saldini 50,
20133 Milano, Italy
E-mail address: `piccarreta@socrates.mat.unimi.it`

Progress in Mathematics, Vol. 196

Cellular approximations using Moore spaces

José L. Rodríguez and Jérôme Scherer

Abstract. For a two-dimensional Moore space M with fundamental group G, we identify the effect of the cellularization CW_M and the fibre $\overline{P}_M$ of the nullification on an Eilenberg–Mac Lane space $K(N, 1)$, where N is any group: both induce on the fundamental group a group theoretical analogue, which can also be described in terms of certain universal extensions. We characterize completely M-cellular and M-acyclic spaces, in the case when $M = M(\mathbf{Z}/p^k, 1)$.

1. Introduction

Let M be a pointed connected CW-complex. The nullification functor P_M and the cellularization functor CW_M have been carefully studied in the last few years (see e.g. [8], [17], [18], [14]). These are generalizations of Postnikov sections and connective covers, where the role of spheres is replaced by a connected CW-complex M and its suspensions. This list of functors also includes plus-constructions and acyclic functors associated with a homology theory, for which M is a universal acyclic space (cf. [2], [13], [22], [24]). Recall that a connected space X is called *M-cellular* if $CW_M X \simeq X$, or, equivalently, if it belongs to the smallest class $\mathcal{C}(M)$ of spaces which contains M and is closed under homotopy equivalences and pointed homotopy colimits. Analogously, X is called *M-acyclic* if $P_M X \simeq *$ or, equivalently, $\overline{P}_M X \simeq X$. It was shown in [14] that the class of M-acyclic spaces is the smallest class $\overline{\mathcal{C}(M)}$ of spaces which contains M and is closed under homotopy equivalences, pointed homotopy colimits, and extensions by fibrations.

Very interesting examples are given by the family of Moore spaces $M(\mathbf{Z}/p, n)$, the homotopy cofibre of the degree p self-map of S^n. For $n \geq 2$, these spaces are the "building blocks" for simply-connected p-torsion spaces. More precisely, it is shown in [3] (see also [11]) that the $M(\mathbf{Z}/p, n)$-cellular spaces are exactly the $(n-1)$-connected spaces X such that $p \cdot \pi_n X = 0$ and $\pi_k X$ is p-torsion for $k > n$. However the methods used in those papers can not handle the non-simply connected case. In this paper we introduce the group theoretical tools that are necessary to deal with this case. They apply to the more general situation when M is a two-dimensional CW-complex with fundamental group G. As we will see in Proposition 3.10 and in the introduction of Section 3, the interesting phenomena

The first author was partially supported by DGICYT grants PB94–0725 and PB97-0202. The second by the grant 81LA-51213 of the Swiss National Foundation for Science and the CRM.

occur when $H_2(M;\mathbf{Z}) = 0$. In that case we say that M is a Moore space of type $M(G,1)$, and we shortly write $M = M(G,1)$.

The *G-socle* of a group N, which we denote by S_GN, is the subgroup of N generated by the images of all homomorphisms from G into N. We introduce the class $\mathcal{C}(G)$ for any group G. It is the smallest class of groups containing G which is closed under isomorphisms and colimits. We construct explicitly the right adjoint C_G to the inclusion of $\mathcal{C}(G)$ in the category of groups and show the following (see also Theorem 3.3).

Theorem 3.9 *Let M be a two dimensional CW-complex and G its fundamental group. Let $X = K(N,1)$ where N is any group. Then we have a natural isomorphism*

$$\pi_1(CW_MX) \cong C_GN.$$

Moreover, the action of C_GN on the higher homotopy groups of CW_MX is trivial.

We further prove the existence of a central extension

$$0 \to A \to C_GN \to S_GN \to 1$$

which is universal in the sense explained in Theorem 3.7.

The proof of such results uses a description of Chachólski, exhibiting CW_MX as the fibre of a map $X \to LX$, where LX is obtained from X by first killing all maps from M, and then applying ΣM-nullification.

This leads us, in the case when $G = \mathbf{Z}/p$, to the following result. We must note that our proof is also valid for $M(\mathbf{Z}/p,n)$ with $n \geq 2$, cases which were previously dealt with in [3] or [11].

Theorem 7.2 *Let $M = M(\mathbf{Z}/p,1)$ be the cofibre of the degree p self-map of S^1 and X be a connected space. Then X is M-cellular if and only if π_1X is generated by elements of order p and $H_n(X;\mathbf{Z})$ is p-torsion for $n \geq 2$.*

In particular, a nilpotent space X is $M(\mathbf{Z}/p,1)$-cellular if and only if π_1X is generated by elements of order p and $\pi_n(X)$ is p-torsion for $n \geq 2$.

Of course, the homotopy groups of non-nilpotent M-cellular spaces need not be p-torsion. For instance, the universal cover of $M(\mathbf{Z}/2,1)$ is S^2. Likewise, a space all whose homotopy groups are generated by p-torsion elements need not be M-cellular, as shown by Example 7.4, where we compute the $M(\mathbf{Z}/2,1)$-cellularization of $K(\Sigma_3,1)$. Theorem 7.5 gives then a general formula for computing the $M(\mathbf{Z}/p,1)$-cellularization of classifying spaces of $\mathbf{Z}/p$-cellular groups in terms of their q-completions.

Nullifications with respect to Moore spaces are better understood. Our aim here is to investigate the homotopy fibre of such nullifications.

Recall that the *G-radical* of a group N, which we denote by T_GN as in [10] or [8], is the smallest subgroup of N such that $\mathrm{Hom}(G, N/T_GN) = 0$. It is known [10] that when M is a two dimensional CW-complex with fundamental group G, then $\pi_1 P_M X \cong \pi_1 X / T_G(\pi_1 X)$. If in addition M is an $M(G,1)$, the space $P_M X$ can be viewed, for a suitable ring R, as the fibrewise R-completion, in the sense of Bousfield and Kan [9], of a covering fibration associated to the G-radical subgroup; see [9], [10], and [12].

We introduce the class $\overline{\mathcal{C}(G)}$, for any group G. It is the smallest class of groups containing G which is closed under isomorphisms, colimits, and extensions. We show in Proposition 4.11 that the fundamental group of any M-acyclic space belongs to this class. We define then $D_G N$ as the fundamental group of $\overline{P}_M K(N,1)$, note that the action of $D_G N$ on the higher homotopy groups of $\overline{P}_M K(N,1)$ is trivial, and prove:

Theorem 4.13 *Let $M = M(G,1)$ be a two-dimensional Moore space. Then D_G is right adjoint to the inclusion of $\overline{\mathcal{C}(G)}$ in the category of groups.*

As in the case of cellularization, there exists a central extension

$$0 \to B \to D_G N \to T_G N \to 1 \tag{1.1}$$

which is universal in the sense explained in Theorem 4.3.

A very enlightening example is given by the acyclic space described by Berrick and Casacuberta in [2, Example 5.3], which turns out to be an $M(G,1)$ for some acyclic group G. In this case $M(G,1)$-nullification is equivalent to Quillen's plus-construction and the G-radical of any group N is its largest perfect subgroup. Thus, in this case, the central extension (1.1) is the usual universal central extension of $T_G N$.

Similar results have been obtained by Mislin and Peschke in [22] in the case when P_M is the plus construction associated to a generalized homology theory. In all these cases CW_M and $\overline{P}_M$ coincide.

Finally, we obtain the following result for $G = \mathbf{Z}/p$ (compare with [3] or [11]).

Theorem 7.1 *Let $M = M(\mathbf{Z}/p,1)$. Then X is M-acyclic if and only if $\pi_1 X$ coincides with its $\mathbf{Z}/p$-radical and $H_n(X;\mathbf{Z})$ is p-torsion for $n \geq 2$.*

In particular, a nilpotent space X is $M(\mathbf{Z}/p,1)$-acyclic if and only if $\pi_1 X$ coincides with its $\mathbf{Z}/p$-radical and $\pi_n(X)$ is p-torsion for $n \geq 2$.

Acknowledgements: This paper was mainly elaborated during the 1998 Topology Semester at the CRM. We are especially indebted to Jon Berrick and Wojciech Chachólski for several enlightening discussions. We warmly appreciate Carles Casacuberta's advice and encouragement. The second author would like to thank the CRM for making the life of a post-doctoral fellow so pleasant.

2. Preliminary results

We give here a short review of the terminology involved in the theory of homotopical localization. We also remind the reader some of the results needed in this paper. More details can be found in [17], [18], [14], [7].

By an idempotent augmented functor in the category of spaces, we mean a functor E from the category of pointed spaces to itself. It preserves weak homotopy equivalences and is equipped with a natural transformation $c : E \to \mathrm{Id}$ from E to the identity functor, such that $c_{EX} : EEX \simeq EX$ for all spaces X.

The most important examples for us are CW_M and $\overline{P}_M$. Let M be a connected CW-complex. A map $Y \to X$ is an M*-cellular equivalence* if it induces a weak equivalence on pointed mapping spaces

$$\mathrm{map}_*(M, Y) \xrightarrow{\sim} \mathrm{map}_*(M, X).$$

There exists then, for each connected space X, an M*-cellular approximation* map $CW_M X \to X$, which is universal (initial) among all M-cellular equivalences to X. The spaces for which $CW_M X \simeq X$ are called M*-cellular.* The class of M-cellular spaces has been identified as the smallest class $\mathcal{C}(M)$ of spaces containing M and closed under weak equivalences and pointed homotopy colimits; see [18, 2.D] and [14, Theorem 8.2]. A connected space Z is said to be M*-null* if $\mathrm{map}_*(M, Z)$ is weakly contractible, that is, $[\Sigma^k M, X] = *$ for all $k \geq 0$. There exists a map $X \to P_M X$, called M*-nullification*, which is universal (terminal) among all maps from X to an M-null space. Finally denote by $\overline{P}_M X \to X$ the homotopy fibre of $X \to P_M X$. A connected space X is called M*-acyclic* if $P_M X \simeq *$, i.e., $\overline{P}_M X \simeq X$. The class of M-acyclic spaces has been identified in [14, Theorem 17.3] as the class $\overline{\mathcal{C}(M)}$. In addition to being closed under weak equivalences and pointed homotopy colimits, it is also closed under extensions by fibrations.

So, every M-cellular space is M-acyclic and furthermore, it is known that each ΣM-acyclic space is M-cellular ([18, 3.B.3]). Hence by universality we have natural maps

$$\overline{P}_{\Sigma M} X \xrightarrow{\alpha} CW_M X \xrightarrow{\beta} \overline{P}_M X. \tag{2.2}$$

Thus $CW_M X$ can be thought of as the fibre of a mixing process between M- and ΣM-nullification. More precisely:

Theorem 2.1. ([14, Theorem 20.5]) *Let M be any connected CW-complex. There is a fibration*

$$CW_M X \longrightarrow X \xrightarrow{\eta} LX$$

where η is the composition of the inclusion $X \to X'$ of X into the homotopy cofibre of the evaluation map $\vee_{[M,X]} M \to X$, followed by $X' \to P_{\Sigma M}(X')$.

Note that the inclusion $X \to X'$ is in fact functorial in the homotopy category, and it is universal (initial) among all maps $X \to Z$ such that $M \to X \to Z$ is homotopically trivial (compare with [2, Corollary 2.2]). Hence, $X \to LX$ is also functorial in the homotopy category.

The fundamental groups of $P_M X$ and LX have a group theoretical meaning in the case when M is a two-dimensional CW-complex, as we next explain. The G*-socle* of a group N, which we denote by $S_G N$, is the subgroup of N generated by the images of all homomorphisms from G into N. The G*-radical* of N, which we denote by $T_G N$ as in [10] or [8], is the smallest subgroup of N such that $\mathrm{Hom}(G, N/T_G N) = 0$. The G-radical of N can be constructed as a (possibly transfinite) direct limit of subgroups T_i where T_1 is the G-socle of N and the quotient $T_i/T_{i-1} = S_G(N/T_{i-1})$. In other words, the groups N such that $T_G N = N$ are precisely the groups which have a normal series whose factors coincide with their G-socles. The first link between the topological functors P_M and L and their discrete analogues is given by the following result. The first isomorphism is obtained in [10, Theorem 3.5] (see also [6, Theorem 5.2]).

Lemma 2.2. *If M is a two-dimensional CW-complex with fundamental group G and X is any space, we have isomorphisms*

$$\pi_1(P_M X) \cong \pi_1 X/T_G(\pi_1 X) \quad \text{and} \quad \pi_1 X' \cong \pi_1 LX \cong \pi_1 X/S_G(\pi_1 X),$$

where X' and LX are as in Theorem 2.1. □

Such a two-dimensional CW-complex M is called a Moore space if the integral homology group $H_2(M;\mathbf{Z}) = 0$. It has type $M(G,1)$ if $\pi_1 M \cong G$. Using Theorem 2.1 it is easy to see that two Moore spaces of type $M(G,1)$ determine the same cellularization and nullification functors.

Bousfield computed in [8, Section 7] the effect of the nullification functor with respect to a two-dimensional Moore space $M(G,1)$ on nilpotent spaces; see also [10, Theorem 4.4] for $G = \mathbf{Z}/p$, and [12, Theorem 2.4] when $G = \mathbf{Z}[1/p]$. Let J be the set of primes p for which G_{ab} is uniquely p-divisible. Define $R = \mathbf{Z}_{(J)}$, the integers localized at J, if G_{ab} is torsion, and $R = \oplus_{p\in J}\mathbf{Z}/p$ otherwise. Let $R_\infty X$ be the Bousfield-Kan R-completion of X; see [9]. Then $P_M X$ can be obtained as the fibrewise R-completion of the covering fibration associated to the G-radical of $\pi_1 X$. That is, we have a diagram of fibrations

$$\begin{array}{ccccc} \tilde{X} & \longrightarrow & X & \longrightarrow & K(\pi_1 X/T_G(\pi_1 X),1) \\ \downarrow & & \downarrow & & \| \\ R_\infty \tilde{X} & \longrightarrow & P_M X & \longrightarrow & K(\pi_1 X/T_G(\pi_1 X),1), \end{array} \tag{2.3}$$

where $R_\infty \tilde{X}$ is simply connected, as the fundamental group of $\tilde{X}$ is R-perfect, i.e., $H_1(\tilde{X};R) = H_1(T_G(\pi_1 X);R) = 0$. Hence $R_\infty \tilde{X}$ coincides with $\tilde{X}^+_{HR}$, the plus-construction with respect to ordinary homology with coefficients in R; see [9, VII.6] and [12]. The universal cover of $P_M X$ is thus equivalent to the three following spaces:

$$P_M \tilde{X} \simeq R_\infty \tilde{X} \simeq \tilde{X}^+_{HR}.$$

Let $A_R X$ denote the homotopy fibre of the plus construction $X \to X^+_{HR}$ (cf. [9, VII, 6.7]). Then, a connected space X is HR-acyclic, i.e., $\tilde{H}_*(X;R) = 0$, if and only if $A_R X \simeq X$. The above remark immediately implies the following.

Proposition 2.3. *Let $M = M(G,1)$ be a two-dimensional Moore space, and X be any connected space. If $\tilde{X}$ denotes the covering of X corresponding to the subgroup $T_G(\pi_1 X)$, then*

$$\overline{P}_M X \simeq A_R \tilde{X} \simeq \overline{P}_M \tilde{X},$$

where R is the ring associated to G as above. □

Corollary 2.4. *Let $M = M(G,1)$ be a two-dimensional Moore space, and X be any connected space. Then, X is M-acyclic if and only if*

$$T_G(\pi_1 X) = \pi_1 X \quad \text{and} \quad H_k(X;R) = 0 \quad \text{for} \quad k \geq 2\,. \qquad \square$$

3. The fundamental group of $M(G,1)$-cellular spaces

In this section we define algebraically a G-cellularization functor C_G in the category of groups. We show that $C_G N$ coincides with the fundamental group of the $M(G,1)$-cellularization of $K(N,1)$. This yields a characterization of $C_G N$ as a certain universal central extension of the G-socle of N. We also prove that the action of $C_G N$ is trivial on the higher homotopy groups of $CW_{M(G,1)}K(N,1)$.

As suggested by Dror-Farjoun, we introduce the closed class of groups $\mathcal{C}(G)$. It is the smallest class of groups containing G, and closed under isomorphisms and taking colimits. In other words, if $F : I \to$ *Groups* is a diagram with $F(i) \in \mathcal{C}(G)$ for any $i \in I$, then $\operatorname{colim}_I F$ should again belong to $\mathcal{C}(G)$.

The following proposition gives the explicit construction of a G-cellularization functor. The existence of such a functor is also ensured by [5, Corollary 7.5].

Proposition 3.1. *Let G be a group. The inclusion $\mathcal{C}(G) \subset$ Groups has a right adjoint $C_G :$ Groups $\to \mathcal{C}(G)$.*

Proof. For any group N, the map $C_G N \to N$ is constructed by induction as follows. First define $C_0 = *_{h:G\to N} G$, the free product of as many copies of G as there are homomorphisms from G to N, and let $h_0 : C_0 \to N$ be the evaluation morphism (so that $h_0(C_0) = S_G N$). Now take the free product $*_{(h',h'')}G$, where (h',h'') is any pair of morphisms $G \rightrightarrows C_0$ coequalized by h_0. Define C_1 as the coequalizer of $*_{(h',h'')}G \rightrightarrows C_0$, and repeat this process (maybe transfinitely). Notice that this inductive construction of $C_G N$ shows that we have a natural epimorphism $C_G N \twoheadrightarrow S_G N$ for any group N. The group $C_G N$ is in $\mathcal{C}(G)$ and the morphism $c : C_G N \to N$ is universal (terminal) with this property as it induces a bijection of sets $\operatorname{Hom}(G,c) : \operatorname{Hom}(G, C_G N) \cong \operatorname{Hom}(G,N)$. □

By analogy to the case of spaces, a group N in $\mathcal{C}(G)$ is called *G-cellular.*

Lemma 3.2. *Let M be a two-dimensional CW-complex with fundamental group G. A group homomorphism $N \to N'$ induces an isomorphism $C_G N \cong C_G N'$ if and only if $CW_M K(N,1) \simeq CW_M K(N',1)$.*

Proof. The pointed mapping space $\operatorname{map}_*(M, K(N,1))$ is weakly equivalent to the discrete set $\operatorname{Hom}(G,N)$. □

Since π_1 commutes with homotopy colimits, the fundamental group of any M-cellular space is $\pi_1 M$-cellular, for any M. Furthermore, the following holds (this could also have been taken as definition of C_G):

Theorem 3.3. *Let M be a two-dimensional CW-complex and G its fundamental group. Let $X = K(N,1)$ where N is any group. Then we have a natural isomorphism*

$$\pi_1(CW_M X) \cong C_G N.$$

Proof. By the previous observation, $\pi_1(CW_M X)$ is G-cellular. It thus only remains to prove that $c : CW_M X \to X$ induces a bijection of sets $\pi_1(c)_* = \mathrm{Hom}(G, \pi_1(c))$. Consider the following commutative diagram (of sets)

$$\begin{array}{ccc} [M, CW_M X] & \xrightarrow{e} & \mathrm{Hom}(G, \pi_1(CW_M X)) \\ \downarrow c_* & & \downarrow \pi_1(c)_* \\ [M, X] & \xrightarrow{e'} & \mathrm{Hom}(G, N). \end{array}$$

Since M is two-dimensional, e is surjective and X being a $K(N,1)$, e' is bijective. On the other hand, c_* is also bijective by the universal property of CW_M. Thus, $\pi_1(c)_*$ is bijective, as desired. □

Lemma 3.4. *Let M be a two-dimensional CW-complex with fundamental group G and let $\tilde{X}$ denote the covering of X corresponding to the subgroup $S_G(\pi_1 X)$. Then*

(i) *$(\tilde{X})'$ is the universal cover of X',*

(ii) *$L\tilde{X}$ is the universal cover of LX,*

where X' and LX are defined in Theorem 2.1.

Proof. The covering fibration $\tilde{X} \to X \xrightarrow{p} K(\pi_1 X / S_G(\pi_1 X), 1)$ induces a bijection $[M, \tilde{X}] \cong [M, X]$ because the map $[M, p]$ is trivial. Apply now Mather-Puppe theorem (see [16, Proposition 6.1]) saying that "the fibre of the push-out is the push-out of the fibres" when the base space is fixed. This produces a new fibration $(\tilde{X})' \to X' \to K(\pi_1 X / S_G(\pi_1 X), 1)$, which is the universal cover fibration by Lemma 2.2. So part (i) holds. For part (ii) we note that the previous fibration has a ΣM-null base, and is therefore preserved under ΣM-nullification. □

The following corollary could also have been proved directly by checking that the covering $\tilde{X} \to X$ is indeed an M-equivalence.

Corollary 3.5. *Let $\tilde{X}$ denote the covering of X corresponding to the subgroup $S_G(\pi_1 X)$ and let M be a two-dimensional CW-complex with fundamental group G. We have a homotopy equivalence $CW_M X \simeq CW_M \tilde{X}$.* □

Lemma 3.6. *Let G be any group, and $0 \to A \to E \to N \to 1$ be a central extension of groups. Then $C_G E \cong C_G N$ if and only if $\mathrm{Hom}(G_{\mathrm{ab}}, A) = 0$ and the natural map $\mathrm{Hom}(G, N) \longrightarrow H^2(G; A)$ is trivial.*

Proof. Apply $\mathrm{map}_*(K(G,1), -)$ to the fibration $K(E,1) \to K(N,1) \to K(A,2)$. This gives a new fibration, whose homotopy sequence

$$0 \to \mathrm{Hom}(G, A) \to \mathrm{Hom}(G, E) \to \mathrm{Hom}(G, N) \to H^2(G; A)$$

is exact as in [9, IX, 4.1]). The lemma is proved. □

We now know enough to describe our first universal central extension. It is nothing but a universal central G-cellular equivalence.

Theorem 3.7. *Let G be any group. Then, for each group N, there is a central extension*

$$0 \to A \to C_G N \to S_G N \to 1$$

such that $\mathrm{Hom}(G_{\mathrm{ab}}, A) = 0$ *and the natural map* $\mathrm{Hom}(G, N) \longrightarrow H^2(G; A)$ *is trivial. Moreover, this extension is universal with respect to these two properties.*

Proof. Let M be a two-dimensional CW-complex with fundamental group G and let $X = K(S_G N, 1)$. The space LX is 1-connected by Lemma 2.2 and define then

$$A = \pi_2 LX \cong \pi_2 X'/T_{G_{\mathrm{ab}}}(\pi_2 X')$$

(this isomorphism follows from [8, Theorem 7.5]). The long exact sequence in homotopy of the fibration $CW_M X \to X \to LX$ produces now the desired central extension, where we identify $C_G(S_G N)$ with $C_G N$. This can be deduced from Corollary 3.5. The G-cellularization of $C_G N \to S_G N$ is also an isomorphism since $\mathrm{Hom}(G, C_G N) \cong \mathrm{Hom}(G, S_G N)$. Hence $\mathrm{Hom}(G_{\mathrm{ab}}, A) = 0$ and the natural map $\mathrm{Hom}(G, N) \longrightarrow H^2(G; A)$ is trivial by Lemma 3.6. The universal property is a direct consequence of the same lemma. □

Corollary 3.8. *Let M be a two-dimensional CW-complex and G its fundamental group. Then $S_G N = N$ if and only if $LK(N,1)$ is 1-connected, and $C_G N \cong N$ if and only if $LK(N,1)$ is 2-connected. In particular*

(1) $\pi_2 LK(N,1) \cong H_2 LK(S_G N, 1)$;
(2) $\pi_3 LK(N,1) \cong H_3 LK(C_G N, 1)$. □

In the next theorem we identify the universal cover of $CW_M K(N,1)$ and remark that the action of the fundamental group is trivial. This could also be seen as a particular case of [20, Proposition A.1] or the even more general [22, Corollary 7.7].

Theorem 3.9. *Let M be a two-dimensional CW-complex and G its fundamental group. Then $CW_M K(N,1) \simeq CW_M K(C_G N, 1)$ and the universal cover fibration is given by*

$$\Omega LK(C_G N, 1) \longrightarrow CW_M K(N,1) \longrightarrow K(C_G N, 1).$$

Moreover the action of $C_G N$ on $\pi_n CW_M K(N,1)$ for $n \geq 2$ is trivial.

Proof. The first part follows from Lemma 3.2. The fibration of Theorem 2.1

$$CW_M K(N,1) \to K(C_G N, 1) \to LK(C_G N, 1)$$

induces a long exact sequence of $C_G N$-modules in homotopy. But $LK(C_G N, 1)$ is a 2-connected space by Corollary 3.8, so that the action of $C_G N$ on the higher homotopy groups of $CW_M K(N,1)$ is trivial. □

We end this section by a result that motivates our study of Moore spaces. Indeed the effect of the cellularization functor for other two-dimensional complexes is completely understood.

Proposition 3.10. *Let M be a two-dimensional CW-complex with fundamental group G. Assume that $H_2(M;\mathbf{Z}) \neq 0$. Then*

$$CW_M K(N,1) \simeq K(C_G N,1).$$

Proof. Choose a presentation $\phi : *\mathbf{Z} \to *\mathbf{Z}$ of G, and realize it as a map f between wedges of circles having M as its homotopy cofibre. Note that a simply connected space Y is ΣM-null if and only if the G-radical of $\pi_2 Y$ is trivial, and $\pi_k Y$ is ϕ_{ab}-local for any $k \geq 3$, i.e., $\mathrm{Hom}(\phi_{\mathrm{ab}}, \pi_k Y)$ is bijective (see [23, Theorem 4.3.6]). When $H_2(M;\mathbf{Z}) \neq 0$, the homomorphism ϕ_{ab} is not injective and any ϕ_{ab}-local group is trivial. So $LK(C_G N,1)$ is the trivial space, as it is already 2-connected (see [23, Corollary 4.3.9]). □

Example 3.11. If $G = C_2$ and N is nilpotent then $C_G N = S_G N$ (by Corollary 7.3 below). However, $C_G N \ncong S_G N$ in general, as shown by the following example, which was suggested by Alejandro Adem:

$$N = \langle a,b,c,d \mid a^2 = b^2 = c^2 = d^2 = 1, abab = cdcd \rangle .$$

In other words N is the push-out of the diagram $C_2 * C_2 \leftarrow \mathbf{Z} \to C_2 * C_2$ where both arrows send the generator of $\mathbf{Z}$ to the commutator. The Mayer–Vietoris sequence shows then that $H_2 N \cong \mathbf{Z} \cong A$. Thus $C_G N$ is an extension of N by $\mathbf{Z}$. This also provides an example of a quotient of a free product of copies of G which is not cellular.

4. The fundamental group of $M(G,1)$-acyclic spaces

We imitate now the preceding section, replacing CW_M by $\overline{P}_M$. First we change our closed class. In addition to being closed under isomorphisms and colimits, the class $\overline{\mathcal{C}(G)}$ is assumed to be closed under taking arbitrary extensions. That is, if $N \hookrightarrow E \twoheadrightarrow Q$ is an extension with $N, Q \in \overline{\mathcal{C}(G)}$, then E belongs to $\overline{\mathcal{C}(G)}$ as well. The right adjoint of the inclusion of $\overline{\mathcal{C}(G)}$ in the category of groups is denoted by D_G and we construct it by topological means, namely as the fundamental group of $\overline{P}_{M(G,1)} K(N,1)$. It could be interesting to have an algebraic description of $D_G N$, similar to that of $C_G N$, in terms of colimits and extensions by short exact sequences.

A well known topological proof of the existence of the universal central extension of a perfect group N uses Quillen's plus-construction. We will follow exactly the same line of proof here, the plus-construction being replaced by a nullification functor with respect to a Moore space. This is a true generalization of this old result in light of [2], where it is proven that the plus-construction is indeed the nullification with respect to a Moore space. Another approach is taken in

[22], where the plus-construction associated to any homology theory determines a universal central extension.

When M is a two-dimensional CW-complex with fundamental group G which is not a Moore space, i.e., $H_2(M;\mathbf{Z}) \neq 0$, the effect of P_M is drastic. As in the proof of Proposition 3.10, one shows that, for any connected space X,

$$P_M X \simeq K(\pi_1 X / T_G(\pi_1 X), 1).$$

Hence $\overline{P}_M X$ is the covering of X corresponding to the subgroup $T_G(\pi_1 X)$. From now on, we will therefore only consider Moore spaces.

Define, for a two-dimensional Moore space $M = M(G,1)$,

$$D_G N := \pi_1(\overline{P}_M K(N,1)).$$

This does not depend on the choice of M by the observation made after Lemma 2.2.

Lemma 4.1. *Let M be a two-dimensional Moore space with fundamental group G and let B denote the group $\pi_2 P_M K(T_G N, 1)$. The space $K(B,2)$ is then M-null, or equivalently,* $\mathrm{Hom}(G_{\mathrm{ab}}, B) = 0 = \mathrm{Ext}(G_{\mathrm{ab}}, B)$.

Proof. The space $P_M K(T_G N, 1)$ is 1-connected by Lemma 2.2. The second Postnikov section $K(\pi_2 X, 2)$ of a simply connected M-null space X is M-null as well, since $\pi_2 P_M X$ only depends on $\pi_2 X$ by [8, Theorem 7.5]. □

The groups B satisfying $\mathrm{Hom}(G_{\mathrm{ab}}, B) = 0 = \mathrm{Ext}(G_{\mathrm{ab}}, B)$ can only be of the two following forms ([8, 7.1]):

Fact 4.2. Let J denote the set of primes p such that G_{ab} is uniquely p-divisible, and J' the complementary set of primes. Then, either G_{ab} is J'-torsion and B is J-local, or G_{ab} is uniquely J-divisible and B has to be Ext-J-complete (in the sense of [9]). In other words, $\mathrm{Hom}(G_{\mathrm{ab}}, B) = 0 = \mathrm{Ext}(G_{\mathrm{ab}}, B)$ if and only if $\mathrm{Hom}(H, B) = 0 = \mathrm{Ext}(H, B)$ where $H = \oplus_{p \in J'} \mathbf{Z}/p$ in the case when G_{ab} is torsion, or $H = \mathbf{Z}[J^{-1}]$ otherwise.

We are now ready to prove the existence of our second universal central extension.

Theorem 4.3. *Let G be the fundamental group of a two-dimensional Moore space, which we denote by M. Then, for each group N, there is a central extension*

$$0 \to B \to D_G N \to T_G N \to 1$$

such that $\mathrm{Hom}(G_{\mathrm{ab}}, B) = 0 = \mathrm{Ext}(G_{\mathrm{ab}}, B)$. *Moreover, this extension is universal (initial) with respect to this property.*

Proof. The idea of the proof is analogous to that of Theorem 3.7. Let X denote $K(T_G N, 1)$. The fibration $\overline{P}_M X \to X \to P_M X$ produces the desired extension using Proposition 2.3 to identify $D_G T_G N$ with $D_G N$. By Lemma 4.1 the group $B = \pi_2 P_M K(T_G N, 1)$ satisfies the property.

We check now the universal property. Let B' be an abelian group having the above property, and $0 \to B' \to E \to T_G N \to 1$ a central extension. Realize it as a fibration $K(E,1) \longrightarrow X \longrightarrow K(B',2)$. As the base space is M-null, there exists a map $P_M X \to K(B', 2)$, unique up to homotopy, inducing a map of fibrations. The map on the fibres gives the desired morphism $D_G N \to E$. □

Example 4.4. Let $G = C_2$ and N be the group described in Example 3.11. Then $D_G N$ is an extension of N by $\mathbf{Z}[1/2]$.

Remark 4.5. The group $T_G N$ is R-perfect and the central extension of Theorem 4.3 is the universal central extension of $T_G N$ with coefficients in R, that is, the one induced from the fibration $A_R K(T_G N, 1) \to K(T_G N, 1) \to K(T_G N, 1)^+_{HR}$. These two extensions coincide by Proposition 2.3. Even though this "R universal central extension" seems to be classical, we do not know any other reference than [22].

Example 4.6. Let $G = \mathbf{Z}[1/p]$, so that $R = \mathbf{Z}/p$. In [22, Proposition 5.4] Mislin and Peschke computed that

$$B \cong \operatorname{Ext}(\mathbf{Z}(p^\infty), H_2(T_G N; \mathbf{Z})) \oplus \operatorname{Hom}(\mathbf{Z}(p^\infty), H_1(T_G N; \mathbf{Z})),$$

where $\mathbf{Z}(p^\infty)$ is the p-torsion subgroup of $\mathbf{Q}/\mathbf{Z}$. Let $G = \mathbf{Z}/p$, so that $R = Z[1/p]$. Then $B = H_2(T_G N; R)$.

An interesting consequence of the previous result is that the functor D_G is idempotent. It is worth noting that it seems rather difficult to prove this fact directly from the definition.

Theorem 4.7. *Let M be a two-dimensional Moore space with fundamental group G. Then $\overline{P}_M K(D_G N, 1) \simeq \overline{P}_M K(N, 1)$ and in particular the functor D_G is idempotent. The universal cover fibration is given by*

$$\Omega P_M K(D_G N, 1) \longrightarrow \overline{P}_M K(N, 1) \longrightarrow K(D_G N, 1).$$

Moreover the action of $D_G N$ on $\pi_n \overline{P}_M K(N, 1)$ for $n \geq 2$ is trivial.

Proof. The functor P_M preserves the fibration

$$K(D_G N, 1) \to K(T_G N, 1) \to K(B, 2)$$

of Theorem 4.3 since $K(B, 2)$ is M-null by Lemma 4.1. Thus so does the functor $\overline{P}_M$. That is, we have $\overline{P}_M K(D_G N, 1) \simeq \overline{P}_M K(T_G N, 1)$. The later space is equivalent to $\overline{P}_M K(N, 1)$ by Proposition 2.3. The statements about the universal cover fibration follow as in Theorem 3.9. □

Remember that the ring R is determined by the group G as follows: $R = \mathbf{Z}_{(J)}$ if G_{ab} is torsion, and $R = \oplus_{p \in J} \mathbf{Z}/p$ otherwise. We say that a group N is *super R-perfect* if $H_1(N; R) = 0 = H_2(N; R)$.

Proposition 4.8. *Let G be the fundamental group of a two-dimensional Moore space M. The following statements are equivalent:*

(1) *$D_G N \cong N$.*
(2) *The space $P_M K(N, 1)$ is 2-connected.*
(3) *$H^2(N; B) = 0$ for any B such that* $\operatorname{Hom}(G_{ab}, B) = 0 = \operatorname{Ext}(G_{ab}, B)$.
(4) *$T_G N = N$ and N is super R-perfect.*

Proof. Theorem 4.3 implies that (1), (2), and (3) are equivalent. We only prove that (4) implies (2). Since N coincides with its G-radical, $P_M K(N,1) \simeq K(N,1)^+_{HR}$ (see diagram (2.3)), and this space is 1-connected. Thus $K(N,1)^+_{HR} \simeq K(N,1)_{HR}$ the HR-homological localization by [22, Proposition 1.6]. Moreover, by [4, Theorem 5.5], $\pi_2 K(N,1)_{HR}$ is an HR-local group. But $H_1(\pi_2 K(N,1)_{HR}; R) = 0$ since N is super R-perfect, so it has to be trivial. □

As a consequence, we obtain the following formulae for the low-dimensional homotopy groups of $P_M K(N,1)$; cf. [1, Corollary 8.6].

Corollary 4.9. *Let M be a two-dimensional Moore space and G its fundamental group. Then*

(1) $\pi_2 P_M K(N,1) \cong H_2 P_M K(T_G N,1)$;
(2) $\pi_3 P_M K(N,1) \cong H_3 P_M K(D_G N,1)$. □

We end this section by proving that this topological construction gives nothing else but the right adjoint of the inclusion of $\overline{\mathcal{C}(G)}$ into the category of groups. We denote the class $\{N \mid D_G N \cong N\}$ by $\mathcal{D}(G)$.

Proposition 4.10. *Let G be the fundamental group of a two-dimensional Moore space M. The class $\mathcal{D}(G)$ is closed under arbitrary extensions and colimits.*

Proof. The class of G-radical groups is closed under colimits and extensions, and so is the class of super R-perfect groups: An easy Hochschild–Serre spectral sequence argument shows that an extension of super R-perfect groups is again super R-perfect, and a Mayer–Vietoris argument shows it for a push-out. Since homology commutes with telescopes, the proposition is proved. □

Proposition 4.11. *Let G be the fundamental group of a two-dimensional Moore space M. Then $\overline{\mathcal{C}(G)} = \mathcal{D}(G)$.*

Proof. By Proposition 4.10, $\overline{\mathcal{C}(G)} \subset \mathcal{D}(G)$. To show the converse we prove that the fundamental group of any space in $\overline{\mathcal{C}(M)}$ is in $\overline{\mathcal{C}(G)}$. But $\overline{\mathcal{C}(M)}$ is the smallest class containing M which is closed under homotopy colimits and extensions by fibrations. Clearly the fundamental group of the homotopy colimit of a diagram all whose values have π_1 in $\overline{\mathcal{C}(G)}$ is again in $\overline{\mathcal{C}(G)}$. So assume we have a fibration $F \to E \to B$ of connected spaces, where the fundamental groups of F and B are in $\overline{\mathcal{C}(G)}$. We have to prove $\pi_1 E \in \overline{\mathcal{C}(G)}$. The cokernel of the boundary morphism $\pi_2 B \to \pi_1 F$ is isomorphic to the coinvariants $(\pi_1 F)_{\pi_2 B} = colim_{\pi_2 B}(\pi_1 F)$ and thus belongs to $\overline{\mathcal{C}(G)}$. Therefore $\pi_1 E$ is an extension of two groups of $\overline{\mathcal{C}(G)}$. □

Corollary 4.12. *Let G be the fundamental group of a two-dimensional Moore space. A group N belongs then to $\mathcal{D}(G)$ if and only if there exists an M-acyclic space X with $\pi_1 X \cong N$.* □

Theorem 4.13. *Let M be a two-dimensional Moore space and G its fundamental group. The augmented functor D_G is then right adjoint to the inclusion of $\overline{\mathcal{C}(G)}$ in* *the category of groups, i.e., we have an isomorphism* $\mathrm{Hom}(L, D_G N) \cong \mathrm{Hom}(L, N)$ *for any group* $L \in \overline{\mathcal{C}(G)}$.

Proof. The map $K(D_G N, 1) \to K(N, 1)$ induces a weak equivalence

$$\overline{P}_M K(D_G N, 1) \simeq \overline{P}_M K(N, 1)$$

by Theorem 4.7. Let $L \in \overline{\mathcal{C}(G)}$. Then

$$\mathrm{map}_*(\overline{P}_M K(L, 1), K(D_G N, 1)) \simeq \mathrm{map}_*(\overline{P}_M K(L, 1), K(N, 1)),$$

i.e., $\mathrm{Hom}(L, D_G N) \cong \mathrm{Hom}(L, N)$. □

5. Acyclic spaces

We illustrate the preceding sections by the case when the Moore space M is acyclic (with respect to ordinary integral homology). We identify the functors $\overline{P}_M$ and CW_M. The motivating example is the universal acyclic group $\mathcal{F}$ of Berrick and Casacuberta [2, Example 5.3]. It satisfies $S_{\mathcal{F}} N = T_{\mathcal{F}} N = \mathcal{P}N$, the maximal perfect subgroup of N. In this case $C_{\mathcal{F}} N = D_{\mathcal{F}} N = \widetilde{\mathcal{P}N}$ the universal central extension of $\mathcal{P}N$ and the two central extensions coincide. If $M = M(\mathcal{F}, 1)$, the functors $\overline{P}_M$ and CW_M coincide with Dror's acyclic functor A, the fibre of Quillen's plus-construction.

We want to consider now an arbitrary acyclic group G, and an acyclic complex M with fundamental group G. This space M is of course not determined by the group. Since $\Sigma M \simeq *$, the fibration of Theorem 2.1 has the form

$$CW_M X \to X \to X'$$

where X' is the homotopy cofibre of the map $\vee_{[M,X]} M \to X$. Let X_N^+ denote the plus-construction of X with respect to a perfect, normal subgroup N of $\pi_1 X$, and let $A_N X$ be the homotopy fibre of the natural map $X \to X_N^+$. The universal property of the plus-construction ensures that $X' \simeq X_S^+$, where $S = S(M, X)$ is the topological socle, i.e., the subgroup generated by the images of all homomorphisms $\pi_1(M) \to \pi_1(X)$ which are induced by maps $M \to X$ (see [2, Section 2]). This subgroup of $\pi_1 X$ is sometimes also called the subgroup swept by M. Arguing similarly with $P_M X$ we deduce the following; cf. [2, Corollary 2.2].

Theorem 5.1. *Let M be an acyclic CW-complex. The map $\beta : CW_M X \to \overline{P}_M X$ of (2.2) is then equivalent to $A_S X \to A_T X$ where S is the subgroup swept by M and T is such that $\pi_1 P_M X \cong \pi_1 X / T$.* □

When M is two-dimensional, $S = S_G N$ and $T = T_G N$ for any space X with fundamental group N. We also have $S = S_G N$ if M is any CW-complex and $X = K(N, 1)$. Therefore, we deduce the following.

Corollary 5.2. *Let X be a space with fundamental group N and M an acyclic CW-complex with fundamental group G. Suppose that M is of dimension two or that X is a $K(N,1)$. Then $\pi_1(CW_M X) \cong C_G N$ is the universal central extension of $S_G N$.* □

6. Nilpotent spaces

When X is a nilpotent space, the homotopy long exact sequence associated to the fibration $\overline{P}_M X \to X \to P_M X$ yields the homotopy groups of the M-acyclic part of X, as follows:

Proposition 6.1. *Let M be a Moore space $M(G,n)$ with cells in dimension n and $n+1$, where $n \geq 1$, and let X be any connected space. Suppose that X is nilpotent if $n = 1$. Let J be the set of primes p such that G_{ab} is uniquely p-divisible and J' be the complementary set of primes. Then $\overline{P}_M X$ is $(n-1)$-connected and for $k \geq n$,*

(I) *if G_{ab} is torsion, then*

$$\pi_k(\overline{P}_M X) \cong \begin{cases} \Pi_{p\in J'}(\mathbf{Z}(p^\infty) \otimes \pi_{k+1}X \ \oplus \ \mathrm{Tor}(\mathbf{Z}(p^\infty), \pi_k X)) & \text{if } k \geq n+1, \\ \Pi_{p\in J'}\mathbf{Z}(p^\infty) \otimes \pi_{n+1}X \ \oplus \ T_G(\pi_n X) & \text{if } k = n; \end{cases}$$

(II) *if G_{ab} is not torsion, then*

$$\pi_k(\overline{P}_M X) \cong \begin{cases} \Pi_{p\in J}(\mathrm{Ext}(\mathbf{Z}[1/p], \pi_{k+1}X) \ \oplus \ \mathrm{Hom}(\mathbf{Z}[1/p], \pi_k X)) & \text{if } k \geq n+1, \\ \Pi_{p\in J}\mathrm{Ext}(\mathbf{Z}[1/p], \pi_{n+1}X) \ \oplus \ D_G(\pi_n X) & \text{if } k = n. \end{cases}$$

Proof. We use from [8, Theorem 7.5] that in the first case we have

$$\pi_k(P_M X) \cong \begin{cases} \pi_k(X) \otimes \mathbf{Z}_{(J')} & \text{if } k \geq n+1, \\ \pi_n X / T_G(\pi_n X) & \text{if } k = n; \end{cases}$$

In the second case we have:

$$\pi_k(P_M X) \cong \begin{cases} \Pi_{p\in J}(\mathrm{Ext}(\mathbf{Z}(p^\infty), \pi_k X) \oplus \mathrm{Hom}(\mathbf{Z}(p^\infty), \pi_{k-1}X)) & \text{if } k \geq n+1, \\ \pi_n X / T_G(\pi_n X) & \text{if } k = n. \end{cases}$$

□

Example 6.2. Let G be a rational group of rank 1 of type $(r_2, r_3, r_5, \ldots)$. That is, G is the additive subgroup of $\mathbf{Q}$ generated by the fractions $1/p^s$, for $s \leq r_p$ (we write $r_p = \infty$ if G is uniquely p-divisible). Note that if $r_p < \infty$ then the G-radical contains the $\mathbf{Z}/p$-radical. Moreover, in the category of abelian groups, the G-socle coincides with the G-radical if and only if $G = \mathbf{Z}[J^{-1}]$ (see [19]).

This allows us to construct two subgroups of $\mathbf{Q}$ having the same set of primes for which they are uniquely p-divisible, but with distinct radical, and thus distinct acyclic approximation. Fix a prime p and define G by $r_p = \infty$ and $r_q = 1$ when $q \neq p$, so that $H = \mathbf{Z}[1/p]$. Let $M = M(G,1)$ and $M' = M(H,1)$. Then we have $\overline{P}_M K(\mathbf{Z}[1/p],1) \simeq *$, while $\overline{P}_{M'} K(\mathbf{Z}[1/p],1) \simeq K(\mathbf{Z}[1/p],1)$.

We next give a description of the class of nilpotent M-acyclic spaces; compare with Corollary 7.9 in [8], see also [21]. The case when $n = 1$ gives a less general result than Corollary 2.4, but gives a characterization of $M(G,1)$-acyclic spaces in terms of their homotopy groups rather than their homology groups.

Proposition 6.3. *Let M be a Moore space $M(G,n)$ with cells in dimension n and $n+1$, where $n \geq 1$, and let X be any connected space. Suppose that X is nilpotent if $n = 1$. Then, X is M-acyclic if and only if X is $(n-1)$-connected, $\pi_n X$ coincides with its G-radical, and $\pi_k(X)$ is J'-torsion for $k \geq n+1$ in the case when G is torsion, or $\pi_k(X)$ is uniquely J-divisible for $k \geq n+1$ otherwise.* □

7. The torsion case

In this section we only deal with the case of the Moore spaces $M(\mathbf{Z}/p^k,1)$, homotopy cofibre of the degree p^k self-map of S^1, for $k \geq 1$. We give a characterization of $M(\mathbf{Z}/p^k,1)$-cellular spaces, which holds even for non-nilpotent spaces. The $M(\mathbf{Z}/p^k,1)$-acyclic spaces have been already identified in Corollary 2.4. The following reformulation only makes use of the fact that an abelian group A is p-torsion if and only if $A \otimes \mathbf{Z}[1/p] = 0$.

Theorem 7.1. *Let $M = M(\mathbf{Z}/p^k,1)$, $k \geq 1$. Then a space X is M-acyclic if and only if $\pi_1 X$ coincides with its $\mathbf{Z}/p$-radical and $H_n(X;\mathbf{Z})$ is p-torsion for $n \geq 2$.* □

Theorem 7.2. *Let $M = M(\mathbf{Z}/p^k,1)$, $k \geq 1$. Then a space X is M-cellular if and only if $\pi_1 X$ is generated by elements of order p^l for $l \leq k$ and $H_n(X;\mathbf{Z})$ is p-torsion for $n \geq 2$.*

Proof. We use again the fact that $CW_M X$ can be obtained as the fibre of the map $X \to P_{M(\mathbf{Z}/p^k,2)}X'$, where X' is the cofibre of $\vee M \to X$. So we have to find a necessary and sufficient condition for $P_{M(\mathbf{Z}/p,2)}X'$ to be trivial. First X' has to be 1-connected, and this is equivalent to $\pi_1 X$ coinciding with its $\mathbf{Z}/p^k$-socle. Knowing that X' and thus $P_{M(\mathbf{Z}/p,2)}X'$ are 1-connected, the triviality of the latest is equivalent to its acyclicity. By Proposition 6.3 the homotopy, or equivalently the reduced integral homology of X', has to be p-torsion. The long exact sequence in homology of the cofibration sequence defining X' shows that this is equivalent to $\tilde{H}_*(X;\mathbf{Z})$ being p-torsion. □

Corollary 7.3. *Let $M = M(\mathbf{Z}/p^k,1)$. A nilpotent space X is M-cellular if and only if $\pi_1 X$ is generated by elements of order p^l for $l \leq k$ and $\pi_n(X)$ is p-torsion for $n \geq 2$.* □

The characterization given in [14, 12.5] or [15] of $M(\mathbf{Z}/2,n)$-cellular spaces (π_n is generated by involutions, and the higher homotopy groups are 2-torsion) is true for $n=1$ if we work in the category of nilpotent spaces. An easy counter-example for non-nilpotent spaces is given by $M(\mathbf{Z}/2,1)$ itself. It is of course an $M(\mathbf{Z}/2,1)$-cellular space, even though $\pi_2 M(\mathbf{Z}/2,1)\cong\mathbf{Z}$. We finally consider the following example.

Example 7.4. The symmetric groups Σ_n are C_2-cellular for $n\geq 2$. Indeed, using the presentation

$$\Sigma_n=\langle \sigma_1,\ldots,\sigma_{n-1}\mid \sigma_i^2=1, \sigma_i\sigma_{i+1}\sigma_i=\sigma_{i+1}\sigma_i\sigma_{i+1}, \sigma_i\sigma_{i+j}\sigma_i=\sigma_{i+j} \text{ for } j\geq 2\rangle\,,$$

one can obtain Σ_n as a push-out of a family of homomorphisms between free products of C_2. For $n-1\geq j\geq 1$, let G_j be the coproduct of $n-1-j$ copies of $C_2 * C_2$, and H_j the coproduct of as many copies of C_2. Define $G_j\to H_j$ to be the coproduct of the fold maps $C_2*C_2\to C_2$. Let K be the coproduct of $n-1$ copies of C_2, and call the generators $x_1,\cdots,x_{n-1}$. Define a map $G_1\to K$ by sending the $(2i-1)$st generator to $x_ix_{i+1}x_i$ and the $2i$th one to $x_{i+1}x_ix_{i+1}$, where $n-1\geq i\geq 1$. For $j\geq 2$, the map $G_j\to K$ is defined by sending the $(2i-1)$st generator to $x_ix_{i+j}x_i$ and the $2i$th one to x_{i+j}. The push-out of the diagram

$$(*_{j=1}^{n-1}H_j)\leftarrow(*_{j=1}^{n-1}G_j)\to K$$

is then the symmetric group Σ_n.

The spaces $K(\Sigma_n,1)$ have therefore C_2-cellular fundamental group, and 2-torsion higher homotopy groups. They are however, for $n\geq 3$, not $M(\mathbf{Z}/2,1)$-cellular by Theorem 7.2, since the integral homology of $K(\Sigma_n,1)$ contains 3-torsion. Actually, we can even compute the cellularization of $K(\Sigma_3,1)$. We know by Theorem 2.1 that it is the fibre of the map $K(\Sigma_3,1)\to P_{M(\mathbf{Z}/2,2)}K(\Sigma_3,1)'$. The later space is simply connected, and is 3-complete. Its mod 3 cohomology is that of $K(\Sigma_3,1)$, so it is by [9, VII, 4.4] the delooping of $S^3\{3\}$, the fibre of the degree 3 self-map of S^3. In other words $CW_{M(\mathbf{Z}/2,1)}K(\Sigma_3,1)$ is a space whose fundamental group is Σ_3 and whose universal cover is $S^3\{3\}$. Thus the universal cover fibration is

$$S^3\{3\}\to CW_{M(\mathbf{Z}/2,1)}K(\Sigma_3,1)\to K(\Sigma_3,1).$$

The action of Σ_3 on $S^3\{3\}$ is trivial by Theorem 3.9.

In the above example we could identify a certain 3-complete space as the delooping of $S^3\{3\}$. In general it is of course not to expect to find nice and well-known spaces as the fibre of the cellularization. However, the same argument as above proves the following result.

Theorem 7.5. *Let π be a finite C_p-cellular group. The universal cover fibration of Theorem 3.9 is then*

$$\Omega K(\pi,1)^{\wedge}_{p'}\to CW_{M(\mathbf{Z}/p,1)}K(\pi,1)\to K(\pi,1)$$

where $X^{\wedge}_{p'}$ denotes the completion away from p, i.e., $X^{\wedge}_{p'}=\prod_{q\neq p}X^{\wedge}_q$. □

References

[1] A. J. Berrick, *An approach to algebraic K-theory,* Research Notes in Math., **56** (1982), Pitman, London.

[2] A. J. Berrick and C. Casacuberta, *A universal space for plus-constructions,* Topology, **38** (1999), no. 3, 467–477.

[3] D. Blanc, *Mapping spaces and M-CW-complexes,* Forum Math., **9** (1997), no. 3, 367–382.

[4] A. K. Bousfield, *The localization of spaces with respect to homology,* Topology, **14** (1975), 133–150.

[5] A. K. Bousfield, *Constructions of factorization systems in categories,* J. Pure Appl. Algebra, **9** (1977), no. 2, 207–220.

[6] A. K. Bousfield, *Localization and periodicity in unstable homotopy theory,* J. Amer. Math. Soc., **7** (1994), no. 4, 831–873.

[7] A. K. Bousfield, *Unstable localization and periodicity,* in: Algebraic Topology: New Trends in Localization and Periodicity, Progress in Math., **136** (1996), Birkhäuser Verlag, Basel, 33–50.

[8] A. K. Bousfield, *Homotopical localizations of spaces,* Amer. J. Math., **119** (1997), no. 6, 1321–1354.

[9] A. K. Bousfield and D. M. Kan, *Homotopy limits, completions and localizations,* Lecture Notes in Math., **304** (1972), Springer-Verlag, Berlin Heidelberg New York.

[10] C. Casacuberta, *Anderson localization from a modern point of view,* in: The Čech Centennial, Contemp. Math., **181** (1995), Amer. Math. Soc., Providence, 35–44.

[11] C. Casacuberta, L. J. Hernández, and J. L. Rodríguez, *Models for torsion homotopy types,* Israel J. Math., **107** (1998), 301–318.

[12] C. Casacuberta and J. L. Rodríguez, *On towers approximating homological localizations,* J. London Math. Soc. (2), **56** (1997), no. 3, 645–656.

[13] C. Casacuberta, J. L. Rodríguez, and D. Scevenels, *Singly generated radicals associated with varieties of groups,* Proceedings of Groups St Andrews 1997 in Bath I, London Math. Soc., Lecture Note Series, **260** (1999), Cambridge University Press, 202–210.

[14] W. Chachólski, *On the functors CW_A and P_A,* Duke J. Math., **84** (1996), no. 3, 599–631.

[15] W. Chachólski, *Desuspending and delooping cellular inequalities,* Invent. Math., **129** (1997), no. 1, 37–62.

[16] W. Chachólski, *A generalization of the triad theorem of Blakers-Massey,* Topology, **36** (1997), no. 6, 1381–1400.

[17] E. Dror Farjoun, *Cellular inequalities,* in: The Čech centennial, Contemp. Math., **181** (1995), Amer. Math. Soc., Providence, 159–181.

[18] E. Dror Farjoun, *Cellular spaces, null spaces and homotopy localization,* Lecture Notes in Math., **1622** (1996), Springer-Verlag, Berlin Heidelberg New York.

[19] T. H. Fay, E. P. Oxford, and G. L. Walls, *Singly generated socles and radicals,* in: Abelian group theory (Honolulu, Hawaii, 1983), Lecture Notes in Math., **1006** (1983), Springer, Berlin Heidelberg New York, 671–684.

[20] J.-C. Hausmann and D. Husemoller, *Acyclic maps,* Enseign. Math. (2), **25** (1979), no. 1-2, 53–75.

[21] L. J. Hernández, *Closed model categories for uniquely S-divisible spaces,* preprint (1996).

[22] G. Mislin and G. Peschke, *Central extensions and generalized plus-construction,* preprint (1998).

[23] J. L. Rodríguez, *On homotopy colimits of spaces with a single homology or homotopy group,* Ph. D. thesis, Universitat Autònoma de Barcelona (1997).

[24] J. Y. Tai, *Generalized plus-constructions and fundamental groups,* J. Pure Appl. Algebra, **132** (1998), no. 2, 207–220.

J. L. Rodríguez
Departamento de Geometría, Topología y Química Orgánica
Universidad de Almería
E-04120 Almería, Spain
E-mail address: `jlrodri@ual.es`

J. Scherer
Institut de Mathématiques
Université de Lausanne
CH–1015 Lausanne
E-mail address: `Jerome.Scherer@ima.unil.ch`

Progress in Mathematics, Vol. 196

Configuration spaces with summable labels

Paolo Salvatore

Abstract. An n-monoid is the appropriate extension of an A_∞-space for the theory of n-fold loop spaces. We define spaces of configurations on n-manifolds with summable labels in partial n-monoids. In particular we obtain an n-fold delooping machinery, that extends the construction of the classifying space by Stasheff. Our configuration spaces cover also symmetric products, spaces of rational curves and spaces of labelled subsets. A configuration space with connected space of labels has the homotopy type of the space of sections of a certain bundle. This extends and unifies results by Bödigheimer, Guest, Kallel and May.

1. Introduction

The interest in labelled configuration spaces in homotopy theory dates back to the seventies. May [15] and Segal [20] showed that the 'electric field map' $C(\mathbb{R}^n; X) \to \Omega^n\Sigma^n(X)$ is a weak homotopy equivalence if X is path connected, and in general is the group completion. Segal showed later [21] that the inclusion $Rat_*(S^2) \hookrightarrow \Omega^2 S^2$ of the space of based rational selfmaps of the spheres into all based selfmaps is the group completion. He used the identification of $Rat_*(S^2)$ with a space of configurations in $\mathbb{C}$ with partially summable labels in $\mathbb{N} \vee \mathbb{N}$, by counting zeros and roots multiplicities. Guest has recently extended his framework in [9] to the space of based rational curves on projective toric varieties. Labelled configuration spaces on manifolds have been studied by Bödigheimer in [3], where the labels are in a based space, and by Kallel in [10], where the summable labels belong to a discrete partial abelian monoid. In both cases the authors have theorems of equivalence between configuration and mapping spaces. We define configuration spaces on a manifold M with labels in A, where A need not to be abelian. It is sufficient that A has a partial sum that is homotopy commutative up to level $dim(M)$. The definition is not trivial and involves tensor products over the Fulton-MacPherson operad. A substantial part of the paper introduces the necessary tools. We generalize the results listed above to the non-abelian setting, and construct a geometric n-fold delooping in one step. Here is a plan of the paper:

Supported by the TMR Grant ERBSMBICT983241 of the European Commission.

In the second section we define the preliminary notion of a partial algebra over an operad and its completion. In the third section we introduce the Fulton-MacPherson operad F_n. We call an algebra over F_n an n-monoid. A 1-monoid is exactly an A_∞-space [12]. In the fourth section we describe the homotopical algebra of topological operads and their algebras. The main results characterize the homotopy type of F_n.

Corollary 4.8. *The unbased Fulton-MacPherson operad is cofibrant.*

Proposition 4.9. *The operad of little n-cubes is weakly equivalent to F_n.*

This implies that the structure of n-monoid is invariant under based homotopy equivalences, and any connected n-monoid has the weak homotopy type of a n-fold loop space. In the fifth section we recall from [14] that a partial compactification $C(M)$ of the ordered configuration space on an open parallelizable n-manifold M is a right module over F_n. We define the configuration space $C(M; A)$ on M with summable labels in a partial n-monoid A, by tensoring $C(M)$ and A over the operad F_n. The definition of $C(M; A)$ is extended to a general open n-manifold M when A is framed, i.e. it has a suitable $GL(n)$-action. In the sixth section we define $C(M, N; A)$ for a relative manifold (M, N) by forgetting the particles in N. Let us denote $B_n(A) = C(I^n, \partial I^n; A)$. For $n = 1$ we obtain the well known construction by Stasheff.

Proposition 6.11. *If A is an A_∞-space, then $B_1(A)$ is homeomorphic to the classifying space $B(A)$ by Stasheff.*

The n-monoid completion of a partial n-monoid A is $C(\mathbb{R}^n; A)$, up to homotopy. We obtain the n-fold delooping of this space in one step:

Theorem 7.3. *If A is framed, then $B_n(A)$ is an n-fold delooping of $C(\mathbb{R}^n; A)$.*

Finally we characterize configuration spaces on manifolds under some conditions.

Theorem 7.6. *If M is a compact closed parallelizable n-manifold and A is a path connected partial framed n-monoid, then there is a weak equivalence*

$$C(M; A) \simeq Map(M; B_n(A)).$$

As corollary we obtain a model for the free loop space on a suspension built out of cyclohedra. This answers a question by Stasheff [22].

Corollary 7.7. *If X is path connected and well pointed, then there is a weak homotopy equivalence $C(S^1; X) \simeq Map(S^1, \Sigma X)$.*

This work is part of the D.Phil. thesis of the author [18], University of Oxford, 1998, written under the direction of U. Tillmann. I am grateful to her for her continuous support, and to M. Markl and J. Stasheff for many valuable suggestions.

2. Partial modules over operads

Let $\mathcal{C}$ be a symmetric closed monoidal category, with tensor product $\otimes$ and unit element e. We assume that $\mathcal{C}$ has small limits and colimits.

Definition 2.1. *A Σ-object X in $\mathcal{C}$ is a collection of objects $X(n)$, for $n \in \mathbb{N}$, such that $X(n)$ is equipped with an action of the symmetric group Σ_n.*

The category of Σ-objects in $\mathcal{C}$ will be denoted by $\Sigma\mathcal{C}$. We observe as in [8] that $\Sigma\mathcal{C}$ is a monoidal category as follows: given two objects A and B, their tensor product $A \otimes B$ is defined by

$$(A \otimes B)(n) = \coprod_{k=0}^{\infty} A(k) \otimes_{\Sigma_k} \Big(\coprod_{\pi \in \mathrm{Map}(n,k)} \bigotimes_{i=1}^{k} B_{\pi^{-1}(i)} \Big),$$

where $Map(n,k)$ is the set of maps from $\{1,\dots,n\}$ to $\{1,\dots,k\}$. Here each element B_S, where S is a set of numbers, is identified to $B_{\#S}$ by the order preserving bijection, and the action of Σ_n is given accordingly. There is a natural embedding functor $j : \mathcal{C} \hookrightarrow \Sigma\mathcal{C}$ considering an object X as a Σ-object concentrated in degree 0, so that

$$j(X)(n) = \begin{cases} X & \text{if } n = 0, \\ \emptyset & \text{if } n \neq 0, \end{cases}$$

with the trivial actions of the symmetric groups. Here $\emptyset$ denotes the initial object of $\mathcal{C}$. The functor j is left adjoint to the forgetful functor $B \mapsto B(0)$ from $\Sigma\mathcal{C}$ to $\mathcal{C}$. More generally we have an embedding functor $j_n : \Sigma_n\mathcal{C} \to \Sigma\mathcal{C}$, that is left adjoint to the forgetful functor $B \mapsto B(n)$.

The unit element ι of $\Sigma\mathcal{C}$ is defined by

$$\iota(n) = \begin{cases} e & \text{if } n = 1, \\ \emptyset & \text{if } n \neq 1. \end{cases}$$

Definition 2.2 ([8]). *An operad in $\mathcal{C}$ is a monoid in the monoidal category $\Sigma\mathcal{C}$. We denote the category of operads in $\mathcal{C}$ by $\mathcal{OP}(\mathcal{C})$.*

This means that an operad (F, μ, η) is a Σ-object F together with composition morphism $\mu : F \otimes F \to F$ and unit morphism $\eta : \iota \to F$, such that the associativity property $\mu(\mu \otimes F) = \mu(F \otimes \mu) : F \otimes F \otimes F \to F$ and the unit property $\mu(F \otimes \mu) = \mu(\mu \otimes F) = id_F : F \to F$ hold. Note that the functor $F \otimes \_$ forms a triple.

Example 2.3. *The category $\mathcal{CH}_R$ of non negatively graded chain complexes over a commutative ring R is monoidal by the tensor product. The operads in $\mathcal{CH}_R$ are called differential graded operads over R.*

Definition 2.4 ([14]). *Given an operad F in $\mathcal{C}$, a left F-module A is a Σ-object A, with a morphism $\rho : F \otimes A \to A$ of Σ-objects such that*

$\rho(F \otimes \rho) = \rho(\mu \otimes A) : F \otimes F \otimes A \to A$ and $\rho(\eta \otimes A) = id_A : A \to A$.

In other words A is an algebra over the triple $F \otimes \_$. Dually we define the notion of right F-module. We denote the category of left F-modules by Mod_F, and the category of right F-modules by ${}_F Mod$.

Definition 2.5. *If F and G are operads in $\mathcal{C}$, then a F-G-bimodule A is a left F-module and a right G-module such that the left F-module structure map is a right G-homomorphism.*

Definition 2.6. *An algebra X over an operad F, or F-algebra, is an object X of $\mathcal{C}$ together with a left F-module structure on $j(X)$.*

We denote the category of F-algebras by Alg_F. Moreover we will denote by $F(Y)$ the free F-algebra generated by the object Y in $\mathcal{C}$. This object is defined by $j(F(Y)) = F \otimes j(Y)$.

Definition 2.7. *A partial left F-module A is a Σ-object A in $\mathcal{C}$ together with a Σ-object $Comp$ in $\mathcal{C}$ called the* object of composables, *a monomorphism $i : Comp \hookrightarrow F \otimes A$ in $\Sigma\mathcal{C}$ and a composition map $\rho : Comp \to A$ such that*

1. *The unit $\eta \otimes A : A \to F \otimes A$ factors uniquely through $\tilde{\eta} : A \to Comp$ and the composition $\rho(\tilde{\eta}) = id_A$ is the identity.*
2. *The pullbacks in $\Sigma\mathcal{C}$ of $i : Comp \hookrightarrow F \otimes A$ along the two maps $(\mu \otimes A)(F \otimes i) : F \otimes Comp \to F \otimes A$ and $F \otimes \rho : F \otimes Comp \to F \otimes A$ coincide. Moreover the compositions of the two pullback maps with $\rho : Comp \to A$ coincide.*

Partial right F-modules are defined dually. A partial F-algebra is an object X of $\mathcal{C}$ such that $j(X)$ is a partial left F-module. We denote the categories of partial left F-modules, right F-modules and F-algebras respectively by $PartMod_F$, ${}_F PartMod$, and $PartAlg_F$.

A morphism $g : (A, Comp_A) \to (B, Comp_B)$ of partial left F-modules is a morphism in $\Sigma\mathcal{C}$ such that $(F \otimes g)i_A : Comp_A \to F \otimes B$ factors through $\tilde{g} : Comp_A \to Comp_B$, and $g\rho_A = \rho_B \tilde{g}$.

We exhibit a functor from partial to total left modules that is left adjoint to the forgetful functor. The analogous construction for right modules is exactly dual.

If A is a partial F-left module, then define $\hat{A}$ by the coequalizer in the category $\Sigma\mathcal{C}$

$$F \otimes Comp \underset{F\otimes\rho}{\overset{(\mu\otimes A)(F\otimes i)}{\rightrightarrows}} F \otimes A \dashrightarrow \hat{A}\,.$$

Proposition 2.8. *There is a left F-module structure on $\hat{A}$.*

Proof. The proof is modelled on Lemma 1.15 in [8]. The coequalizer above is reflexive because the input arrows admit the common section $F \otimes \tilde{\eta} : F \otimes A \to F \otimes Comp$. Now $\hat{A}$ admits the structure of left F-module, because by Lemma 2.3.8 in [17] $F \otimes \_$ preserves reflexive coequalizers. Moreover $\hat{A}$ is the coequalizer of the pair above in the category of left F-modules. □

Proposition 2.9. *The completion $A \mapsto \hat{A}$ induces a functor that is left adjoint to the forgetful functor $U : Mod_F \to PartMod_F$.*

Definition 2.10. *For any right F-module C with structure map $\sigma : C \otimes F \to C$ and a partial left F-module $(A, Comp, i)$ we define the tensor product $C \otimes_F A$ as coequalizer in $\Sigma\mathcal{C}$*

$$C \otimes Comp_A \underset{C\otimes\rho}{\overset{(\sigma\otimes A)(C\otimes i)}{\rightrightarrows}} C \otimes A \dashrightarrow C \otimes_F A\,.$$

Dually we define the tensor product of a partial right F-module and a left F-module.

Proposition 2.11. *Given a partial right F-module A, an F-G-bimodule B, and a partial left G-module C, there is a natural isomorphism*

$$(A \otimes_F B) \otimes_G C \cong A \otimes_F (B \otimes_G C).$$

The isomorphism holds because the tensor product is a left adjoint and preserves colimits.

3. The Fulton-MacPherson operad

The category $\mathcal{CG}$ of compactly generated weak Hausdorff topological spaces is a closed monoidal category with all limits and colimits, hence it satisfies the assumptions of the previous section. We note however that in general the forgetful functor to the category of sets does not preserve colimits. Operads and modules in $\mathcal{CG}$ shall be called simply topological operads and topological modules.

The key topological operads in this paper are the Fulton-MacPherson operads, that are suitable cofibrant versions of the little cubes operads. They were introduced in [8]. We recall their definition. Consider the differential-geometric blow-up of $(\mathbb{R}^n)^k$ along the small diagonal $\Delta = \{x_1, \dots, x_k \mid x_1 = \cdots = x_k\}$. The blow-up is explicitly obtained if we replace the diagonal by its normal sphere bundle. The fiber of the trivial normal bundle at the origin is $F = \{y_1, \dots, y_k \mid \sum_{i=1}^{i=k} y_i = 0\}$ and the sphere bundle $PF = (F - 0)/(\mathbb{R}^+)$ can be seen as the space of closed half-lines in F. Then the blow-up is

$$\mathrm{Bl}_\Delta((\mathbb{R}^n)^k) = \{(x, y) \in (\mathbb{R}^n)^k \times PF \,|\, x - \pi_\Delta(x) \in y\},$$

where the orthogonal projection is $\pi_\Delta(x_1, \dots, x_k) = (\sum_{i=1}^{i=k} x_i/k, \dots, \sum_{i=1}^{i=k} x_i/k)$. For any set $S \subseteq \{1, \dots, k\}$ let us denote by $\mathrm{Bl}_\Delta((\mathbb{R}^n)^S)$ the blow-up of $(\mathbb{R}^n)^S$ along its small diagonal.

Let $C_k^0(\mathbb{R}^n) \subset Map(\{1, \dots, k\}, \mathbb{R}^n)$ be the space of ordered pairwise distinct k-tuples in $\mathbb{R}^n$. There is a natural right Σ_k-action on this space, and we consider it as left Σ_k-space by the opposite action. As $C_k^0(\mathbb{R}^n)$ does not intersect any diagonal, there is a natural embedding

$$j : C_k^0(\mathbb{R}^n) \to \prod_{S \subseteq \{1,\dots,n\},\, \#S \geq 2} \mathrm{Bl}_\Delta((\mathbb{R}^n)^S).$$

Definition 3.1. *The Fulton-MacPherson configuration space $C_k(\mathbb{R}^n)$ is the closure of the image of j.*

We note that $GL(n)$ acts diagonally on each blowup, j is a $GL(n)$-equivariant map and therefore $C_k(\mathbb{R}^n)$ is a $GL(n)$-space.

In a similar way we define the Fulton-MacPherson configuration space $C_k(M)$ of a smooth open manifold M. In this case one builds the differential-geometric blowups of M^k along the diagonal Δ_M by gluing together $M^k - \Delta_M$ and the normal sphere bundle via a tubular neighbourhood of Δ_M in M^k. It turns out that $C_k(M)$ is a manifold with corners Σ_k-equivariantly homotopy equivalent to its interior, the ordinary configuration space $C_k^0(M)$ of ordered pairwise distinct k-tuples in M.

There is a blow-down map $b : C_k(M) \to M^k$ such that the composite $C_k^0(M) \xhookrightarrow{j} C_k(M) \xrightarrow{b} M^k$ is the inclusion. We will say that the blow-down map gives the *macroscopic locations* of the particles.

There is a characterization of the Fulton-MacPherson configuration space by means of trees due to Kontsevich. For us a *tree* is an oriented finite connected graph with no cycles such that each vertex has exactly one outcoming edge. An *ordered tree* is a tree together with an ordering of the incoming edges of each vertex. The ordering is equivalent to the assignation of a planar embedding. The only edge with no end vertex is the *root*, the edges with no initial vertex are the *twigs*, and all other edges are *internal.* A tree on a set I is a tree together with a bijection from the set of its twigs to I. The *valence* of a vertex is the number of incoming edges. Let $G(n)$ be the group of affine transformations of $\mathbb{R}^n$ generated by translations and positive dilatations.

Proposition 3.2 ([12]). *Let M be an open manifold. Then each element in $C_k(M)$ is uniquely determined by:*

1. *Distinct macroscopic locations $P_1, \dots, P_l \in M$, with $1 \leq l \leq k$.*
2. *For each $1 \leq i \leq l$ a tree T_i with f_i twigs, so that $\sum_{i=1}^{l} f_i = k$, and for each vertex in T_i of valence m an element in $C_m^0(\tau_{P_i}(M))/G(n)$, where $\tau_{P_i}(M)$ is the tangent plane at P_i.*
3. *A global ordering of the k twigs of the trees.*

Definition 3.3. *If $b : C_k(\mathbb{R}^n) \to (\mathbb{R}^n)^k$ is the blowdown map, then the Fulton-MacPherson space is $F_n(k) = b^{-1}(\{0\}^k)$.*

This space contains all configurations macroscopically located at the origin.

Proposition 3.4 ([8]). *The space $F_n(k)$ is a manifold with corners, and it is a compactification of $C_k^0(\mathbb{R}^n)/G(n)$.*

The *faces* of $F_n(k)$ are indexed by trees on $\{1, \ldots, k\}$, and the codimension of a face is equal to the number of internal edges of the indexing tree.

Proposition 3.5 ([14]). *The spaces $F_n(k)$ $k \geq 0$ assemble to form a topological operad. Moreover $F_n(k)$ is a Σ_k-equivariant deformation retract of $C_k(\mathbb{R}^n)$.*

The composition law is easily described in terms of trees: each element in F_n is described by a single tree by Proposition 3.2. If $a \in F_n(k)$ and $b_j \in F_n(i_j)$ for $j = 1, \ldots, k$ then $a \circ (b_1, \ldots, b_k) \in F_n(i_1 + \cdots + i_k)$ corresponds to the tree obtained by merging the j-th twig of the tree of a with the root of the tree of b_j for $j = 1, \ldots, k$, and assigning the new twigs the induced order. This operation on trees will be called *grafting*. Note that $F_n(1)$ is a point, the unit ι of the operad, and is represented by the trivial tree. We assume that $F_n(0)$ is a point, the empty configuration. We stress the fact that Getzler and Jones in [8] consider the unpointed version $\tilde{F}_n$ such that $\tilde{F}_n(k) = F_n(k)$ for $k > 0$ and $\tilde{F}_n(0) = \emptyset$. Their paper focuses on the differential graded operad $e_n = H_*(\tilde{F}_n; \mathbb{Q})$, the rational homology of $\tilde{F}_n$. They denote $H_*(F_n; \mathbb{Q})$ by e_n^+. The deformation retraction $r : C_k(\mathbb{R}^n) \times I \to C_k(\mathbb{R}^n)$ such that $r(C_k(\mathbb{R}^n) \times \{1\}) = F_n(k)$ is defined for $t \neq 0$ and $x \in C_k^0(\mathbb{R}^n)$ by $r(x,t) = xt$.

Definition 3.6. *We call an algebra over F_n an n-monoid, and an algebra over $\tilde{F}_n$ an n-semigroup.*

Example 3.7 ([12]). *The 1-monoids are the A_∞-spaces.*

In fact $F_1(i) = K_i \times \Sigma_i$, where K_i denotes the associahedron by Stasheff [22], so F_1 is the symmetric operad generated by the non-symmetric Stasheff operad K. But an A_∞ space is by definition an algebra over K.

4. Homotopical algebra and the little discs

We describe the closed model category structure of the categories of topological operads and their algebras.

Definition 4.1 ([6]). *A cofibrantly generated model category is a closed model category* [16], *together with a set I of* generating cofibrations, *and a set J of* generating trivial cofibrations, *so that the fibrations and the trivial fibrations are respectively the maps satisfying the right lifting property with respect to the maps in J and I.*

Consider the free operad functor $\mathbb{T} : \Sigma(\mathcal{CG}) \to \mathcal{OP}(\mathcal{CG})$, left adjoint to the forgetful functor $\mathbb{U} : \mathcal{OP}(\mathcal{CG}) \to \Sigma(\mathcal{CG})$. Let $\mathcal{S}_n$ be the family of subgroups of Σ_n. The simplicial version of the following proposition is 3.2.11 in [17].

Proposition 4.2 ([18]). *The category of topological operads is a cofibrantly generated model category, with the following structure:*

1. *The set of generating cofibrations is $I = \{\mathbb{T}(\partial I^i \times H \setminus \Sigma_n \hookrightarrow I^i \times H \setminus \Sigma_n) \mid i, n \in \mathbb{N},\ H \in \mathcal{S}_n\}$;*

2. *The set of generating trivial cofibrations is*
$$J = \{\mathbb{T}((I^{i-1} \times \{0\}) \times H \setminus \Sigma_n \hookrightarrow I^i \times H \setminus \Sigma_n) \mid i, n \in \mathbb{N},\ H \in \mathcal{S}_n\};$$
3. *A morphism f is respectively a weak equivalence or a fibration if for any $n \in \mathbb{N}$ and $H \in \mathcal{S}_n$ the restriction f_n^H of f_n to the H-invariant subspaces is respectively a weak homotopy equivalence or a Serre fibration.*

There is a functorial cofibrant resolution for operads, introduced in [1]. Let A be a topological operad. Let M_k be the set of isomorphism classes of *ordered* trees on $\{1, \dots, k\}$, and for each tree t let $V(t)$ be the set of its vertices and $E(t)$ the set of its internal edges. For each vertex $x \in V(t)$ let $|x|$ be its valence.

Definition 4.3. *The space of ordered trees on $\{1, \dots, k\}$ with vertices labelled by elements of A, and with internal edges labelled by real numbers in $[0,1]$ is*
$$T_k(A) = \coprod_{t \in M_k} \Big(\prod_{x \in V(t)} A(|x|) \times [0,1]^{\#E(t)} \Big).$$

Let T_t be the summand indexed by a tree $t \in M_k$. For each internal edge $e \in E(t)$ the operad composition induces a map $\partial_e : T_t \to T_{t-e}$, where $t - e$ is obtained from t by collapsing e to a vertex. If e goes from x to y, $|y| = n$, and e is the $i-th$ incoming edge of y, then $\partial_e(x)$ is the multiplication of the composition $\theta_i : A(|x|) \times A(|y|) \longrightarrow A(|x| + |y| - 1)$ by the identity maps of the vertices in $V(t) - \{x \cup y\}$.

Definition 4.4. *The space $WA(k)$ is the quotient of $T_k(A)$ under the following relations:*

1. *Suppose that $t \in T_k(A)$, v is a vertex of t of valence n labelled by $\alpha \in A(n)$, the subtrees stemming from v are $t_1 < \dots < t_n$, and $\sigma \in \Sigma_n$. Then t is equivalent to the element obtained from t by replacing α by $\sigma^{-1}\alpha$ and by permuting the order of the subtrees to $t_{\sigma_1} < \dots < t_{\sigma_n}$.*
2. *If $t \in T_k(A)$ has an edge e of length 0, then t is equivalent to the labelled tree obtained by collapsing e to a vertex, and composing the labels of its vertices.*
3. *If $t \in T_k(A)$ has a vertex w of valence 1 labelled by the unit $\iota \in A(1)$ of the operad A, then t is equivalent to the labelled tree obtained by removing w. If w is between two internal edges of lengths s and t, then we assign length $s + t - st$ to the merged edge.*

There is an action of Σ_k on $WA(k)$ induced by permuting the labelling of the twigs of elements in $T_k(A)$.

Proposition 4.5 ([1]). *There is an operad structure on WA, defined by grafting trees, and by assigning length 1 to the new internal edges. A natural ordering of the twigs of the composite is induced. The trivial tree consisting of an edge with no vertices is the identity of WA.*

For us a cofibration of topological spaces is the retract of a generalized CW-inclusion [6]. We say that a pointed space (X, x_0) is well-pointed if the inclusion $\{x_0\} \hookrightarrow X$ is a cofibration. The following proposition is essentially proved in [1].

Proposition 4.6. *Let A be a topological operad such that $(A(1), \iota)$ is well-pointed and each $A(n)$ is a cofibrant space. Then WA is a cofibrant resolution of A.*

Proposition 4.7. *There is an isomorphism of topological operads $W(\tilde{F}_n) \cong \tilde{F}_n$.*

Proof. We observe that $W(\tilde{F}_n)(i)$ is obtained by gluing together for each *face* S of $F_n(i)$ of codimension d a copy of $S \times [0,1]^d$. This is true because the codimension of a face of the manifold with corners $F_n(i)$ is equal to the number of internal edges of the associated tree. Then $W(\tilde{F}_n)(i)$ admits the structure of manifold with corners diffeomorphic to $F_n(i)$, and the composition maps of both operads are described by grafting of trees. □

Corollary 4.8. *The operad $\tilde{F}_n$ is cofibrant.*

Let D_n be the operad of little n-discs. The space $D_n(k)$ is the space of k-tuples of direction preserving affine selfembeddings of the unit n-disc with pairwise disjointed images. The operad structure is defined by multicomposition of the embeddings. There is a sequence of Σ_k-equivariant homotopy equivalences $D_n(k) \to C_k^0(I^n) \hookrightarrow C_k^0(\mathbb{R}^n) \hookrightarrow C_k(\mathbb{R}^n) \to F_n(k)$. The first map sends the little discs to their centers, the last is a deformation retraction. The inclusion $C_k^0(I^n) \hookrightarrow C_k^0(\mathbb{R}^n)$ is a Σ_k-equivariant homotopy equivalence because the inclusion $I^n \hookrightarrow \mathbb{R}^n$ is isotopic to a homeomorphism. The image of the composite r_k is the interior of the manifold with corners $F_n(k)$. It follows that the Σ-map $r = \{r_k\}$ is not an operad map because all elements in the boundary of $F_n(k)$ are composite.

Proposition 4.9. *The operad D_n is weakly equivalent to F_n.*

Proof. We build an extension $R : WD_n \to F_n$ of $r : D_n \to F_n$ that is a map of operads and a weak equivalence.

An element a in $WD_n(k)$ is represented by a labelled tree $\tau \in T_k(D_n)$ on $\{1, \dots, k\}$. If the i-tuple $(f_1^v, \dots, f_i^v)$ labels a vertex v of valence i, then for each j we associate the embedding f_j^v to the j-th *incoming* edge $e_j(v)$ of v. The equivalence relation defining WD_n preserves this association. We observe incidentally that the multicomposition of the labels of τ is the k-tuple of embeddings $(g_1, \dots, g_k)$ such that for each j g_j is the composition of the embeddings associated to the edges along the unique path from the j-th twig to the root. Suppose that the internal edges of τ are labelled by numbers in $(0,1)$. Let $l(e)$ denote the length of an edge e and if $r \in (0,1]$ let δ_r be the dilatation of the n-disc by r. Consider the labelled tree τ' obtained from τ by replacing for each vertex v and for each $j = 1, \dots, |v|$ the embedding f_j^v by the rescaling $f_j^v \circ \delta_{l(e_j(v))}$. Let b be the multicomposition of the labels of τ', and set $R_k(a) = r_k(b)$. We have defined R_k on a dense subspace of $WD_n(k)$. The map R_k extends to $WD_n(k)$ and R is an operad map, because the boundary and the composition of F_n are described by a limit procedure. Let $i_k : D_n(k) \to WD_n(k)$ be the inclusion such that $i_k(a)$ is represented by the tree on $\{1, \dots, k\}$ with a single vertex labelled by a. The map $R_k : WD_n(k) \to F_n(k)$ is a Σ_k-equivariant homotopy equivalence for each

k because i_k is such [1] and $R_k i_k = r_k$. In particular R is a weak equivalence of topological operads, and $D_n \simeq F_n$. □

The simplicial analogue of the following proposition is 3.2.5 in [17].

Proposition 4.10 ([18] [19]). *Let F be a topological operad. Then the category Alg_F is a cofibrantly generated model category with the following structure:*

1. *The set of generating cofibrations is $I = \{F(\partial I^i) \hookrightarrow F(I^i) \mid i \in \mathbb{N}\}$.*
2. *The set of trivial generating cofibrations is*
$$J = \{F(I^{i-1} \times \{0\}) \hookrightarrow F(I^i) \mid i \in \mathbb{N}\}.$$
3. *A F-homomorphism is a weak equivalence or a fibration if it is respectively a weak homotopy equivalence or a Serre fibration.*

Under mild conditions there is a functorial cofibrant resolution of topological algebras over operads, introduced in [1]. Let A be a topological operad. Consider the Σ-space W^+A defined similarly as WA, except that relation 3 is not applied if w is the root vertex. It turns out that W^+A is an $A - WA$-bimodule, by action of A on the label of the root, and by grafting trees representing elements of WA. Let X be an A-algebra. It has a WA-algebra structure induced by the projection $\varepsilon : WA \to A$. We define the A-algebra $U_A(X) = W^+A \otimes_{WA} X$. The projection $W^+A \to WA$, obtained by extending relation 3 to the root vertex, induces an A-homomorphism $\pi : U_A(X) \to X$, that is a deformation retraction, see p. 51 of [1].

Proposition 4.11 ([18]). *If X is a cofibrant space then $U_A(X)$ is a cofibrant A-algebra.*

Definition 4.12. *If A is a topological operad, and X, Y are A-algebras, then a homotopy A-morphism from X to Y is an A-homomorphism from $U_A(X)$ to Y.*

Proposition 4.13. *If F is a topological operad, X is the retract of a generalized CW-space, and $Ho(Alg_F)$ is the homotopy category, then $Ho(Alg_F)(X,Y) = Alg_F(U_F(X),Y)/\simeq$.*

Proof. The set of right homotopy classes $[X, Y]$ in the sense of [16] is the set of F-homomorphisms from a cofibrant model of X to a fibrant model of Y modulo right homotopy. Now $U_F(X)$ is a cofibrant resolution of X, and Y is fibrant because every object is such. It is easy to see that the the right homotopy classes of F-homomorphisms from $U_F(X)$ to Y are ordinary homotopy classes, because Y^I is a path object. □

This result is consistent with the formulation of the homotopy category of F-algebras in [1].

Proposition 4.14. *If Z is an n-semigroup and $p : Y \to Z$ is a homotopy equivalence, then Y has a structure of n-semigroup such that p extends to a homotopy $\tilde{F}_n$-morphism.*

Proof. It is sufficient to observe that $W\tilde{F}_n$ is homeomorphic to $\tilde{F}_n$, and apply the homotopy invariance theorem 8.1 in [1]. □

5. Modules and configuration spaces with summable labels

Proposition 5.1 ([14]). *For any parallelizable open manifold M of dimension n, the space of configurations $C(M) = \coprod_{k\in\mathbb{N}} C_k(M)$ is a right F_n-module.*

Proof. We choose a trivialization of the tangent bundle $\tau(M) \cong M \times \mathbb{R}^n$. Then the composition $C(M) \otimes F_n \to C(M)$ is described by grafting of trees representing elements as in 3.2. □

Markl in [14] gives a similar picture for generic open manifolds by introducing the framed Fulton-MacPherson operads.

Definition 5.2. *Let G be a topological group, and let F be a topological operad such that $F(i)$ is a $G \times \Sigma_i$-space for each i, and the structure map μ of F is G-equivariant. The semidirect product $F \rtimes G$ is the operad defined by $(F \rtimes G)(i) = F(i) \times G^i$, with structure map*

$$\tilde{\mu}((x, g_1, \ldots, g_k); (x_1, g_1^1, \ldots, g_1^{m_1}), \ldots, (x_k, g_k^1, \ldots, g_k^{m_k})) = \\ = (\mu(x; g_1x_1, \ldots, g_kx_k), g_1g_1^1, \ldots, g_kg_k^{m_k}) .$$

Definition 5.3. *The framed Fulton-MacPherson operad is the semidirect product $fF_n = F_n \rtimes GL(n)$.*

Definition 5.4. *Let M be an open n-manifold. The $GL(n)$-bundle of frames on M induces a $GL(n)^k$-bundle $fC_k(M)$ on $C_k(M)$, acted on by Σ_k, that we call the* framed configuration space *of k frames in M.*

Proposition 5.5 ([14]). *The Σ-space $fC(M)$ of framed configurations is a right module over fF_n.*

An element of the framed configuration space $fC_k(M)$ is uniquely determined by labelled trees as in Proposition 3.2, and by additional k frames of the tangent planes associated to the k twigs. A smooth embedding $i : M \hookrightarrow N$ of open n-manifolds induces a right fF_n-homomorphism $fC(i) : fC(M) \hookrightarrow fC(N)$.

Remark 5.6. *If M is a Riemannian n-manifold, then we can define for each k a $O(n)^k$-bundle $f^OC_k(M)$ over $C_k(M)$, so that $f^OC(M)$ is a right $F_n \rtimes O(n)$-module. If M is oriented then we can define a $SO(n)^k$-bundle $f^{SO}C_k(M)$ on $C_k(M)$ so that $f^{SO}C(M)$ is a right $F_n \rtimes SO(n)$-module.*

Definition 5.7. *We call an algebra over fF_n a framed n-monoid.*

Hence a framed n-monoid is an n-monoid equipped with an action of $GL(n)$, that is compatible with the n-monoid structure map.

Definition 5.8. *A partial framed n-monoid is a partial n-monoid with an action of $GL(n)$, such that $GL(n)$ preserves the space of composables and respects the partial composition.*

Definition 5.9. *Let $fD_n(k)$ be the space of k-tuples of of affine selfembeddings of the unit n-disc that preserve angles and have pairwise disjointed images. The multicomposition gives fD_n the structure of an operad, that we call the operad of framed little n-discs.*

Remark 5.10. *Consider the iterated loop space $\Omega^n(X, x_0)$ as the space of maps from the closed unit n-disc to X, sending the boundary to the base point x_0. This space is an algebra over fD_n.*

Proposition 5.11. *The operad of framed little n-discs fD_n is weakly equivalent to the framed Fulton-MacPherson operad fF_n.*

Proof. We apply the same proof of Proposition 4.9 to show that $fD_n \simeq F_n \rtimes O(n)$, and conclude by the homotopy equivalence $O(n) \hookrightarrow GL(n)$. □

If we restrict to the suboperad $\bar{f}D_n \subset fD_n$ containing orientation preserving embeddings, then we obtain a weak equivalence $\bar{f}D_n \simeq F_n \rtimes SO(n)$.

Definition 5.12. *Let A be a partial n-monoid, and let M be an open parallelizable manifold of dimension n. Then the space of configurations in M with partially summable labels in A is $C(M; A) := C(M) \otimes_{F_n} A$.*

An element of $C(M) \otimes A = \coprod_k C_k(M) \times_{\Sigma_k} A^k$ consists by 3.2 of a finite set of trees based at distinct points in M, with vertices labelled by F_n and twigs labelled by A. The equivalence relation defining $C(M; A)$ says that if some twigs labelled by $a_1, \ldots, a_k$ are departing from a vertex labelled by $c \in F_n(k)$ in $t \in C(M) \otimes A$ and $\rho(c; a_1, \ldots, a_k)$ is defined, then we identify t with the forest obtained from t by cutting such twigs, and by replacing their vertex by a twig labelled by $\rho(c; a_1, \ldots, a_k)$. Furthermore if the i-th twig departing from a vertex labelled by c in t is labelled by the base point a_0 then we identify t to forest obtained by cutting the twig and by replacing the label c by $s_i(c)$, where $s_i : F_n(k) \to F_n(k-1)$ is the projection induced by forgetting the i-th coordinate.

If A is an n-monoid, then by iterated identifications any element in $C(M; A)$ has a unique representative consisting of a finite set of trivial trees in M, or points, with labels in $A - \{a_0\}$.

We denote by $| \quad | : CG \to \mathbf{Set}$ be the forgetful functor.

Proposition 5.13. *Suppose that the inclusion $Comp \hookrightarrow F(A)$ is a cofibration, and A is well-pointed. Then*

1. $|C(M; A)| = |C(M)| \otimes_{|F_n|} |A|$;
2. *the space $C(M; A)$ has the weak topology with respect to the filtration $C_k(M; A) = Im(\coprod_{i \le k} C(M)_i \times_{\Sigma_i} A^i), \ k \in \mathbb{N}$.*

Proof. If A is a proper n-monoid, then we have relative homeomorphisms

$$(C_k(M), \partial C_k(M)) \times_{\Sigma_k} (A, a_0)^k \longrightarrow (C_k(M; A), C_{k-1}(M; A))$$

for $k \geq 1$, and we conclude by 8.4, 9.2 and 9.4 in [23]. If A is a partial n-monoid, then we denote by $R_i \subset C_i(M) \times_{\Sigma_i} A^i$ the space of reducible elements that are equivalent to an element of some $C_j(M) \times_{\Sigma_j} A^j$ with $j < i$. For example,

$$R_1 = M \times \{a_0\};$$
$$R_2 = (C_2(M) \times_{\Sigma_2} (A \vee A)) \cup (M \times Comp_2);$$
$$R_3 = (C_3(M) \times_{\Sigma_3} (A \vee A \vee A)) \cup (C_2(M) \times_{\Sigma_2} (Comp_2 \times A)) \cup (M \times Comp_3).$$

We have relative homeomorphisms $(C_i(M) \times_{\Sigma_i} A^i, R_i) \to (C_i(M; A), C_{i-1}(M; A))$, and we argue similarly. □

Definition 5.14. *Suppose that M is an open n-dimensional smooth manifold, and A is a partial framed n-monoid. Then the space of configurations in M with labels in A is $C(M; A) := fC(M) \otimes_{fF_n} A$.*

Note that if M is parallelizable then the definition is consistent with 5.12. In fact the framed configurations in M are given by $fC(M) = C(M) \otimes_{F_n} fF_n$ and by 2.11

$$fC(M) \otimes_{fF_n} A = C(M) \otimes_{F_n} fF_n \otimes_{fF_n} A = C(M) \otimes_{F_n} A.$$

Proposition 5.15. *Let A be a partial framed n-monoid with base point a_0 such that the inclusions $Comp \hookrightarrow fF_n(A)$ and $\{a_0\} \hookrightarrow A$ are cofibrations of $GL(n)$-spaces. Let M be an open n-manifold. Then*

1. $|C(M; A)| = |fC(M)| \otimes_{|fF_n|} |A|$;
2. *the space $C(M; A)$ has the weak topology with respect to the filtration $C_k(M; A) = Im(\coprod_{i \leq k} fC(M)_i \times_{\Sigma_i} A^i)$, $k \in \mathbb{N}$.*

We give some examples of configuration spaces with summable labels. Let us denote by $\hat{A}^n$ the completion of a partial n-monoid A.

Proposition 5.16. *If A is a partial n-monoid, then there is a strong deformation retraction $w_A : C(\mathbb{R}^n; A) \to \hat{A}^n$.*

Proof. It is sufficient to observe that there is a deformation retraction of right F_n-modules $w : C(\mathbb{R}^n) \to F_n$. If an element $x \in C(\mathbb{R}^n; A)$ is represented by a finite number of labelled trees $\tau_1, \ldots, \tau_k$ based at distinct points $P_1, \ldots, P_k \in \mathbb{R}^n$, then $w_A(x)$ is represented by the single tree obtained by connecting $\tau_1, \ldots, \tau_k$ to a root vertex labelled by the class $[P_1, \ldots, P_k] \in C_k^0(\mathbb{R}^n)/G(n) \subset F_n(k)$. □

Example 5.17. *If M is a discrete partial monoid, then $\hat{M}^1$ has the homotopy type of its monoid completion. If M is abelian then $\hat{M}^\infty$ has the homotopy type of its abelian monoid completion.*

Definition 5.18. *Let A be a partial abelian monoid and M an n-manifold. We denote by $C^0(M;A)$ the quotient of $\coprod_k C_k^0(M) \times_{\Sigma_k} A^k$ under the following relation $\sim$: if $(m_1,\ldots,m_k) \in C_k^0(M)$, $a_1,\ldots,a_k \in A$, $m_1 = m_2$ and $a_1 + a_2$ is defined, then*

$$(m_1,\ldots,m_k; a_1,\ldots,a_k) \sim (m_2,\ldots,m_k; a_1 + a_2,\ldots,a_k).$$

Lemma 5.19. *If A is a partial abelian monoid, and M is an n-dimensional open manifold, then the inclusion $C^0(M;A) \hookrightarrow C(M;A)$ is a weak equivalence.*

Proof. The proof makes use of the fact that a copy of the manifold with corners $C_k(M)$ lies inside its interior $C_k^0(M)$, so the retraction $r : C_k(M) \to C_k^0(M)$ is a Σ_k-equivariant homeomorphism onto its image. We compare via this retraction the pushout diagram for $C_k^0(M;A)$

$$\begin{array}{ccc} C_k^0(M) \times_{\Sigma_k} T_k(X,x_0) \cup (C^0(M) \times_\tau Comp(A))_k & \longrightarrow & C_{k-1}^0(M;A) \\ \Big\downarrow & & \Big\downarrow \\ C_k^0(M) \times_{\Sigma_k} A^k \cup (C^0(M) \times_\tau Comp(A))_k & \longrightarrow & C_k^0(M;A), \end{array}$$

and the pushout diagram for $C_k(M;A)$

$$\begin{array}{ccc} C_k(M) \times_{\Sigma_k} T_k(X,x_0) \cup (C^0(M) \times_\tau Comp(A))_k & \longrightarrow & C_{k-1}(M;A) \\ \Big\downarrow & & \Big\downarrow \\ C_k(M) \times_{\Sigma_k} A^k & \longrightarrow & C_k(M;A). \end{array}$$

Here we denote by $(C^0(M) \times_\tau Comp(A))_k$ the subspace of $C(M)_k \times_{\Sigma_k} A^k$ of those labelled configurations such that several points are concentrated in the same macroscopic location if and only if their labels are summable. The inclusion of the space on the left hand top corner of the first diagram into that of the second diagram is a homotopy equivalence, because r induces a common retraction onto a copy of the second space. The same holds for the spaces on the left hand bottom corner. We conclude by induction and the gluing lemma [4].

□

If we regard a pointed space (A, a_0) as a partial abelian monoid with $x + a_0 = x$ as the only defined sums, for $x \in A$, then $C^0(M;A)$ is the configuration space with labels studied in [3].

Corollary 5.20. *Let (A, a_0) be a well-pointed space. Then for any open n-manifold M there is a weak equivalence $C^0(M;A) \simeq C(M;A)$.*

Proof. The space A is a partial n-monoid by $Comp = \coprod_k F_n(k) \times_{\Sigma_k} \vee_{i=1}^k A$. □

For some background about toric varieties we refer to [7].

Corollary 5.21. *If V is a projective toric variety such that $H_2(V)$ is torsion free, then there exists a partial discrete abelian monoid Δ_V, such that the union of some components of $(\hat{\Delta_V})^2$ is homotopy equivalent to the space $Rat(V)$ of based rational curves on V.*

Proof. Guest has shown in [9] that if Δ_V is the fan associated to the variety V [7] then the union of some components of $C^0(\mathbb{R}^2; \Delta_V)$ is homeomorphic to $Rat(V)$. The corollary follows from the theorem and from proposition 5.16. □

Remark 5.22. *It is possible to define labelled configurations with support in a manifold with corners M. It is sufficient to choose an embedding $M \hookrightarrow M'$, with M' open, consider the right F_n-submodule $C(M) \hookrightarrow C(M')$ of configurations macroscopically located at points of M, and carry through the discussion as for open manifolds.*

6. The relative case

We define relative labelled configuration spaces on relative manifolds.

Let (X, x_0) be a pointed topological space. Let M be a manifold with corners and $N \hookrightarrow M$ a cofibration such that $M - N$ is an open manifold. We obtain easily from 3.2 that each element $c \in C(M; X)$ is uniquely determined by a finite set $S(c) \subset M$, and for each $P \in S(c)$ a labelled tree T_P as in 3.2, with the only difference that the twigs of the tree are labelled by $X - x_0$.

Definition 6.1. *The based space $C(M, N)(X)$ is the quotient $C(M; X)/\sim$ by the equivalence relation such that $a \sim a'$ if and only if $S(a) \cap (M-N) = S(a') \cap (M-N)$ and the trees indexed by these intersections coincide. The base point is the class $[a]$ such that $S(a) \subset N$.*

If we regard pointed spaces as partial n-monoids, then the n-monoid completion induces a monad (F_n^*, η_*, μ_*) on the category of pointed compactly generated spaces $\mathcal{CG}_*$. Each element in the completion $F_n^*(X) = \hat{X}^n$ is represented by a tree with vertex labels in F_n and twigs labels in $X - x_0$. The product μ_* is given by grafting of trees, and the unit η_* sends an element x to the trivial tree labelled by x.

Proposition 6.2. *If M is a parallelizable n-manifold, and $N \hookrightarrow M$ is a cofibration such that $M - N$ is open, then the functor $C(M, N)$ has a structure of right algebra over F_n^*.*

Proof. We need to exhibit a natural transformation $\lambda : C(M,N)F_n^* \to C(M,N)$ such that $\lambda \circ C(M,N)\eta_*$ is the identity and the diagram

$$\begin{array}{ccc} C(M,N)F_n^*F_n^* & \xrightarrow{C(M,N)\mu_*} & C(M,N)F_n^* \\ \downarrow{\scriptstyle \lambda F_n^*} & & \downarrow{\scriptstyle \lambda} \\ C(M,N)F_n^* & \xrightarrow{\lambda} & C(M,N). \end{array}$$

commutes. The morphism λ is obtained by grafting of trees. □

Definition 6.3. *If (A,ρ) is an n-monoid, and M,N are as before, then the space $C(M,N;A)$ of configurations in (M,N) with summable labels in A is the coequalizer*

$$C(M,N)F_n^*(A) \overset{C(M,N)\rho}{\underset{\lambda A}{\rightrightarrows}} C(M,N)A \dashrightarrow C(M,N;A)\,.$$

Definition 6.4. *A partial n-monoid A is good if the inclusion $Comp(A) \to F_n(A)$ is a cofibration, and the partial composition $\rho : Comp(A) \to A$ induces a map on the quotient $Comp^*(A) \subset F_n^*(A)$ of $Comp(A)$.*

The definition of a good framed partial n-monoid is similar. ¿From now on we will assume implicitly that all partial (framed) n-monoids are good.

By means of the framed Fulton-MacPherson operad we can define similarly $C(M,N;A)$, if M is an n-dimensional manifold with corners, $N \hookrightarrow M$ is a cofibration, and A is a good partial framed n-monoid, and as in 5.13 we obtain:

Proposition 6.5. *Define a filtration so that $[a] \in C_k(M,N;A)$ if and only if k is the number of twigs of trees in $S(a) \cap (M-N)$. Then $C(M,N;A)$ has the weak topology with respect to the filtration and it is compactly generated.*

Definition 6.6. *If A is a partial framed n-monoid, then $B_k(A) = C((I^k, \partial I^k) \times I^{n-k}; A)$ for $i = 1, \ldots, n$.*

If A is a partial abelian monoid and (M,N) is any pair then we define the relative labelled configuration space $C^0(M,N;A)$ as quotient of $C^0(M;A)$, by identifying configurations that coincide on $M-N$. We state the relative version of 5.19.

Proposition 6.7. *If M is a manifold, $N \hookrightarrow M$ is a cofibration, and $M-N$ is open, then there is a weak equivalence $C^0(M,N;A) \simeq C(M,N;A)$.*

Proof. The proof is similar to that of 5.19. In this case we use for each k a Σ_k-equivariant retraction $r_k : C_k(M) \to C_k^0(M)$ such that r_k preserves $b^{-1}(\overline{M^k - N^k})$, where $b : C_k(M) \to M^k$ is the blowdown. □

Corollary 6.8. *If V is a projective toric variety such that $H_2(V)$ is torsion free, with torus T and fan Δ_V, then there is a weak equivalence $B_2(\Delta_V) \simeq V \times_T ET$.*

Proof. Guest has shown in [9] that $V \times_T ET$ is homotopy equivalent to $C^0(I^2, \partial I^2; \Delta_V)$. □

The relative version of 5.20 is:

Corollary 6.9. *For any well-pointed space X there is a weak equivalence* $C^0(M, N; X) \simeq C(M, N; X)$.

Corollary 6.10. *Let X be a well-pointed space considered as partial n-monoid. Then there is a weak equivalence* $\Sigma^n(X) \xrightarrow{\simeq} B_n(X)$.

Proof. The space of open configurations $C^0(I^n, \partial I^n; X)$ retracts onto $\Sigma^n(X)$, considered as space of configurations of a single labelled point in $(I^n, \partial I^n)$. The retraction is achieved [5] by pushing radially the particles away onto the boundary. But the inclusion $C^0(I^n, \partial I^n; X) \hookrightarrow C(I^n, \partial I^n; X) = B_n(X)$ is a weak equivalence by 6.9. □

By means of configuration spaces we obtain the classifying space constructed by Stasheff.

Proposition 6.11. *Let (A, a_0) be a well-pointed A_∞ space. The quotient map* $C(I, \{0\}; A) \to C(I, \partial I; A) = B_1(A)$ *is canonically homeomorphic to the universal arrow* $E(A) \to B(A)$.

Proof. It is sufficient to carry out the discussion in the non-symmetric case: in fact $C_k([0,1]) = S_k([0,1]) \times \Sigma_k$, where $S_k([0,1])$ compactifies the space of strictly ordered maps from $\{1, \dots, k\}$ to $I = [0,1]$.

Let $S_k(I)\{0,1\} \subseteq S_k(I)$ be the closure of the subspace of maps $\alpha : \{1, \dots, k\} \to I$ such that $\alpha(1) = 0$, $\alpha(k) = 1$. Its elements are described by appropriate trees as in 3.2. For $k \geq 2$, we have homeomorphisms $r : S_k(I)\{0,1\} \leftrightarrows S_k(0) : j$, where $S_k(0)$ is the space of configurations in $\mathbb{R}$ macroscopically concentrated at the point 0.

If $\alpha_i \to \alpha \in S_k(I)\{0,1\}$, $\alpha_i \in C_k^0(I)$, then $r(\alpha) = \lim_i \frac{\alpha_i - \alpha_i(0)}{i(\alpha_i(1) - \alpha_i(0))}$.

If $\beta_i \to \beta \in S_k(0)$, $\beta_i \in C_k^0(\mathbb{R})$, then $j(\beta) = \lim_i \frac{\beta_i - \beta_i(0)}{\beta_i(1) - \beta_i(0)}$.

We have seen in 3.2 that $S_k(0) = K_k$ is the associahedron. Under the identification $K_k \cong S_k(I)\{0,1\}$ the Stasheff space $B(A)$ is defined to be the quotient of $\coprod S_k(I)\{0,1\} \times A^{k-2}$, seen as space of forests labelled by A, under the following steps:

1. We replace a tree on i twigs by a point having as label the action of the tree on its twigs via $K_i \times A^i \to A$.
2. We can cut twigs labelled by a_0.
3. We identify any two labelled forests coinciding outside 0 and 1.

But this quotient is exactly $B_1(A) = C(I, \partial I; A)$. In a similar way one shows that $E(A) = \coprod S_k(I)\{0,1\} \times A^{k-1} / \sim$ is homeomorphic to $C(I, \{0\}; A)$. In this case in 3 we identify forests coinciding outside 0. □

7. Approximation theorems

We say that a partial framed n-monoid A has homotopy inverse if the H-space $\hat{A}^n$ has homotopy inverse.

Lemma 7.1. *Let M be a connected compact n-manifold, $M' \subset M$ a compact n-submanifold, $N \subset M$ a closed submanifold, and A a partial framed n-monoid. Suppose that either A has a homotopy inverse or the pair $(M', N \cap M')$ is connected. Then there is a quasifibration*

$$C(M', N \cap M'; A) \longrightarrow C(M, N; A) \xrightarrow{\pi} C(M, M' \cup N; A).$$

This holds in particular if A is path connected.

Proof. We follow the proof of proposition 3.1 in [5]. The space $C(M, M' \cup N; A)$ has a filtration by $C_k := C_k(M, M' \cup N; A)$. There is a homeomorphism $\alpha_k : \pi^{-1}(C_k - C_{k-1}) \cong C(M', N \cap M'; A) \times (C_k - C_{k-1})$ such that $\pi\alpha_k^{-1}$ is the projection onto the factor $C_k - C_{k-1}$. Choose a collared neighbourhood U of M' in M and a smooth isotopy retraction $r : U \to M'$ such that $r(U \cap N) \subset N$. For each k there is an open neighbourhood U_k of C_k in C_{k+1} such that r induces a smooth isotopy retraction $r_k : U_k \times I \to C_k$, and a smooth isotopy retraction $\tilde{r}_k : \pi^{-1}(U_k) \times I \to \pi^{-1}(C_k)$ covering r_k. For any point $P \in U_k$ we need to show that the restriction $t : \pi^{-1}(P) \to \pi^{-1}(r_1(P))$ of $\tilde{r}_1$ is a weak homotopy equivalence. If we identify domain and range of t to $C(M', N \cap M'; A)$ by α_k, then t pushes the labelled particles away from N, and adds a finite set of trees T in proximity to N. But if the pair $(M', N \cap M')$ is connected, then the trees in T can be moved continuously to N, where they vanish, and t is homotopic to a homeomorphism. On the other hand, if A has a homotopy inverse, then t has a homotopy inverse that pushes the particles away from N and adds some homotopy inverses of the trees in T in proximity to N. □

Proposition 7.2. *Let A be a partial framed n-monoid. Then for $i = 1, \dots, n$ there are maps $s_i : B_{i-1}(A) \longrightarrow \Omega B_i(A)$, such that s_i is a weak homotopy equivalence for $i > 1$, and s_1 is a weak homotopy equivalence if A has a homotopy inverse.*

Proof. Note that $B_0(A)$ is homotopic to the framed n-monoid completion of A. For each i the base point of $B_i(A)$ is the empty configuration. The translation $\tau_1(t) : I^n \to \mathbb{R} \times I^{n-1}$ of the first coordinate by t induces a map $\pi_{\tau_1(t)} : B_0(A) \to B_1(A)$, composite of the induced map $C(I^n; A) \xrightarrow{C(\tau_i(t);A)} C(\mathbb{R} \times I^{n-1}; A)$ and the projection $C(\mathbb{R} \times I^{n-1}; A) \to C((I, \partial I) \times I^{n-1}; A)$. Then the 'scanning' map s_1 is defined for $x \in B_0(A) = C(I^n; A)$ by $s_1(x)(t) = \pi_{\tau_1(2t-1)}(x) \in B_1(A)$. For $i > 0$ the translation of the $(i+1)$-th coordinate by t induces similarly a map $\pi_{\tau_{i+1}(t)} : B_i(A) \to B_{i+1}(A)$, and $s_{i+1} : B_i(A) \to \Omega B_{i+1}(A)$ is given by $s_{i+1}(x)(t) = \pi_{\tau_{i+1}(2t-1)}(x)$. We define $M = I^k \times [0,2] \times I^{n-k-1}$, $N = (\partial I^k \times [0,2] \times I^{n-k-1}) \cup (I^k \times 0 \times I^{n-k-1})$, and we identify $B_k(A)$ to $C(I^k \times [1,2] \times I^{n-k-1}, \partial I^k \times [1,2] \times I^{n-k-1}; A)$

via $\tau_{k+1}(1)$. We consider for $1 \leq k \leq n-1$ a commutative diagram

$$\begin{array}{ccccc} B_k(A) & \longrightarrow & C(M,N;A) & \longrightarrow & B_{k+1}(A) \\ \downarrow{\scriptstyle s_{k+1}} & & \downarrow{\scriptstyle s} & & \| \\ \Omega B_{k+1}(A) & \longrightarrow & PB_{k+1}(A) & \rightarrow & B_{k+1}(A)\,. \end{array}$$

The top row is a quasifibration and the bottom row a fibration. The scanning map s is defined on the total space $C(M,N;A)$ by $s(x)(t) = \pi_{\tau_{k+1}(2t)}(x)$ and is consistent with s_{k+1}. Now the space $C(M,N;A)$ is contractible. In fact by excision $C(M,N;A) \cong C(M',N';A)$, with $M' = \mathbb{R}^k \times (-\infty,2] \times \mathbb{R}^{n-k-1}$ and $N' = M' - (M-N)$. Moreover there is a smooth isotopy $H_t : (M,N) \to (M',N')$, such that H_0 is the inclusion and $H_1(M) \subset N'$. For example define H_t as the dilatation by $3t$ centered in $(\frac{1}{2},\ldots,\frac{1}{2},3,\frac{1}{2},\ldots,\frac{1}{2})$, with 3 at the $(k+1)$-st position. We conclude by comparing the long exact sequences in homotopy and by induction on k. □

The spaces $B_0(A) = C(I^n;A)$ and $\Omega^n B_n(A)$ are both fD_n-algebras. The map $s : B_0(A) \to \Omega^n B_n(A)$ constructed by looping and composing the scanning maps in proposition 7.2 can be extended to a homotopy fD_n-morphism by rescaling suitably the scanning maps on the labels of trees in $U_{fD_n}(B_0(A))$. By 7.2 we obtain:

Theorem 7.3. *If A is a partial framed n-monoid, then $s : B_0(A) \to \Omega^n B_n(A)$ is the group completion. If A has homotopy inverse, then s is a weak homotopy equivalence.*

Actually s is the group completion in the homotopy category of fD_n-algebras.

Corollary 7.4 ([15]). *If X is a well-pointed space, then $s : C^0(\mathbb{R}^n;X) \to \Omega^n\Sigma^n X$ is the group completion. If X is path connected, then s is a weak homotopy equivalence.*

Proof. Consider X as a partial n-monoid as in corollary 5.20. Now $B_0(A) = C(I^n;A) \simeq \hat{A}^n$ by the same argument of proposition 5.16 . Moreover $\hat{A}^n \simeq C(\mathbb{R}^n;A)$ by 5.16, $C(\mathbb{R}^n;A) \simeq C^0(\mathbb{R}^n;A)$ by 5.20 and $\Sigma^n X \simeq B_n(X)$ by 6.10. Now we can apply the theorem. □

Corollary 7.5 ([9]). *If V is a projective toric variety such that $H_2(V)$ is torsion free, then $s : Rat(V) \to \Omega^2(V)$ is the group completion.*

Proof. Apply corollaries 5.21 and 6.8, and restrict to the relevant components. □

Given an n-manifold M, and its tangent bundle τ, there is a bundle $\gamma = C(\tau,\partial\tau;A)$ on M with fiber $B_n(A) = C(I^n,\partial I^n;A)$, consisting of relative fiberwise configurations in the fiberwise one-point compactification modulo the section at infinity $(\hat{\tau},\infty)$. Whether ∂M is empty or not we can define a map s :

$C(M, \partial M; A) \to \Gamma(M; B_n A)$ to the space of sections of γ. Note that if M is parallelizable then $\Gamma(M; B_n A) = Map(M; B_n A)$. The scanning map s is constructed by the exponential map: if $x \in C(M, \partial M; A)$, then $s(x)$ sends a point $P \in M$ to the restriction of x to a small disc neighbourhood of P modulo its boundary.

Theorem 7.6. *Let A be a partial framed n-monoid. Let M be a compact connected n-manifold with boundary. Then the scanning map $s : C(M, \partial M; A) \to \Gamma(M; B_n A)$ is a weak homotopy equivalence. If A has homotopy inverse and N is a compact connected n-manifold without boundary then $s : C(N; A) \to \Gamma(N; B_n A)$ is a weak homotopy equivalence.*

Proof. We follow the proof of 10.4 in [10]. There is a finite handle decomposition of M with no handles of index n. If M' is obtained from M'' by attaching a handle H of index i , then we apply lemma 7.1 and we obtain a quasifibration $C(H, \overline{\partial H - \partial H \cap M''}; A) \to C(M', \partial M'; A) \to C(M'', \partial M''; A)$. On the other hand we have a fibration $\Gamma(H/(H \cap M''); B_n A) \to \Gamma(M'; B_n A) \to \Gamma(M''; B_n A)$. But $C(H, \overline{\partial H - \partial H \cap M''}; A) \cong B_{n-i}(A)$, and $\Gamma(H/(H \cap M''); B_n A) \simeq \Omega^i B_n(A)$. We compare the two sequences by the scanning maps and we conclude by proposition 7.2 and induction on the number of handles. In the case of N we have even a handle of index n and we apply the second part of proposition 7.2. □

Corollary 7.7. *If X is a well-pointed path connected space then $s : C(S^1; X) \to Map(S^1, \Sigma X)$ is a weak homotopy equivalence.*

Proof. We consider X as partial framed 1-monoid as in corollary 5.20. By corollary 6.10 $B_1(X) \simeq \Sigma X$. We apply the second part of the theorem, and note that $\Gamma(S^1, \Sigma X) \simeq Map(S^1, \Sigma X)$ because S^1 is parallelizable. □

This answers a question raised by Stasheff in [22] p. 10. The analogous result for $C^0(S^1; X)$ is in [3].

Any partial framed n-monoid gives an approximation theorem for mapping spaces, and the homotopy theorist is tempted to discover new examples. It might be worth considering colimits of abelian monoids in the category of n-monoids.

References

[1] J.M. Boardman, *Homotopy structures and the language of trees*, Collection: Algebraic Topology, Vol. XXII, AMS, 1971, 37–58.

[2] J.M. Boardman and R.M. Vogt, Homotopy invariant algebraic structures on topological spaces, LNM 347, 1973.

[3] C-F. Boedigheimer, *Stable splittings of mapping spaces*, LNM 1286, 1987, 174–187.

[4] R. Brown, A geometric account of general topology, homotopy types and the fundamental groupoid. Ellis Horwood ltd., Chichester, 1988.

[5] D. McDuff, *Configuration spaces of positive and negative particles*, Topology 14 (1975), 91–107.

[6] W. G. Dwyer, P. S. Hirschhorn and D. M. Kan, Model categories and more general abstract homotopy theory, preprint.

[7] W. Fulton, Introduction to toric varieties, Ann. of Math. Stud. 131, Princeton Univ. Press, Princeton, N.J., 1993.

[8] E. Getzler and J.D.S. Jones,*Operads, homotopy algebra, and iterated integrals for double loop spaces*, Preprint hep-th/9403055.

[9] M. Guest, *The topology of the space of rational curves on a toric variety*, Acta Math. 174 (1995), no. 1, 119–145.

[10] S. Kallel, *Spaces of particles on manifolds and generalized Poincaré dualities*, Preprint math/9810067.

[11] F. Kato, On spaces realizing intermediate stages of the Hurewicz map, Master's thesis, Department of Mathematics, Shinshu University, 1996.

[12] M. Kontsevich, *Feynman diagrams and low-dimensional topology*, First European Congress of Mathematicians, Vol. II, 97–121, Birkhäuser 1994.

[13] S. MacLane, Categories for the working mathematician, Springer, 1971.

[14] M. Markl, *A compactification of the real configuration space as an operadic completion*, Journal of Algebra 215 (1999), 185–204.

[15] J.P. May, The geometry of iterated loop spaces, LNM 271, 1972.

[16] D. Quillen, Homotopical algebra, LNM 43, 1967.

[17] C. Rezk, Spaces of algebra structures and cohomology of operads, Ph. D. Thesis, MIT, 1996.

[18] P. Salvatore, Configuration operads, minimal models and rational curves, D. Phil. Thesis, Oxford, 1998.

[19] R. Schwänzl and R.M. Vogt, *The categories of $A_\infty-$ and E_∞-monoids and ring spaces as closed simplicial and topological categories*, Arch. Math. 56 (1991), 405–411.

[20] G. Segal, *Configuration spaces and iterated loop spaces*, Invent. Math. 21 (1973), 213–221.

[21] G. Segal, *The topology of rational functions*, Acta Math. 143 (1979), 39–72.

[22] J. Stasheff, *From operads to "physically" inspired theories*, Proceedings of Renaissance Conference, Cont. Math. 202, 53–81, AMS, 1997.

[23] N.E. Steenrod, *A convenient category of topological spaces*, Mich. Math. J. 14 (1967), 133–152.

Mathematisches Institut,
Universität Bonn,
Beringstrasse 1
53115 Bonn, Germany
E-mail address: salvator@math.uni-bonn.de

Progress in Mathematics, Vol. 196

Kaleidoscoping Lusternik–Schnirelmann category type invariants

Hans Scheerer and Daniel Tanré

Abstract. A general LS-category type invariant is defined as a function of two variables. It specializes to (old and new) relative invariants and generates strong category notions. Each invariant is determined by an axiom scheme whose form has been established by Lusternik and Schnirelmann. We also discuss the corresponding absolute invariants and formulate, in a general category $\mathcal{C}$, a method to define LS-category and strong LS-category concepts. They specialize to the usual ones and it turns out that some relative invariants are absolute invariants in the new sense.

1. Introduction

The original concept of category invented by Lusternik and Schnirelmann in [12] was one for maps that can be characterized by a simple axiom scheme. Meanwhile many variations of the first concept have been introduced (we refer to the survey articles [10], [11]). Today, the original invariant is called a *relative* one.

In the present paper, we give a "general" invariant which is a function on the class of pairs of maps (f, g) having a fixed target B. In fact, its definition is based on the question when does g factor through f', if f' is obtained from f by successively applying certain constructions. We show in a kaleidoscoping manner that this invariant specializes to many known and also some new invariants. And we like to emphasize that each variant is determined by its axiom scheme. We also introduce relative strong category notions which are governed by slightly different axiom systems. It has to be observed, moreover, that any other factorization problem similarly leads to corresponding other "general invariants".

We then discuss absolute versions of the relative concepts. On the way we are led to formulate a general method to define LS-category and strong LS-category concepts in any category $\mathcal{C}$. And it turns out that some relative invariants are absolute ones in that sense.

The paper is organized as follows. In Section 2 we set up the general LS-invariant. A first specialization to a relative invariant is given in Section 3. A second one related to the category in the sense of Clapp and Puppe [1] appears in Section 4 as well as a notion of strong relative category. In Section 5 we discuss the transition to the absolute case and a general method to obtain invariants in any category $\mathcal{C}$.

In an Appendix, we consider a mapping theorem for some of the relative category concepts which specializes to the mapping theorem of [7], [8].

In Sections 2, 3 and 4, we work in the category $\mathcal{T}_*$ of pointed spaces; the symbols $mor(\mathcal{T}_*)$, $mor(-,B)$ denote the class of morphisms of $\mathcal{T}_*$, resp. the class of morphisms of $\mathcal{T}_*$ with target B. We recall that $\mathcal{T}_*$ is a closed model category with homotopy equivalences as weak equivalences and Hurewicz fibrations as fibrations. Moreover it is also a J-category in the sense of [3]. All we do in these sections could also be performed in any other model category (with J-structure if necessary). Any map $f\colon X \to Y$ of $\mathcal{T}_*$ factors as a weak equivalence followed by a fibration which we call the associated fibration $f'\colon X' \to Y$.

2. The general invariant

We fix $B \in \mathcal{T}_*$.

Notation 2.1. Let $f\colon X \to B$, $g\colon Y \to B$ be elements of $mor(-,B)$. We write $f \leq g$ if there exists a homotopy commutative diagram

$$\begin{array}{ccc} X & \longrightarrow & Y \\ {\scriptstyle f}\searrow & & \swarrow{\scriptstyle g} \\ & B. & \end{array}$$

Let $\mathcal{K}$, ψ be classes of constructions, i.e., functions

$$mor(-,B) \to \{\text{subclasses of } mor(-,B)\}\,.$$

Note that classes of constructions can be composed in the sense of correspondences.

Definition 2.2. Given f, $g \in mor(-,B)$ we set

$$(\mathcal{K},\psi)\text{-}cat(f,g) = \inf\{r \mid \exists f' \in \mathcal{K}^r(\psi(f)) \text{ such that } g \leq f'\}\,.$$

Definition 2.3. A class of constructions $\mathcal{K}$ is *reasonable* if for all f, $g \in mor(-,B)$ with $f \leq g$ and $f' \in \mathcal{K}(f)$ there is $g' \in \mathcal{K}(g)$ with $f' \leq g'$.

Proposition 2.4. (1) *If $g_1 \leq g_2$, then $(\mathcal{K},\psi)\text{-}cat(f,g_1) \leq (\mathcal{K},\psi)\text{-}cat(f,g_2)$.*

(2) *Let $\mathcal{K}$ be reasonable, let f, $g \in mor(-,B)$ and $g' \in \mathcal{K}(g)$. Then*

$$(\mathcal{K},\psi)\text{-}cat(f,g') \leq (\mathcal{K},\psi)\text{-}cat(f,g) + 1.$$

(3) *Let $\mathcal{K}$, ψ be reasonable and $f_1 \leq f_2$. Then*

$$(\mathcal{K},\psi)\text{-}cat(f_2,g) \leq (\mathcal{K},\psi)\text{-}cat(f_1,g).$$

Proof. (1) Suppose g_2 factors through a map in $\mathcal{K}^r(\psi(f))$. Then so does g_1.

(2) If g factors through a map in $\mathcal{K}^r(\psi(f))$, then g' factors through a map in $\mathcal{K}^{r+1}(\psi(f))$.

(3) Left as exercise. □

Theorem 2.5. *Let $\mathcal{K}$ be reasonable. Then the function $(\mathcal{K},\psi)$-cat satisfies the following three properties and is the maximum of all functions $n\colon mor(-,B)\times mor(-,B)\to \mathbb{N}\cup\{\infty\}$ satisfying these:*

(i) *$n(f,g)=0$ if there is $\alpha\in\psi(f)$ such that $g\le\alpha$;*
(ii) *for all f and $g_1\le g_2$, one has $n(f,g_1)\le n(f,g_2)$;*
(iii) *for all f, $g\in mor(-,B)$ and $g'\in\mathcal{K}(g)$, $n(f,g')\le n(f,g)+1$ holds.*

Proof. By definition and Proposition 2.4, $(\mathcal{K},\psi)$-cat satisfies (i), (ii), (iii).

Let now $(\mathcal{K},\psi)\text{-}cat(f,g)=r$ and choose $\alpha\in\psi(f)$, $f'\in\mathcal{K}^r(\alpha)$ such that $g\le f'$. Then $n(f,g)\le n(f,f')\le n(f,\alpha)+r=r$. □

Remark 2.6. Property (ii) is the analogue of the "domination property" for the usual LS-category: If Y dominates X, i.e., there are maps $X\xrightarrow{i}Y\xrightarrow{\pi}X$ with $\pi\circ i\sim id_X$, then $cat(X)\le cat(Y)$.

Convention. *We will always assume that $\mathcal{K}$ is reasonable.*

Here is a trivial example:

Example 2.7. Let $\mathcal{K}=id$ and $\psi=id$. Then

$$(\mathcal{K},\psi)\text{-}cat(f,g)=\begin{cases}0 & \text{if } g\le f,\\ \infty & \text{else.}\end{cases}$$

Remark 2.8. Consider the following factorization problem: Given a homotopy commutative square, simply denoted by $(f;g)$,

$$\begin{array}{ccc} E & \overset{v}{\leftarrow} & Y\\ {\scriptstyle f}\downarrow & & \downarrow{\scriptstyle g}\\ B & \underset{u}{\leftarrow} & X,\end{array}$$

does there exists a map $w\colon X\to E$ (called *solution*) such that $f\circ w\sim u$ and $w\circ g\sim v$?

View $mor(-,B)$ as the category of objects over B and let $\mathcal{K}$ be a class of functors $T\colon mor(-,B)\to mor(-,B)$ admitting natural transformations $id\to T$. Then we may define

$$\mathcal{K}\text{-}cat(f;g)=\inf\{r\mid \exists f'\in\mathcal{K}^r(f) \text{ with a solution } w' \text{ for the problem } (f';g)\}$$

where $(f';g)$ is the square obtained from $(f;g)$ using repeatedly the natural transformations as follows:

$$\begin{array}{ccccc} \tilde{E} & \leftarrow & E & \leftarrow & Y\\ {\scriptstyle T(f)}\downarrow & & {\scriptstyle f}\downarrow & & \downarrow{\scriptstyle g}\\ B & = & B & \leftarrow & X.\end{array}$$

The axiomatic description of this invariant takes a slightly different form:

The function $\mathcal{K}$-cat on the class of squares $(f;g)$ satisfies the following properties and is the maximum of all functions satisfying these:

(i) $n(f;g) = 0$ if the factorization problem $(f;g)$ has a solution;
(ii) $n(f';g) \geq n(f;g) - 1$ for all $f' \in \mathcal{K}(f)$.

As an application we just mention the invariant on $mor(-, B)$ given by $f \mapsto \mathcal{K}\text{-}cat(f; f)$ where $\mathcal{K}$ consists of the one functor assigning to f its join with $* \to B$ over B. This invariant is used in [2].

3. First specialization to a notion of relative LS-category

We choose $\psi = id$ and delete ψ from the notation, i.e., we obtain a function $\mathcal{K}\text{-}cat(f, g)$.

Definition 3.1. Given $f \in mor(-, B)$, we set $\mathcal{K}\text{-}cat(f) = \mathcal{K}\text{-}cat(f, id_B)$.

We leave it as an exercise to characterize the function $\mathcal{K}\text{-}cat$ on $mor(-, B)$ by the axiomatization scheme.

Example 3.2. Let $\mathcal{L}$ be a class of maps in $mor(-, B)$. Denote by $\mathcal{L}*$ the class of constructions $f \mapsto \{f *_B \rho \mid \rho \in \mathcal{L}\}$ (for a definition of the join $*_B$ see the Appendix).

(i) Take $\mathcal{L} = \{* \to B\}$. Then $\mathcal{L} * \text{-}cat(* \to B) = cat(B)$ (where $cat(B)$ is calculated according to Ganea [9]). To see this we note that performing the construction r times starting from $* \to B$ gives the r-th Ganea space of B.

(ii) Let $\mathcal{L}$ be the class of spherical fibrations over B. Then

$$\mathcal{L} * \text{-}cat(f) = \begin{cases} 0 & \text{if } f \text{ has a section}, \\ 1 & \text{else.} \end{cases}$$

Example 3.3. Let $\mathcal{G} \subset mor(\mathcal{T}_*)$ be a subclass and for each $B \in \mathcal{T}_*$ let $\mathcal{G}(B) = \{f \in \mathcal{G} \mid target(f) = B\}$. Suppose moreover that $\mathcal{G}$ satisfies the following property:

$$\boxed{\text{M}} \text{ For all } g\colon B \to C \in mor(\mathcal{T}_*), \quad g_* (\mathcal{G}(B)) \subset \mathcal{G}(C).$$

To such a class $\mathcal{G}$ we shall associate a **modified Ganea construction:** Given $f\colon X \to B$, let $f'\colon E \to B$ be an associated fibration with fibre F, choose $\alpha \in \mathcal{G}(F)$ and define $g\colon E \cup_F C(\alpha) \to B$ by $g|E = f$, $g|C(\alpha) = *$ (where $C(\alpha)$ denotes the mapping cone of α).

Define now the class of constructions $\mathcal{K}(\mathcal{G})$ by

$$\mathcal{K}(\mathcal{G})(f) = \{g \mid g \text{ obtained by a modified Ganea construction from } f\}.$$

The condition $\boxed{\text{M}}$ ensures that $\mathcal{K}(\mathcal{G})$ is reasonable. We thus have the function $\mathcal{K}(\mathcal{G})\text{-}cat(f, g)$ and obtain the relative invariant $\mathcal{K}(\mathcal{G})\text{-}cat(f)$.

Remark 3.4. The important examples for such $\mathcal{G}$ are obtained as follows: Let $\mathcal{A} \subset Obj(\mathcal{T}_*)$ be a subclass and define

$$\mathcal{G}_{\mathcal{A}}(B) = \{f\colon A \to B \mid f \in mor(A, B),\ A \in \mathcal{A}\}.$$

We think that $\mathcal{K}(\mathcal{G}_{\mathcal{A}})\text{-}cat(f)$ may be an interesting invariant, just as the corresponding absolute invariant $X \mapsto \mathcal{K}(\mathcal{G}_{\mathcal{A}})\text{-}cat(* \to X)$ (compare Section 5). Indeed, if $\mathcal{A} = \mathcal{T}_*$, then $\mathcal{K}(\mathcal{G}_{\mathcal{A}})\text{-}cat(* \to X) = cat(X)$.

4. Relative invariants according to Clapp and Puppe

4.1. General case

We assume now that the function ψ is constant, i.e., we may look at ψ as being a subclass of $mor(-, B)$, B fixed again.

Definition 4.1. Given $g \in mor(-, B)$, we set $(\mathcal{K}, \psi)\text{-}cat(g) = (\mathcal{K}, \psi)\text{-}cat(id_B, g)$.

Example 4.2. As in Example 3.2, let $\mathcal{L} \subset mor(-, B)$ be a class. Then $(\mathcal{L}*, \mathcal{L})\text{-}cat(g)$ is the relative invariant defined by Clapp and Puppe in [1] for the case that $\mathcal{L} = \mathcal{G}_{\mathcal{A}}(B)$ (for $\mathcal{G}_{\mathcal{A}}$ see example 3 above).

Example 4.3. Let $\mathcal{G}$ be as in Example 3.3. Then we have the invariant

$$(\mathcal{K}(\mathcal{G}), \mathcal{G}(B))\text{-}cat(f).$$

In particular, $(\mathcal{K}(\mathcal{G}), \mathcal{G}(B))\text{-}cat(id_B)$ is the absolute invariant considered in [13] in case $\mathcal{G} = \mathcal{G}_{\mathcal{A}}$.

4.2. Strong relative category

Notation 4.4. Let g_1, $g_2 \in mor(-, B)$. Then we write $g_1 \sim g_2$ if there is a homotopy equivalence h with $g_2 \circ h \sim g_1$.

Definition 4.5. Given $g \in mor(-, B)$, set

$$(\mathcal{K}, \psi)\text{-}Cat(g) = \inf\{r \mid g \sim g' \text{ for some } g' \in \mathcal{K}^r(\psi)\}.$$

This time we will write down the axiom scheme for $(\mathcal{K}, \psi)\text{-}cat$ and for $(\mathcal{K}, \psi)\text{-}Cat$ to emphasize that it is the analogue of the corresponding scheme for usual cat and Cat (see e.g. [13]).

Theorem 4.6. *The function $(\mathcal{K}, \psi)\text{-}cat$ on $mor(-, B)$ satisfies the following three properties and is the maximum of all functions $n\colon mor(-, B) \to \mathbb{N} \cup \{\infty\}$ satisfying these:*

(i) *$n(g) = 0$ if $g \leq \alpha$ for some $\alpha \in \psi$;*
(ii) *if $g_1 \leq g_2$, then $n(g_1) \leq n(g_2)$;*
(iii) *for any $g' \in \mathcal{K}(g)$ one has $n(g') \leq n(g) + 1$.*

Proof. Apply Theorem 2.5. □

Theorem 4.7. *The function $(\mathcal{K}, \psi)\text{-}Cat$ on $mor(-, B)$ satisfies the following three properties and is the maximum of all functions $n\colon mor(-, B) \to \mathbb{N} \cup \{\infty\}$ satisfying these:*

(i) *$n(g) = 0$ if $g \sim \alpha$ for some $\alpha \in \psi$;*
(ii) *if $g_1 \sim g_2$, then $n(g_1) = n(g_2)$;*
(iii) *for any $g' \in \mathcal{K}(g)$ one has $n(g') \leq n(g) + 1$.*

Proof. We show maximality: Let $(\mathcal{K}, \psi)\text{-}Cat(g) = r$ and choose $g' \in \mathcal{K}^r(\psi)$, $g \sim g'$. Then $n(g') \leq n(\alpha) + r = r$, for some $\alpha \in \psi$. □

Remark 4.8. Obviously $(\mathcal{K},\psi)\text{-}cat \leq (\mathcal{K},\psi)\text{-}Cat$ and the problem arises to say anything about the difference.

Example 4.9. Let $\mathcal{K} = id$ and ψ be the class of spherical fibrations over B. Then $(\mathcal{K},\psi)\text{-}cat(g) = 0$ if g factors through a spherical fibration and

$$(\mathcal{K},\psi)\text{-}Cat(g) = \begin{cases} 0 & \text{if } g \sim g' \text{ with } g' \text{ spherical,} \\ \infty & \text{else.} \end{cases}$$

Hence, $(\mathcal{K},\psi)\text{-}cat(B \times (S^1 \vee S^1) \to B) = 0$ and its strong category is infinite.

5. Absolute invariants

An absolute LS-invariant should be a function $Obj(\mathcal{T}_*) \to \mathbb{N} \cup \{\infty\}$ which can be characterized by a simple axiom system in our favorite form. To obtain an absolute invariant from a relative one we need classes of constructions $\mathcal{K},\psi\colon mor(\mathcal{T}_*) \to$ {subclasses of $mor(\mathcal{T}_*)$} such that

$$\boxed{*} \text{ for } f \in mor(-,B),\ \mathcal{K}(f),\ \psi(f) \subset mor(-,B).$$

We then obtain classes of constructions

$$\mathcal{K}(B),\psi(B)\colon mor(-,B) \to \{\text{subclasses of } mor(-,B)\}$$

as we need to define

$$(\mathcal{K},\psi)\text{-}cat(B) = (\mathcal{K}(B),\psi(B))\text{-}cat(* \to B, id_B).$$

The choice of the pair $(* \to B, id_B)$ does make sense in the familiar examples.

We will give a solution to the problem of finding conditions on $\mathcal{K}$, ψ such that this invariant can be described by a simple axiom scheme (as an absolute invariant). However, in analogy to the relative situation we may set up a general method to define absolute LS-category notions. This should be done in arbitrary categories, for it will then be possible to interprete some relative invariants in $\mathcal{T}_*$ as absolute ones in $(\mathcal{T}_*)_B$. It is this scheme we apply in 5.2.

5.1. Strong category and category in a category $\mathcal{C}$

Let $\mathcal{C}$ be a category, and ψ_0, $\mathcal{K}_0$ be classes of constructions

$$Obj(\mathcal{C}) \to \{\text{subclasses of } Obj(\mathcal{C})\}.$$

Definition 5.1. Given $X \in Obj(\mathcal{C})$, we set

$$(\mathcal{K}_0,\psi_0)\text{-}Cat(X) = \inf\{r \mid \exists Y \in \mathcal{K}_0^r(\psi_0(X)),\ Y \cong X\}.$$

Suppose that in addition there is given a class γ of morphisms (called *special*).

Definition 5.2. Given $X \in Obj(\mathcal{C})$, we set

$$(\mathcal{K}_0,\psi_0,\gamma)\text{-}cat(X) = \inf\{r \mid \exists Y \in \mathcal{K}_0^r(\psi_0(X)) \text{ and } \exists(\sigma\colon X \to Y) \in \gamma\}.$$

The most common choice of special morphisms is the class of morphisms $i\colon X \to Y$ such that there exists $r\colon Y \to X$ with $r \circ i = id_X$. In that case we drop γ from the notation.

Remark 5.3. Suppose that in Definition 5.2 we let γ be the class of isomorphisms then we just obtain Definition 5.1; thus we may consider the definition of Cat as a particular case of that of cat.

Example 5.4. Let $\mathcal{C} = Ho\text{-}\mathcal{T}_*$, $\psi_0 = \{*\}$ and $\mathcal{K}_0(X)$ the class of mapping cones of $f\colon Y \to X$, $Y \in \mathcal{T}_*$, then $(\mathcal{K}_0, \psi_0)\text{-}cat$ and $(\mathcal{K}_0, \psi_0)\text{-}Cat$ agree with the usual LS-invariants.

Example 5.5. Let $\mathcal{C} = Ho\text{-}\mathcal{T}_*$, let $\mathcal{A} \subset Obj\,(\mathcal{T}_*)$, $\psi_0 = \mathcal{A}$, $\mathcal{K}_0(X) = \{$double mapping cylinders $X \leftarrow Y \to A,\ A \in \mathcal{A}\}$, then $(\mathcal{K}_0, \psi_0)\text{-}cat$ and $(\mathcal{K}_0, \psi_0)\text{-}Cat$ agree with the invariants introduced by Clapp and Puppe [1, Propositions 5-4 and 5-5].

Example 5.6. Let $\mathcal{C} = Ho\text{-}\mathcal{T}_*$, let $\mathcal{A} \subset Obj\,(\mathcal{T}_*)$, $\psi_0 = \mathcal{A}$, $\mathcal{K}_0(X)$ the class of mapping cones of $f\colon A \to X$, $A \in \mathcal{A}$, then $(\mathcal{K}_0, \psi_0)\text{-}cat$ and $(\mathcal{K}_0, \psi_0)\text{-}Cat$ agree with the invariants introduced in [13, Definition 3].

Example 5.7. Let $\mathcal{C} = Ho\text{-}\mathcal{T}_*$, let $\psi_0 = \{*\}$, let $\mathcal{K}_0(X)$ be the class of total spaces of spherical fibrations over X. Then e.g.

$$(\mathcal{K}_0, \psi_0)\text{-}cat\,(U(n)) = (\mathcal{K}_0, \psi_0)\text{-}Cat\,(U(n)) = n$$

(where $U(n)$ denotes the unitary group in n variables).

Example 5.8. Let $\mathcal{C}$ be the category of modules, ψ_0 the class of free modules and $\mathcal{K}_0 = id$, then, for a module M, the two invariants are infinite except

$(\mathcal{K}_0, \psi_0)\text{-}Cat(M) = 0$ if M is free,
$(\mathcal{K}_0, \psi_0)\text{-}cat(M) = 0$ if M is projective.

Example 5.9. Let $\mathcal{C}$ be the category of differentiable manifolds, $\psi_0 = \{*\}$, $\mathcal{K}_0(X) = X \times \mathbb{R}$, $\gamma = \{$embeddings$\}$. Then $(\mathcal{K}_0, \psi_0, \gamma)\text{-}cat(X) \le m$ if X embeds in $\mathbb{R}^m$.

Example 5.10. Let $\mathcal{C} = Ho\text{-}\mathcal{T}_*$. Set $\mathcal{K}_0(X) = \psi_0(X) = \{\Omega X, \Sigma X\}$ (where Ω, Σ denote the loop space and reduced suspension respectively). By [5], there are non-trivial spaces X with $(\mathcal{K}_0, \psi_0)\text{-}Cat(X) < \infty$.

We state characterization schemes for $(\mathcal{K}_0, \psi_0, \gamma)\text{-}cat$ and $(\mathcal{K}_0, \psi_0)\text{-}Cat$ under some hypotheses (which are fulfilled in all examples of this section except the last one) and leave the proof to the kaleidoscope viewer.

Definition 5.11. The triple $(\mathcal{K}_0, \psi_0, \gamma)$ is *reasonable* if

(i) for each $(\sigma\colon X \to Y) \in \gamma$ the relation $\psi_0(X) \supset \psi_0(Y)$ holds and if $X' \in \mathcal{K}_0(X)$, there exists $Y' \in \mathcal{K}_0(Y)$ and $(\sigma'\colon X' \to Y') \in \gamma$, and
(ii) for each X and $X' \in \mathcal{K}_0(X)$ one has $\psi_0(X') \supset \psi_0(X)$, and
(iii) for each X and $X' \in \psi_0(X)$ there is $Z \in \psi_0(X')$ and $X' \to Z$ in γ, and
(iv) γ is closed under composition.

Theorem 5.12. *Let $(\mathcal{K}_0, \psi_0, \gamma)$ be reasonable. Then the function $(\mathcal{K}_0, \psi_0, \gamma)$-cat satisfies the following three properties and is the maximum of all functions $n\colon Obj(\mathcal{C}) \to \mathbb{N} \cup \{\infty\}$ satisfying these:*

(i) *$n(X) = 0$ if there is $(\sigma\colon X \to Y) \in \gamma$ with $Y \in \psi_0(X)$;*
(ii) *given $(\sigma\colon X \to X') \in \gamma$, then $n(X) \leq n(X')$;*
(iii) *given $X' \in \mathcal{K}_0(X)$, then $n(X') \leq n(X) + 1$ holds.*

Theorem 5.13. *Suppose that the triple $(\mathcal{K}_0, \psi_0, \iota)$ where ι is the class of isomorphisms is reasonable. Then the function $(\mathcal{K}_0, \psi_0)$-Cat satisfies the following properties and is the maximum of all functions $n\colon Obj(\mathcal{C}) \to \mathbb{N} \cup \{\infty\}$ satisfying these:*

(i) *$n(X) = 0$ if $X \cong Y \in \psi_0(X)$;*
(ii) *if $X \cong X'$, then $n(X) = n(X')$;*
(iii) *given $X' \in \mathcal{K}_0(X)$, then $n(X') \leq n(X) + 1$ holds.*

5.2. From relative category notions towards absolute ones

Let $\mathcal{K}$, ψ be classes of constructions $mor(\mathcal{T}_*) \to \{\text{subclasses of } mor(\mathcal{T}_*)\}$ as at the beginning of this section. We obtain classes of constructions $\mathcal{K}_0$, $\psi_0\colon Obj(\mathcal{T}_*) \to \{\text{subclasses of } Obj(\mathcal{T}_*)\}$ as follows.

Definition 5.14. Set $\psi_0(X)$ equal to the class of domains A where $\alpha\colon A \to X$ belongs to $\psi(X)(* \to X)$. For $X \in \mathcal{T}_*$, let $\mathcal{K}_0(X)$ be the class of X' such that there exist $f\colon X \to Y$ and $f'\colon X' \to Y$ belonging to $\mathcal{K}(Y)(f)$.

Thus we have the invariants $(\mathcal{K}_0, \psi_0)$-*cat* and $(\mathcal{K}_0, \psi_0)$-*Cat* as above (recall that we continue to work in Ho-$\mathcal{T}_*$). The invariant $(\mathcal{K}, \psi)$-$cat(* \to X, id_X)$ will be abbreviated by $\widetilde{cat}(X)$.

Proposition 5.15. *The inequality $(\mathcal{K}_0, \psi_0)$-$cat(X) \leq \widetilde{cat}(X)$ holds for all $X \in \mathcal{T}_*$.*

Proof. Let $\widetilde{cat}(X) = n$, $n < \infty$. Then there exists $f' \in \mathcal{K}^n(\psi(* \to X))$ with homotopy section. By definition, the domain X' of f' has $(\mathcal{K}_0, \psi_0)$-$Cat(X') \leq n$, hence $(\mathcal{K}_0, \psi_0)$-$cat(X) \leq n$. □

To get the reverse inequality we need many more assumptions on $\mathcal{K}$ and ψ:

Definition 5.16. We call $\mathcal{K}$, ψ *reasonable in the extended sense* if the following conditions are satisfied:

(i) Given $u\colon X \to Y$ and $\alpha\colon A \to X$ in $\psi(X)(* \to X)$, then there exists $\beta\colon B \to Y$ in $\psi(Y)(* \to Y)$ such that $u \circ \alpha \leq \beta$.
(ii) Similarly, given $u\colon X \to Y$, $f\colon U \to X$ and $g\colon V \to Y$ with $u \circ f \leq g$, given $f' \in \mathcal{K}(X)(f)$, then there exists $g' \in \mathcal{K}(Y)(g)$ such that $u \circ f' \leq g'$.

Theorem 5.17. *Let $\mathcal{K}$, ψ be reasonable in the extended sense. Suppose that for all $X \in \mathcal{T}_*$ the following is true:*

(i) *For all $A \in \psi_0(X)$, $\widetilde{cat}(A) = 0$;*
(ii) *for all $f\colon X \to Y$ and $(f'\colon X' \to Y) \in \mathcal{K}(Y)(f)$, $\widetilde{cat}(X') \leq \widetilde{cat}(X) + 1$.*

Then $(\mathcal{K}_0, \psi_0)$-$cat = \widetilde{cat}$.

Proof. It remains to show that the equation $(\mathcal{K}_0, \psi_0)\text{-}cat(X) = n$ implies that $\widetilde{cat}(X) \leq n$. If $(\mathcal{K}_0, \psi_0)\text{-}cat(X) = n$, then X is a retract up to homotopy of X' with $(\mathcal{K}_0, \psi_0)\text{-}Cat(X') = n$, i.e., $X' \sim X''$ where X'' can be constructed from some $A \in \psi_0(X)$ in a particular way. By condition (i), (ii), we obtain $\widetilde{cat}(X'') \leq n$. Now the condition "reasonable in the extended sense" implies that $\widetilde{cat}$ satisfies the domination property, hence $\widetilde{cat}(X) \leq \widetilde{cat}(X') = \widetilde{cat}(X'') \leq n$. □

Example 5.18. The conditions of the theorem hold in the following situations:

(a) $\psi = id$ and $\mathcal{K} = \mathcal{K}(\mathcal{G}_\mathcal{A})$ or

(b) $\psi = \mathcal{G}_\mathcal{A}$ and $\mathcal{K} = \mathcal{K}(\mathcal{G}_\mathcal{A})$ or
$\psi = \mathcal{G}_\mathcal{A}$ and $\mathcal{K} = \mathcal{G}_\mathcal{A}*$ (compare 4.1).

5.3. Back to relative category notions

Choose an object $B \in \mathcal{C}$ and consider $\mathcal{C}_B$ the category of maps with target B. For ψ, $\mathcal{K}$: $Obj(\mathcal{C}_B) \to$ {subclasses of $Obj(\mathcal{C}_B)$}, we obtain $(\mathcal{K}, \psi)\text{-}Cat(f)$, $f\colon X \to B$. Various choices of classes of special morphisms then give corresponding "relative" category notions.

As an example let us consider the case where $\mathcal{C}$ is the homotopy category of $\mathcal{T}_*$, $B \in \mathcal{T}_*$, and assume that ψ is constant. Then $(\mathcal{K}, \psi)\text{-}Cat(f)$, $f \in mor(-, B)$, coincides with the corresponding notion of Section 4. Moreover, with $\gamma = mor\,(\mathcal{C}_B)$ the function $(\mathcal{K}, \psi, \gamma)\text{-}cat$ coincides with the category notion of Clapp and Puppe of Section 4. However, the invariants of Section 3 cannot be obtained that way.

6. Appendix

This section is devoted to the definition of the join of two maps with the same target and to a mapping theorem for some relative category notions. We work in the category $\mathcal{T}_*$ and do not distinguish maps from homotopy classes of maps.

Definition 6.1. Let $f\colon A \to B$ and $g\colon C \to B$ be two maps in $\mathcal{T}_*$ with associated fibrations $f'\colon A' \to B$ and $g'\colon C' \to B$. The pullback of f' and g' is called a *homotopy pullback* of f and g and denoted by $A \times_B C$.

Dually, by considering the pushout of the associated cofibrations, we have the notion of *homotopy pushout* of maps $B \to X$, $B \to Y$, denoted by $X \vee_B Y$.

Definition 6.2. Let $f\colon A \to B$ and $g\colon C \to B$ be maps in $\mathcal{T}_*$ and let $P = A \times_B C$, $J = A \vee_P C$. The universal properties of pullback and pushout give a canonical map $J \to B$, called the *join of f and g over B* and denoted by $f *_B g$.

Recall one of the main properties of the join construction [4, Theorem 3.1]:

Theorem 6.3. *Consider two homotopy pullbacks*

$$\begin{array}{ccccc} A & \longrightarrow & B & \longleftarrow & C \\ \downarrow & & \downarrow & & \downarrow \\ A' & \longrightarrow & B' & \longleftarrow & C'. \end{array}$$

Then the join construction gives rise to a new homotopy pullback:

$$\begin{array}{ccc} A *_B C & \longrightarrow & B \\ \downarrow & & \downarrow \\ A' *_{B'} C' & \longrightarrow & B'. \end{array}$$

Given $h\colon B \to C \in \mathcal{T}_*$, $\mathcal{L}_B \subset mor(-,B)$ and $\mathcal{L}_C \subset mor(-,C)$, we recall from Example 4.2 that $(\mathcal{L}_B*, \mathcal{L}_B)$-$cat$ is a function on $mor(-,B)$ and $(\mathcal{L}_C*, \mathcal{L}_C)$-$cat$ is one on $mor(-,C)$.

Theorem 6.4. *Suppose that $h\colon B \to C$, $\mathcal{L}_B$ and $\mathcal{L}_C$ verify the following property: for all $\alpha \in \mathcal{L}_C$ there exists $\beta \in \mathcal{L}_B$ and a homotopy pullback*

$$\begin{array}{ccc} A_1 & \to & B_1 \\ {\scriptstyle\beta}\downarrow & & \downarrow{\scriptstyle\alpha} \\ B & \xrightarrow{h} & C \end{array}$$

Let

$$\begin{array}{ccc} X & \to & Y \\ {\scriptstyle f}\downarrow & & \downarrow{\scriptstyle g} \\ B & \xrightarrow{h} & C \end{array}$$

be a homotopy pullback. Then $(\mathcal{L}_B*, \mathcal{L}_B)$-$cat(f) \le (\mathcal{L}_C*, \mathcal{L}_C)$-$cat(g)$.

Proof. Given $\alpha_1, \alpha_2, \ldots, \alpha_k \in \mathcal{L}_C$ and the corresponding $\beta_1, \beta_2, \ldots, \beta_k \in \mathcal{L}_B$, by the previous join theorem (Theorem 5.17) there is a homotopy pullback

$$\begin{array}{ccc} G_k(f, \beta_1, \ldots, \beta_k) & \longrightarrow & G_k(g, \alpha_1, \ldots, \alpha_k) \\ \downarrow & & \downarrow \\ B & \xrightarrow{h} & C \end{array}$$

where $G_k(f, \beta_1, \ldots, \beta_k)$ denotes the iterated joins over B of the indicated maps. Hence, if the right arrow of the diagram above has a section, the left one has a section. □

Corollary 6.5. (1) *Let $F \xrightarrow{i} B$ be the homotopy fibre of $h\colon B \to C$. Let $\mathcal{L}_B = \{i\}$ and $\mathcal{L}_c = \{* \to C\}$. Then $(\mathcal{L}_B*, \mathcal{L}_B)$-$cat(i) \le (\mathcal{L}_C*, \mathcal{L}_C)$-$cat(*) = cat(C)$.*

(2) *If moreover $i \sim *$, then $cat(B) \le cat(C)$.*

Proof. (1) Apply Theorem 6.3 to the homotopy pullback

$$\begin{array}{ccc} F & \to & Y \\ \downarrow & & \downarrow \\ B & \xrightarrow{h} & C. \end{array}$$

(2) The existence of homotopy commutative diagrams

$$\begin{array}{ccc} G_n(i, \ldots, i) & \longrightarrow & G_n(B) = G_n(*, \ldots, *) \\ & \searrow \quad \swarrow & \\ & B & \end{array}$$

implies $cat(B) \le (\mathcal{L}_B*, \mathcal{L}_B)$-$cat(i)$. □

References

[1] Clapp M., Puppe D., Invariants of the Lusternik–Schnirelmann type and the topology of critical sets, *Trans. Amer. Math. Soc.* **298** (1986), 603–620.

[2] Cornea O., Some properties of the relative Lusternik–Schnirelmann category. In: Stable and Unstable Homotopy. Fields Inst. Comm., Amer. Math. Soc., 1998, 67–72.

[3] Doeraene J.-P., LS-category in a model category, *J. Pure Appl. Algebra* **84** (1993), 215–261.

[4] Doeraene J.-P., Homotopy pullbacks, homotopy pushouts and joins, *Bull. Soc. Math. Belg.* **5** (1998), 15–37.

[5] Eggar M. H., On compositions of the loop and suspension functors, *Proc. Amer. Math. Soc.* **123** (1995), 597–606.

[6] Félix Y., *La dichotomie elliptique-hyperbolique en homotopie rationnelle,* Astérisque **176**, Soc. Math. France, 1989.

[7] Félix Y., Lemaire J.-M., On the mapping theorem for Lusternik–Schnirelmann category, *Topology* **24** (1985), 41–43; **27** (1987), 177.

[8] Félix Y., Lemaire J.-M., On the mapping theorem for Lusternik–Schnirelmann category II, *Canad. J. Math.* **XL** (1988), 1389–1398.

[9] Ganea T., Lusternik–Schnirelmann category and strong category, *Illinois J. Math.* **11** (1967), 417–427.

[10] James I. M., On category in the sense of Lusternik–Schnirelmann, *Topology* **17** (1978), 331–348.

[11] Lemaire J.-M., Lusternik–Schnirelmann category: an introduction. In: Algebra, algebraic topology and their interactions. Proceedings of a conference held in Stockholm, August 3–13, 1983, and later developments. Editor: J.-E. Roos. *Lecture Notes in Math.* **1183**, Springer-Verlag, 1986, 259–276.

[12] Lusternik L., Schnirelmann L., *Méthodes topologiques dans les problèmes variationnels*, Hermann, Paris, 1934.

[13] Scheerer H., Tanré D., Variation zum Konzept der Lusternik–Schnirelmann Kategorie, *Math. Nachr.* **207** (1999), 183–194.

Freie Universität Berlin
Mathematisches Institut
Arnimallee 2–6
D–14195 Berlin, Germany
E-mail address: `hanschee@zedat.fu-berlin.de`

U.F.R. de Mathématiques
UMR CNRS 8524
Université des Sciences et Technologies de Lille
59655 Villeneuve d'Ascq Cedex, France
E-mail address: `Daniel.Tanre@agat.univ-lille1.fr`

Progress in Mathematics, Vol. 196

Essential category weight and phantom maps

Jeffrey Strom

Abstract. The purpose of this paper is to study the relationship between maps with infinite essential category weight and phantom maps (there is a brief summary of the main results on essential category weight in the appendix to this paper). It is not hard to see that any map with $E(f) = \infty$ is a phantom map. We give examples to show that the converse is not always true: there are phantom maps f with $E(f) = 1$. We also show that if ΩX is homotopy equivalent to a finite dimensional CW complex then every phantom map $f\colon X \longrightarrow Y$ has $E(f) = \infty$. We are able to adapt many of the results of the theory of phantom maps to give us results about maps with $E(f) = \infty$. Finally, we use the connections between essential category weight and phantom maps to answer a question (asked by McGibbon) about phantom maps.

The purpose of this paper is to study the relationship between maps with infinite essential category weight and phantom maps.

A map (all maps and spaces in this paper are pointed) $f\colon X \longrightarrow Y$ has **essential category weight** at least N if $f \circ g \simeq *$ whenever $g\colon Z \longrightarrow X$ and Z is a CW complex with Lusternik–Schnirelmann category at most N (or, equivalently, if the Lusternik–Schnirelmann category of g is at most N). We write $E(f) \geq N$; observe that if $f \not\simeq *$, then $E(f) < \mathrm{cat}(X)$. The appendix to this paper contains a brief summary of the main results on essential category weight.

A map f has infinite essential category weight if $E(f) \geq N$ for all N. Write $E(X, Y)$ to denote the set of pointed homotopy classes of maps $f\colon X \longrightarrow Y$ with $E(f) = \infty$. It follows from Theorem 9 of the Appendix that $E(X, Y)$ is functorial in both X and Y.

Rudyak [11] has made the observation that if $E(f) = \infty$, then f must be a phantom map. Recall that a map $f\colon X \longrightarrow Y$ is a phantom map if $f \circ g \simeq *$ whenever $g\colon Z \longrightarrow X$ and the dimension of Z is finite (this is a phantom map of the first kind in McGibbon's article [7]). To prove this, it suffices to look at each component of the domain; since connected finite dimensional spaces have finite category, the observation is immediate.

Let $\mathrm{Ph}(X, Y)$ denote the set of pointed homotopy classes of phantom maps $f\colon X \longrightarrow Y$. Thus, $E(X, Y) \subseteq \mathrm{Ph}(X, Y)$. It is not hard to see that this inclusion can be proper. For example, there is a phantom map $f\colon \mathbb{CP}^\infty \longrightarrow S^3$ whose suspension is nontrivial [5]. Thus, Σf is a phantom map, but $E(\Sigma f) = 1$. For another example, if G is any one of the groups $Sp(2), Sp(3), G_2, F_4$, then there are stably nontrivial phantom maps $\Omega G \longrightarrow K$ for certain finite type CW complexes K [7]. The suspensions of these maps are phantom maps with essential category weight 1.

This leads us to a natural question: for what spaces is it true that $E(X,Y) = \mathrm{Ph}(X,Y)$? Our first theorem is a partial answer to this question.

Theorem 1. *Let X be a space whose loop space ΩX is homotopy equivalent to a finite dimensional CW complex. Then for any Y*

$$E(X,Y) = \mathrm{Ph}(X,Y).$$

Proof. Let $f\colon X \longrightarrow Y$ be a phantom map; we need to show that $E(f) = \infty$. By Theorem 10 in the Appendix, it suffices to show that $f|_{B_N\Omega X} \simeq *$ for each $N > 0$. This is the case because, since ΩX is homotopy equivalent to a finite dimensional CW complex, so is $B_N\Omega X$. □

It is trivial that $E(X,Y) = \mathrm{Ph}(X,Y)$ if $\mathrm{Ph}(X,Y) = *$. But this is far from the situation in Theorem 1, as we now show.

According to McGibbon (Example 3.12 in [7]), if G is a compact Lie group, then the universal phantom map out of BG is nontrivial. The same argument works equally well for any space X whose loop space ΩX is homotopy equivalent to a connected finite dimensional CW complex. Thus, we have the following corollary.

Corollary 2. *If ΩX is homotopy equivalent to a connected finite dimensional CW complex, then there is a space Y such that $E(X,Y) \neq *$. It follows that* $\mathrm{cat}(X) = \infty$.

The universal phantom map [6] is a useful tool in the study of $\mathrm{Ph}(X,Y)$. There is an analogous map out of X which is weakly universal with respect to the property of having infinite essential category weight.

For each CW complex X, we may form the cofiber sequence

$$\bigvee B_N\Omega X \xrightarrow{\ i\ } B\Omega X \simeq X \xrightarrow{\ \Phi\ } \bigvee \Sigma B_N\Omega X$$

in which i is the wedge of the inclusions of the $B_N\Omega X$.

Theorem 3. *Every map $f\colon X \longrightarrow Y$ with $E(f) = \infty$ has a factorization*

$$\begin{array}{ccc} X & \xrightarrow{\quad f \quad} & Y \\ & {\scriptstyle\Phi}\searrow \quad \nearrow & \\ & \bigvee \Sigma B_N\Omega X. & \end{array}$$

This factorization is not unique in general.

The proof is a straighforward adaptation of Gray and McGibbon's Theorem 1 in [6], in which we replace the filtration

$$X_0 \subseteq X_1 \subseteq \cdots \subseteq X_n \subseteq \cdots \subseteq X$$

by the filtration

$$B_1\Omega X \subseteq \cdots \subseteq B_N\Omega X \subseteq \cdots \subseteq B\Omega X \simeq X$$

and use Theorem 10 to interpret essential category weight in terms of this filtration.

This filtration of X can also be used to give an algebraic computation of $E(X,Y)$.

Theorem 4. *For CW complexes X and Y,*

$$E(X,Y) \cong \lim{}^1[\Sigma B_N\Omega X, Y].$$

Proof. According to Bousfield and Kan [1], there is a short exact sequence

$$* \longrightarrow \lim{}^1[\Sigma B_N\Omega X, Y] \longrightarrow [X,Y] \xrightarrow{\ j\ } \lim[B_N\Omega X, Y] \longrightarrow * .$$

Theorem 10 allows us to identify $E(X,Y)$ with $j^{-1}(*)$. □

Some other results of [6] carry over as well. For example, we have the following corollary, which is analogous to Theorem 2 in [6].

Corollary 5. *Let X be a CW complex; then $E(X,Y) = *$ for every Y if and only if ΣX is dominated by $\bigvee \Sigma B_N\Omega X$.*

Gray and McGibbon [6] use the universal phantom map to show that if f and g are composable phantom maps, then $f \circ g \simeq *$; the same is true *a fortiori* if $E(f) = E(g) = \infty$. However, we can say much more.

Proposition 6. *If g is a phantom map and $E(f) > 1$, then $f \circ g \simeq *$. This is the case, in particular, if $E(g) = \infty$.*

Proof. We have the following commutative diagram

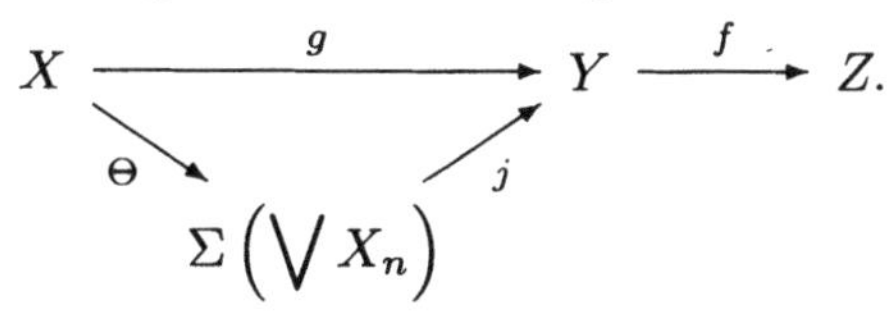

Since cat $(\Sigma(\bigvee X_n)) = 2$ and $E(f) \geq 2$, we conclude that $f \circ j \simeq *$, which finishes the proof. □

We have seen how to apply the theory of phantom maps to the study of maps with $E(f) = \infty$. Our final result goes in the other direction: we will use the theory of essential category weight to answer a question of McGibbon's.

Following Roitberg [9], McGibbon observes that if X and Y have finite type and Y is grouplike, then $\mathrm{Ph}(X,Y)$ has an abelian group structure which is natural in X. He asks in Question 6 of [7] whether the assumption of finite type is necessary. The exact sequence due to Bousfield and Kan shows that it is not. McGibbon's intention was to ask for a geometric argument that shows why the commutator of two phantom maps should be trivial. This is the question we resolve in our proof of Theorem 7.

Theorem 7. *If X and Y are CW complexes and Y is grouplike, then* $\mathrm{Ph}(X, Y)$ *has an abelian group structure which is natural in X.*

Proof. We have to show that the commutator of any two phantom maps is trivial.

Let $\chi\colon Y \times Y \longrightarrow Y$ denote the commutator map. Since $\chi|_{Y\vee Y} \simeq *$, there is a factorization

$$\begin{array}{ccc} Y \times Y & \xrightarrow{\quad\chi\quad} & Y \\ & {\scriptstyle\wedge}\searrow \quad \nearrow & \\ & Y \wedge Y & \end{array}$$

in which $\wedge$ is the usual quotient map. By Theorem 8 in the Appendix, $E(\wedge) > 1$. Using Theorem 9 in the Appendix, we see that $E(\chi) > 1$.

Now let $f, g\colon X \longrightarrow Y$ be phantom maps; it follows that their product

$$f \times g\colon X \times X \longrightarrow Y \times Y$$

is also phantom. Since the commutator $[f, g]$ is equal to $\chi \circ (f \times g) \circ d$, where $d\colon X \longrightarrow X \times X$ is the diagonal map, the theorem follows directly from Proposition 6. □

A final note before moving on to the Appendix. This paper was written while I held a visiting position at Wayne State University. My thanks to the Mathematics department there, and especially to Chuck McGibbon and Bob Bruner, for providing a helpful and stimulating working environment.

Appendix: Essential category weight

This is a summary of results from [13]. Essential category weight is a homotopy invariant version of category weight, which was introduced by Fadell and Husseini in [2]. Essential category weight has been studied independently by Rudyak ([11], [12]), who called it **strict category weight**.

Many of the most important lower bounds on the category of a space X take the form of a product formula. Theorem 8 can be considered the basic product formula from which all others follow.

Theorem 8. *Let $f\colon X \longrightarrow K$ and $g\colon Y \longrightarrow L$. In the diagram*

$$\begin{array}{ccc} X \times Y & \xrightarrow{f \times g} & K \times L \\ & {\scriptstyle\widehat{f\times g}}\searrow & \downarrow{\scriptstyle\wedge} \\ & & K \wedge L \end{array}$$

we have $E(\widehat{f \times g}) \geq E(f) + E(g)$.

Proof. Write $E(f) = p$ and $E(g) = q$, and suppose $h\colon Z \longrightarrow X \times Y$ with $\mathrm{cat}(Z) \leq p + q$. Then we may write $Z = A \cup B$ with $\mathrm{cat}(A) \leq p$ and $\mathrm{cat}(B) \leq q$. Thus $(f \circ h)|_A \simeq *$ and $(g \circ h)|_B \simeq *$. We can use homotopy extension to lift $(f \times g) \circ h$ into $K \vee L$, which shows that $\wedge \circ (f \times g) \circ h \simeq *$ and completes the proof. $\square$

When f and g represent cohomology classes (in any cohomology theory), this result becomes a stronger version of the classical cup length lower bound for $\mathrm{cat}(X)$. When f and g are maps to a grouplike space, we obtain a strengthening of Whitehead's theorem [16] that the nilpotence length of $[X, Y]$ is bounded above by the category of X.

Next we give a formula for the essential category weight of a composition of two maps.

Theorem 9. *Let $f\colon X \longrightarrow Y$ and $g\colon Y \longrightarrow Z$. Then*

$$E(g \circ f) \geq E(g) \cdot E(f).$$

Proof. Write $E(f) = p$ and $E(g) = q$ and suppose $h\colon W \longrightarrow X$ with $\mathrm{cat}(W) \leq pq$. Then we may write

$$W = A_1 \cup \cdots \cup A_q$$

with $\mathrm{cat}(A_i) \leq p$. Thus $(f \circ h)|_{A_i} \simeq *$ for each i, and so $f \circ h$ has a factorization

$$\begin{array}{ccccccc} W & \xrightarrow{h} & X & \xrightarrow{f} & Y & \xrightarrow{g} & Z \\ & \searrow & & \nearrow_{i} & & & \\ & & W' & & & & \end{array}$$

in which $W' = W \cup (\coprod CA_i)$. Since $\mathrm{cat}(W') \leq q$ and $E(g) = q$, we conclude that $g \circ i \simeq *$, which proves the theorem. $\square$

A very important consequence of this theorem is that $E(g \circ f) \geq E(g)$ and $E(g \circ f) \geq E(f)$.

If f represents a cohomology class u and g represents a cohomology operation θ, we find that $E(\theta(u)) \geq E(\theta) \cdot E(u)$. Thus, Fadell and Husseini's result [2] that $E(P^I(u)) \geq 2$ if $|u| = e(I)$ can be instantly improved to $E(P^I(u)) \geq 2E(u)$. Nearly complete calculations of $E(\theta)$ for ordinary cohomology operations can be found in [13].

Finally, we address the question of how to compute the essential category weight of a given map. Recall that if X is a CW complex, then we can form a filtration

$$B_1\Omega X \subseteq \cdots \subseteq B_N\Omega X \subseteq \cdots \subseteq B\Omega X \simeq X$$

by the usual construction of classifying spaces [8], [14], [3].

Theorem 10. *Let $f\colon X \longrightarrow Y$ be any map. Then $E(f) \geq N$ if and only if the composite*

$$B_N\Omega X \hookrightarrow B\Omega X \simeq X \xrightarrow{f} Y$$

is nullhomotopic.

Proof. Recall from [14] that a map $f\colon Z \longrightarrow BG$ factors through a space Q with $\mathrm{cat}(Q) \le N$ if and only if f factors through $B_N\Omega X$. Thus, if $f|_{B_N\Omega X} \simeq *$, then $E(f) \ge N$.

It follows from the construction of $B_N\Omega X$ that $\mathrm{cat}(B_N\Omega X) \le N$; thus, if $E(f) \ge N$, then $f|_{B_N\Omega X} \simeq *$. □

The filtration of X by $B_N\Omega X$ gives rise to a Rothenberg–Steenrod spectral sequence (also called a Moore spectral sequence) [10]; the connection between this spectral sequence and $\mathrm{cat}(X)$ has been studied by Whitehead [17], Ginsburg [4] and Toomer [15]. Many of their results can be derived as instant corollaries of Theorem 10. Furthermore, we can see that the E_∞-term of this spectral sequence is precisely the graded module on $H^*(X)$ associated to the filtration by essential category weight.

References

[1] A. K. Bousfield and D. M. Kan, *Homotopy Limits, Completions, and Localizations*, Lecture Notes in Math. **304**, Springer, 1972.

[2] E. Fadell and S. Y. Husseini, *Category weight and Steenrod operations*, Bol. Soc. Mat. Mexicana **37** (1992) 151–161.

[3] T. Ganea, *Lusternik–Schnirelmann category and strong category*, Illinois J. Math. **11** (1967), 417–427.

[4] M. Ginsburg, *On the Lusternik–Schnirelmann category*, Ann. of Math. **77** (1963), 538–551.

[5] B. Gray, *Spaces of the same n-type for all n*, Topology **5** (1966), 241–243.

[6] B. Gray and C. A. McGibbon, *Universal phantom maps*, Topology **32** (1993), 371–394.

[7] C. A. McGibbon, *Phantom maps*, a chapter in *The Handbook of Algebraic Topology*, I. M. James, ed., North Holland, 1995.

[8] J. Milnor, *Construction of universal bundles, II*, Ann. of Math. **63** (1956), 430–436.

[9] J. Roitberg, *Weak identities, phantom maps, and H-spaces*, Israel J. Math. **66** (1989), 319–329.

[10] M. Rothenberg and N. Steenrod, *The cohomology of classifying spaces of H-spaces*, Bull. Amer. Math. Soc. **71** (1965), 872–875.

[11] Y. Rudyak, *On category weight and its applications*, Topology **38** (1999), 37–56.

[12] Y. Rudyak, *Category weight: New ideas concerning Lusternik–Schnirelmann category*, Homotopy and Geometry, Banach Center Publications **45**, Warszawa (1998), 47–61.

[13] J. Strom, *Category Weight and Essential Category Weight*, University of Wisconsin-Madison Ph.D. Thesis (1997).

[14] A. Svarc, *The genus of a fiber space*, Amer. Math. Soc. Translations **55** (1966), 49–140.

[15] G. Toomer, *Lusternik–Schnirelmann category and the Moore spectral sequence*, Math. Z. **138** (1974), 123–143.

[16] G. W. Whitehead, *On mappings into group-like spaces*, Comment. Math. Helv. **28** (1954), 320–328.

[17] G. W. Whitehead, *The homology suspension* in: Colloque de Topologie Algébrique, Louvain (1956), 89–95.

Department of Mathematics,
Dartmouth College,
Hanover, NH 03755, USA
E-mail address: jeffrey.strom@dartmouth.edu

MIX
Papier aus verantwortungsvollen Quellen
Paper from responsible sources
FSC® C105338

If you have any concerns about our products,
you can contact us on
ProductSafety@springernature.com

In case Publisher is established outside the EU,
the EU authorized representative is:
Springer Nature Customer Service Center GmbH
Europaplatz 3, 69115 Heidelberg, Germany

Printed by Libri Plureos GmbH
in Hamburg, Germany